MOLECULAR NEUROBIOLOGY

GORDON GUROFF

Laboratory of Developmental Neurobiology
National Institute of Child Health and Human Development
National Institutes of Health
Bethesda, Maryland

MARCEL DEKKER, INC.   New York and Basel

Library of Congress Cataloging in Publication Data

Guroff, Gordon [Date]
   Molecular neurobiology.

   Includes indexes.
   1. Neurochemistry. I. Title.
QP356.3.G87   612'.8042   79-22812
ISBN 0-8247-6862-0

MARCEL DEKKER, INC.
270 Madison Avenue, New York, New York 10016

Current printing (last digit):
10 9 8 7 6 5 4 3 2 1

PRINTED IN THE UNITED STATES OF AMERICA

To Marjorie

FOREWORD

Neurochemistry as a specialty, and the biochemical approach to behavior and brain function, have clearly come of age and into general acceptance, but they still represent a young discipline whose validity was questioned by many not so long ago. Now the importance of understanding brain function in biological terms is unequivocally recognized. In spite of this coming of age and recognition, it is not an easy task to construct a single volume on the subject. Many metabolic reactions that are important for neural function are the same in the brain as in most other organs; therefore their detailed discussion is not justified in a neurochemical text. On the other hand, because the different metabolic pathways become specific in terms of structure, development, or malfunction, their discussion often needs to involve subjects usually dealt with in books of anatomy, embryology, or pathology. The few writers of books on neurochemistry of manageable size have had to make their own approach; no traditional form for the treatment of this subject exists. The author needs to treat the subject in an interdisciplinary manner, but he has to avoid being too diffuse in his approach. The treatment of the subject has to be precise; molecular biology is too advanced to be ignored. Events in the nervous system can now be understood in precise molecular biological terms.

Dr. Guroff has had to follow this nontraditional tradition to create his own format, and he has succeeded admirably. The book recognizes that in the nervous system more than in any other organ, metabolism, structure, and function have to be discussed with a unified outlook, and this is the way this book was written; this is why it starts with cell types and discusses metabolism especially in terms of malfunctions, and the discussion of even the most complex mechanisms is precise and based on a sound molecular approach.

The reason neurochemistry is now recognized is that recent results from many excellent contributors have shown that function can be understood, malfunction perhaps corrected, as one better understands cerebral metabolic interrelationships. These contributions have created an exciting period in this discipline, with hopes for great advances; the field is indeed in a phase of rapid development. This excitement, and hope for the future, can be eagerly felt by the reader of the present book.

The book also shows the fruits of recent research in a concise way. The logical road from basic knowledge to understanding of pathological changes leads to new methods of treatment and is by far the best road to take. These aspects also add to the excitement and hope that the reader should gain.

Abel Lajtha
Director, New York State Research Institute
for Neurochemistry and Drug Addiction
Ward's Island, New York

PREFACE

This book is intended to serve as an outline for a course in contemporary neuro-chemistry for students with a background in general biochemistry. It seemed, when I was asked to undertake this venture, that such an outline was not available. As the book progressed it became, in addition, an attempt to relate the biochemical and structural properties of the brain to its functions and to introduce the reader, in a systematic way, to the many disorders that result from alterations in the fundamental biochemistry of the brain.

The book, then, is designed to progress from structure through chemistry to function. It is, hopefully, up to date, but some effort has been made to provide a background to recent observations and to acknowledge the contributions of specific individuals whose work has brought us to this point. The clinical descriptions, while not detailed, may serve to alert students considering neuro-biology as a future to the remarkable gaps which exist in our ability to understand the misfires of the human brain. Perhaps these excursions from the bench to the bedside are a unique feature of the construction of this book.

I apologize if the treatment leans too heavily on areas of my own particular research interest. Possibly, by way of recompense, it will convey some of the excitement and wonder one frequently feels in the laboratory. In any case, I hope it will provide a comprehensive statement of the level at which we understand, or fail to understand, the brain.

The construction of this book required the cooperation and encouragement of family, friends, and colleagues, who read and corrected various portions of the manuscript. My special thanks to Mrs. Kitty Kunkle who typed the manuscript and to Mrs. Rose Smith who prepared the illustrations.

Gordon Guroff

CONTENTS

Metabolism and Function

BIOCHEMICAL CYTOLOGY OF NERVE AND BRAIN

1
CELL TYPES

The nervous system consists, broadly speaking, of three types of cells. Within these general types, however, there are several subtypes and even considerable uncertainty about some of the classifications. In fact, there are a number of reports of the identification of new cell types in the nervous system which are as yet unsubstantiated. Considering only the major classifications, there are the neuronal, the neuroglial, and the endothelial cells. The endothelial cells are present in the linings of the brain, in the structural elements or connective tissue matrix, and in the blood vessels. These cells resemble endothelial cells in other parts of the body, and those in the brain seem to be in no way unique. The truly neural elements, then, are the neurons and the glia.

Neurons

The nerve cells, or neurons, are the communicating elements of the nervous system. They receive information from the environment or from other neurons and pass it on to receptor cells such as those of muscles or glands. Although each neuron is truly unique by virtue of its position in the nervous system and the pattern of its connections to other cells, it is possible to group the neurons in a number of different ways [1]. As we shall see later, the neurons can be classified on the basis of the exact chemistry of their communicating mechanism. In the present context, neurons can be classified according to their function: (1) sensory, accepting information from the sense organs; (2) motor, passing impulse to effector or action cells; or (3) associative, communicating information from one neuron to another. Morphologically, neurons have been grouped as: (1) unipolar, cells with but one extension or process; (2) bipolar, cells with two processes, one axon and one dendrite; or (3) multipolar, cells with one axon and many dendrites. Other classifications are of value as well. For example, classification on the basis of the action of the neuron, i.e., excitatory, inhibitory; or classification based on the length of the axonal process, i.e., long, short, or arborized (branched). But

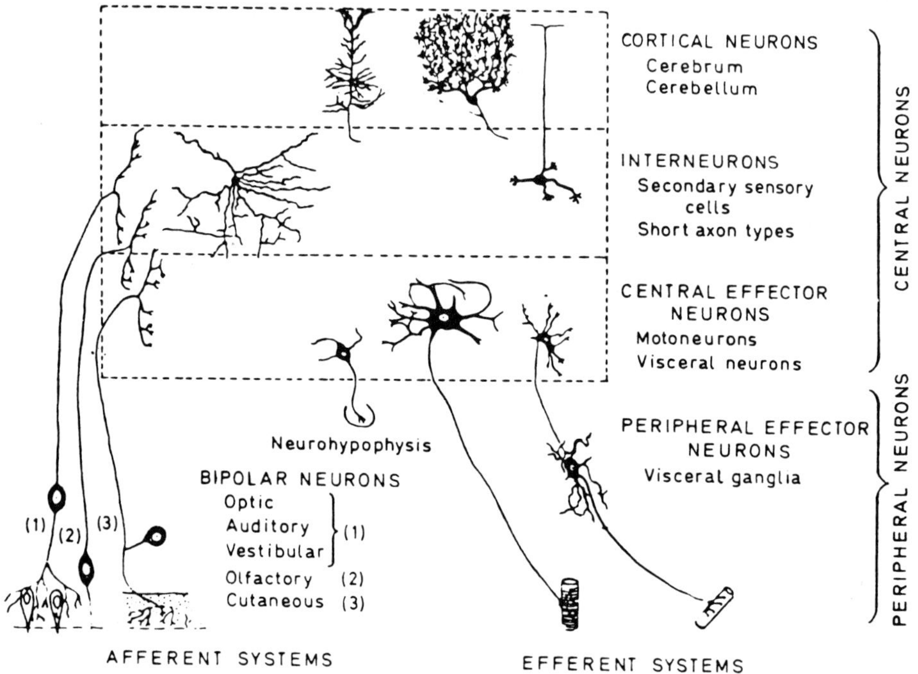

Figure 1.1   Diagram of some representative species of neuron in the mammalian nervous system. [Modified from D. Bodian, *Cold Spring Harbor Symp. Quant. Biol. 17:*1 (1952).]

classification of the neurons is difficult at best because the neuron is the most polymorphous of all the cell types in the body (Fig. 1.1). And since its pattern of connections to other cells is truly reflected in its form, it has been said that the shape of a neuron is its most properly neural feature.

There are variously estimated to be 10^{10} to 10^{12} neurons in the normal human brain. These cells can make several hundred or even several thousand connections with other neurons. Thus, the neuronal network is stupifyingly complex. As an example, Hydén has estimated that the dendritic assembly of one Purkinje cell in the cerebellum may pass through the same space as the axons from 270,000 other neurons [2]. Accepting the complexity of the neuronal network and the anatomical uniqueness of each neuron, some few things can be said, in general, about the morphology of the neuron. Neurons are generally polarized into a receiving or dendritic zone and a transmitting or axonal zone (Fig. 1.2). The cell body, or perikaryon, generally lies in the dendritic zone. The dendrites

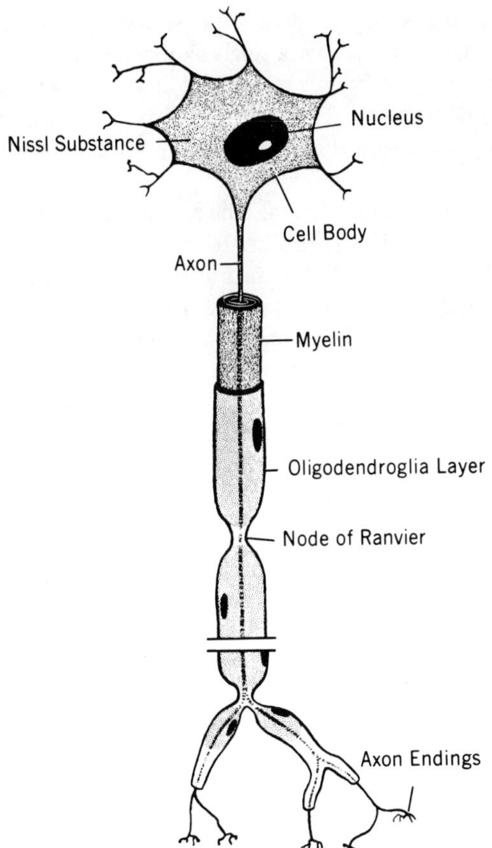

Figure 1.2 The morphology of a typical motoneuron showing the soma with its large central nucleus, Nissl granular material, dendritic branches, and long axon with terminal teledendrites enclosed (except for the proximal axon) by oligodendroglia and myelin. The diagram is based on conventional descriptions derived from light microscopy. (Redrawn and modified from A. W. Ham, *Histology*, 3rd ed., Pitman, London, 1957.)

usually are many in number and are generally covered with synaptic inputs from other cells. They are short, usually less than 1 mm long, and are very fine, on the order of 5 μm in diameter. The perikaryon, or soma, is ordinarily a small part of the total volume of the cell but can range up to 100 μm in diameter. Many cell bodies, on the other hand, are smaller than a red blood cell. The volume of the perikaryon of various neurons varies from the large spinal motoneurons at 2.5×10^{-7} cm^3, to the sympathetic ganglion cells at 1×10^{-8} cm^3, and down to the

spinal internuncial cells which are smaller still [3]. Most of the surface area of
the cell body is occupied by dendrites. Neurons have a single axon which is much
thicker than the dendrites. The axon may be up to a meter long in some cases and
can be on the order of 20 to 25 μm in diameter. The single axon can be arborized
at the end. Each terminus of the axon is enlarged into a synaptic bouton before
it encounters the receiving element of the next cell.

The neurons arise embryologically from the neural tube and are thus ectodermal in origin. Upon differentiation they lose the ability to divide. They constitute about 10% of the cell population of the adult brain, but the ratio of neurons to glia may vary somewhat from species to species and certainly decreases as
the animal matures. Neuronal densities also vary to some extent from a low of
10^7 neurons per cubic centimeter in the human motor cortex to a high of 10^8 per
cubic centimeter in the mouse visual cortex [4].

The neurons function by receiving electrochemical stimuli at the dendritic
end, conducting them through the cell body and down the axon by means of a
membrane-associated propagated electrical response, and communicating this information across the narrow space which forms the junction (the synapse) between two excitable cells to a receptor cell, via packets of specific chemical transmitters. Facilitating this process is a unique property of the neuronal membrane,
its ability to remain in a polarized state. The inside of the neuronal membrane is
some 70 to 90 mV more negative than the outside. As we shall see later, this
unique cell property is the basis for what is probably the sole function of the
neuron, that is cell-to-cell communication.

Glia

There are three generally accepted kinds of glial cells in the central nervous system. These are the oligodendrocytes, the astrocytes, and the microglia. The
oligodendrocytes and the astrocytes are sometimes grouped together as macro-
glia and are ectodermal in origin, being derived, as are the neurons, from the
neural tube. The microglia are mesodermal in origin and are derived from blood-
borne macrophages which enter the brain during the neonatal period. The glia in
general retain the property of cell division by mitosis even in the mature state
and proliferate in the adult nervous system expecially in response to injury. The
neuroglia are so called because they were at first thought to be a noncellular kind
of interstitial substance holding the neurons in place, thus "nerve glue." It was
not until somewhat later that they were recognized as discrete cellular entities.
Even now it is not clear that all existing types of glia have been described or
whether the arbitrary classifications of oligodendrocytes, astrocytes, and microglia truly describe the existing cell types. However, classification aside, it is generally felt that the glia, as a whole, function in any of a number of ways to support and sustain the neurons.

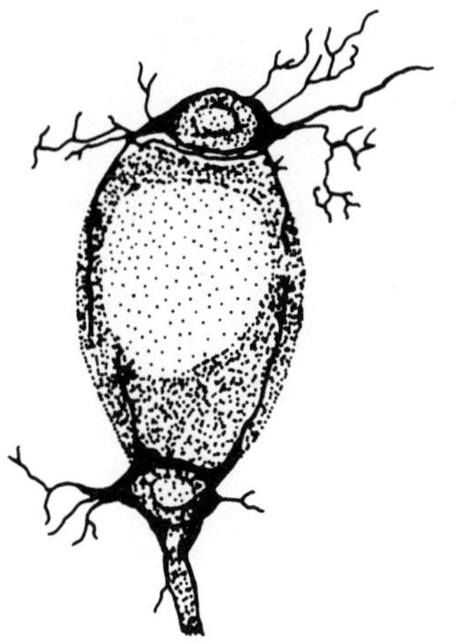

Figure 1.3   Oligodendroglia adhering to a neuron. [Redrawn from P. Glees and K. Meller, in *The Structure and Function of Nervous Tissue*, Vol. I (G. Bourne, ed.), Academic, New York, 1969.]

The oligodendrocytes are generally smaller than neurons and have short processes. They rarely lie in contact with the vascular bed but rather are in contact with the axons of neurons (Fig. 1.3). They envelope the axons; several axons from several cells can be covered by a single oligodendrocyte. They occur primarily in the white matter and are responsible for the myelin sheath that surrounds the axons. The oligodendrocytes envelope the axon, probably rotating around it, and covering it as they go with layer upon layer of their own membrane substance, which becomes the sheath. In the peripheral nerves, the elaboration of this myelin sheath is performed by a similar type of cell, known as the Schwann cell. Although the two cell types seem to fulfill a similar function it is known that they are antigenically different. If an animal is immunized with peripheral nerve, the Schwann cells are attacked and "experimental allergic neuritis" develops. Immunization with tissue from the central nervous system, on the other hand, leads to degeneration of the oligodendrocytes specifically, and a different condition, "experimental allergic encephalitis," ensues [5]. Both cell types, however, envelop axons by wrapping around them over and over again, elaborating layer upon layer of membrane material as they go.

The astrocytes have highly irregular cell bodies but are in general shaped like neurons (Fig. 1.4), except that they have no axon. They are found between the capillary vessel and the neuron or its axon and form somewhat of a supporting framework for the neurons. Their processes are slender and filamentous, but they enlarge at the ends to form astrocyte expansions, or "feet." These feet constitute a membrane around the external surfaces of the brain and around the blood vessels which nourish the brain. Because of their characteristic location between capillaries and neurons, it has been suggested that they also provide nutrition for the neurons, transporting gas, water, ions, and nutrients, from the

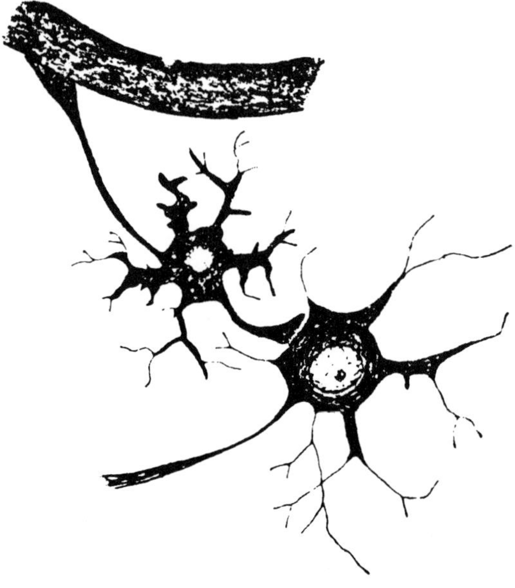

Figure 1.4 Protoplasmic astrocyte establishing a cellular connection between a vessel and a neuron. [Redrawn from P. Glees and K. Meller, in *The Structure and Function of Nervous Tissue,* Vol. I (G. Bourne, ed.), Academic, New York, 1919.]

blood vessels to the nerve cells. More recent evidence, to be discussed later, does not support this function for the astrocytes. Although the astrocyte membrane is known to be polarized to about the same extent as that of the neuron, the work of Kuffler and his group [6] has indicated that this polarization is of an entirely different nature. The astrocyte cannot be induced to carry an action potential and, thus, is not thought to have a signaling function like that of the neuron.

The microglia are scattered seemingly at random throughout the central nervous system. They account for perhaps 10% of the glia cells. Microglia are much smaller than astrocytes and have a scanty perikaryon. They have wavy processes which branch and give off spinelike projections (Fig. 1.5). They are present in both white and gray matter but are more abundant in the gray. The microglia are inactive in the normal brain, but are activated by inflammation or injury to the tissue. Upon injury they proliferate rapidly, retract their processes, and move toward the lesion. They ingest cellular debris by phagocytosis and transport the debris to the capillaries. They are thought, then, to be scavenger cells. Some reservation about the function of the microglia and, in fact, about their very existence has recently been aroused since, although they have been

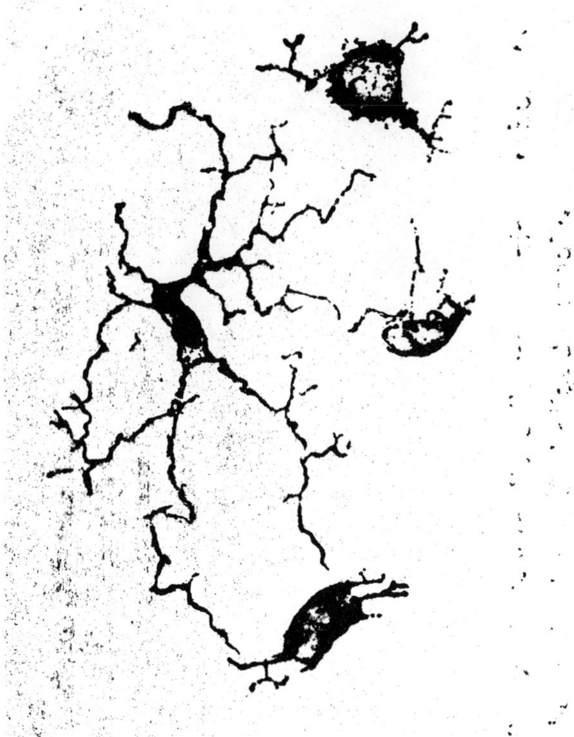

Figure 1.5   Microglia. [Redrawn from P. Glees and K. Meller, in *The Structure and Function of Nervous Tissue,* Vol. I (G. Bourne, ed.), Academic, New York, 1969.]

seen repeatedly and their action defined clearly by light microscopy, they have yet to be observed with the electron microscope [7].

Extracellular Space

The extracellular space of the nervous system is different from that of other tissues (Fig. 1.6). Between the relatively small cell bodies is an intricate feltwork made up of axons, dendrites, and various glial processes intermingled and interconnected. This dense and complicated web is known as the neuropil. The exact amount of extracellular space in the brain has been a controversial subject for some time. Early electron micrographs of brain uniformly and convincingly showed that little or no extracellular space existed. This concept was fostered

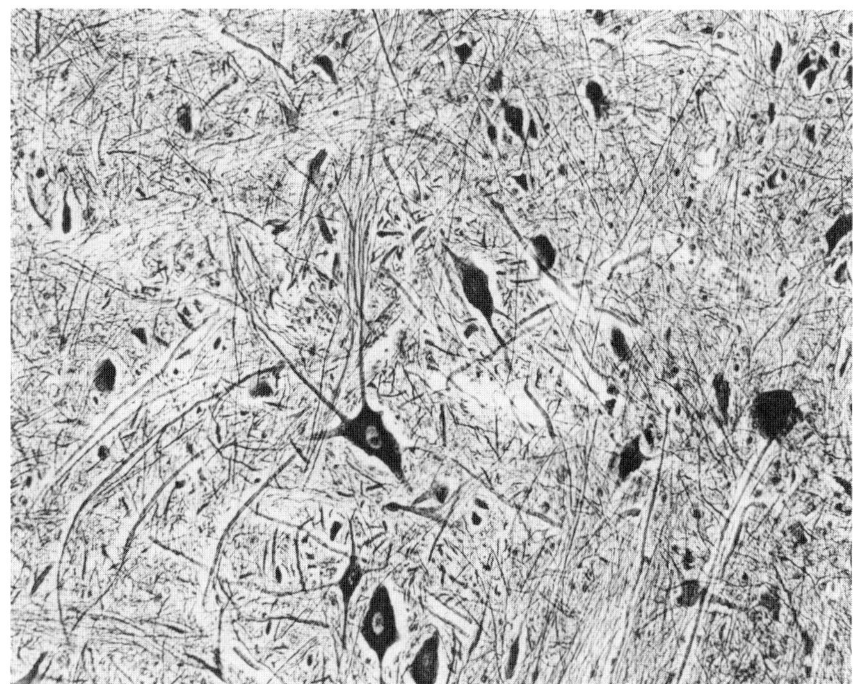

Figure 1.6 Micrograph of neural tissue. Cluster of neurons in the gray matter of the spinal cord of the monkey. Paraffin section, 15μm. Bodian silver stain ×150. [Figure courtesy of Dr. David Bodian, from D. Bodian, in *The Neurosciences, A Study Program* (G. C. Quarton, T. Melnechuk, and F. O. Schmitt, eds.), Rockefeller University Press, New York, 1967.]

by its adoption as an attractive explanation for the phenomenon of the "blood-brain barrier," which is discussed below [8]. That is, if there is no extracellular space, things which enter the brain must pass through cellular material and this, of course, might be much slower than passage, as in other tissues, through extra-cellular fluid. It has now been shown that electron microscopy can yield vastly different estimates for the extracellular space of brain, depending on the techniques used to prepare the tissue for observation. Conventional dehydration methods gave estimates ranging from 0 to 5%, much lower than the spaces estimated for nonneural tissues. Recent studies using more rapid methods, in which the tissue is frozen within seconds of its sectioning and then fixed with osmium tetroxide in acetone, lead to much larger values, approximating those found with other tissues. The explanation probably lies in the observation that brain cells, notably the glia, imbibe fluids at a remarkable rate once the circulation has been arrested. The low estimates from early electron micrographs probably were due to the time which elapsed during the preparation of the tissue. Indeed, if the tissue is allowed to stand for just a few minutes before the freezing and osmium tetroxide methodology is applied, the electron micrographs again look as if no extracellular space is present.

More conventional physiological approaches to the study of brain extracellular space have also led to conflicting estimates, in many cases because the special properties of the extracellular fluid of brain were not appreciated. The use of the classic extracellular markers such as thiocyanate, inulin, sucrose, sulfate, chloride, sodium, or iodide by conventional administration to the intact animal are fraught with uncertainties as far as brain is concerned. In general, these difficulties include the following. The protracted time necessary for these markers to reach equilibrium with the brain may allow the measurements to be influenced by a number of other factors. Then, the turnover of the cerebrospinal fluid may prevent a true equilibrium from being attained. Finally, it is not at all clear which fluid should be used to serve as a reference for these measurements, since, unlike other tissues, the extracellular fluid of brain does not resemble an ultrafiltrate of plasma but is more like the cerebrospinal fluid.

Specific problems of interpretation with each of these markers are as follows. Sodium, chloride, and thiocyanate are known to be intracellular to some extent in brain. Sodium and chloride concentrations, in addition, are known to be much higher in cerebrospinal fluid than in plasma. Thiocyanate is known to be bound to proteins. Inulin, sucrose, and sulfate are known to be removed rapidly from the cerebrospinal fluid by bulk flow, and thiocyanate and iodide are actively transported out of the cerebrospinal fluid by specific carrier mechanisms. So, none of these markers can be expected to reach a true equilibrium with the extracellular fluid but merely to come to some steady-state concentration. Thus, using the standard techniques, measurements will yield high values if the markers can also slowly penetrate the cells. More to the point, however, is the newer realization that the concentration of these markers can be vastly affected by the outward pumping mechanisms associated with the cerebrospinal fluid, either the bulk flow turnover or the specific active transport mechanisms. That is, in the brain, the concentration of such a marker depends on the equilibrium between its rate of permeation of the extracellular fluid and the rate of its excretion via the cerebrospinal fluid back into the blood. Some of these problems have been overcome using newer and more sophisticated methods in which an equilibrium is attained by combined ventriculocisternal perfusion along with an intravenous infusion. When such methodologies are applied, the best figures for extracellular space seem to be 14 to 24% obtained with sulfate, inulin, or sucrose, and 21 to 30% obtained with sodium, chloride, thiocyanate, or iodide.

The latter figures seem the more reliable since these markers achieve a more rapid entry into the brain.

Measurements of extracellular space using the electrical impedance of the tissue as an index have given an additional insight into this problem. The assumption is that only the extracellular ions carry the applied current since the cell membranes provide a much greater barrier to current flow. In the rat, the cat, and the rabbit, cortical impedance measurements yield a figure for the extracellular space of between 18 and 25%.

Thus, the controversy has raged and the various methodologies have been evaluated and criticized [4,9-11]. At the moment the thought is that the extracellular space of brain is not too different from that of other tissues. That is, the extracellular space of brain is certainly quite substantial and a consensus now seems to be grouped about a true brain extracellular space of between 15 and 25%. It is of interest that this extracellular compartment is subject to change by a variety of abnormal conditions. For example, the extracellular space decreases markedly upon asphyxia, whether measured by impedance or observed in the electron microscope. Cerebral edema caused by trauma, on the other hand, leads to a marked increase in the sulfate or thiocyanate space. Clearly, these changes can be interpreted as being due to rapid shifts of ions and water between the extracellular and the intracellular compartments.

Biochemical Comparison Between Neurons and Glia

Biochemical comparisons of the chemical and enzymatic composition of neurons and glia have been attempted using several different experimental approaches. The problem is a difficult one, certainly, because of the intricate intertwining of the various cell types. To date there has been only a limited amount of work done, for example, on the comparative biochemistry of the various types of glia. Even the several methods used to study the metabolic and structural differences between the neurons and the glia are open to serious criticism. Some information of a more or less general sort is available and has been presented in a number of reviews [2,12-15].

The major approaches to the problem include: (1) the isolation of viable cells by microdissection, as pioneered by Hydén and his group and later by Giacobini; (2) the use of quantitative histochemistry on minute, frozen sections from plugs of brain, primarily, if not exclusively, by Lowry and his collaborators; and (3) the preparation of fractions enriched in either neurons or glia by means of centrifugation of the disaggregated tissue through stepwise gradients by Rose and others. Some information has been obtained by comparing gray matter with white, since gray is thought to contain primarily neurons and the white matter mostly oligodendroglia. The objections to these measurements are that the white matter contains large amounts of myelin sheath and axons as well as glia, and the gray matter, although containing only neurons, may contain a specialized type of neuron and may not be representative of neurons in general. That is, the "purer" the gray tissue, the more specialized its constituent cells may be. Investigation of tumors of the nervous system as representatives of pure cell types, e.g., gliomas or neuroblastomas, have been reported, but these studies are also suspect. Obviously, it is not known to what extent the tissue is altered upon becoming malignant. Studies of certain nonmammalian preparations have been carried out. These studies are possible since certain species possess neurons or glia of sufficient size to be isolated intact by microdissection. Among the preparations available are the neurons of the snail (*Helix*) and of the sea-slug (*Aplysia*), or of the puffer fish (*Spheroides maculatus*), the stretch receptor of the crayfish (*Orcinectes virilis*), and the Mauthner neuron of the goldfish (*Carassius auratus*). For glial studies, people have used the glia of the leech central nervous system or of the amphibian optic nerve. But all these data are questionable even if they can be extrapolated to the cells of higher organisms. The fact that these cells are so big as to be available to microsurgery suggests that they are not typical. Cells in culture have been studied to provide comparative information, but a discussion of the behavior of these cells is reserved for a later section. The three techniques which have been most thoroughly exploited to date, then, are the microdissection methods of Hydén, the quantitative histochemistry of Lowry, and the centrifugal techniques of Rose, and these will now be discussed in detail.

By sectioning brain or spinal cord and lightly staining the sections with methylene blue it is possible to visualize the neurons. Viable cells have been removed under a dissecting microscope using glass or steel knives or with a glass rod with a nylon loop attached to the end. Using such micromethods, Hydén and his coworkers have been able to obtain several dozen neurons in less than an hour. Cells so prepared respire for many hours, carry out oxidative phosphorylation, and maintain a membrane potential. For comparative purposes these studies involved the concomitant removal of an equivalent volume of the glial cell mass surrounding the neuron under study. Ingenious analytical microtechniques have then been applied to these minute bits of tissue so that their metabolic activities and chemical compositions could be compared. For studying the oxidative activities, a microdiver technique was developed. This technique involves sealing the cell or cells, suspended in media, in a capillary tube. Changes in the gas pressure cause the capillary to rise or sink, and thus the appropriate enzymatic activity can be evaluated. In the work done by Hydén's group it was shown that the glycolytic and oxidative activities of the two cell types are roughly comparable (Table 1.1) [16-18]. Using similar methods, Giacobini has shown that marked differences exist between neurons and glia with regard to certain specific enzymes. Carbonic anhydrase has been found to be localized almost exclusively in glial cells (Table 1.2) [19,20]. In some samples, the glia associated with a

Table 1.1 Distribution of Some Enzyme Activities in Neurons and Glia in the Rabbit

Enzyme or system	Source	Neuron body	Associated glia	Ratio N/G
		μl O_2 consumed per hour, $\times 10^{-4}$		
Cytochrome oxidase	Spinal ganglia	4.5	1.6	2.8
	Dieters' cell	4.2	11.5	0.4
Succinoxidase	Spinal ganglia	3.4	5.1	0.7
	Dieters' cell	2.2	4.2	0.5
α-Ketoglutarate oxidation	Dieters' cell	2.2	2.1	1.1
Glutamate oxidation	Dieters' cell	2.2	1.1	2.0
Pyruvate and malate oxidation	Dieters' cell	0.8	1.9	0.4
Anaerobic glycolysis	Dieters' cell	9.1[a]	7.4[a]	1.2

[a]These values represent μl CO_2 produced per hour, $\times 10^{-4}$.

Table adapted from the work of H. Hydén: H. Hydén, S. Lovtrup, and A. Pigon, *J. Neurochem.* 2:304 (1958); A. Hamberger and H. Hydén, *J. Cell Biol.* 16:521 (1963); A. Hamberger, *Acta Physiol. Scand.,* Suppl. 203, 58 (1963).

Table 1.2 Distribution of Carbonic Anhydrase and the Cholinesterases in Neurons and Glia

Enzyme	Source	Neuron body	Associated glia	Ratio N/G
		μl of CO_2 or O_2 per hour, $\times 10^{-4}$		
Carbonic anhydrase	Dieters' cell	6.4	385	0.02
Acetylcholinesterase	Anterior horn cell	10-50	0	—
	Spinal ganglia	2-20	0	—
	Sympathetic ganglia	2-30	0	—
Nonspecific cholinesterase	Anterior horn cell	0	4.1	—
	Spinal ganglia	5	7	0.7
	Sympathetic ganglia	3	9	0.3

Table adapted from the work of E. Giacobini: E. Giacobini, *Science 134:*1524 (1961); E. Giacobini, *J. Neurochem.* 9:169 (1962); E. Giacobini, *Acta Physiol. Scand.,* Suppl. 156, 45(1959).

Table 1.3   Content and Composition of RNA in Neurons and Glia of the
Dieters' Nucleus of the Rat

RNA	Neuron body	Associated glia
Content, pg/cell or equivalant weight of glia	1545	123
Composition (%)		
Adenine	20.5	25.3
Guanine	33.7	29.0
Cytosine	27.4	26.5
Uracil	18.4	19.2

Table adapted from the work of H. Hydén: H. Hydén and A. Pigon, *J. Neurochem.* 6:57
(1960); H. Hydén and B. McEwen, *Proc. Natl. Acad. Sci. USA* 55:354 (1966); E. Egyhazi
and H. Hydén, *J. Biophys. Biochem. Cytol.* 10:403 (1961); H. Hydén and E. Egyhazi, *Proc.
Natl. Acad. Sci. USA* 48:1366 (1962); H. Hydén and E. Egyhazi, *Proc. Natl. Acad. Sci.
USA* 49:618 (1963).

particular neuron contained up to 120 times as much carbonic anhydrase as the
neuron itself. Carbonic anhydrase, in fact, is now generally accepted as a glial
marker. The reason for this surprising distribution is not known, but an explanation in terms of a function for carbonic anhydrase in chloride transport has been
proposed [12]. Similar methods have been used to evaluate the distribution
among neurons and glia of the enzymes which hydrolyze acetylcholine. The
specific acetylcholinesterase has been found to be present only in neurons while
the glia contain most, if not all, of the nonspecific butyrylcholinesterase (Table
1.2) [21]. More recent work by Hydén and his coworkers has centered on the
content and composition of the proteins and nucleic acids of neurons and their
associated glia. In their hands, the brain-specific protein, S-100, was found primarily in the glia [22]. Also, it was observed that the neurons contain more
RNA and that its composition is different than that of glia (Table 1.3) [23-26].
The observation that there is more RNA is not unexpected since it has been
known for many years that neurons stain more or less selectively with dyes which
bind to RNA. The compositional differences can be explained on much the same
basis, i.e., the extra RNA of the neurons is probably due to the presence of more
ribosomes in these cells. Thus, the gross RNA of the glia probably resembles the
DNA composition to a greater extent, since the RNA of the ribosomes does not.
However, these differences in composition may reflect a difference in the nature
of the proteins synthesized by the two cell types and, hence, be a true reflection
of the differences in function. Among the most surprising and interesting findings from these studies is that the activities of various enzymes and the content
and composition of RNA in the cells changes under different physiological and
environmental conditions. A consideration of these data is presented in Chap. 26.
    Several points should be made with regard to this work. First, the methods

used are difficult and expensive to perform, and few, if any, similar data have been presented by other groups. Second, it is generally acknowledged that the neurons isolated are stripped of both axons and dendrites and, although resealed, represent only the neuronal cell body and do not reflect the metabolism of the processes. Third, the only neurons amenable to study are, perforce, the very large ones which can be dissected out and may not be at all typical. Although the above points certainly should be considered in evaluating the data obtained by such methods, these investigations have provided some extremely important facts.

The work of Lowry and his collaborators has involved the use of frozen samples and the application of intricate metabolic techniques for measuring minute quantities of enzymes, cofactors, and metabolites. The general approach of these studies has been to prepare "plugs" of nervous tissue which encompass many cell layers or tracts. These plugs are then freeze-dried and sectioned into disks of some 10 μg in weight [27]. Since certain layers are enriched in cell bodies, others in dendrites, others in synapses, and so on, it is possible to correlate various biochemical properties with various types or parts of cells [28-30]. By using alternate sections for microassay and for microscopy these correlations can be made rather precisely. In other studies, single cell bodies were microdissected from the freeze-dried sections [31,32]. Data on the distribution of a great number of enzymes have been obtained from such studies (Table 1.4). Also, accurate data on the content of several lipids, γ-aminobutyric acid, DNA, riboflavin, chloride, and phosphate in various cells or cell layers have been presented. By using the product of one reaction as cofactor for another, enormous amplification of the effects of various enzymes or other constituents has been obtained, enabling final measurements of trace quantities to be at conventional levels. Perhaps the best illustration of the enormous sensitivity of these tools is the measurement of the activity of malic dehydrogenase in samples of 0.01 μg dry weight [31].

In general, Lowry found that the enzyme activities of the major energy-yielding systems are usually at a lower level in nerve cell bodies than in the average brain tissue. Further, the cell layers containing neural processes had a much higher content of these enzymes than the layers containing cell bodies. Thus, Lowry, among others, has suggested that a major portion of the metabolism of the brain is dendritic.

Although a number of groups have presented data on the macroseparation of glia and neurons by centrifugal techniques, by far the most complete work has been done by Rose and his coworkers. Not only has the methodology been presented in detail [33,13,14], and carefully modified [34], but rather thorough comparative studies have been done by this group in an attempt to evaluate their methods in comparison with the other techniques available for this purpose [35, 36]. Basically, the Rose technique allows the preparation of fractions enriched in glial or neuronal elements through the use of discontinuous density centrifuga-

Table 1.4   Enzyme Activities in Single Spinal Ganglion Cell Bodies and Ganglion Cell Capsules of the Rabbit[a]

Enzyme	Ganglion cell body	Ganglion glial capsule	Ratio N/G
Hexokinase	7.2	2.2	3.3
Phosphoglucoisomerase	49	24	2.0
Lactic dehydrogenase	50	28	1.8
Glucose-6-phosphate dehydrogenase	2.15	4.8	0.4
6-Phosphogluconate dehydrogenase	1.29	2.22	0.6
Isocitric dehydrogenase	4.5	6.2	0.7
Malic dehydrogenase	231	90	2.6
Glutamate dehydrogenase	2.5	1.6	1.6
Glutamate-oxalacetate transaminase	35	6.5	5.4

[a]Moles of substrate converted per kilogram of lipid-free dry weight per hour per cell or equivalent volume.

Table adapted from the work of O. H. Lowry: O. H. Lowry, N. R. Roberts, and M. L. Chang, *J. Biol. Chem. 222*:97 (1956); O. H. Lowry, in *Metabolism of the Nervous System* (D. Richter, ed.), Pergamon, New York, 1964.

tion in Ficoll, a sucrose polymer of approximately 4×10^5 molecular weight. The cleansed tissue is "teased" through a nylon mesh into 0.1 M KCl containing 10% Ficoll. This suspension is then filtered through a stainless steel grid. Alternatively, the separation can be done using glass bead filtration [34]. The filtrate is layered onto 30% Ficoll which in turn covers a layer of 1.45 M sucrose in a centrifuge tube. The tube is then spun in a swinging bucket rotor. After a rather brief centrifugation the various interfaces are collected. The undisrupted tissue and the myelin fraction remain at the top, and the red cells and free nuclei pellet at the bottom. The glia fraction is at the interface between the 10% and the 30% Ficoll and the neuronal fraction at the interface between the 30% Ficoll and the sucrose (Fig. 1.7). From 20 rat brains a neuronal fraction can be obtained which contains (parts of) up to 10% of the neurons originally present. This fraction has on the order of 3×10^7 "cells" and about 100 mg of wet weight. The contamination with glia, evaluated using carbonic anhydrase as a glial marker, is about 11%. The glia fraction itself yields about 200 mg of wet weight. The entire procedure takes about 3 hr.

Criticisms of this methodology by its developers and by others have centered around the questions of the state of the cells isolated and the levels of contamination of the various fractions. There is no question that the "neuronal"

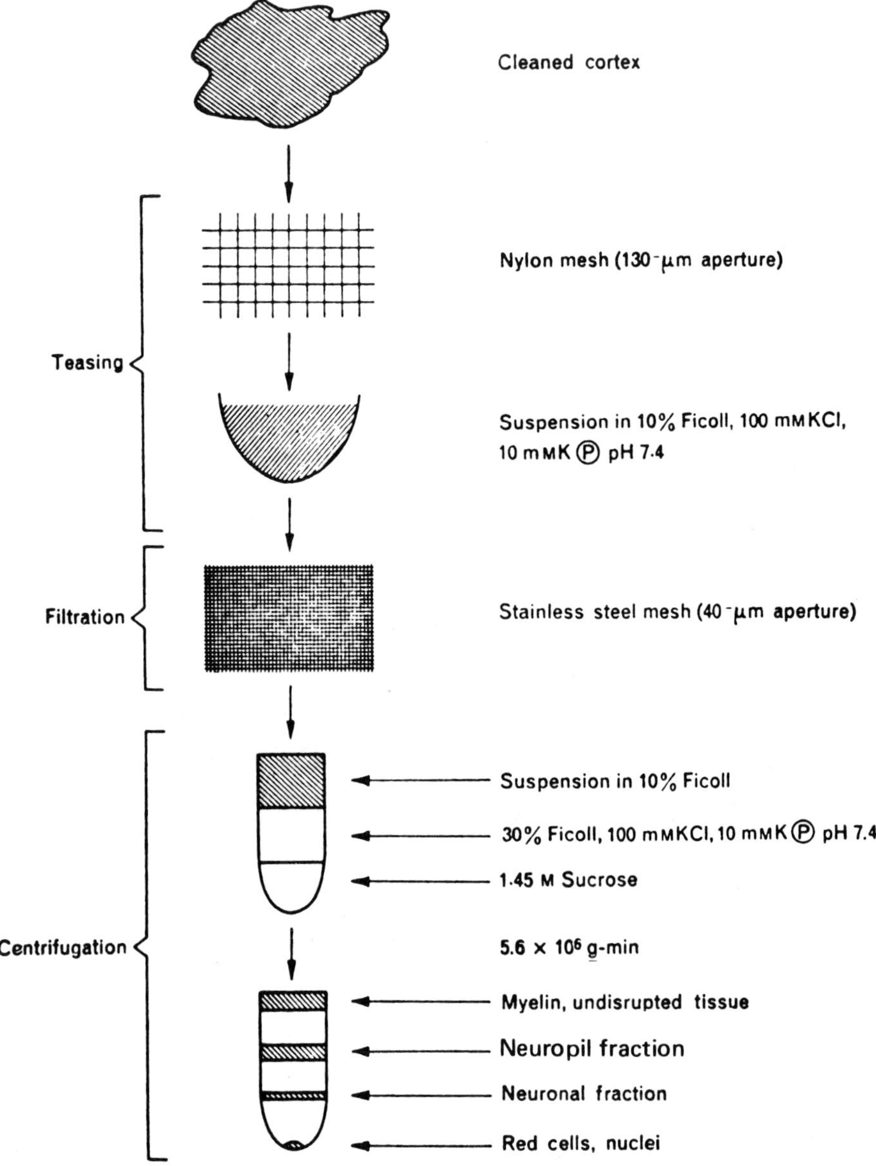

Figure 1.7   Bulk separation of neurons and glia. [From S. P. R. Rose, *Biochem. J. 102*:33 (1967); also in *Applied Neurochemistry* (A. N. Davison and J. Dobbing, eds.), F.A. Davis Company, Philadelphia, 1968.]

fraction is, in fact, composed mainly of cell bodies of neurons which, although resealed and probably mostly covered with complete cell membrane, are devoid of
axons and dendrites. This is a serious drawback since it is becoming quite clear
that these cell processes have an active metabolism which may be quite different
from the metabolism of the cell body itself. And the neuronal fraction, although
only 11% contaminated by glia, is also known to contain endothelial cells and
some cell debris. The "glial" fraction contains no neurons, but does have a substantial contamination with synaptic elements, dendrites, axons, and subcellular
particles. Thus, these fractions can be considered enriched for glia or neuronal
cell bodies, bur are certainly not pure. The fractions isolated, on the other hand,
are most surely metabolically active, and their properties suggest that the data
obtained have substantial physiological validity. Rose has shown that the neuronal
elements behave like osmometers, so they are intact [37]. Consistent with this is
the finding that these bodies will not allow soluble enzymes to escape into the
medium [33] and will take up amino acids and ions actively [37,38]. The preparations respire at about 70 to 80% of the rate of brain slices and generate CO_2,
lactate, and amino acids from glucose. The cells also manufacture ATP and proteins. In Rose's words they behave "like a slightly damaged slice." The slice, of
course, behaves unfortunately like a more or less damaged brain. Nonetheless,
cell fractions from young animals can be grown in tissue culture, and although
membrane potentials have not been detected in these cells, this may be merely because most of them are extremely small for such experimental probing. In any
case, the macrotechniques remain the only way to obtain substantial amounts of
material for biochemical study.

Using the macromethods, a number of comparisons between the neuronal-
enriched and the glial-enriched fractions have been made (Table 1.5). Rose has
found that the neuronal fraction has a higher content of certain free amino acids
than the glial. For example, the neuronal fraction contains substantially more
glutamate, glutamine, γ-aminobutyrate, alanine, and aspartate [37]. Many enzymes, on the other hand, appear to be in about equal concentration in the two
fractions. The lactic and succinic dehydrogenase, aspartate transaminase, glutamate decarboxylase, and triphosphoinositide phosphomonoesterase content of
the two fractions are similar [33,37]. Cytochrome oxidase is about twofold and
glutamate dehydrogenase about fourfold higher in the neuronal fraction [33,39].

Generally speaking, then, it is known that the overall armamentarium of
metabolic enzymes is present in both glia and neurons. Although the various techniques give widely varying quantitative results, both cell types can oxidize glucose
and manufacture ATP. The only clear qualitative enzymatic differences are the
presence of carbonic anhydrase only in the glia and the specific acetylcholinesterase only in the neurons. Certain brain-specific proteins seem to be localized in
one or the other of the cell types. That is, the S-100 protein seems to be primarily a glial component and the 14-3-2 protein occurs only in neurons. These

Table 1.5 Comparison of the Contents and Properties of Neuron-Enriched (Neuronal) and Glial-Enriched (Neuropil) Fractions

Constituent	Neuronal	Neuropil	Ratio Neuronal/Neuropil
Protein (%)	0.9	14.3	0.06
DNA (μg/mg protein)	67.6	8.6	7.9
RNA (μg/mg protein)	71.2	16.4	4.3
ATP (nmol/mg protein)	5.18	1.72	3.0
ATP (nmol/mg protein), after 60 min incubation with glucose	12.31	7.36	—
Glutamic acid (nmol/mg protein)	19.9	15.1	2.7
Aspartic acid (nmol/mg protein)	10.5	10.9	0.96
Glutamine (nmol/mg protein)	47.2	33.0	1.43
Alanine (nmol/mg protein)	41.7	30.3	1.38
GABA (nmol/mg protein)	21.1	11.3	1.87
Glutamate dehydrogenase (μmol substrate/mg protein/hour)	1.99	0.50	4.0
Aspartate aminotransferase (μmol substrate/mg protein/hour)	5.73	5.75	1.0
Glutamate decarboxylase (μmol substrate/mg protein/hour)	0.24	0.20	1.2
Glutamine synthetase (μmol substrate/mg protein/hour)	1.40	0.79	1.8

Table adapted from the work of S. P. R. Rose: S. P. R. Rose, *J. Neurochem. 16:*1319 (1969); S. P. R. Rose, *J. Neurochem. 15:*1415 (1968).

proteins are considered in detail in Chap. 15. The rate of RNA and protein synthesis of neurons is generally thought to be higher than that of the glia, and in the mature brain, DNA synthesis occurs only in the glia. The composition of the total RNA in the cells is quite different, the glial RNA being much higher in guanine and lower in adenine than the neuronal and, in general, resembling the composition of the DNA. And, although the neurons maintain a greater free amino acid pool than do the glia, the rate of conversion of glucose to amino acids proceeds twofold to threefold faster in the glia. For all the ingenious and delicate techniques now available it is clear that the methodology for the separation of the many different cell types in the brain is woefully inadequate and that even more refined methods are badly needed.

References

1. T. R. Shantha, S. L. Manocha, G. H. Bourne, and J. A. Kappers, in *The Structure and Function of Nervous Tissue,* Vol. II (G. H. Bourne, ed.), Academic, New York, 1969.
2. H. Hydén (ed.), in *The Neuron,* Elsevier, Amsterdam, 1967.
3. B. Wyke, *Principles of General Neurology,* Elsevier, Amsterdam, 1969.
4. B. G. Cragg, in *Applied Neurochemistry* (A. N. Davison and J. Dobbing, eds.), F.A. Davis, Philadelphia, 1968.
5. B. H. Waksman and R. D. Adams, *Am. J. Pathol. 41:*135 (1962).
6. S. W. Kuffler, J. G. Nicholls, and R. K. Orkand, *J. Neurophysiol. 19:*768 (1966).
7. A. Peters, S. L. Palay, and H. deF. Webster, *The Fine Structure of the Nervous System,* Harper & Row, New York, 1970.
8. R. Edström, *Acta Psychiatr. Neurol. Scand. 33:*403 (1958).
9. J. Dobbing, in *Applied Neurochemistry* (A. N. Davison and J. Dobbing, eds.), F.A. Davis, Philadelphia, 1968.
10. S. W. Kuffler and J. G. Nicholls, *Ergebn. Physiol. Biol. Chem. Exp. Pharmakol. 57:*1 (1966).
11. H. A. Pappius, in *Handbook of Neurochemistry,* Vol. II (A. Lajtha, ed.), Plenum, New York, 1969.
12. E. Giacobini, in *Morphological and Biochemical Correlates of Neural Activity* (M. M. Cohen and R. S. Snider, eds.), Hoeber, New York, 1964.
13. S. P. R. Rose, in *Applied Neurochemistry* (A. N. Davison and J. Dobbing, eds.), F.A. Davis, Philadelphia, 1968.
14. S. P. R. Rose, in *Handbook of Neurochemistry,* Vol. 11 (A. Lajtha, ed.), Plenum, New York, 1969.
15. M. M. Brand and G. M. Lehrer, in *Handbook of Neurochemistry,* Vol. Va (A. Lajtha, ed.), Plenum, New York, 1971.
16. H. Hydén, S. Lovtrup, and A. Pigon, *Neurochem. 2:*304 (1958).
17. A. Hamberger and H. Hydén, *J. Cell Biol. 16:*521 (1963).
18. A. Hamberger, *Acta Physiol. Scand.,* Suppl. 203, *58* (1963).

19. E. Giacobini, *Science 134:*1524 (1961).
20. E. Giacobini, *J. Neurochem. 9:*169 (1962).
21. E. Giacobini, *Acta Physiol. Scand.,* Suppl. 156, *45* (1959).
22. H. Hydén and B. McEwen, *Proc. Natl. Acad. Sci. USA 55:*354 (1966).
23. H. Hydén and A. Pigon, *J. Neurochem. 6:*57 (1960).
24. E. Egyhazi and H. Hydén, *J. Biophys. Biochem. Cytol. 10:*403 (1961).
25. H. Hydén and E. Egyhazi, *Proc. Natl. Acad. Sci. USA 48:*1366 (1962).
26. H. Hydén and E. Egyhazi, *Proc. Natl. Acad. Sci. USA 49:*618 (1963).
27. O. H. Lowry, *J. Histochem. Cytochem. 1:*420 (1953).
28. O. H. Lowry, N. R. Roberts, K. Y. Leiner, M. L. Wu, and A. L. Farr, *J. Biol. Chem. 207:*1 (1954).
29. O. H. Lowry, N. R. Roberts, K. Y. Leiner, M. L. Wu, A. L. Farr, and R. W. Albers, *J. Biol. Chem. 207:*39 (1954).
30. O. H. Lowry, N. R. Roberts, M. L. Wu, W. S. Hixon, and E. J. Crawford, *J. Biol. Chem. 207:*19 (1954).
31. O. H. Lowry, N. R. Roberts, and M. L. Chang, *J. Biol. Chem. 222:*97 (1956).
32. O. H. Lowry, in *Metabolism of the Nervous System* (D. Richter, ed.), Pergamon, New York, 1956.
33. S. P. R. Rose, *Biochem. J. 102:*33 (1967).
34. S. P. R. Rose and A. K. Sinha, *Life Sci. 9:*907 (1970).
35. S. P. R. Rose, *J. Neurochem. 17:*809 (1970).
36. A. K. Sinha and S. P. R. Rose, *Brain Res. 33:*205 (1971).
37. S. P. R. Rose, *J. Neurochem. 16:*1319 (1969).
38. H. F. Bradford and S. P. R. Rose, *J. Neurochem. 14:*373 (1967).
39. S. P. R. Rose, *J. Neurochem. 15:*1415 (1968).

2
CELLULAR ORGANIZATION

Cell Body

The cell body, or perikaryon, of the neuron is the metabolic center of the cell. The cytoplasm, or karyoplasm, of the neuron is extremely viscous, much more so than the cytoplasm of other somatic cells of the body. This fluid is so viscous that it does not diffuse out upon injury to the membrane.

The cell body is surrounded by the cell membrane, which is covered with dendrites. The dendrites are extensions of the cell membrane and may number 200,000 or more per cell. The dendrites are the receptive surfaces of the cell and are generally short, may be arborized, and are covered with spines or granules.

Nucleus

As in most cells, the cytoplasm of the cell body contains several different organelles (Fig. 2.1). There is generally only one large, centrally located nucleus. This nucleus is pale and roughly spherical. It comprises about one-fourth to one-fifth of the volume of the cell body in newborn rats, for example, and as the brain matures this ratio drops to about one-tenth. In general, the nucleus is on the order of 1/100 or less of the total cell volume, considering all the processes.

The nucleus contains one or more formed bodies known as nucleoli. The nucleolus is frequently at the center of the nucleus and contains most of the nuclear RNA. It is an irregular, spongy matrix with no limiting membrane. The nucleolus contains a number of enzymes, including acid phosphatase, phosphodiesterase, lactic acid dehydrogenase, succinic acid dehydrogenase, glucose-6-phosphate dehydrogenase, monoamine oxidase, ATPase, and glucose-6-phosphatase. Probably, the nucleolus moves back and forth to the nuclear membrane, discharging RNA into the cytoplasm.

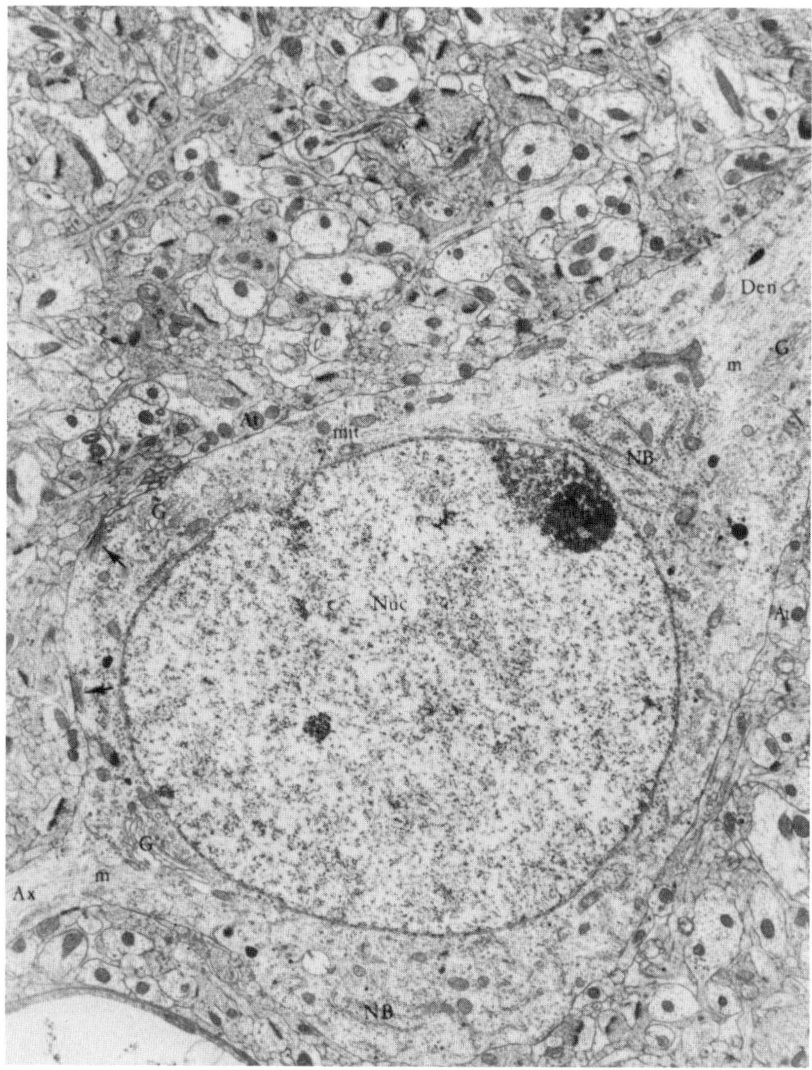

Figure 2.1 The cell body of a pyramidal neuron. Occupying the center of the field is the perikaryon of a pyramidal neuron from the superficial layers of the cortex. Most of the perikaryon is occupied by a large, rounded nucleus (Nuc), which contains a nucleolus (ncl) and homogeneously dispersed karyoplasm. The cytoplasm surrounding the nucleus is confined to a thin rim containing a few Nissl bodies (NB). One relatively large Nissl body lies at the base of the apical dendrite (Den). Some of the cisternae of the granular endoplasmic reticulum are continuous with subsurface cisternae (arrows). Also prominent within the

Golgi

In the cytoplasm near the nucleus is the Golgi apparatus (Fig. 2.2a). Upon staining this organelle appears as a network or a spheroidal system of vesicles, vacuoles, sacs, and filaments disposed around the nucleus like a garland or shell. Rather surprisingly, the Golgi contains all the thiamine pyrophosphatase in the cell, and this enzyme is used as a histological marker to identify the Golgi. The Golgi apparatus is frequently found extending into the dendrites, but not into the axon. It may function in forming vesicles or lysosomes.

Nissl Bodies

The most characteristic feature of the perikaryon of neurons is the presence of a basophilic substance called Nissl bodies (Fig. 2.2b). This substance has been shown to be a special and concentrated form of the endoplasmic reticulum which exists in parallel rows and is lined with ribosomes. These structures contain ribonucleoprotein and possibly polysaccharides and sphingomyelin, and are the principal site of protein synthesis in the cell. The Nissl substance is so intense in the neuron that its staining with basic dyes is the classic method for detecting neurons histologically, and of differentiating them from the surrounding glia.

Mitochondria

Mitochondria occur in large numbers randomly distributed throughout the cell body (Fig. 2.2c). The mitochondria are smaller than in other tissues and appear rodlike or filamentous in the neuron. The cristae of these mitochondria are sometimes found to run the length of the structure instead of across, and the mitochondria do not contain the dense granules usually found in mitochondria from other tissues.

Other Inclusions

The cell body also contains a number of other particles in varying amounts. Lysosomes are found, usually as spherical or oval bodies from 0.3 to 2 μm in diameter.

Figure 2.1 (continued)
perikaryal cytoplasm are elements of the Golgi apparatus (G), mitochondria (mit), and microtubules (m) that stream into both the apical dendrite (Den) and the emerging axon (Ax). A number of axon terminals (At) synapse upon the surface of this neuron. Cerebral cortex from an adult rat, X7200. (Figure courtesy of Dr. Alan Peters, from A. Peters, S. L. Palay, and H. deF. Webster, in *The Fine Structure of the Nervous System,* Harper & Row, New York, 1970.)

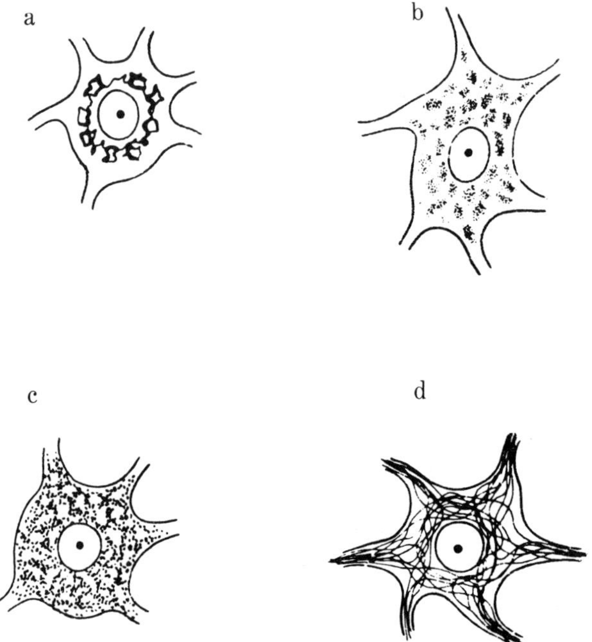

Figure 2.2 Various inclusions in the neuronal cell body. (a) Golgi apparatus of a motoneuron in the spinal cord; osmium tetroxide inpregnation. (b) Motoneuron in the abducens nucleus showing the distribution of Nissl bodies; thioin. (c) Motoneuron in the abducens nucleus showing the distribution of mitochondria; Altmann-Kull method. (d) Motoneuron in the spinal cord, showing the distribution of neurofibrils; Cajal silver stain.

These lysosomes, as in other tissues, contain many of the hydrolytic activities of the cell, such as acid phosphatase, and their number and activity increase following injury to the tissue. Melanin granules are found primarily in the substantia nigra and the locus ceruleus. They contain large amounts of protein, lipid, and carbohydrate, small amounts of RNA, and traces of various metals. The lipopigment granules (or lipofuchsin) can be seen in neurons examined by electron microscopy. They contain lipid, carbohydrate, and protein and can be readily visualized with lipid stains. Although they have been termed the "senile pigment" this is something of a misnomer since, in humans, they appear during the first and second decade of life. The appearance of these particles may be related to the autooxidation of fatty acids since a similar pigment (ceroid) appears as a consequence of prolonged vitamin E deficiency. Lipid droplets or particles are frequently seen in the cell body. In young individuals these particles may be made up of phospholipid and some lipoprotein. With age some neutral lipid appears in the particle. Glycogen particles are sometimes found in small amounts and seem to increase during hibernation and decrease with fever or starvation. Iron-containing inclusions are observed in some cell bodies. Finally, the perikaryon is crowded with filamentous or membranous fibers about 100 Å in diameter, called neurofilaments or neurofibrils, which are fine strands of protein and which make up a meshwork of unknown function (Fig. 2.2d).

Dendrites and Axons

The dendrites contain Nissl bodies and mitochondria, and a series of rodlike channels 200 to 250 Å in diameter known as neurotubules, which are the microtubules of neural tissue. These are different from the neurofilaments. These channels are almost certainly involved in the orderly flow of materials from the cell bodies to the extremities of the processes, a phenomenon known as "axonal flow." The axon arises from the perikaryon through a raised area known as the axonal hillock. The axon in mature neurons is surrounded and enveloped by the myelin sheath elaborated by specific glial cells. The axon contains mitochondria, neurofilaments, and neurotubules, but not Nissl substance.

Synaptic Endings

The axon terminus is usually expanded or bulbous and is known as the synaptic bouton. The bouton is crowded with mitochondria and with a large number of synaptic vesicles, which differ in size and shape depending on the type of neuron from which they originate. The bouton also contains a small amount of RNA. These vesicles are mostly membrane bound and contain the transmitter substance. The vesicles may be manufactured in the Golgi apparatus and transported to the bouton by axonal flow, although they have never been observed in the normal axon. On the other hand, they probably originate in the bouton itself, being made directly from the synaptic membrane.

The neurons communicate across the synaptic cleft, or synapse. This cleft is formed by the apposition of the axonal bouton of one cell and the dendritic surface of another (Fig. 2.3). Generally, the synaptic cleft is between 200 and 400 Å wide. The surfaces at the synapse have sticky, adherent plaques. The cleft is permeated by cytoplasmic filaments, and these filaments may form a dense plate midway between the surfaces.

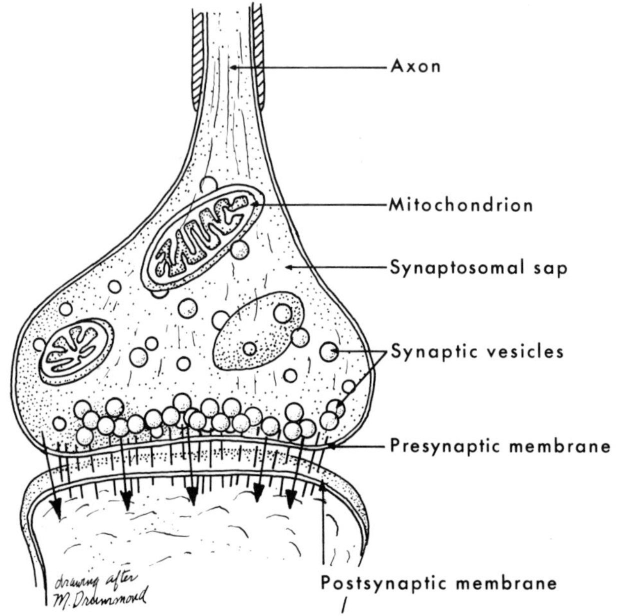

Figure 2.3   Schematic section of a synapse (drawing by M. Duncan). [Figure courtesy of Dr. Patrick L. McGeer, from P. L. McGeer, "The Chemistry of Mind," *Am. Sci. 59:*221 (1971).]

Glial Organization

Substantially less information is available about the details of the cellular organization of the various types of glia. The astrocytes have large numbers of intracellular organelles and an oval nucleus with two or more nucleoli. There are many free ribosomes but not much granular endoplasmic reticulum. Relatively small numbers of mitochondria appear, but a well-developed Golgi apparatus is present. The astrocytes contain glycogen granules and filaments about 60 to 100 Å thick. The oligodendrocytes have a small, more or less round nucleus in which the nucleolus is not as apparent a feature as it is in neuronal nuclei. The ribosomes are densely packed, some being bound and some free. These cells contain a variable number of mitochondria and have a well developed Golgi apparatus. They have conspicuous neurotubules, but few glycogen granules are found. Generally, the nucleus and the cytoplasm of the oligodendrocytes is denser than those of the astrocytes. Unfortunately, the information about cellular organization which is available about the various classes of glia does not add much to the quest for the exact biochemical functions of these cell types.

3
FRACTIONATION AND CHARACTERIZATION OF SUBCELLULAR PARTICLES

Differential Centrifugation

Isolation of the various subcellular particles in relatively homogeneous form can be accomplished by various centrifugal techniques. Nearly all the useful methodology is based on a combination of classic differential centrifugation in buffered sucrose solutions followed by centrifugation of the appropriate fraction through discontinuous density gradients made up of sucrose or sucrose-Ficoll solutions [1]. Generally, the cells are homogenized in 0.32 M sucrose in a Potter-Elvehjem homogenizer with a Teflon pestle. The clearance between the pestle and the wall of the homogenizer should be on the order of 0.025 cm. The number of strokes or passes used to homogenize the tissue is usually specified for a given procedure. The sucrose solution is generally buffered with Tris at pH 6.2 to 6.7 or with histidine, and may also contain small amounts of Ca^{2+} or Mg^{2+} (1 mM), possibly some K^+, and, in some few procedures, a low concentration of a nonionic detergent to aid in the solubilization of the cytoplasmic membranes.

Differential centrifugation of such an homogenate can yield fractions similar to those isolated from liver, but more extensive purification is required to give homogeneous preparations because the particles behave differently than do hepatic particles, and because structures other than those found in liver are present in brain. But, by differential centrifugation of brain, one can obtain a crude nuclear pellet at low speeds, a crude mitochondrial pellet by moderate-speed centrifugation of the nuclear supernatant, a crude microsomal pellet by high-speed centrifugation of the mitochondrial supernatant, and a high-speed supernatant containing the soluble materials of the cell (Fig. 3.1). Discontinuous density gradient centrifugation of these crude fractions is advantageous for a number of reasons. First, a number of components can be separated at once. Second, the density of the steps in the gradient, the time of centrifugation, and the speed of the centrifugation can all be varied to provide great discretion in the separations achieved. And, third, the layering of the sample on top of such

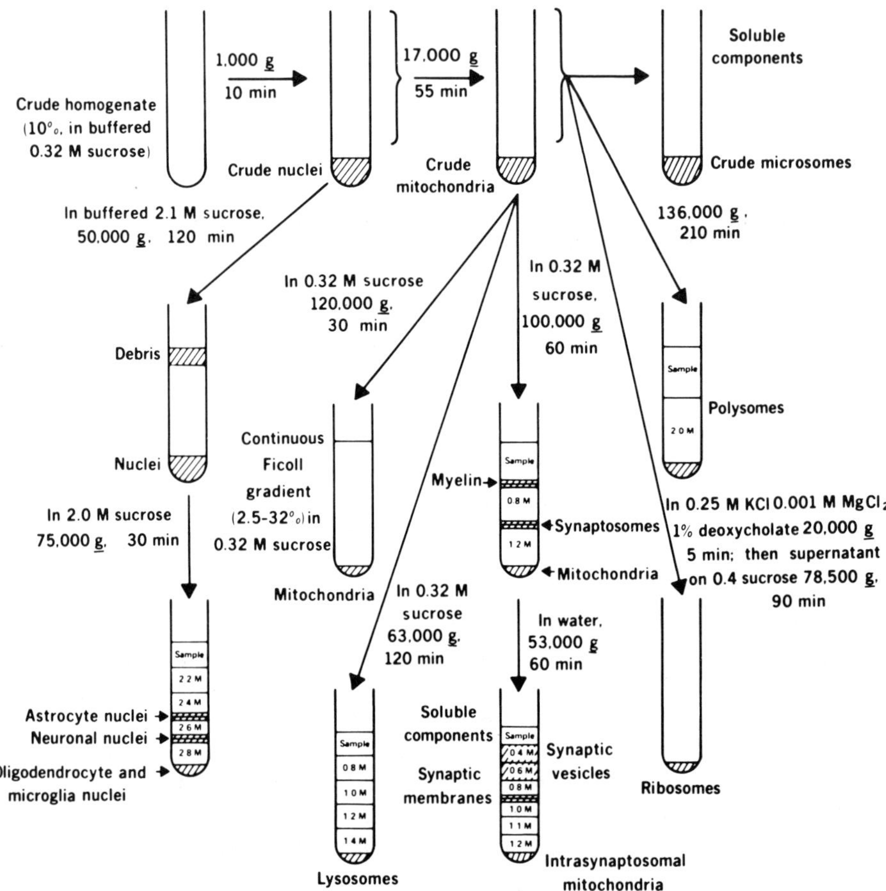

Figure 3.1   Flow sheet for the separation of subcellular particles from brain.

a gradient allows complete removal of the desired component from the initial sample, which may contain substantial amounts of material entrapped from the lighter fractions in the original differential separation.

Nuclei

Purified nuclei can be obtained from the crude nuclear pellet by taking advantage of their very high density and centrifuging them through buffers containing 1.5 to 2.5 M sucrose [1,2] (Fig. 3.2). Alternatively, the original homogenate can

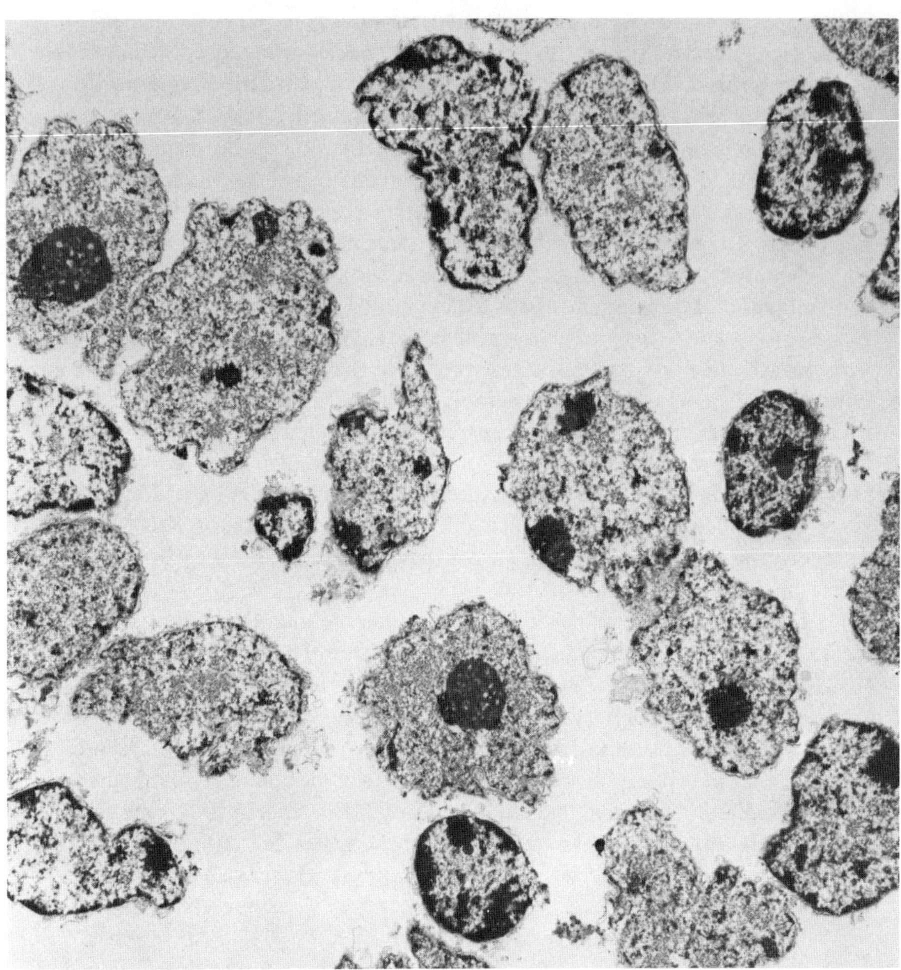

Figure 3.2   Electron micrograph of nuclei from adult mouse brain, ×6700. Fresh brain was homogenized in 0.25 M sucrose-5 mM MgCl$_2$, and successively washed in 1M, 1.4 M, and 1.8 M sucrose-3 mM MgCl$_2$. The nuclear pellet was further centrifuged in 2.3 M sucrose-3 mM MgCl$_2$ in an SW-41 rotor at 22,000 rpm for 30 min. For electron microscopy, the pellet was fixed in 3.5% glutaraldehyde in 0.07 M phosphate buffer (pH 7.0) followed by osmication, dehydration through graded ethanols, and embedding in epon. Sections were stained with uranyl acetate and lead citrate. Note the numerous well-defined nuclei and the lack of cytoplasmic contaminants. [Figure courtesy of Dr. David Choie, from D. D. Choie, E. R. Friedberg, S. R. van den Berg, and M. M. Herman, *J. Neurochem.* *29*:811 (1977).]

be made in 2.2 M sucrose and spun at moderate speeds to pellet the nuclei [3].
In sucrose concentrations of this magnitude the nuclei form a pellet while all the
other components float. By repeated sedimentation under these conditions a
fairly homogeneous nuclear population can be obtained. Stepwise density gradients can be used to separate the nuclei even further into populations representing neuronal and glial nuclei [4,5]. By these techniques brain nuclei have been
found to be variable in size and shape and rather more fragile than liver nuclei.
The purity of these nuclear fractions is best assessed by phase contrast microscopy. Various marker enzymes can be searched for to determine the amount of
contamination. For example, assays for cytochrome oxidase, succinic dehydrogenase, or glutamate dehydrogenase will estimate the mitochondrial population,
while measurement of acetylcholinesterase or of the Na^+,K^+-ATPase will give an
estimate of the contamination by cytoplasmic membranes. Various cautions
have been observed to ensure the preparation of pure nuclear fractions. A pH
of 6.3 to 6.5 appears to be optimal [2]. The inclusion of 1 mM $MgCl_2$ gives the
best preservation of the nuclear structure [4,6]. The addition of K^+ ions prevents swelling [6]. And the presence of a small amount of Triton X-100 aids in
dissolving the cytoplasmic membranes of the cells, but has little apparent effect
on the nuclear membrane [2].

The nuclei from brain, as from other tissues, contain DNA and RNA polymerases [2,7-9]. Also, the nuclei of brain have been shown to hold certain of
the glycolytic enzymes [7,10]. In addition, the enzymes ATPase, adenylate
kinase, NAD-pyrophosphorylase, 5′-nucleotidase, polynucleotide phosphorylase,
and poly(C)-synthetase and poly (A)-synthetase have been reported. It is interesting that NAD-pyrophosphorylase has been shown to be about 10-fold higher in
neuronal nuclei than in glial nuclei [11]. Many other reports of enzyme activities
in the nuclear "fraction" have appeared, but, given the uncertainties involved in
the preparation of clean nuclei, most of these reports must await confirmation
in other laboratories.

Mitochondria

The crude mitochondrial pellet obtained by differential centrifugation of brain
homogenates is heavily contaminated with myelin and with nerve endings. Purified brain mitochondria can be obtained by centrifuging this crude fraction
through density gradients of sucrose or, better still, of sucrose-Ficoll combinations. Several methods of purification are available using either continuous [12]
(Fig. 3.3) or discontinuous gradients [1,13,14].

The brain is a particularly rich source of mitochondria. About 15% of the
protein of the brain is mitochondrial in origin. This is not too surprising, since
brain uses on the order of 25% of the oxygen and glucose consumed by the body
and, of course, must have an extremely active and plentiful energy metabolism.

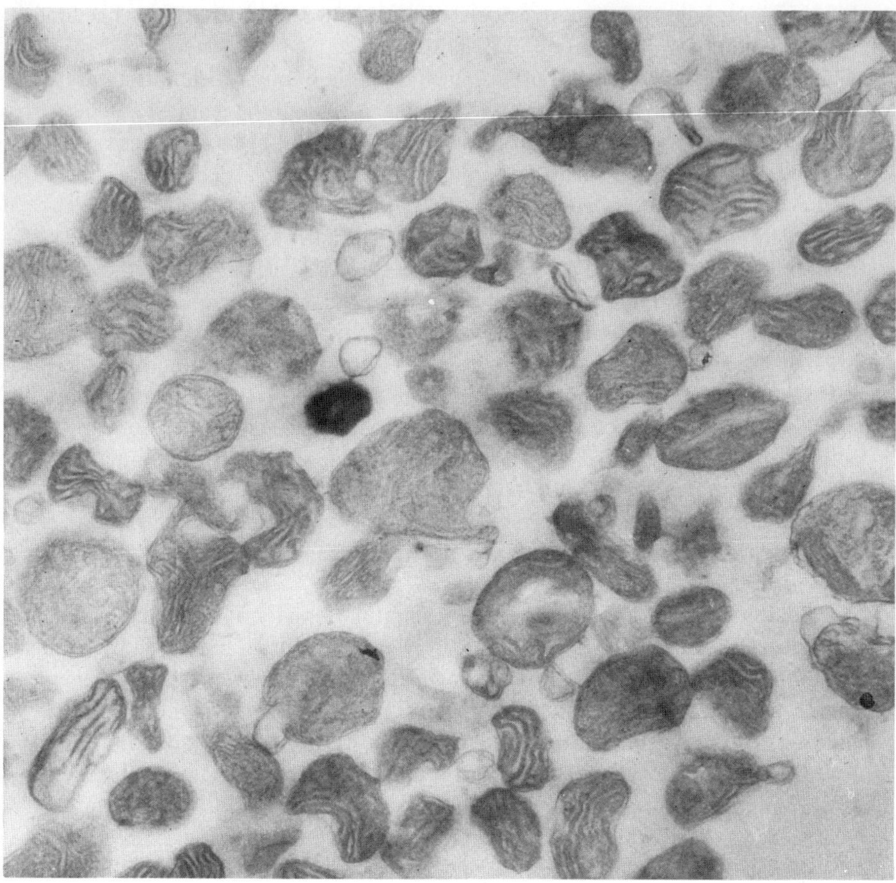

Figure 3.3 Electron micrograph of mitochondria isolated from rat brain by Ficoll density gradient centrifugation, ×40,000. [Figure courtesy of Dr. Leo Abood, from L. G. Abood, in *Handbook of Neurochemistry*, Vol. II (A. Lajtha, ed.), Plenum, New York, 1969.]

Mitochondria are most concentrated in the neuronal elements and particularly in those portions of the cell devoted to excitation, e.g., the synapses and the axon hillock.

Purified mitochondria from brain are grossly comparable to mitochondria isolated from liver or heart. There are some few qualitative differences and quite a number of quantitative differences. They are generally somewhat smaller in size and appear as rods or rounded granules when isolated. They are about 35%

lipid, 57% protein, and 5% ash, calculated on a dry weight basis [15]. The lipid portion contains on the order of 75% phospholipid, 20% cholesterol, and 5% diglyceride and triglyceride. Cardiolipin is found in brain mitochondria, but not in the other subcellular fractions of the brain. Although the detailed studies which have been done with heart mitochondria have not been as extensively pursued with brain, it is clear that brain mitochondria can also be dissociated into a number of submitochondrial particles [15]. Upon examination by electron microscopy it can be seen that brain mitochondria are largely composed of a knoblike structure called the elementary particle. These particles, which may number around 100,000 per mitochondria, have a diameter of about 100 Å and are attached to the membranes of the mitochondria by individual stems. When isolated mitochondria are treated with a variety of disrupting techniques, such as with dilute surfactants, ultrasonics, or repeated freezing and thawing, the mitochondria disintegrate into a number of submitochondrial fractions. Some of the soluble enzymes associated with the intact mitochondria, such as glutamate dehydrogenase and β-hydroxybutyrate dehydrogenase, are lost. A membrane fraction and a particulate which carries on both electron transport and oxidative phosphorylation can be prepared, as well as a fraction of membranes with this respiratory complex attached.

A discussion of the details of the respiratory activity of brain appears in Chap. 21. It should be mentioned here, however, that brain mitochondria contain, of course, the Krebs cycle dehydrogenases, and all the requisite machinery for electron transport and oxidative phosphorylation. They also incorporate radioactive amino acids into protein. Characteristically, brain mitochondria have, quantitatively, a different substrate range than liver or kidney mitochondria. Brain mitochondria have a very high level of α-glycerophosphate dehydrogenase, and they contain most of the hexokinase and monoamine oxidase of the cell. The hexokinase found associated with mitochondria from the brain is bound to the external membrane and acts only on substrates present in the cytosol. It is not known whether this enzyme is bound to mitochondria in situ or whether it binds during the isolation, but the finding that hexokinase ends up in the mitochondrial compartment is unique to the brain. Monoamine oxidase or succinic dehydrogenase activities can be used as markers for brain mitochondria, and their measurement is frequently used to determine mitochondrial contamination in other fractions isolated from the cell.

Synaptosomes

Some of the early studies on brain mitochondria revealed anomalous properties. Many of these results are now known to be due to the fact that, in brain, the mitochondrial fraction contains a particle unique to brain known as a "synaptosome." Before this was recognized, the properties of the synaptosomes were included in many of the characteristics attributed to brain mitochondria.

The work of Whittaker and of deRobertis and their colleagues has provided a large body of information about the synaptosomes and about the structural components of the synaptic region [16,17]. This particle is formed when neural tissue is homogenized. During the shearing process associated with homogenization, the postsynaptic cell breaks just beyond the synaptic cleft and the presynaptic cell breaks just beyond the terminal bouton of the axonal process. The axonal process then reseals to form a particle which represents the "pinched-off" nerve ending. These particles, which sediment in the mitochondrial fraction, have been extensively investigated and have provided useful information in such diverse areas as the synthesis, storage, and release of presumed transmitter substances, the identification of new candidate transmitters, the composition, turnover, and energy metabolism of the axonal cytoplasm (axoplasm), the rates of axonal flow, and the characteristics of the unmyelinated neuronal membrane.

A number of procedures are available for the isolation of synaptosomes [1]. Invariably they involve a very precise homogenization of the tissue in a Potter-Elvehjem homogenizer with a prescribed tolerance of 0.25 mm between the pestle and the side of the glass tube mortar. The tissue is homogenized in nine volumes of 0.32 M sucrose and then spun, first at 1000 g for 10 min to remove the nuclear pellet, and then at 20,000 g to obtain the crude mitochondrial fraction. The synaptosomes are obtained from the crude mitochondrial fraction either by the use of discontinuous gradients of sucrose [18-20] (Fig. 3.4) or of sucrose-Ficoll mixtures [13,21]. The sucrose procedure involves the preparation of gradients made up of layers of 0.8 and 1.2 M sucrose. The crude mitochondrial fraction, reconstituted in 0.32 M sucrose, is layered on top of these discontinuous gradients and the tubes are centrifuged at 53,000 g for 2 hr. The mitochondria are spun to the bottom in a pellet, the contaminating myelin floats at the top of the 0.8 M layer, and the synaptosomes are found in a diffuse layer at the interface between the 0.8 and the 1.2 M sucrose. The Ficoll techniques involve discontinuous gradients between 5 and 20% Ficoll in 0.32 M sucrose. After centrifuging for 45 min at 64,000 g, the synaptosomes are found at the interface between 7.5 and 13% Ficoll. Although objections to the use of Ficoll have been raised based on the slow establishment of centrifugal equilibrium due to its great viscosity, and the lack of uniformity of this polymer from batch to batch, both methods yield morphologically intact and metabolically active materials. All methods yield preparations somewhat contaminated with unspecified membrane fragments.

The morphology of synaptosomes can be examined by a number of procedures [16]. The particles have an average diameter of about 0.5 μm and a volume of 0.1 μm^3. They contain, of course, the terminal axoplasm, some mitochondria, and a population of synaptic vesicles, and are bounded by an outer membrane (Fig. 3.5). Many of the synaptosomes, especially those prepared in mild concentrations of detergents such as Triton X-100, also retain portions of the postsynaptic membrane. The treatment of intact synaptosomes with Triton-X-100 under specified conditions can lead to the isolation of these junctional

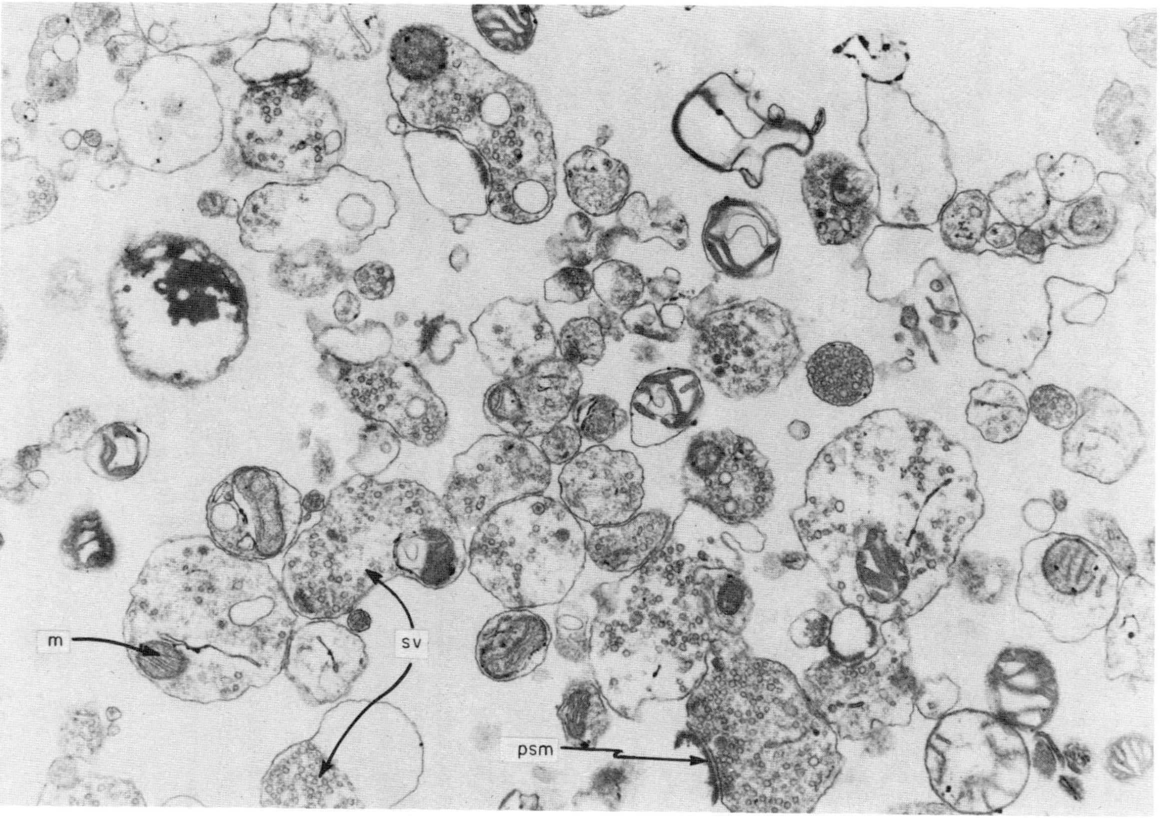

Figure 3.4   Synaptosomes.   Note synaptic vesicles (sv) and mitochondria (m) enclosed within membrane to form a nerve-ending particle which may have a thickened portion of the postsynaptic membrane (psm) adherent to it.   (Figure courtesy of Dr. V. P. Whittaker.)

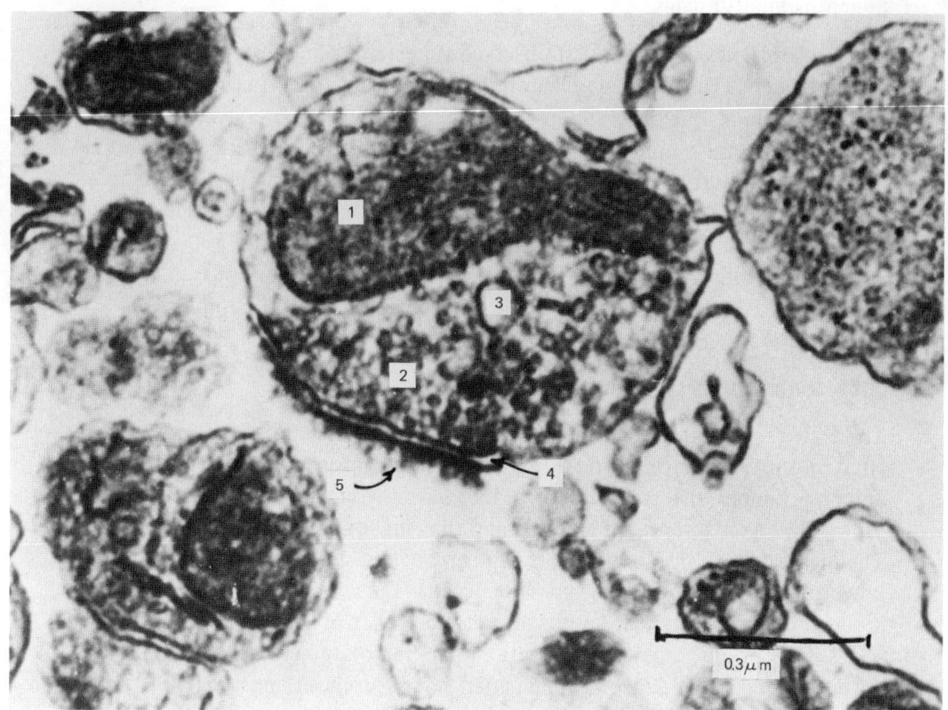

Figure 3.5   A synaptosome, showing: (1) a mitochondrion, (2) synaptic vesicles, (3) larger vesicles, (4) synaptic cleft containing electron-dense material, and (5) fibrillar postsynaptic thickening ("web"); ×60,000. (Figure courtesy of Dr. V. P. Whittaker.)

complexes containing the synaptic cleft intact, with its synaptic filaments and the cytoplasmic tufts comprising the subsynaptic web. The synaptosome itself contains something less than 100 vesicles. About 4% of the volume of the synaptosome is taken up by vesicles, about 24% by mitochondria, about 8% by the external membrane, and about 64% is cytoplasm [16]. When carefully carried out the technique has a fairly good yield. It has been estimated that the yield of synaptosomes from 1 g of guinea pig cortex is on the order of 4×10^{11} particles. This is comparable to the total number of nerve terminals estimated to be in this tissue [22]. Various estimates of the yield range between 10 and 50%. The preparations of synaptosomes from many species and from various areas of the central nervous system have fairly similar characteristics. Attempts to isolate synaptosomes from peripheral tissue have had limited success, primarily due to the very low yields obtained using these techniques.

Synaptosomal Vesicles

Synaptosomal vesicles were first observed via electron microscopy. There is excellent evidence that these vesicles are involved in synaptic transmission. This evidence includes the observation that degeneration of the nerve leads to a lysis of the vesicles and a concomitant alteration in the ability of the nerve to transmit impulses. Further studies showed that stimulation of the nerve leads to a depletion of the vesicles, and chemically induced convulsions produce almost complete loss of the vesicles. The vesicles are between 400 and 500 Å in diameter and are generally thought now to be identical to the "quanta" of transmitter substance released into the synapse when a nerve cell transmits an impulse.

Subsynaptosomal Fractionation

The vesicles, as well as the other components of the synaptosome, can be isolated by subsynaptosomal fractionation. The synaptosomes can be disrupted by any of a number of techniques, including warming, mechanical agitation, ultrasonics, detergents, repeated freezing and thawing, or treatment with cobra venom. The method of choice, however, is exposure of the synaptosomes to solutions of low osmotic strength. This is possible because the vesicles and the mitochondria are much more resistant to hypoosmotic shock than are the synaptosomes themselves. Practically, subsynaptic fractions can be isolated either from the crude mitochondrial pellet or from the purified synaptosomes themselves [1]. In either case, the particles are suspended in water and disrupted by homogenization or merely by passing them through a pipette several times. This suspension is then layered on a discontinuous sucrose gradient and centrifuged. The soluble constituents of the synaptosomes remain on the top, and the vesicles equilibrate between 0.4 and 0.6 M sucrose. The synaptosomal membranes are found in layers at the interface between 0.8 and 1.0 M sucrose, and the synaptosomal mitochondria form a pellet at the bottom. Thus, four fractions can be isolated from the synaptosomes: the vesicles, the mitochondria, the synaptic membranes, and the soluble constituents. The vesicles can be obtained in about 15% yield by such techniques.

The properties of the synaptosomes reflect, to a great extent, the properties of their constituent components. They respire, carry on oxidative phosphorylation, transport cations and choline, and incorporate radioactive amino acids into proteins. They contain all the candidate transmitter substances, including acetylcholine, norepinephrine, dopamine, serotonin, γ-aminobutyric acid, histamine, and glycine. The enzymes involved in the metabolism of these presumed transmitter substances are also found in synaptosomes. Acetylcholinesterase is present in this fraction and seems to be bound to the outside surface of the synaptosomal membrane. Choline acetylase is also present but is part of the soluble components.

In addition, glutamic acid decarboxylase, for the synthesis of γ-aminobutyric acid, and 5-hydroxytryptophan decarboxylase, for the synthesis of serotonin, are found in synaptosomes. Presumably, the synaptosome fraction represents a mixed population of nerve endings. Therefore, the presence of the various transmitters and their respective metabolic machinery is confined to given subpopulations of this fraction. That is, there must be a fraction of synaptosomes representing the cholinergic nerves, a fraction representing the adrenergic nerves, and so on. The attempts to subfractionate synaptosomes into these populations has met with only limited success, and even the few encouraging results which have been published are suspect [16].

The vesicles contain, most importantly, a large portion of the transmitter substance. In the case of acetylcholine, somewhere between 50 and 70% is present in the vesicles. The remainder is found in the soluble portion of the synaptosome. This observation explains the data obtained by many groups showing that there are two pools of acetylcholine in synaptosomes. The labile pool, and the one which is rapidly labeled with radioactive choline, is the one in the soluble portion. The stable pool, which is resistant to osmotic release and poorly or slowly labeled with radioactive material, is the one contained in the vesicles. The vesicle pool of acetylcholine, somewhere between 900 and 2000 molecules per vesicle, is estimated to be at a concentration of about 0.1 to 0.2 M. The vesicles are also the location of the other transmitter substances as well, but considerably less work has been done to characterize the distribution and properties of the pools of these other transmitters. Certainly, several different kinds of vesicles are present in the vesicular fraction. Presumably, each synaptosome contains a homogeneous population of vesicles containing, in turn, only the transmitter substance used by the cell from which the synaptosome was derived. The vesicles also contain a high content of ATP. The only enzyme proved to be in the vesicles is an ATPase activated by Mg^{2+} or Ca^{2+}. It is thought that the ATP and the ATPase are intimately involved in the release of transmitter during the transmission process. About 45% of the dry weight of the vesicle is lipid, and the lipid composition is unique. There is less cholesterol and cerebroside than in the synaptic membranes and very little ganglioside. The vesicles, on the other hand, are relatively enriched for phospholipid. The insoluble fraction of the vesicles has been shown to contain about 16 different proteins. One of these is a proteolipid. The origin of the vesicles is still not known for sure, but they may originate from the neurotubules. It is known that the protein of the vesicles has a half-life on the order of 21 days, so it is thought that the vesicles are used over and over again for the release of transmitter substance.

The mitochondria of the synaptosomes are much like other mitochondria of neural origin. They contain the normal mitochondrial enzymes and a small amount of RNA. Some few quantitative differences in the amino acid metabolizing enzymes, the regulation of mitochondrial metabolism by substrates, and the sensitivity of pyridine nucleotide dehydrogenases was found [23]. But, in general, the properties of these mitochondria are in no way unusual.

The soluble portion of the synaptosome contains the glycolytic enzymes, and the enzymes for transmitter synthesis. Also, enzymes for protein synthesis have been found. This is consistent with the observation that the synaptosomes incorporate radioactive amino acids into protein.

The external membrane of the synaptosome is quite different than the membrane of the vesicles. A number of enzymes are present in the synaptosome membrane. Among them are acetylcholinesterase, a Na^+,K^+-activated ATPase thought to be involved in cation transport and quite different than the ATPase of the vesicles, glutamine synthetase, and adenylate cyclase. The membranes also contain RNA and gangliosides. It is to be expected that the postsynaptic membranes contain the receptor sites to which the transmitter substances bind. This area of investigation is considered in Chap. 24. It can be noted now that the synaptosomal membranes, most certainly those portions which were originally postsynaptic, do bind inhibitors of the transmission process and that some work indicates that the receptor, at least for acetylcholine, can be isolated from these membranes.

The discovery of the synaptosomes and their components has elicited an enormous outpouring of research. Indeed, the study of the synaptosomes has been one of the most interesting aspects of recent neurochemistry, since these particles presumably represent an important portion of the transmitting mechanism of the communicating cell.

Microsomes

Brain microsomes can be prepared by high-speed centrifugation of the postmitochondrial supernatant. Two main fractions can be separated, the smooth microsomes, which are poor in RNA, and the granular microsomes, which are relatively richer in ribosomes and RNA. The smooth and the granular microsomes can be obtained in fairly homogeneous form.

Polyribosomes and Ribosomes

Polyribosomes have been isolated from brain by centrifugation through high-density sucrose solutions [24,25] (Fig. 3.6). The presence of relatively high concentrations of Mg^{2+} during the preparation of these polysomes is necessary to protect against disaggregation into ribosomes. The sensitivity of brain polysomes to dissociation has led to the suggestion that these brain aggregates are more sensitive to changes in the experimental, or indeed the physiological, environment than are polysomes from other tissues [26]. As in other tissues, polyribosomal structures can be found either free in the cytoplasm or bound to the membranes of the

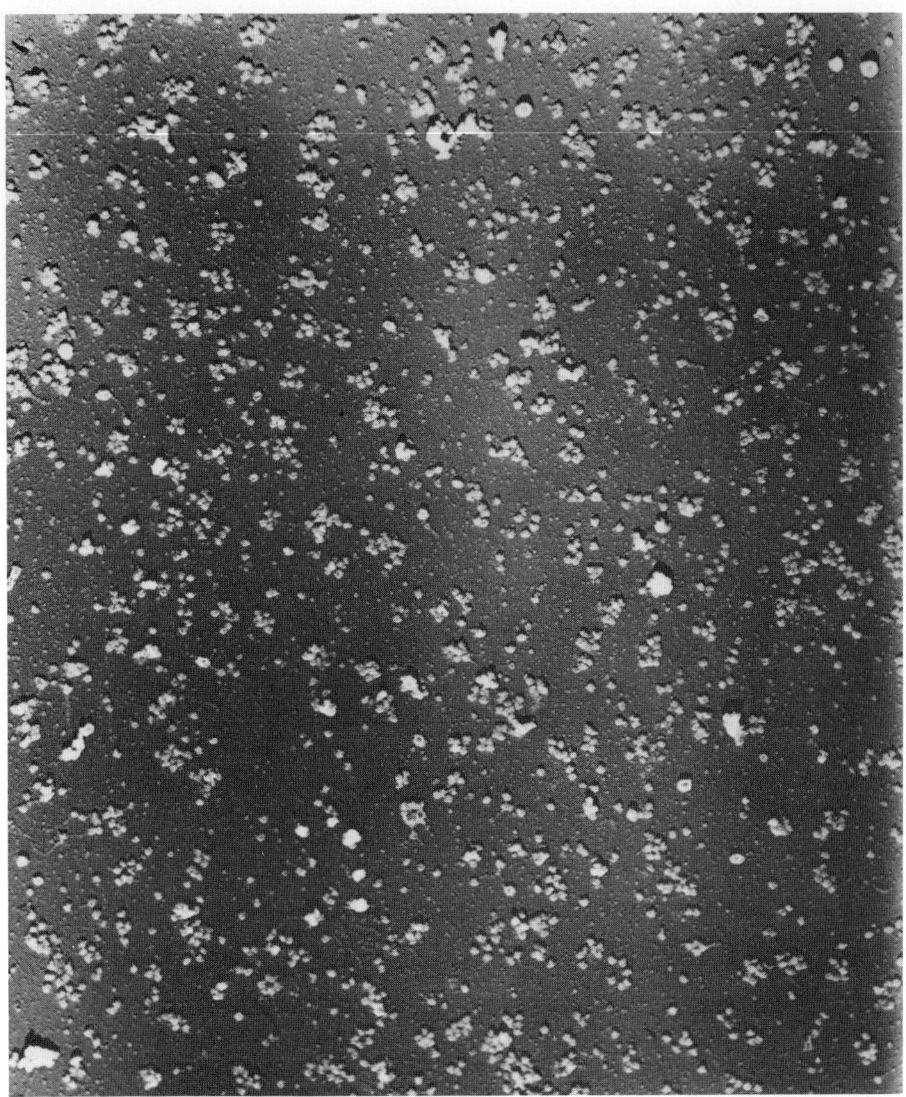

Figure 3.6   Electron micrograph of polysomes from rat brain, ×63,000.   [Figure courtesy of Dr. H. R. Mahler.   Reprinted from A. T. Campagnoni and H. R. Mahler, *Biochemistry* 6:956 (April 1967).   Copyright 1967 by the American Chemical Society.   Reprinted by permission of the copyright owner.]

cell. And, as in other cells, the free polyribosomes seem to be involved in the synthesis of proteins for use in the interior of the cell and the membrane-bound polyribosomes in the synthesis of proteins designed to be secreted by the cell [27]. A great deal of work has been done in an attempt to characterize the various types of polysomal structures in brain cells [27]. Special interest in these studies is warranted since it has been found that the proportion of free and bound polysomes changes with age and with different physiological and pathological conditions. The results of such studies indicate that the ribosomes from the various different polysome structures are all the same. They cannot be distinguished from one another even though one population comes from the bound polysomes, another from the free, and a third from free cytoplasmic ribosomes. The polysomes seem to acquire structural and functional specificity by virtue of their interactions with the membranes of the cell or with the strands of the various messenger RNAs which hold them together [27].

Treatment of polyribosomes with deoxycholate will dissociate them into ribosomes [28-31]. Other treatments, such as storage in low Mg^{2+} or exposure to RNase under defined conditions, will also liberate ribosomes. The method of choice, however, is deoxycholate treatment. Complete removal of Mg^{2+} by exposure of the ribosomes to ethylenediaminetetraacetate (EDTA) breaks the particles down to their constituent subunits [32-34].

As mentioned above, brain polyribosomes may be more sensitive to changes in the environment than polyribosomes from other tissues. The ribosomes themselves, however, have been shown to be much like ribosomes from other tissues. They are dense bodies measuring 150 to 250 Å across. They are 40% RNA, and the rest is protein. No DNA, lipid, or hexose has been found. The RNA is most certainly bound to protein in the form of ribonucleoprotein since, although the RNA from the ribosomes is eminently RNase sensitive when extracted, the ribosomes themselves are quite resistant to RNase. The intact brain ribosome has a Svedberg sedimentation constant of 80S, as do ribosomes from other tissues. And the subunits released by treatment with EDTA are 60S and 40S, as in other tissues.

The ribosomal RNA consists of a 28S fraction and an 18S fraction in a 2:1 ratio. Also found are a 5S RNA and traces of transfer RNA sedimenting at 4S. The composition of the ribosomal RNA in brain is in no way unique compared to ribosomal RNA of other tissues. The protein of the ribosomes is basic and resembles ribosomal protein from other mammalian sources. The polyamines spermine and spermidine have been found in brain ribosomes. The only enzymes known to be present in the ribosomes, in addition to the enzymes required for protein synthesis, are some tightly bound ribonucleases, an active one with a pH optimum at 5.4 and a less active one with a pH optimum at 7.9, and a tightly bound phosphomonoesterase. These enzymes seem to be an intimate part of the ribosome structure, but can be released from the particle by treatment with solutions of high salt concentration.

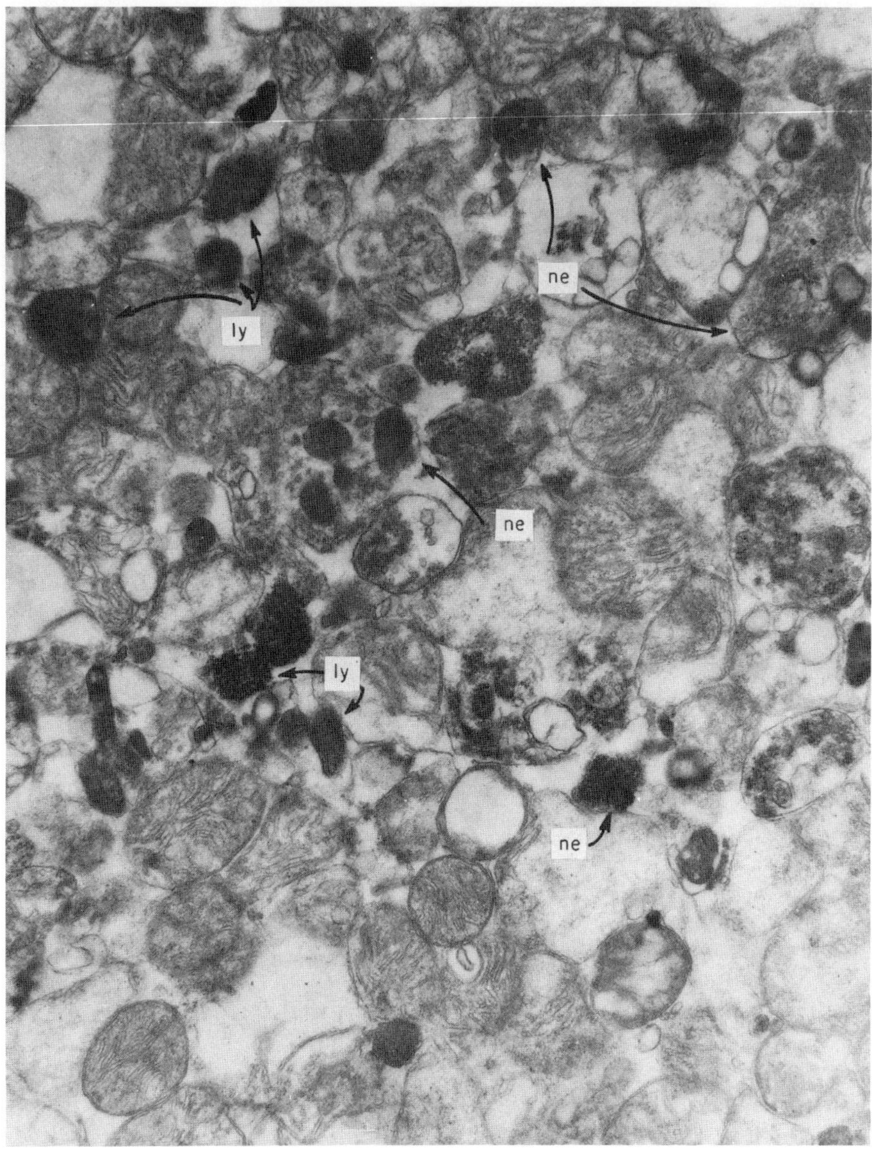

Figure 3.7   Brain lysosomes.   Lysosome-rich fraction from brain.   Note lysosomes (ly) and nerve-ending particles (ne).   Figure courtesy of Dr. H. Koenig, *J. Neurochem.* *11*:729 (1964).

Lysosomes

Brain lysosomes are a heterogeneous population of dense membrane-limited, granular structures which contain a great number of the hydrolytic activities of the cell (Fig. 3.7). They are irregular in shape and heterogeneous in size, but, on the average, measure about 0.5 μm across. They contaminate virtually every fraction isolated from the tissue because of their heterogeneous morphology and can be isolated from either the crude mitochondrial fraction or the crude microsomal fraction by centrifugation through a discontinuous sucrose gradient. Lysosome-enriched pellets can be obtained from either fraction by sedimentation through a 1.4 M sucrose layer [35]. No homogeneous preparations of brain lysosomes are as yet available.

The lysosomes from brain have been demonstrated to contain acid phosphatase, β-glucuronidase, β-galactosidase, N-acetyl-β-D-glucosaminidase, acid RNase, acid DNase, aryl sulfatase, cathepsin, sialidase, and cerebroside galactosidase. Strong evidence for the presence of a number of other hydrolytic activities has been presented [36]. In fact, these particles contain activities which hydrolyze virtually every macromolecule in the cell and vast numbers of smaller molecules. Invariably, the enzymes have pH optima between 3.8 and 5.5. In addition to the enzymatic components, the lysosomes also contain large amounts of lipoprotein.

The lysosomes can be identified by a number of histological techniques. Most of these are based on the activity of the various hydrolyses. They also are identifiable by their unique autofluorescence [37]. Irradiation of the lysosomes at about 360 nm produces a yellow fluorescence which seems to be due to the presence of pyridine nucleotides and flavins and perhaps other, as yet unidentified constituents.

The enzymes in the lysosomes are inactive or "latent" until they are released. The structural basis for this latency is not known for sure but it seems due either to the limiting membrane which surrounds the particle or to the binding and inactivation of the enzymes by the polyanionic lipoproteins of the matrix, probably the latter. The enzymes become active if the lysosome is ruptured by any of a number of techniques, including freezing and thawing, sonication, hypoosmotic shock, or treatment with detergents or organic reagents. Certain other milder treatments will release some of the enzymes without totally disrupting the structure. The function of the lysosomes is believed to be in the digestion of material brought into the cell (heterophagy) or in the digestion of endogenous material (autophagy). Certainly it is known that lysosomal enzymes become active when the cell is injured or dies. As is discussed in Chap. 16, the accumulation of various macromolecules in the lysosomal structures is a major characteristic of certain of the degenerative diseases involving neural tissue.

References

1. S. H. Appel, E. D. Day, and D. D. Mickey, in *Basic Neurochemistry* (R. W. Albers, G. J. Siegel, R. Katzman, and B. W. Agranoff, eds.), Little, Brown, Boston, 1972.
2. D. A. Rappoport, P. Maxcy, Jr., and H. F. Daginawala, in *Handbook of Neurochemistry*, Vol. II (A. Lajtha, ed.), Plenum, New York, 1969.
3. P. Mandel, A. R. Dravid, and N. Pete, *J. Neurochem. 14:*301 (1967).
4. H. Løvtrup-Rein and B. S. McEwen, *J. Cell Biol. 30:*405 (1966).
5. T. Kato and M. Kurokawa, *J. Cell Biol. 32:*649 (1967).
6. M. Sporn, T. Wanko, and W. Dingman, *J. Cell Biol. 15:*109 (1962).
7. D. A. Rappoport, R. R. Fritz, and A. Moraczewski, *Biochim. Biophys. Acta 74:*42 (1963).
8. S. H. Barondes, *J. Neurochem. 11:*663 (1964).
9. S. C. Bondy and H. Waelsch, *J. Neurochem. 12:*751 (1965).
10. L. G. Abood, R. W. Gerard, J. Banks, and R. D. Tschirgi, *Am. J. Physiol. 168:*728 (1952).
11. M. Kurokawa, T. Kato, and H. Inamura, *Proc. Jpn. Acad. 42:*1217 (1966).
12. R. Tanaka and L. G. Abood, *J. Neurochem. 10:*571 (1963).
13. A. A. Abdel-Latif, *Biochim. Biophys. Acta 121:*403 (1966).
14. A. A. Abdel-Latif and L. G. Abood, *J. Neurochem. 11:*9 (1964).
15. L. G. Abood, in *Handbook of Neurochemistry*, Vol. II (A. Lajtha, ed.), Plenum, New York, 1969.
16. V. P. Whittaker, in *Handbook of Neurochemistry*, Vol. II (A. Lajtha, ed.), Plenum, New York, 1969.
17. E. deRobertis and G. R. Lores Arnaiz, in *Handbook of Neurochemistry*, Vol. II (A. Lajtha, ed.), Plenum, New York, 1969.
18. E. G. Gray and V. P. Whittaker, *J. Physiol. (London) 153:*35P (1960).
19. E. G. Gray and V. P. Whittaker, *J. Anat. 96:*79 (1962).
20. E. deRobertis, A. P. deIraldi, G. R. Lores Arnais, and C. Gomez, *J. Biophys. Biochem. Cytol. 9:*229 (1961).
21. L. A. Autilio, S. H. Appel, P. Pettis, and P. Gambetti, *Biochemistry 7:*2615 (1968).
22. F. Clementi, V. P. Whittaker, and M. N. Sheridan, *Z. Zellforsch. Mikrosk. Anat. 72:*126 (1966).
23. L. Salganicoff and E. deRobertis, *J. Neurochem. 12:*287 (1965).
24. C. E. Zomzely, S. Roberts, D. M. Brown, and C. Provost, *J. Mol. Biol. 19:*455 (1966).
25. A. T. Campagnoni and H. R. Mahler, *Biochemistry 6:*956 (1967).
26. S. Roberts, in *Handbook of Neurochemistry*, Vol. Va (A. Lajtha, ed.), Plenum, New York, 1971.
27. M. R. V. Murthy, in *Protein Metabolism of the Nervous System* (A. Lajtha, ed.), Plenum, New York, 1970.
28. M. R. V. Murthy and D. A. Rappoport, *Biochim. Biophys. Acta 95:*132 (1965).
29. S. Yamagami, Y. Kawakita, and S. Naka, *J. Neurochem. 12:*607 (1965).

30. C. E. Zomzely, S. Roberts, D. M. Brown, and D. Rapaport, *Biochim. Biophys. Acta 103:*529 (1965).
31. C. E. Zomzely, S. Roberts, and D. Rapaport, *J. Neurochem. 11:*567 (1964).
32. V. Hanzon and G. Toschi, *Exp. Cell Res. 16:*256 (1959).
33. R. K. Datta and J. J. Ghosh, *J. Neurochem. 10:*611 (1963).
34. S. Yamagami, M. Masui, and Y. Kawakita, *J. Neurochem. 10:*849 (1963).
35. H. Koenig, D. Gaines, T. McDonald, R. Gray, and J. Scott, *J. Neurochem. 11:*729 (1964).
36. H. Koenig, in *Handbook of Neurochemistry,* Vol. II (A. Lajtha, ed.), Plenum, New York, 1969.
37. H. Koenig, *J. Histochem. Cytochem. 11:*556 (1963).

4
AXONAL FLOW

The neuron, unlike any other cell in the body, may extend over a substantial distance. The physical distance between the metabolic and synthetic center of the cell, the perikaryon, and the transmitting element, the axonal bouton, can be up to a meter or more. For the axon and the synapse to function, the terminal portions must be nourished and their enzymatic machinery must be refreshed and replenished. Simply put, the axons must either make everything they need locally or there must be some orderly mechanism to provide material for them from the cell body. Although there is some evidence for local synthesis of some of the axonal constituents, it is known that much of the requirement for axonal nourishment is provided by synthesis of the constituents in the cell body and transport to the nerve terminal by a process known as "axonal flow."

In the late twenties, Ramon y Cajal observed that if axons were mechanically constricted, a swelling occurred on the side of the constriction nearest the cell body. This observation led to the formulation of the concept of axonal flow by Weiss in 1943 [1]. Basically, this is the movement of cytoplasm from the cell body down the axon at a regular rate, carrying with it proteins and nutrients from the cell body to the terminus. Through the further work of Weiss and of many others, this concept is now widely accepted as the mechanism by which axons remain viable.

The original constriction experiments have been extended [2,3] to show that (1) the swelling of the axon is proportional to the distance from the constriction, (2) the swelling increases with time, (3) the swelling on the side nearest the cell body is matched by a decrease in volume on the other side, (4) multiple constrictions lead to multiple swellings, and (5) release of the constriction leads to a wave of swelling which passes down the axon at a regular rate. Other work has clearly shown that the levels of various oxidative enzymes, and of certain phospholipids, rise in the swollen axonal segment [4,5]. Particulate matter, including mitochondria and catecholamine-containing granules, also accumulates.

A number of other techniques have been used to investigate this phenomenon [5]. Transection experiments have shown that certain materials are elaborated at a fixed rate from the cut end. Also, when the cut ends seal, the stumps appear to swell, and certain of the cytoplasmic constituents increase in concentration. These experiments are of limited value for this purpose since it is known that extensive changes take place in the cell body when the axon is cut and so no analogy to the normal situation is possible. Axonal flow has also been observed in intact axons by autoradiography [6]. Upon administration of radioactive amino acid, it was shown that radioactive protein appeared in cell bodies of the Purkinje cells of the cerebellum within 4 hr. After 1 day the level of radioactive protein in these cell bodies had decreased. It was not until several days later, however, that the radioactive protein had reached the white matter containing the axonal processes of these cells. Another valuable approach has been to examine the appearance of materials in the synaptosomes. Upon administration of a radioactive amino acid the total brain protein becomes labeled within a few minutes and the labeling is completed within an hour. The radioactivity thereafter declines. In contrast, the radioactivity of the proteins in the synaptosomes increases for several days. This increase is most certainly due to the transport of radioactive proteins from the cell body to the axonal bouton. Similar conclusions have been reached from studies in which the mitochondrial enzyme, monoamine oxidase, was irreversibly inhibited and then the return of its activity via new protein synthesis was followed in whole brain and in synaptosomes [7-10]. Studies of nerve cells in tissue culture have indicated that certain particles in the axon can move in an abrupt or leaping fashion ("saltatory" transport). It is difficult to extrapolate these observations to the physiological situation since the culture system is quite artificial. The ligation experiments suggest that the flow is not completely unidirectional ("retrograde axonal flow"). Evidence for accumulation of materials on the side of the ligation away from the cell body has been presented. No information is available on the flow of materials into the dendrites but it can be expected to involve a similar process.

Not all materials are transported. As a matter of fact, the process of axonal flow seems to be fairly selective. It is known that labeled proteins, phospholipids, and catecholamines move down the axon. Mitochondria and vesicles also are transported, but ribosomes are not. Certain specific enzymes such as monoamine oxidase are transported, possibly by virtue of their mitochondrial localization. Although there is some data to the contrary [4], cholinesterase appears to be made in all parts of the axon and not transported by axonal flow. This conclusion is based on elegant work using irreversible inhibition of the enzyme and following its reappearance in the various parts of the axon [11]. It is interesting that the same experimental design used on two different enzymes, cholinesterase and monoamine oxidase, gives opposite results, clearly demonstrating the selectivity of the transport. Another clear demonstration of the selectivity, and, if anything, an even more surprising one, is given by the data showing that glutamate but not xylose moves down the axon [12].

It is clear that there is more than one rate of movement of materials down the axon. It is appropriate to talk about fast and slow axoplasmic transport. The slow rate is easily defined at about 1 to 2 mm/day. This is the rate at which axons regenerate after transection and probably the rate at which they grow normally. Materials which are shown to move at this rate are probably merely moving along with the growing axon. Fast flow is more difficult to define. Phospholipid has been reported to move at 70 mm/day [13], protein at 200 mm/day [14], and various amino acids at rates ranging all the way from two to several hundred millimeter per day. Rapid rates for the transport of catecholamine-containing granules have been observed [15,16]. It is best, at the present state of our information, to conclude that each component has its own characteristic rate of axoplasmic movement and that there are many rates of transport, not just slow and fast, even in the same axon.

Several suggestions as to the mechanism have been advanced over the years. The transport was, at one time, thought to be due to the hydrodynamic pressure generated in the cell body by the continued synthesis of the cell constituents. Evidence against this mechanism includes the observation that axoplasmic movement continues between two ligations of an axon [15]. It has been suggested that a peristaltic movement of the surrounding glial cells drives materials down the axon. This seems unlikely in view of the fact that some forms of axoplasmic flow have been observed in denuded axons in culture, and further, because of the various rates of transport shown by various molecules. The best evidence now favors the participation of neurotubules and their constituent protein. These tubules are some 200 to 250 Å in diameter and extend almost the whole length of the axon process. They are made up of a globular protein which binds colchicine and is thought to have contractile properties. This protein is discussed in detail in Chap. 15. For the moment it is only necessary to say that the protein is almost certainly involved in axonal flow, probably by virtue of its contractile properties. The most compelling pieces of evidence for this are the various observations showing that colchicine inhibits axonal transport [17-19]. Whatever the mechanism, it is known that the flow depends upon metabolic energy, but not on continued protein synthesis [9], that it is probably not affected by nerve stimulation [16], and that it seems to decrease as the animal ages [6,9]. It is important to note that an understanding of the mechanism of axonal movement may have dividends beyond simply the information it provides on neuronal nutrition. Certain viruses and toxins have been found to invade the nervous system by a systematic progression through the neuronal network [4].

References

1. P. Weiss, *Arch. Surg. 46:*525 (1943).
2. P. Weiss and H. B. Hiscoe, *J. Exp. Zool. 107:*315 (1948).
3. P. Weiss, in *Regional Neurochemistry* (S. S. Kety, ed.), Pergamon, Oxford, 1961.
4. R. L. Friede, *Topographic Brain Chemistry,* Academic, New York, 1966, Chap. XV.
5. S. H. Barondes, in *Handbook of Neurochemistry,* Vol. II (A. Lajtha, ed.), Plenum, New York, 1969.
6. B. Droz and C. P. Leblond, *J. Comp. Neurol. 121:*325 (1963).
7. S. H. Barondes, *Science 146:*779 (1964).
8. S. H. Barondes, *J. Neurochem. 13:*721 (1966).
9. S. H. Barondes, *J. Neurochem. 15:*343 (1968).
10. S. H. Barondes, *Commun. Behav. Biol.* A, *1:*179 (1968).
11. E. Koenig and G. B. Koelle, *J. Neurochem. 8:*169 (1961).
12. G. A. Kerkut, in *Axoplasmic Transport, Report of a Work Session,* Neurosciences Research Program Bulletin, Vol. 2, MIT Press, Cambridge, Mass., 1967.
13. G. W. Kreutzberg and W. Wechsler, *Acta Neuropathol. 2:*349 (1963).
14. R. J. Lasek, in *Axoplasmic Transport, Report of a Work Session,* Neurosciences Research Program Bulletin, Vol. 2, MIT Press, Cambridge, Mass., 1967.
15. A. Dahlstrom, *Acta Physiol. Scand. 69:*158 (1967).
16. A. Dahlstrom, in *Axoplasmic Transport, Report of a Work Session,* Neurosciences Research Program Bulletin, Vol. 2, MIT Press, Cambridge, Mass., 1967.
17. K. A. C. James, J. J. Bray, I. G. Morgan, and L. Austin, *Biochem. J. 117:* 767 (1970).
18. K. A. C. James and L. Austin, *Biochem. J. 117:*773 (1970).
19. J. Sjostrand, M. Frizell, and P. O. Hasselgren, *J. Neurochem. 17:*563 (1970).

5
THE MYELIN SHEATH

In both the central and peripheral nervous systems, the axons of most neurons
are covered with a white sheath, much like the insulation on a wire. This myelin
sheath is formed by specific glial cells, the oligodendrocytes in the central nervous
system, and the Schwann cells in the peripheral nervous system. It is, in fact,
made up of successive layers of the plasma membrane of these glia, and is formed,
probably, by the action of the glial cell rotating in a spiral around the axon process (Fig. 5.1). In the central nervous system, on the other hand, it seems that
one oligodendrocyte can myelinate several axons, and so a simple picture of the
glia revolving around the axon, leaving a membrane spiral as it goes, does not
seem quite adequate, but it is the picture most fitting the pattern of myelination.
In any case, the myelin sheath is made up of the membrane of the glia and is the
most condensed and concentrated membrane structure found in the body.

Myelin constitutes about 50% of the white matter of the brain and is mainly responsible for the chemical and biochemical differences between white matter
and gray matter. The myelin component accounts for the glistening white appearance, the low water content, and the high lipid content of the white matter.

Structure

The structure of the myelin sheath accurately reflects its origin as a condensed
membrane. The plasma membrane is probably organized as pictured by Davson
and Danielli [1] as a lipid bilayer bounded on both sides by protein. The myelin
sheath, then, is a spiral of these layers, with the protein coating of adjacent layers
tightly fused. And truly, the sheath, as observed by optical and X-ray techniques,
and more recently in the electron microscope, appears as a series of mixed layers
of lipid and protein. The early work of Schmitt and his colleagues [2] established that myelin was a concentric, layered structure, and that it consisted of a
repeating unit. It was possible for Schmitt to write in 1939 that, "The proteins
occur as thin sheets wrapped concentrically about the axon with two bimolecular

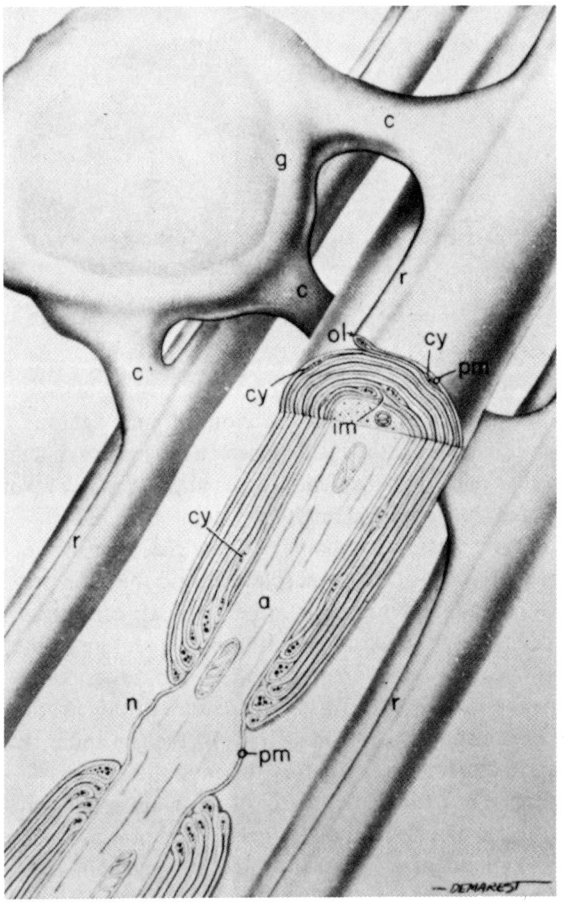

Figure 5.1   Central nervous system myelin-glial relationships: The plasma membrane (pm); the inner mesaxon (im); trapped glial cytoplasm (cy); loops of the plasma membrane (ol); internal ridge (r); glial cell (g) and its connection (c) to the axon (a); and the node are indicated. [Figure courtesy of Dr. Hans Ris and the Rockefeller University Press from M. B. Bunge, R. P. Bunge, and H. Ris, *J. Biophys. Biochem. Cytol. 10*:67 (1961).]

layers of lipoids interspersed between adjacent protein layers" [3] . This description is quite consistent with our current concept of the myelin structure. The more recent studies with the electron microscope have added refinements to the thinking about the structure but have not radically altered Schmitt's description (Figs. 5.2, 5.3, 5.4).

The myelin sheath does not cover the entire axon. It is segmented, and each segment is contributed by a single glial cell. The intervals of uncovered axon between the segments are called the nodes of Ranvier. The function of the myelin sheath is not to prevent short circuits or leakage as was first imagined. It is, on the contrary, to facilitate conduction, by permitting or forcing the current to leap between these nodes. This so-called saltatory conduction which occurs in myelinated axons is much faster, on the order of sixfold faster, than conduction in axons of comparable size without a myelin sheath. The speed with which the current moves from node to node through the extraaxonal space is much faster than it could move through the axonal substance, and this seems to be a major reason for the presence of the segmental myelin sheath.

The nodes also appear to be centers of high metabolic activity because the glial cells impinge at these points and the glial mitochondria collect near the points of contact. Branching of a myelinated fiber always takes place at the node. And in the central nervous system a node may sometimes participate in the formation of a synapse.

Composition

Although the structure of the myelin sheath has been known for decades, the chemical and enzymatic composition of myelin was not investigated until methods for its isolation became available in the early 1960s. Now a number of techniques are available [4,5], but all of them take advantage of the low density of the myelin structure which is a consequence of its high lipid content. Merely sedimenting a brain homogenate through a series of sucrose layers of increasing density will lead to myelin-enriched fractions in the top layers. It is generally necessary to treat the myelin-enriched fractions with distilled water to remove the sheath from the axonal fragment to which it is attached. When this is done, the myelin will peel away from the axon and will form myelin "vesicles." If this is not done the myelin preparations will contain axonal fragments. Absolutely pure myelin is difficult to prepare even with the most detailed techniques because, by the very nature of the formation of myelin, the layers include small amounts of cellular material from the glial cell of origin. For the best myelin preparations it has proved advisable to start with dissected white matter rather than whole brain. Even with this additional caution it has proved quite difficult to get pure myelin from young animals, in which the amounts of myelin present are relatively small. Most recently, procedures have been devised specifically for this purpose [5].

The purity of a given myelin preparation is best assessed by electron microscopy. The myelin particles retain their characteristic layered appearance. Even so it is difficult to detect a small amount of particulate contamination in a field of myelin vesicles. Other criteria of purity include the solubility of the prepara-

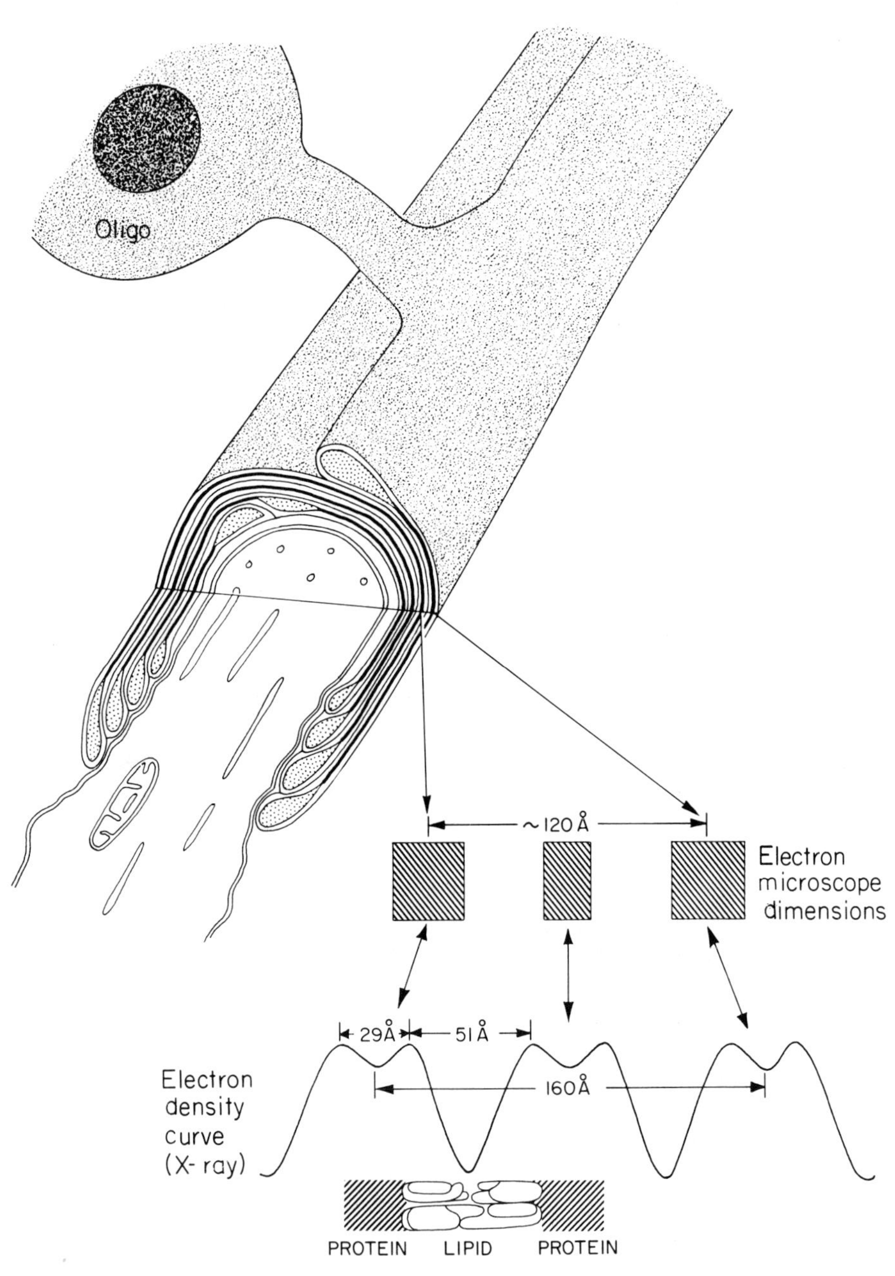
Oligo
~120 Å
Electron
microscope
dimensions
29Å
51Å
160Å
Electron
density
curve
(X-ray)
PROTEIN   LIPID   PROTEIN

tion in chloroform-methanol (2:1) and the absence of enzymatic markers for
other subcellular fractions. The best preparations have been shown to be on the
order of 95% soluble in the chloroform-methanol solvent and to have very low
levels of succinic dehydrogenase, Na^+,K^+-activated ATPase, glucose-6-phosphate
phosphatase, 5'-nucleotidase, and nucleic acid.

By now the composition of myelin is thoroughly known, and except for
some trace constituents, general agreement exists as to its chemical and biochemical makeup [6,7]. X-ray diffraction studies have shown that the water content
of tightly wound myelin is about 40%. Lipid makes up about two-thirds of the
solid matter and protein about one-third. No polysaccharides have been found,
although some carbohydrate may be present in the form of glycoproteins. The
myelin from different species has approximately the same composition although
some species variation is known to occur (Table 5.1).

The lipid component of myelin contains, as major constituents, phospholipid, glycolipid, and sterol in the proportions 2:1:1 by weight. The molar ratios
of these components is generally found to be sterol, 4; phospholipid, 3; and glycolipid (galactolipid), 2. The sterol, as implied above, is almost all cholesterol.
Some small amount of desmosterol has been found. The presence of cholesterol
esters, rather than cholesterol, in myelin, is characteristic either of immature
myelin or of some pathologic state. The major lipids of myelin, then, are cholesterol, cerebrosides, ethanolamine phosphatides (as plasmalogens), and lecithin.
The structures of these components is presented in Chap. 17. None of the brain
lipids have been shown to be absent from myelin, with the possible exception of
cardiolipin which is uniquely mitochondrial in origin. And no lipids are known
which are present only in myelin.

At least three different kinds of proteins occur in myelin. About 50 to
60% of the protein is in the form of "proteolipid" described first by Folch and
his coworkers [8,9]. Although all the proteins of myelin are soluble in organic
solvents when the lipids of myelin are present, these proteolipids remain lipid

Figure 5.2   A composite diagram summarizing some of the ultrastructural data on
central nervous system (CNS) myelin. At the top an oligodendrocyte is shown
connected to the sheath by a process. The cutaway view of the myelin and axon
illustrates the relationship of these two structures at the nodal and paranodal regions. Only a few myelin layers have been drawn for the sake of clarity. The
lower part of the figure shows roughly the dimensions and appearance of one
myelin repeating unit as seen with fixed and embedded preparations in the electron microscope. This is contrasted with the dimensions of the electron density
curve of CNS myelin obtained by X-ray diffraction studies in fresh nerve. The
components responsible for the peaks and troughs are sketched below. (From
E. S. Goldensohn and S. H. Appel, eds., *Scientific Approaches to Clinical Neurology,* Lea & Febiger, Philadelphia, 1977.)

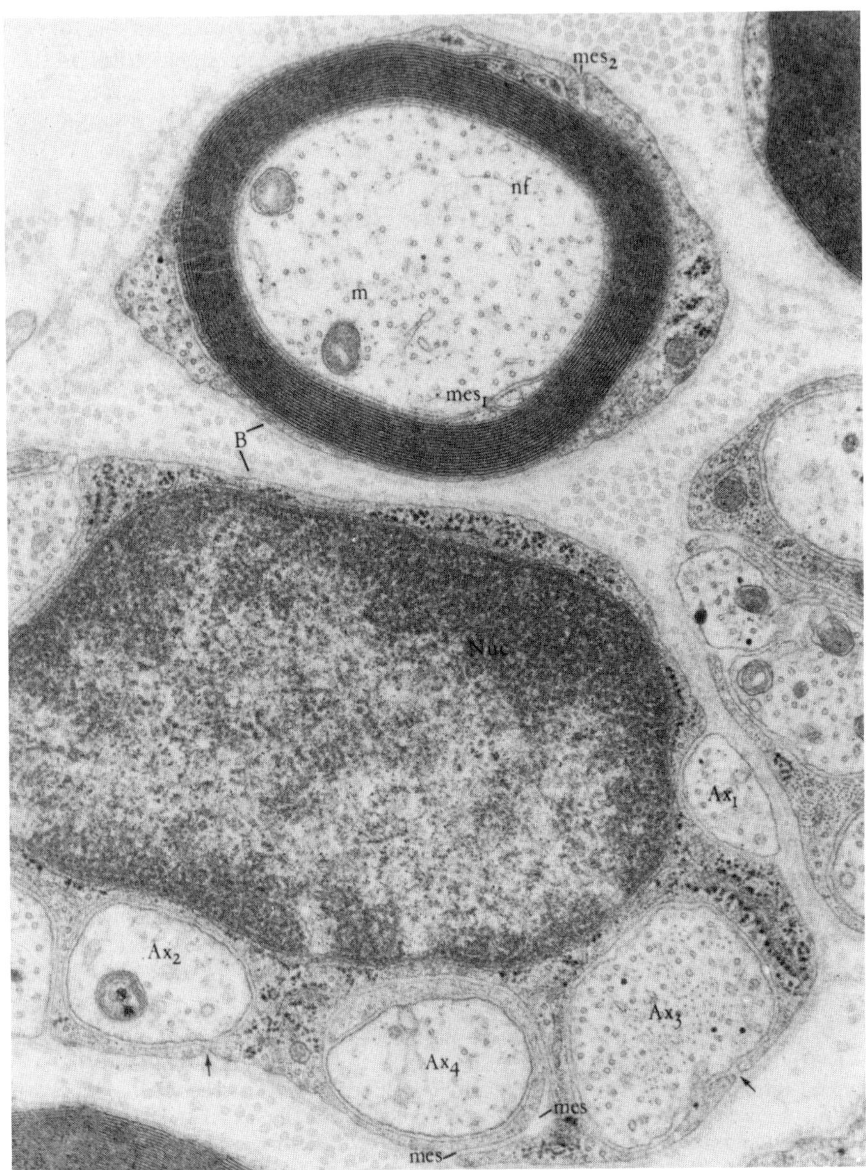

Figure 5.3a   Myelinated nerve fiber, adult peripheral nerve. In the lower half of the picture is a Schwann cell sectioned transversely at its nucleus (Nuc). Six unmyelinated axons are enclosed to varying degrees in the peripheral cytoplasm of this cell. One axon (AX₁) is covered on one side only by basal lamina (B).

Table 5.1   Gross Composition (mg/g of dry myelin) of Myelin from
Various Species

Constituent	Human	Cow	Rabbit	Rat	Frog
Protein	456	295	380	479	390
Glycolipid	136	266	171	113	69
Phospholipid	238	319	310	290	364
Sterol	178	170	160	122	156

Table adapted from L. C. Mokrasch, in *Handbook of Neurochemistry* (A.
Lajtha, ed.), Vol. I, Plenum, New York, 1969.

soluble even after removal of the lipid, and are, thus, distinct from the class of
lipoproteins which contain lipid but are water-soluble. The proteolipid may be
extracted from myelin with $CHCl_3$:CH_3OH (2:1). The molecular weight of the
protein in this fraction is between 20,000 and 30,000, and there is probably one
molecular species present or, at most, a few. The protein portion of the proteolipid has a remarkably high content of nonpolar amino acids, which probably
accounts for its unusual solubility properties. About 60% of the residues are
either aliphatic or aromatic, mostly glycine, alanine, leucine, isoleucine, and
tryptophan. The lipid is bound noncovalently to the protein, at least partly by
salt links. Proteolipids are also found in other membranes of the body, although
there is some immunological evidence to suggest that the myelin proteolipid is
a unique entity [10]. The function of the myelin proteolipid appears to be to
stabilize the myelin windings by providing a continuous hydrophobic milieu between layers of protein. Very little proteolipid is found in the myelin of peripheral nerve.

A second protein found in myelin is a small, basic material comprising
about 30% of myelin protein and which has been intensively investigated because of its antigenic properties. As shown by Kies and her coworkers [11-13],
and by others, this material, when administered subcutaneously, produces a demyelinating condition known as experimental allergic encephalomyelitis and so
is known as the EAE antigen. Interest in this material stems from the fact that
the condition has some similarities to multiple sclerosis. Most species have a

Figure 5.3a (continued)
The others are completely enclosed by mesaxons (mes). At the top is a myelinated axon. The spiral of myelin starts at the internal mesaxon (mes_1) and terminates at the external mesaxon (mes_2). Microtubules (m) and neurotubules (nf)
can be seen. Sciatic nerve from an adult rat, ×33,800.

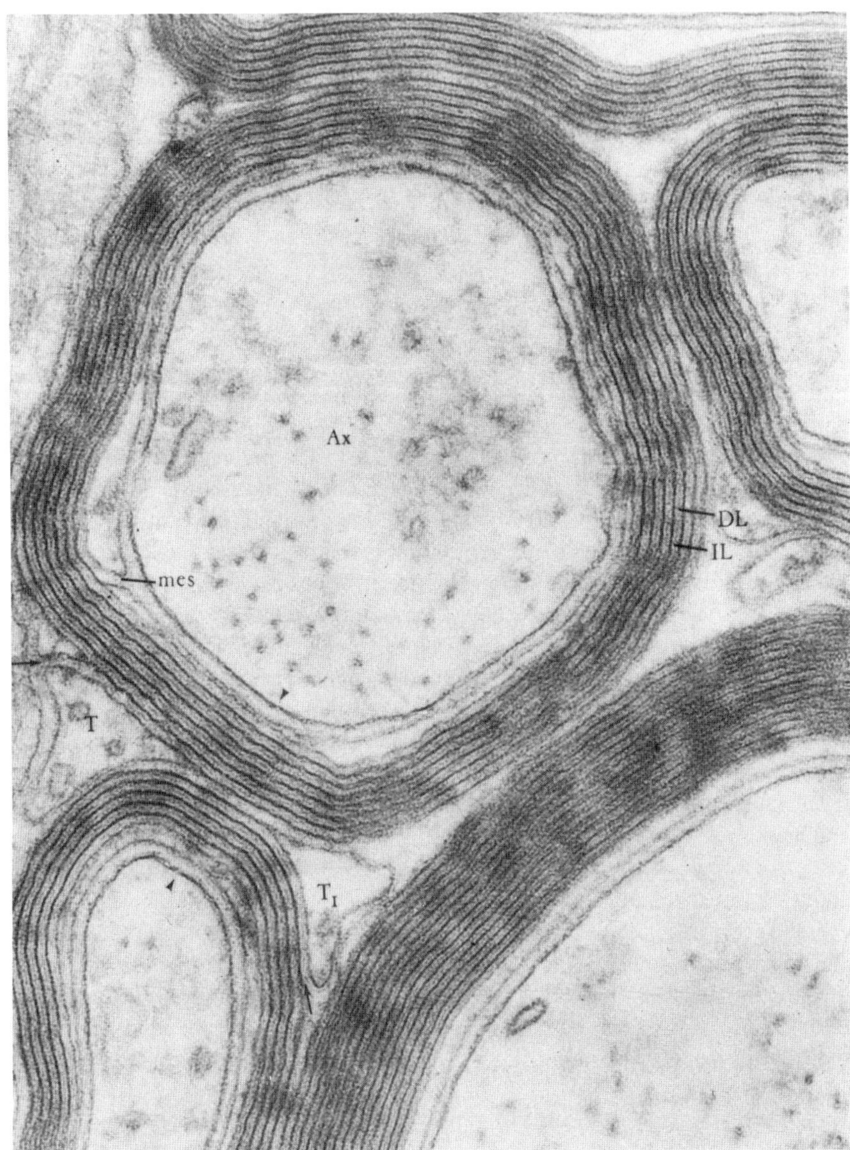

Figure 5.3b Myelinated nerve fibers, adult central nervous system. The picture includes the entire transverse section of one myelinated axon (AX) and portions of four others. The spiraled lamellae of the myelin sheath start at the inner mesaxon (mes). Here the intraperiod line (IL) is produced by apposition of the outer faces of the plasma membrane belonging to the oligodendrocyte process

single such antigen, but in the rat and the mouse there are two proteins with antigenic activity. The antigens have molecular weights between 15,000 and 20,000. Specifically, the bovine antigen has a molecular weight of 18,000, 169 amino acid residues, an isoelectric pH greater than 12, and no tertiary structure. The lack of tertiary structure makes the protein stable to ordinary denaturing conditions. These proteins are found only in myelin and appear to be complexed with phospholipid in the native state. When isolated free of their lipid component they are quite water soluble. The antigen has been studied in detail, and the structures of antigens from several species are completely known [14]. The bovine antigen has 54% polar amino acids and contains no cysteine and only one tryptophan. Only nine residues around the single tryptophan are necessary for the antigenic activity to be expressed in the guinea pig, and modification of the single tryptophan residue abolishes the activity. The antigenic determinants for activity of the bovine antigen in other species seem to be somewhat different. The characteristics of experimental allergic encephalomyelitis is discussed later in this chapter.

Another group of proteins found in myelin are called the Wolfgram proteolipids [15]. They differ from the proteolipid described by Folch and Lees in that they are soluble in acid. They have relatively high molecular weights, and more than 50% of their amino acid residues are polar. This fraction is heterogeneous and contains at least two or three major components.

Thus, when myelin proteins are examined by electrophoretic methods, the major components detected are a proteolipid band, two or three Wolfgram bands, and one or two bands due to basic proteins (EAE antigen). Myelin also contains a small amount of glycoprotein [16,17] and a proteolipid-like material(s) known as the DM 20 protein [18]. The materials referred to in the older literature as neurokeratin, trypsin-resistant protein residue (TRPP), and neurosclerin, are now thought to be artifacts arising from the denaturation of the proteolipid during preparation and are not considered to be true components of myelin [19].

Figure 5.3b (continued)
that forms the sheath. The major dense line (DL), which alternates with the intraperiod line, is formed by the apposition of the cytoplasmic surfaces of the same plasma membrane. The sheath surrounding the axon (AX) contains six lamellae. These terminate on the outside of the sheath at the external tongue process (T). The major dense line terminates as the cytoplasmic faces of the plasma membrane separate from each other to surround the tongue process, and the intraperiod line ends where the plasma membrane of the tongue process turns away from the outside of the sheath (arrows). Optic nerve from an adult rat, X121,550. (Figures 5.3a and b courtesy of Dr. Alan Peters, from A. Peters, S. L. Palay, and H. deF. Webster, *The Fine Structure of the Nervous System,* Harper & Row, New York, 1970.)

Figure **5.4** Electron micrograph of the myelin sheath of a peripheral nerve fiber of the rabbit. Note the double structure of the interperiod line, $\times 252{,}000$. [From W. T. Norton, in *Basic Neurochemistry* (R. W. Albers, G. J. Siegel, R. Katzman, and B. W. Agranoff, eds.), Little, Brown, Boston, 1972.]

Enzymatic Activity

Several reports of enzyme activity associated with myelin have proved to be incorrect. The probable cause of such reports is the inevitable presence of small amounts of glial cellular material trapped between the layers of the glial membrane making up the myelin. The presence of such cellular material in myelin preparations has also led to reports on the presence of RNA, amino acids, pyridine nucleotides, and various ions as myelin constituents. The chances are that these materials, as well as the number of enzymes found in trace quantities, are present as contaminants. There is compelling evidence that one enzyme is truly a constituent of myelin. That enzyme is 2′,3′-cyclic nucleotide-3′-phosphohydrolase [20,21]. About 60% of the total activity found in brain is present in the myelin fraction. The activity of this enzyme increases as myelination progresses and is markedly deficient in strains of mice with aberrations in the myelination process. There is evidence that myelin also contains a neutral proteinase and that peripheral myelin has alkaline phosphatase activity. The occurrence of leucine aminopeptidase activity and esterase activity has been reported. But these last two have not been thoroughly substantiated. It seems that the only enzyme completely accepted as a myelin constituent is the 2′,3′-cyclic nucleotide-3′-phosphohydrolase. The reason for the presence of this enzyme is not known.

Developmental Aspects

The process of myelination coincides with maturation of brain function. In species such as rats, where myelination takes place after birth, the newborn are quite helpless. In other species, such as horses, cows, and guinea pigs, myelination takes place before birth and the newborn animals are capable of active, independent function immediately upon birth. In humans, myelination is partially complete at birth, and almost complete at 2 years of age, but continues at a low rate into the second decade.

Not only does the amount of myelin in the brain increase dramatically during myelination, but the composition also changes to some extent. Probably the glial membrane is deposited and then the myelin is processed through a series of continuing enzymatic reactions to yield mature myelin. In the early stages of myelination a "myelin-like" fraction can be isolated which may represent this immature membrane system. The maturation process involves several changes. With regard to the protein, the ratio of the various components changes. In the human, the ratio between the EAE antigen and the proteolipid increases, With regard to the lipids, the galactolipids increase, the lecithin decreases. The desmosterol decreases, and the other components remain fairly stable (Table 5.2) [5,7]. More subtle changes in lipid composition also occur, such as increases in the length and degree of unsaturation of the fatty acids [22].

Table 5.2 The Composition of Rat Myelin During Development [a]

Age (days)	15	20	30	60	144	190	425
Lipid, % dry wt	74.2	70.4	72.1	72.5	70.4	70.1	68.0
Sterol	25.1	25.7	25.7	25.9	27.8	27.9	27.4
Desmosterol	3.0	1.8	1.5	1.0	0.3	0.2	0.2
Galactolipid	21.4	23.6	26.2	30.5	33.3	30.7	31.7
Phospholipid[b]	50.4	48.3	48.0	44.0	43.7	43.5	44.9
PE	18.4	18.1	17.5	16.8	15.1	17.1	17.7
PC	16.4	15.6	13.6	12.4	10.7	10.7	11.5
Sph	3.4	3.5	3.3	3.4	3.2	3.1	3.1
PI	1.4	1.4	1.0	1.3	1.2	1.1	1.2
PS	7.0	6.9	7.4	6.5	6.7	7.1	7.6
Plasmalogen	14.1	13.1	13.7	14.4	13.9	13.7	14.3

[a]The figures for lipid are in percentage of dry weight and for desmosterol are in percentage of total sterol. All other figures are wt% of total lipid.

[b]PE, phosphatidylethanolamine; PC, phosphatidylcholine (lecithin); Sph, sphingomyelin; PI, phosphatidylinositol; PS, phosphatidylserine.

Table from the work of W. T. Norton, in *Chemistry and Brain Development* (R. Paoletti and A. N. Davison, eds.), Advances in Experimental Medicine and Biology, Vol. 13, Plenum, New York, 1971.

Metabolic Aspects

Metabolically, mature myelin is one of the most stable structures in the body. In contrast to the rapid incorporation of materials into the myelin sheath during early growth, very little incorporation or degradation occurs after the sheath matures. On the other hand, myelin is not completely inert [22]. Suggestions that myelin is deposited and degraded as a unit or bloc structure [23] have now been contradicted by observations that the various lipid constituents turn over at quite different rates. The most active materials in the myelin appear to be the phosphate moieties of the diphosphoinositides and triphosphoinositides. The monophosphoinositides also turn over rapidly. The mechanism of the turnover and the reason for it are unknown, but it is known that the various phosphoinositides are intimately bound to the myelin proteins. Among the other phospholipids, the rate of turnover of the choline and ethanolamine phospholipids is known to be more rapid than that of the serine phospholipids. Cholesterol, cerebrosides, sulfatides, and sphingomyelins turn over quite slowly. The protein metabolism of mature myelin has not been well studied. The unusual stability of mature myelin is most probably due to the fact that the myelin structure is completely and thoroughly separated from the metabolic apparatus of the glial cell from which it arose.

Myelin Abnormalities

When myelin does not form, or does not form properly, or when already formed myelin breaks down, the consequences to the organism are, of course, quite severe. There are a number of different ways of classifying the known myelin abnormalities, but perhaps the simplest is to group the conditions as follows [24]. Demyelination involves the loss of already formed myelin and can be either primary or secondary. That is, the loss can be the first thing that happens while the axon remains intact, or it can follow some damage to the relevant axon. Dysmyelination is the formation of defective myelin, usually due to some inborn metabolic error. The defective myelin is unstable and breaks down at some stage of development. Hypomyelination, on the other hand, is the formation of adequate myelin but in reduced amounts, due to arrest at some stage, or to some chronic abnormality which slows the production of myelin over the whole course of the early life of the subject.

All the demyelinating conditions seem to have certain biochemical events in common. As the myelin breaks down the water content increases. The various myelin constituents decrease, and almost invariably, as the cholesterol content decreases, cholesterol esters appear. The demyelinating tracts are invaded by glial elements and by phagocytes. The volume formerly occupied by the myelin is filled with other tissue or with fluid. Since these same changes occur in many different demyelinating conditions, whether primary or secondary, they are probably produced by degradative processes which are not specific to the initial cause of the conditions.

Although a number of primary demyelinating diseases are known, the most common and thoroughly investigated is multiple sclerosis. This is a condition of the central nervous system in which scattered, sharply defined demyelinated areas or "plaques" appear in the white matter. These plaques have a well-defined border and a "rubbery-firm, gelatinous, grayish-brown, semi-opaque center" [25]. These areas appear throughout the white matter, but the favored point of attack seems to be the border between the white matter and the gray. The localized demyelination seems to be the primary lesion since, especially in early plaques, the axon appears to be untouched. Irreversible axonal damage is seen in some of the mature plaques. Also, some evidence of remyelination is observed, albeit infrequently. The disturbance appears to come from outside the sheath, because the myelin seems to degenerate from the outside surface.

Biochemically, extensive loss of all lipid components occurs in the area of the plaque. Cholesterol disappears and cholesterol esters are found. Cholesterol and various phospholipids appear in the cerebrospinal fluid. In about two-thirds of the cases the γ-globulin content of the cerebrospinal fluid rises, sometimes 60 to 70%, but no diagnostically reliable alterations have been observed. Analysis of the total myelin of multiple sclerosis victims gives a normal picture. No abnormal myelin is formed. The composition of the myelin does not reflect the local degeneration at the plaques because the contribution of the degenerating borders to the gross analysis is quite small.

No decisive information is available as to the cause of multiple sclerosis, although several theories have been advanced. The suspicion that multiple sclerosis is an autoimmune disease has not been experimentally confirmed. No hard evidence is available to suggest a viral cause, although the possibility of a "slow" or latent virus receives continuing consideration. A most exciting lead has been provided by the finding of a myelinolytic antibody in the γ-globulin fraction of the plasma proteins from some multiple sclerosis patients [26-28]. Unfortunately, this myelinolytic fraction has been found, as well, in a number of other neurological disorders, and is considered a possible consequence of the disease, rather than a cause [25]. Although the incidence of the condition has been shown to be high in some families and in certain races, no hard genetic correlations have been possible. Multiple sclerosis is more prevalent in cold climates than in hot. The incidence of the disease in the United States is on the order of 10 to 75 per 100,000 population, or about 0.05%. There are, then, 75,000 to 100,000 cases at any one time [29]. Again, the cause of this condition is unknown.

Multiple Sclerosis

A 43-year-old married woman was admitted complaining of weakness on the right side, slurred speech, loss of the ability to read, and failing memory. A history revealed an episode of weakness and numbness in the right leg 11 years before, lasting about 1 week, a ten-day period of unsteady walking and blurred vision 9 years before, and a recurrent numbness of the right leg, 8 years before. Further such episodes had been experienced in the 2 or 3 years prior to admission.

Examination showed a right-handed woman who laughed and cried without reason and, at times, spoke incomprehensibly because of slurring and jargon. She was unable to remember three digits and had spatial disorientation. Hearing was unaffected. She had moderately severe right hemiplegia, bilateral pyramidal signs, right hemianesthesia, and mild left-sided cerebeller ataxia. Laboratory tests showed cerebrospinal fluid cells, 14 per cubic millimeter, protein 70 mg/100 ml.

Subsequently she experienced disorientation, and eventually, bilateral blindness. Her mental state deteriorated, and she was subject to right-sided epileptic attacks. Eighteen months after admission this patient showed clinical evidence of widespread demyelination in both hemispheres, including the occipital lobes.

Except in some small percentage of cases in very young individuals, the course of multiple sclerosis is a long one. The appearance of the disease is

usually between the ages of 20 and 40. Life spans of 20 to 30 years after the onset are not unusual, and death generally occurs from causes not primarily related
to the disease itself. Extensive neurological damage, then, can occur in this disease and exist for long periods without being fatal.

Experimental Allergic Encephalomyelitis

Primary demyelination can be caused experimentally in animals in a number of
ways. Among these are the administration of cyanide, azide, malononitrile, or
diphtheria toxin. The most interesting is the condition known as experimental
allergic encephalomyelitis because it has conceptual and observable similarities
to multiple sclerosis. The injection of homologous or heterologous material
from the central nervous system, or of the purified basic protein from myelin,
causes this demyelinating condition. Suspending these preparations in Freund's
adjuvant aids in the production of the condition. There are some species differences in the structure of the basic protein from myelin and even multiple proteins in some species [30]. Even so, most species tested are susceptible to injections from other species, with some differences in the nature of the response depending on the particular combination of donor and recipient [31].

The primary lesion is a destruction of the myelin, presumably due to formation of an antibody to the injected material followed by an antigen-antibody reaction within the myelin sheath. In the early stages the axon remains unaffected,
but during the course of the condition a marked inflammatory reaction develops
with consequent axonal damage. The animals exhibit severe weight loss 10 to 20
days after the injection and suffer paralysis of the hind legs, loss of the righting
reflex, and incontinence. In dogs and monkeys lesions are seen histochemically
which are somewhat similar to the lesions in multiple sclerosis, but in rats and
mice demyelinating lesions are not an obvious feature. In spite of the lack of
prominent lesions in some species, experimental allergic encephalomyelitis is considered surely to be an autoimmune disease of the myelin because of the specific
and well-documented nature of the basic protein antigen. The antibody produced
has been shown to be a cellular, rather than a circulating one, and the condition
can be passively transferred using appropriately sensitized lymphocytes. It has
been suggested that the reason that the basic protein is antigenic and not recognized as a "self-protein" is that it appears in the myelin after the immunocom-
petent cells of the body have developed [30].

The relationship of experimental allergic encephalomyelitis to multiple
sclerosis has been considered in detail by many authors [24,25]. In some species
the lesions of allergic encephalomyelitis seem quite similar to the plaques found
clinically. Also, the circulating myelinoclastic factor observed in multiple sclerosis is at least conceptually similar to the antibody produced in allergic encephalomyelitis. But, despite these similarities, doubt remains as to whether allergic

encephalomyelitis is a valid model of multiple sclerosis. One of the crucial dissimilarities is that the course of EAE is continuous and does not resemble the episodic nature of multiple sclerosis. In fact, it is not yet firmly established that multiple sclerosis is an autoimmune disease. If this could be established without doubt, it might be possible to deal with multiple sclerosis using some of the therapeutic leads developed in studies on the treatment and suppression of experimental allergic encephalomyelitis [24].

Wallerian Degeneration

Even in conditions which are primarily demyelinating in nature, there is eventual damage to the axon. This leads to the concept that the glial cells not only form the myelin and are necessary to maintain it, but are necessary for maintaining the axon as well. The reverse certainly is true; that is, the axon must be intact for the myelin sheath to remain viable. Thus, there are a number of conditions in which demyelination is secondary to axonal damage. Such secondary demyelination is known as Wallerian degeneration and was first described by Waller in 1850. This term is used to describe degeneration of the myelin sheath after damage to the axon, and the sequence of events following axon crushing or cutting have been studied in great detail.

After surgical, experimental, or accidental cutting of an axon there is a progressive destruction of the myelin sheath around the part of the nerve peripheral to the lesion. The myelin begins to fragment after 1 day, and a splitting of the myelin layers can be observed in the early stages. After 3 to 4 days swellings begin to appear in the myelin, and the glial cells around the sheath begin to proliferate. Fatty droplets appear in the myelin at the end of the first week, and the myelin begins to be removed by invading phagocytic cells by the end of the second week.

The disintegrating myelin remains biochemically the same as normal myelin for a time until the structure of the sheath breaks down and enzymatic processes begin to degrade the myelin constituents. Sphingomyelin and cerebroside disappear completely in a period of about 3 months. Cholesterol levels drop, and cholesterol esters appear. Tissue protein content decreases and the nucleic acid content increases as glial cells and other tissue elements infiltrate the site of the lesion. In the later stages the picture becomes somewhat confused because it is difficult to separate the changes in the sheath itself from the concomitant changes in the surrounding tissue. Overall, the structure of the sheath breaks down during the first week or 10 days after sectioning of the axon. Then the chemical degradation of the myelin proceeds for the next several weeks. Only after this time is there any evidence of remyelination or repair.

Among the earliest events observed in such degenerating systems is an increase in proteolytic activity. As early as the first day an increase in acid proteinase activity occurs [32]. This may be due to release of activity from the axonal mitochondria or the axonal lysosomes. Somewhat later the neutral proteinase activity also increases. This latter enzyme, which may be a constituent of the myelin sheath itself, is probably in a latent or inactive state normally and only becomes "unmasked" when the myelin structure breaks down. A number of other enzymatic activities increase much later in the sequence of events, e.g., acid phosphatase and β-glucuronidase activities rise after about 2 weeks. Most measurements, however, indicate that the increased proteolytic activity is the initial enzymatic event. There is a general feeling that this enzyme may digest the protein matrix of the myelin and thereby trigger the complex series of changes seen.

Dysmyelination

In contrast to demyelinating conditions where normal myelin is formed in normal amounts and then is broken down because of some condition external to the myelin itself are the dysmyelinating states in which abnormal myelin is formed and then is broken down or is unable to function properly because of its own faulty chemistry. Two such conditions are known clinically. Since they are both genetic lesions in the metabolism of lipids they are discussed in detail in Chap. 17. In the present context it can be mentioned that the two diseases are known as metachromatic leukodystrophy and Refsum's disease. Both are caused by a complete lack of one of the degradative enzymes of lipid metabolism. The first involves the absence of a sulfatase which converts cerebroside sulfate to cerebroside. Thus, cerebroside sulfate accumulates and there is a virtual absence, in the later stages of the disease, of normal myelin. In the second, Refsum's disease, an enzyme which normally oxidizes branched-chain fatty acids is missing, and phytanic acid, which is a normal constituent of plant material, accumulates in the phospholipids of the myelin. In both cases the myelin formed is grossly abnormal and cannot fulfill the normal metabolic or structural role. Both these conditions are equally well classified as lipid-storage diseases.

Hypomyelination

The third class of myelin diseases, the hypomyelinating conditions, occur when myelin formation is normal but is arrested either partially or completely during the period of myelination. Partial arrest is found in a number of dissimilar states in which the lesion is far removed from myelin metabolism and hypomyelination occurs as a distant consequence of some other aberration in metabolism. Many of the hereditary amino acidurias fall into this category, the best studied perhaps being phenylketonuria. This condition is considered in detail in Chap. 9. It is pertinent to mention here that in this inherited disorder of aromatic amino acid metabolism there is a generalized failure of myelination in the central nervous system. A few relevant studies have been done with cell-free systems which synthesize myelin components. The results suggest that the myelin defect is caused by the toxic effects of some of the abnormal amino acid metabolites which may accumulate in the brain. Whatever the cause it does not seem to be a primary effect of the condition, but rather, a secondary consequence of a lesion in an entirely different area of metabolism. The arrest of myelin formation is only partial, and the myelin formed seems normal in quality but much reduced in quantity. Such hypomyelinations are rather common in the amino acidurias.

Myelination Mutants in Mice

Of the several hundred different mutations which have been observed in inbred mice, almost 100 of them are classified as neurological mutants [33]. Excluding those which are obviously due to malformation of the inner ear, and some others about which very little is known, there are about 30 which appear to have metabolic disturbances at the root of the neurological problems (Table 5.3). Of these, three are of interest in the preceding discussion on the formation of myelin. These three are the quaking, the Jimpy, and the myelin synthesis deficient (MSD).

The quaking mouse, first described in 1964 [34], is the clearest animal analog to the hypomyelinating state. Genetically the characteristic is an autosomal recessive. Physically it is manifest as a tremor of the hindquarters when the mouse is placed on a hard, flat surface, and especially when the mouse is moving around. The tremor develops between 10 and 12 days of life and is full-blown by the time the mouse is 3 weeks old. Otherwise the animals function normally. They eat, drink, swim, and reproduce normally, and live a normal life span. Where learning has been measured the responses of the quaking mice have been indistinguishable from the controls. The central nervous system of these mice is severely deficient in myelin. The deficiency is obvious from the earliest stages of myelination and never catches up. The little myelin that is present appears normal, but recent studies suggest that it has a different composition from mature myelin from controls. The peripheral nerves are fully myelinated. There is no evidence of myelin destruction, merely a severe lack of myelin formation. Biochemically, as one would expect, the levels of the components of myelin are markedly depressed in whole brain measurements. And although some enzymatic differences have been shown to exist between the quaking mice and controls [35], it is not at all clear whether these differences are related to the primary lesion causing the hypomyelination.

The Jimpy mouse, first described in 1954 [36], is not so clearly categorized. In this mutation there seems to be both hypomyelination and demyelination [33]. Genetically this condition is transmitted as a sex-linked recessive. The affected males exhibit a generalized tremor by 10 to 12 days of age, which

Table 5.3 Characteristics of Some of the Neurological Mutants of the Mouse

Name	Gene symbol	Genetic characteristics	Biochemical or physical defect	Behavioral abnormality
Absent corpus callosum	ac	Recessive	Partial or complete absence of the corpus callosum	None noticed
Ataxia	ax	Recessive	Defect in the growth of nerve cells; spinal cord white matter reduced; fewer glia and less glial RNA	Weakness of extremities or trunk, followed by paralysis and stiffening of the hind limbs
Cerebral degeneration	cb	Recessive	Hydrocephalus of the cerebral hemispheres and olfactory lobes; white matter degeneration	Sterile; short life span
Dilute-lethal	dl	Recessive	Low liver phenylalanine hydroxylase; demyelination	Diluted coat color; convulsions; death at 3 weeks
Dystonia musculorum	dt	Recessive	Nerve fiber degeneration; axonal swelling in peripheral nerves; muscle spindles shrunken	Uncoordination; moves by writhing, not walking; insensitive to pain
Ducky	du	Recessive	Not known	Waddling gait; toeing out of the hindfeet; seizures
Epilepsy	el	Dominant	No lesions noted	5-minute convulsions after gentle tossing
Eyeless	ey	Not clear	Absence of eyes or reduction of their size	
Fidget	fi	Recessive	Embryonic eye and inner ear disorder	Head shaking; some run in circles; gyrate in water; eyes may be closed throughout life
Jittery	ji	Recessive	Possible defect in aromatic amino acid metabolism; no apparent pathology	Fall down while trying to run; aberrant righting reaction; squatting; convulsions; death by 30 days

Table 5.3 (continued)

Name	Gene symbol	Genetic characteristics	Biochemical or physical defect	Behavioral abnormality
Jimpy	jp	Recessive (sex-linked)	Hypomyelination; demyelination	Tremor; spontaneous seizures; death by 30 days
Leaner	la	Recessive	Small cerebellum; areas of neuron loss	Ataxia; stiffness; retarded motor activity
Lethargic	lh	Recessive	Not studied	Poor gait; seizures
Leukencephalosis	—	Recessive (probably extinct)	Necrosis of the forebrain; hydrocephalus	—
Myelin synthesis deficient	MSD	Recessive (sex-linked)	Probable dysmyelination due to deposition of abnormal lipids	Tremor; seizures; death in 18-23 days
Pink-eyed sterile	ps	Recessive	Possible defect in tyrosine metabolism	Defective spermiogenesis; uncoordinated behavior; pink-eyed
Quaking	qk	Recessive	Hypomyelination	Tremor of the hindquarters; seizures
Reeler	rl	Recessive	Structural abnormalities of the cerebellum, cerebral cortex; increased concentration of cholinesterase	Instability of posture and gait; tremor; weakness; death at 3-4 weeks
Staggerer	sg	Recessive	Reduction in cerebellar size; reduced number of granule cells	Shuffling, hesitant gait; mild tremor
Spastic	spa	Recessive	Disease mediated by central nervous system tissues below the isocortex; spinal cord necessary	Rapid tremor; stiff posture; noise-induced spasms

Shaker with synd-actylism	sy	Recessive	Inner ear abnormalities; malformation of the cerebellum	Fused toes; head tossing and shaking; deafness; death in 1–2 months
Torpid	td	Recessive	—	Slow movements; clumsy walk; mild seizures
Tottering	tg	Recessive	—	Wobbly gait; intermittant seizures
Trembler	tr	Dominant	No lesions in myelin; no difference in electrocorticograms	Paresis; convulsions; die at weaning
Varitint-waddler	va	Semidominant	Dystrophic degeneration in several areas	Heterozygotes waddle and are spotted; homozygotes are white, violently hyperactive, convulsive
Writher	wh	Recessive	—	Convulsions; unable to stand; die before weaning
Wabbler-lethal	wl	Recessive	Myelin degeneration in specific tracts; some cellular loss, probably secondary; increased succinic dehydrogenase in the cerebellum and brainstem	Whole body tremor; increased tone; convulsions; die at 30 days
Weaver	wv	Recessive	Small, immature cerebellum; absence of granule cells	Instability of gait; fine tremor; leaping when agitated; convulsions

Data abstracted from R. L. Sidman, M. C. Green, and S. H. Appel, *Catalog of Neurological Mutants of the Mouse*, Harvard University Press, Cambridge, Mass., 1965.

is most marked in the hindquarters. They are afflicted with seizures by the fourth week of life and die at around 30 days. The hypomyelination is severe, and in addition, evidence for myelin destruction has been found. Characteristically, these animals accumulate nonpolar lipids in droplets in the white matter. The lipid is presumably cholesterol esters, the hallmark of myelin destruction. Again, the myelination of the peripheral nerves seems normal. Biochemically, the myelin which is formed appears normal. Myelin constituents are, of course, found in greatly reduced amounts. A number of differences in the enzymatic machinery of the brains of these mice from those of normals have been found. However, it is thought that none of these is the primary lesion and may represent, on the other hand, the consequences of a malfunction in the glial cells, which is the true mutation causing the Jimpy state.

The third mutant, observed more recently [37], is the MSD, or "myelin synthesis deficient." The characteristic is transmitted genetically as a sex-linked recessive. Physically the animals exhibit tremors and seizures much like those seen in the Jimpy mice. The life span of these mice is substantially shorter, however, being on the order of 18 to 23 days. It also appears that the two genes are different. The central nervous system of these mice is severely myelin deficient, but the peripheral nerves are fully myelinated. Myelin constituents are present in substantially reduced amounts; for example, the levels of sulfatides and sphingosine are about one-third those of the controls. In addition, some qualitative differences in the composition of the cerebrosides and the sulfatides have been detected via thin-layer chromatography. And the ratio of saturated to unsaturated fatty acids in the myelin of the mutant is quite different than that of the controls (1.30 in the mutant, 0.96 in the controls). There is no evident demyelination, but the finding of abnormalities in the composition of the myelin makes it uncertain whether this mutation should be classified as a dysmyelination or a hypomyelination.

A comparison of the three mutants reveals that the quaking and the MSD have somewhat similar pathology and very different genetic characteristics. The Jimpy and the MSD, which are similar genetically, are apparently different pathologically. And, in terms of the severity of the lesion, clearly the quaking is the mildest, with the MSD being somewhat more drastic than the Jimpy.

References

1. H. Davson and J. F. Danielli, *The Permeability of Natural Membranes*, 2nd ed., Cambridge University Press, Cambridge, Eng., 1952.
2. F. O. Schmitt, in *The Biology of Myelin* (S. R. Korey, ed.), Hoeber, New York, 1959.
3. F. O. Schmitt and R. S. Bear, *Biol. Rev. 14:*27 (1939).
4. S. H. Appel, E. D. Day, and D. D. Mickey, in *Basic Neurochemistry* (R. W.

Albers, G. J. Siegel, R. Katzman, and B. W. Agranoff, eds.), Little, Brown, Boston, 1972.

5. W. T. Norton, in *Chemistry and Brain Development* (R. Paoletti and A. N. Davison, eds.), Advances in Experimental Medicine and Biology, Vol. 13, Plenum, New York, 1971.

6. L. C. Mokrasch, in *Handbook of Neurochemistry*, Vol. 1 (A. Lajtha, ed.), Plenum, New York, 1969.

7. W. T. Norton, in *Basic Neurochemistry* (R. W. Albers, G. J. Siegel, R. Katzman, and B. W. Agranoff, eds.), Little, Brown, Boston, 1972.

8. J. Folch and M. Lees, *J. Biol. Chem. 191:*807 (1951).

9. J. Folch, I. Ascoli, M. Lees, J. A. Meath, and F. N. LeBaron, *J. Biol. Chem. 191:*833 (1951).

10. H. C. Agrawal, K. Fujimoto, W. T. Shearer, B. K. Hartman, and F. Margolis, *Trans. Am. Soc. Neurochem. 7:*108 (1976).

11. W. M. Kies and E. C. Alvord, *Allergic Encephalomyelitis,* Thomas, Springfield, Ill., 1959.

12. W. M. Kies, J. B. Murphy, and E. C. Alvord, in *Chemical Pathology of the Newvous System* (J. Folch-Pi, ed.), Pergamon, Oxford, 1961.

13. M. W. Kies, R. H. Laatsch, O. L. Silva, and E. C. Alvord, *Z. Immun. Allergieforsch. 126:*1 (1964).

14. E. H. Eylar, *Proc. Natl. Acad. Sci. USA 67:*1425 (1970).

15. F. Wolfgram, *J. Neurochem. 13:*461 (1966).

16. J. M. Matthieu, R. O. Brady, and R. H. Quarles, *Brain Res. 86:*55 (1975).

17. R. H. Quarles, in *The Nervous System*, Vol. I (D. B. Tower, ed.), Raven, New York, 1975.

18. H. C. Agrawal, R. M. Burton, M. A. Fishman, R. F. Mitchell, and A. L. Prensky, *J. Neurochem. 19:*2083 (1972).

19. F. N. LeBaron, in *Brain Lipids, Lipoproteins and Leucodystrophies* (J. Folch and H. Bauer, eds.), Elsevier, Amsterdam, 1963.

20. T. Kurihara and Y. Tsukada, *J. Neurochem. 14:*1167 (1967).

21. T. Kurihara and Y. Tsukada, *J. Neurochem. 15:*827 (1968).

22. F. N. LeBaron, in *Handbook of Neurochemistry,* Vol. III (A. Lajtha, ed.), Plenum, New York, 1970.

23. A. N. Davison and A. Peters, *Myelination,* Thomas, Springfield, Ill., 1970.

24. P. Morell, M. B. Bornstein, and W. T. Norton, in *Basic Neurochemistry* (R. W. Albers, G. J. Siegel, R. Katzman, and B. W. Agranoff, eds.), Little, Brown, Boston, 1972.

25. C. W. M. Adams and S. Leibowitz, in *The Structure and Function of Nervous Tissue,* Vol. III (G. H. Bourne, ed.), Academic, New York, 1969.

26. M. B. Bornstein and S. H. Appel, *J. Neuropathol. Exp. Neurol. 20:*141 (1961).

27. M. B. Bornstein, *Natl. Cancer Inst. Monogr. 11:*197 (1963).

28. S. H. Appel and M. B. Bornstein, *J. Exp. Med. 119:*303 (1964).

29. H. H. Merritt, *A Textbook of Neurology,* 4th ed., Lea and Febiger, Philadelphia, 1967, chap. 9.

30. E. R. Einstein and L. P. Chao, in *Protein Metabolism in the Nervous System* (A. Lajtha, ed.), Plenum, New York, 1970.
31. M. W. Kies, in *Protein Metabolism in the Nervous System* (A. Lajtha, ed.), Plenum, New York, 1970.
32. G. Porcellati, in *Protein Metabolism in the Nervous System* (A. Lajtha, ed.), Plenum, New York, 1970.
33. R. L. Sidman, M. C. Green, and S. H. Appel, *Catalog of the Neurological Mutants of the Mouse*, Harvard University Press, Cambridge, Mass., 1965.
34. R. L. Sidman, M. M. Dickie, and S. H. Appel, *Science 144:*309 (1964).
35. J. N. Kanfer, in *Biochemistry of Brain and Behavior* (R. E. Bowman and S. P. Datta, eds.), Plenum, New York, 1970.
36. J. R. S. Phillips, *Z. Indukt. Abstamm. Vererbungs. 86:*322 (1954).
37. H. Meier and A. D. MacPike, *Exp. Brain Res. 10:*512 (1970).

6
THE BLOOD-BRAIN BARRIER

It has been known for some years that many small molecules which penetrate
from the blood into most tissues of the body with ease, do not appear to enter
the brain or enter much more slowly. A number of observations of this nature
have led to the concept that there is a barrier of some sort between the blood and
the brain. The early observations of Ehrlich and the somewhat more systematic
observations of Goldmann [1] in 1909 indicated that blood-borne basic dyes,
and in particular, trypan blue, stained all the tissues of the body but did not enter
the brain. The possibility that the brain did not react with the dye was ruled out
by Goldmann's second experiment in which trypan blue was administered via the
cerebrospinal fluid. Under these conditions the brain became heavily stained.
Thus, the brain seemed to be protected in some way, at least from trypan blue.
These original experiments have been followed by a large body of work on the
phenomenon [2-4]. Briefly, the conclusions from these experiments are that:
(1) some materials, such as trypan blue and proteins, which enter other tissues
from blood, do not enter brain; (2) increased blood levels of some normal metabolites, such as glutamic acid, are not accompanied by increased brain levels; (3)
most materials appear in brain much more rapidly if introduced into the cerebrospinal fluid than if introduced into the blood; and (4) most materials enter tissues
such as liver and muscle much more readily than they do brain. It is now clear
that no single unitary phenomenon can explain all the diverse experimental findings and that the blood-brain barrier concept is based on a number of different
properties of the brain which make passage into it by almost any molecule different than entry into other tissues of the body.

The choice of trypan blue for the original experiments was, in a sense, for-
tuitous. This dye is now known to be rapidly and tightly bound to plasma proteins so that it was not the trypan blue itself which was excluded, but the trypan
blue-protein complex. It may be that the extremely low level of protein in the
cerebrospinal fluid allows the dye to remain free and to enter more readily from
this route. On the other hand, it has recently been shown, as is discussed in the

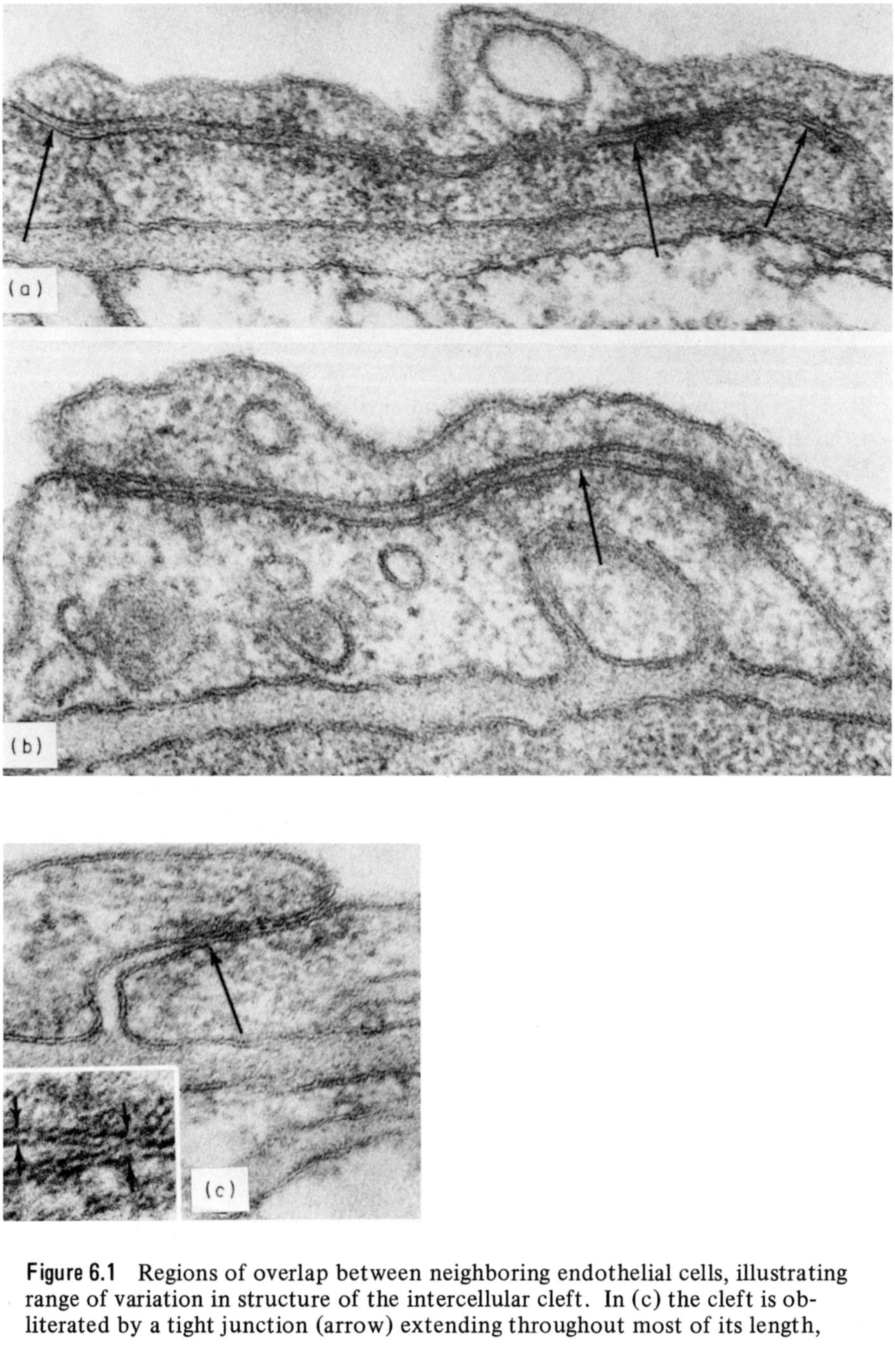

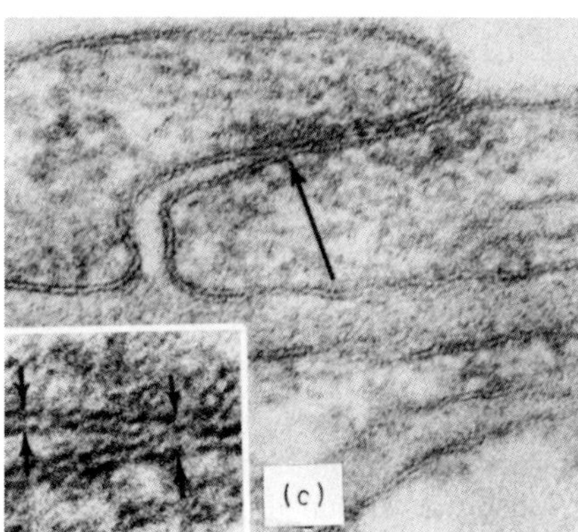

Figure 6.1   Regions of overlap between neighboring endothelial cells, illustrating range of variation in structure of the intercellular cleft. In (c) the cleft is obliterated by a tight junction (arrow) extending throughout most of its length,

next few paragraphs, that proteins themselves enter the brain more readily from
the cerebrospinal fluid than from the blood.

In discussing the blood-brain barrier hypothesis it is necessary to examine
the many different kinds of materials which enter, or do not enter brain, and to
try to uncover the reasons for the observations. The most elegant experiments
are those of Reese and Karnovsky [5] using peroxidase. This small enzyme
(molecular weight, 43,000), when administered through the bloodstream, will
escape from the capillaries of most tissues and can be detected histologically in
the intercellular clefts constituting the extracellular space. The passage of this
marker into brain is much slower than into any other tissue. The barrier to
peroxidase, and presumably to other molecules of similar or greater size, was
shown to be the "tight" junctions which exist between the endothelial cells
making up the brain capillaries (Fig. 6.1). As can be seen in the figure, the cells
overlap, and at points along the overlap the opposed membranes approach each
other very closely. Compared to other tissues the junctions of the endothelial
cells of the brain capillaries are much tighter. In other words, the junctions in
such tissues as heart muscle are "close," whereas those in brain are "closed" [5].
These overlapped cells with their tight junctions form a continuous belt or a
"zonulae occludentes" and provide a complete barrier to peroxidase between
the capillary contents and the extracellular space of the brain. In areas that have
been shown to have no blood-brain barrier to peroxidase, or to the trypan blue-
protein complex, such as the area postrema, the junctions between the cells are
not tight. And there are no tight junctions in the ependyma or the pia-glial surface of the brain, and so proteins pass easily from the cerebrospinal fluid to the
extracellular space of the brain.

An understanding of the blood-brain barrier to small molecules is a more
complex and less precise exercise. First, a number of extracellular markers, such
as inulin, sucrose, or sulfate, enter brain much less readily than they enter other
tissues. That is, after long periods of time following injections of these markers,
their volume of distribution in brain is much less than in other tissues. Some
years ago this was taken as supportive evidence for the concept, based on electron

---

Figure 6.1 (continued)
while in (b) a short tight junction (arrow) occupies only a small part of the cleft.
(a) is most typical; several tight junctions (arrows) are present along the cleft,
and much of the remaining cleft is very narrow. The region of the intercellular
cleft indicated by the arrow in (c) is shown in inset at higher magnification to
illustrate that the total width of the tight junction (between arrows at right) is
less than twice the width of the adjacent plasma membrane (between arrows at
the left). Normal mouse; uranyl acetate, block stain (a) ×121,500; (b) ×171,000;
(c) ×153,000; (c) inset ×378,000. [Figure courtesy of Dr. T. S. Reese, *J. Cell
Biol.* 34: 207 (1967).]

microscopic studies, that the brain had very little extracellular space. However, as discussed in Chap. 1, these electron microscopic data are no longer accepted and so another explanation is necessary. It now seems likely that the cause of the low distribution of these extracellular markers is a combination of the tight junctions which reduce the rate of penetration and the "sink" action of the cerebrospinal fluid. As is mentioned in Chap. 7, the cerebrospinal fluid and the extracellular fluid of the brain are in fairly rapid equilibrium. The cerebrospinal fluid, in addition, is being continually formed and removed by bulk flow. This bulk flow carries away these extracellular markers at a rate comparable to the rate at which they penetrate the capillary wall. Thus, the final steady-state volume of distribution will be a resultant of these two opposing forces, not just a product of the rate of penetration as in other tissues, and not a valid measure of the extracellular space.

Even more complicated is the consideration of the ability of metabolites to enter the brain. In addition to the slow rate of passive permeation of these materials across the brain capillaries, and the bulk flow properties of the constantly renewing cerebrospinal fluid, consideration must be given to the active transport mechanisms known to exist in brain. Not only are there specific transport processes to take in most of the metabolites and ions needed for brain function, but there are specific pumping mechanisms which remove certain materials from brain via the cerebrospinal fluid and actively pump them into the blood. The specificity of these transport systems is demonstrated by the uptake of various sugars. D-Glucose is taken up very readily, but L-glucose is not taken up at all. Mannose and maltose are taken up quickly, galactose rather slowly, and fructose at a very slow rate. The blood-brain barrier to water-soluble small molecules, and especially to ionized molecules, then, is the slow passive movement of these substances across the lipid membranes of the brain capillaries and the brain cells, coupled with the very rapid uptake of related molecules for which specific active transport systems exist. An interesting case is the uptake of glutamic acid. For a long time it was felt that there was an absolute barrier to the penetration of glutamate into the brain, because when the level of glutamate in the blood was raised by injection, the level in the brain was not increased. This was contradicted later by the use of radioactive glutamic acid which, of course, did not raise the absolute level of glutamate in the blood, but did appear rapidly in the brain. So the brain substance is permeable to glutamate, and the reason that the concentration in brain does not rise in response to a rise in the blood concentration is that there is a rapid pumping of glutamate out of the brain via the cerebrospinal fluid, keeping the brain concentration constant and maintaining the homeostatic relationship. Certain anions, some organic acids, and, in particular, penicillin, appear to be in this category. That is, they are transported very rapidly out of the brain via the cerebrospinal fluid and, thus, appear to be completely excluded.

Finally, it is clear that there is no barrier to certain substances. Ethanol,

for example, equilibrates with the brain substance faster than measurements can
be made. Other lipid-soluble materials get into brain at a very fast rate, and many
studies show that, as with other cellular membranes, the entry into brain for a
homologous series of un-ionized compounds for which there are no specific transport mechanisms, has a direct relationship to their lipid solubility. That is, the
more lipid soluble a material is, the more quickly it will enter the brain. Certain
gases, such as CO_2, O_2, N_2O, Xe, and Kr, and many of the anesthetics, get into
brain so rapidly that the only factor limiting their entry is the rate at which they
are presented to the brain substance by the cerebral blood flow. Also, water
equilibrates very quickly from blood to brain.

So it seems that discussions of the blood-brain barrier must specify the substance being studied. No single anatomical structure or functional mechanism
will explain all the effects. The importance of this combination of effects, however, is easy to appreciate. The "barrier" phenomena serve as an insulating and
stabilizing influence, so that the environment of the neural cells is constant, even
in the face of a rapidly changing blood composition.

There are some portions of the brain which are not protected by these
barrier phenomena. These include the area postrema, the median eminence of
the hypothalamus, the line of attachment of the choroid plexus, and the pineal.
It may be that these areas of the brain must sample the blood directly in order
to regulate the levels of its various constituents.

Many conditions are known which cause a breakdown of the barrier mechanisms. For example, treatment with mercurials, various toxins, or high levels
of CO_2 are reported to reduce the barrier function [2, 4]. In addition, injuries
of various sorts also breach the barrier to large molecules. That is, after electroconvulsive shock, convulsions of other sorts, accidental head injuries, or freezing,
large molecules which were previously excluded from the brain can be shown to
enter. In certain tumors of the brain, a proportion of cerebral infarctions, and
many inflammatory states such as meningitis, the blood-brain barrier appears to
be breached, and materials normally excluded from brain enter readily. In fact,
the diagnostic procedures for the localization of tumors and infarctions is based
on the abnormal entry of tracers at the site of the lesion.

It is widely held that the blood-brain barrier is absent or, at least, less effective in very young individuals. There seems, however, to be very little hard
evidence for this conclusion [2]. Many of the characteristics of the barrier systems are present in young animals. The tight junctions in the capillary wall can
be seen. Dyes such as trypan blue are excluded. On the other hand, some metabolites, such as phosphate and cholesterol, do enter the brains of young animals much faster than they enter the brains of adults. This has been taken as
evidence for reduced or undeveloped barrier function. But it is known that the
overall metabolism of the brain increases dramatically during specific phases of
brain growth in young individuals. Many of the observations on increased up-

take of metabolites during these phases may be due to the markedly increased
metabolism or to the greater activity of the various transport mechanisms [2]
and not to a lack of the blood-brain barrier. In support of this it is reported that
the uptake is low in very young animals before the phase of rapid brain growth.
In any case, considerations of the blood-brain barrier phenomena are fundamental to all studies on the relationship of the brain to the rest of the body.

References

1.   E. E. Goldmann, *Beitr. Klin. Chir. 64:*192 (1909).
2.   J. Dobbing, in *Applied Neurochemistry* (A. N. Davison and J. Dobbing, eds.),
      F.A. Davis, Philadelphia, 1968.
3.   H. Davson, in *Chemistry and Brain Development* (R. Paoletti and A. N.
      Davison, eds.), Advances in Experimental Medicine and Biology, Vol. 13,
      Plenum, New York, 1971.
4.   R. Katzman, in *Basic Neurochemistry* (R. W. Albers, G. J. Siegel, R. Katz-
      man, and B. W. Agranoff, eds.), Little, Brown, Boston, 1972.
5.   T. S. Reese and M. J. Karnovsky, *J. Cell Biol. 34:*207 (1967).

7
CEREBROSPINAL FLUID

The brain is bathed in a clear, colorless liquid of low viscosity known as the cerebrospinal fluid. This fluid circulates within and around the brain and spinal cord and fills the subarachnoid spaces, the cerebral ventricles, and the perivascular spaces. This fluid is elaborated principally by the choroid plexuses which are highly vascular formations of modified ependymal cells from the pia. These plexuses are reddish in color due to their intense vascularization and to the large bore of their capillaries. A small proportion of the cerebrospinal fluid is derived directly from the tissues of the brain and the spinal cord. The fluid is formed continuously and also is removed continuously, through valvelike arrangements, into the bloodstream. The rate of formation and the rate of drainage are, of course, equal and amount to about 0.5 ml/min in young adult humans. Since there are on the order of 140 ml of cerebrospinal fluid in such an individual, the rate of formation and removal is about 0.35% per minute, which leads to a complete renewal of the fluid 2 or 3 times a day.

The cerebrospinal fluid is in rapid equilibrium with the extracellular fluid of the central nervous system. On the other hand, it is not in equilibrium with the blood, and its composition is quite different from that of an ultrafiltrate of plasma (Table 7.1). The special nature of the blood-brain barrier, discussed in Chap. 6, as well as the mechanisms of fluid formation discussed below, make the composition of the fluid quite unique and unlike that of the extracellular fluids of the other tissues of the body. In fact, the cerebrospinal fluid is a specific secretion of the brain and not in any sense an interstitial fluid.

There are considered to be four general mechanisms involved in the formation of the cerebrospinal fluid [1]. Active transport from the blood by the choroid plexus accounts for the appearance of Na^+ in the fluid. Also, it is known that the choroid plexus can actively transport materials in the opposite direction. For example, Diodrast and phenolsulfophthalein [2], and possibly the biologically active amines, are actively cleared from the fluid and pumped into the blood by active transport systems. The plexus has a high rate of oxygen uptake, a high

Table 7.1   Composition of Adult Cerebrospinal Fluid (Lumbar) Compared with that of Blood Plasma

	Cerebrospinal fluid	Blood plasma
Constituent	(mg/100 ml)	(mg/100 ml)
Sodium	325	320
Chloride	445	360
Potassium	11	18
Calcium	5	10
Magnesium	2.5	2.7
Bicarbonate	65	175
Inorganic phosphate	2	3.4
Proteins	20-30	7500
Glucose	60-70	90
Urea	20	30
Uric acid	0.5	2.5
Creatinine	1.2	1.3
Free amino acids	1.6	5.8
Lactic acid	15	12
Cholesterol	0.5	160
Ascorbic acid	1.2	0.7
General Properties		
Total solids (g/100 ml)	1.0	8.7
Density (g/ml at 37°C)	1.0010	1.0144
pCO_2 (mmHg)	44	44
pO_2 (mmHg)	25	97
pH	7.31	7.45

Data taken from B. Wyke, *Principles of General Neurology*, Elsevier, Amsterdam, 1969.

rate of glucose consumption, and a large number of mitochondria, and is quite capable of sustaining active transport systems. Diffusion probably accounts for the fluid content of K^+, Ca^{2+}, PO_4^{3-}, HCO_3^-, Cl^-, urea, and water. Pinocytosis may be involved in the transport of small amounts of protein into the fluid. And, since the capillary pressure in the choroid plexus is somewhat greater than the pressure of the cerebrospinal fluid, the diffusion component may be aided by a filtration pressure gradient. The selectivity of the active transport mechanisms, together with the barriers to free diffusion and the constant renewal of the fluid, combine to produce the unique composition. Considering all the above factors, it is not surprising that the composition of the cerebrospinal fluid is unique.

Generally, all the materials present in cerebrospinal fluid are also found in blood plasma, but the concentrations of these materials are quite different (Table

7.1). On the other hand, many things present in the plasma are not constituents of the cerebrospinal fluid. The most striking difference is the low solids content. The plasma has 8.7% solids while the cerebrospinal fluid has about 1% solids. The protein content is also much lower, being on the order of 40 mg/100 ml in cerebrospinal fluid, which is less than 1% of that in plasma. The proteins that are present are the albumin and the α-, β-, and γ-globulins found in plasma. But the total concentrations are much lower, and most of the protein is albumin. All the amino acids normally found in plasma are also found in cerebrospinal fluid, with the possible exception of cysteine, but in much lower amounts. The concentration of glucose is about 70% that of the plasma. Most of the ions found are in lower concentration, but the concentrations of Na^+ and of Cl^- are somewhat higher. Curiously, only two of the vitamins have been found in the cerebrospinal fluid, vitamin C and vitamin B_{12}. Of these, the concentration of vitamin C is twice as high as it is in plasma. A few enzyme activities have been routinely observed in normal fluid. Commonly, lactic acid dehydrogenase and glutamic-oxalacetic transaminase are found. It has also been reported that acetylcholinesterase and acetaldehyde oxidase activities are present. Not present in cerebrospinal fluid are such things as fibrinogen and prothrombin, the bile pigments, various lipids, and most of the hormones. In infants, a few cells are found in the fluid. These may number about $100/mm^3$ and are mostly lymphocytes. No cells are present in normal adult cerebrospinal fluid. The circulation of the fluid from the sites of formation to the sites of removal, and the contribution of various anatomical structures along the way to the formation of the fluid, account for the fact that the composition varies from one site in the system to the other.

Physically, the fluid has a lower viscosity and a lower specific gravity than the plasma. The total osmolality of the fluid is higher than that of plasma. And normally, the pH of the cerebrospinal fluid is a bit lower than that of plasma.

The first ideas as to the function of the fluid centered on its mechanical aspects and its ability to provide insulation against physical shock. While doubtless true, it is now known that the fluid serves other important functions [1,3,4]. Obviously, because of the many, carefully regulated mechanisms governing its formation, its composition is quite stable. It provides a suitable, unvarying environment for the central nervous system, even in the face of alterations in blood composition and changes in the metabolic activity of the nervous tissue itself. Despite wide and rapid fluctuations in the blood pH, the pH of the cerebrospinal fluid remains within narrow boundaries. The K^+ component of the fluid is felt to be a buffering reservoir for the activity of the neurons, in that when the activity of the cells causes an outflow of K^+, it is removed by the circulating fluid, and when K^+ is required, it comes from the fluid. This contention is supported by the finding that the K^+ content of the fluid rises markedly during convulsions. Another prime function appears to be to remove proteins and other materials from the tissue when they are produced by normal cell death or due to such conditions as infarction or injury. That is, since the central nervous system has no lymphatic system to drain the tissue, any cellular material or any material leaking into the tissue from the bloodstream must be removed by the circulating cerebrospinal fluid. This is another aspect of the so-called sink action of the cerebrospinal fluid. Thus, the fluid serves as mechanical and chemical protection for the central nervous system, and provides a channel for the disposal of waste materials.

Because it is relatively accessible, the composition of the cerebrospinal fluid has proved to be of substantial clinical interest. Measurements of the concentrations of cerebrospinal fluid constituents can give indications of neural malfunction. The most valuable constituents for diagnostic purposes have been the enzymes lactic acid dehydrogenase and glutamic-oxalacetic transaminase. These enzymes are derived from the tissue itself, and in certain pathological states involving tissue destruction they appear in high levels in the cerebrospinal fluid. In neurosyphilis, or in multiple sclerosis and other demyelinating diseases, the γ-globulin content of the cerebrospinal fluid rises and the activities of these two enzymes go up markedly. When an infarction or a tumor causes a breakdown in barrier function, passive diffusion of albumin from the blood causes an expansion of the cerebrospinal fluid compartment and results in dangerous edema and intracranial pressure. Subsequent tissue degeneration leads to a rise in the dehydrogenase and transaminase activities in the cerebrospinal fluid. The composition of the cerebrospinal fluid, then, can be an indication of the integrity of the blood-brain barrier systems and of the state of the neural tissue itself. This is true because the fluid is in such rapid equilibrium with the actual extracellular compartment of the central nervous system.

References

1. B. Wyke, *Principles of General Neurology*, Elsevier, Amsterdam, 1969.
2. J. R. Pappenheimer, S. R. Heisey, and E. F. Jordan, *Am. J. Physiol. 200:*1 (1961).
3. H. Davson, in *Handbook of Neurochemistry*, Vol. II (A. Lajtha, ed.), Plenum, New York, 1969.
4. H. Davson, *Physiology of the Cerebrospinal Fluid*, Churchill, London, 1967.

8
TISSUE CULTURE

The study of brain tissue in culture has provided a great deal of useful and unique information and holds promise of a great deal more. Basically, there are two techniques for investigating brain in culture. One technique involves the explant, in which minute fragments of tissue are removed from the animal and maintained in nutrient medium for days, weeks, or even months. The second is the use of dissociated cells, in which tissues are treated with a proteolytic enzyme to separate them one from the other and then are suspended, much like bacteria, and grown either as monolayers or as aggregates. Each technique has several variants and certain unique advantages. But any methodology for the growth of brain tissue in vitro is difficult, time consuming, and sometimes elusive. Thus, it has been written in one review of the field that, "It is recommended that those wishing to undertake a program of organized nerve-tissue culture first arrange an apprenticeship . . . in some established laboratory" [1]. Or, in another, ". . . invariably taken 2 to 3 years to get over teething troubles" [2].

Explant Culture

The use of explants involves the dissection of small pieces of tissue from prenatal or neonatal brain. The amounts which can successfully be explanted are on the order of 1 mm^3. Tissue from adult brain does not survive for long under the conditions of explant. The explants are most successful at a time when morphogenesis is complete and when cytodifferentiation is just beginning. The exact period of time depends upon the species to be used. Under these conditions the tissues recover rapidly from the explanting and develop to maturity in vitro. Many of the events associated with tissue differentiation can be seen in the developing explant, as will be discussed below. But since the development in the explant is somewhat slower than comparable events in the animal, it is better to study rapidly developing animals such as the mouse, rather than more slowly

developing species such as the human. In fact, the mouse and the chicken have
been the species of choice.

The conditions used for the explant culture of brain are not radically different than those appropriate to other tissues in vitro. These have been discussed
in a number of reviews [1-3]. The media used are balanced salt solutions containing fetal or placental serum and embryo extract. A high concentration of
glucose is included. The pH and the temperature must be carefully controlled,
and indeed, different tissue responses have been seen due to alterations of one
degree in the temperature [2]. The vessels used for such cultures are generally
Maximov-type, double cover-slip arrays with a coating of collagen.

Explant techniques have been used in most elegant fashion to observe and describe the various neuronal types in brain [4] (Fig. 8.1a-d). In addition, explants
have been found amenable to a number of different kinds of measurements. Time-
lapse photography, fluorescence or electron microscopy, metabolic studies, and
microelectric recordings have been used to study the responses of these explants
to various conditions and stimuli.

Dissociated Cells

The technique for dissociating cells is a powerful tool for the study of tissues in
general, and for brain in particular. It is of special benefit for neural tissue because it allows the separation of the many types of cells in brain and the eventual
isolation and study of a single cell type. The technique involves the excision of
the tissue, followed by the mincing of the sample into very small fragments, on
the order of the cubic millimeter used for explant. Then the tissue is treated with
a dilute solution of trypsin and, after the trypsinization, drawn through a pipette
several times to separate the cells completely. The cell suspension is then plated
in a culture dish in nutrient medium.

Whether in explant or in dispersed cell culture, a number of general statements can be made about brain tissue in vitro. Certainly, mature tissue does not
do well in culture, and mature neurons do not multiply. Studies of the energy
metabolism of cultured nervous tissue reveal that the two major energy-producing
systems, glycolysis and the Krebs cycle, as well as the hexose monophosphate
shunt, are operative. The exact balance between the various pathways depends
on the tissue cultured and the stage of development at which it is cultured. Although little or no DNA synthesis has been observed, cells in culture synthesize
proteins and RNA and carry on the metabolism associated with the biosynthesis
and interconversion of the nucleotides. Lipid metabolism associated with the
synthesis of myelin has been repeatedly found in vitro. So it is clear that nervous
tissue in culture carries on the fundamental metabolic activities that most cells
exhibit in vitro or in vivo. The more interesting question, of course, is what properties do they exhibit which are unique to brain.

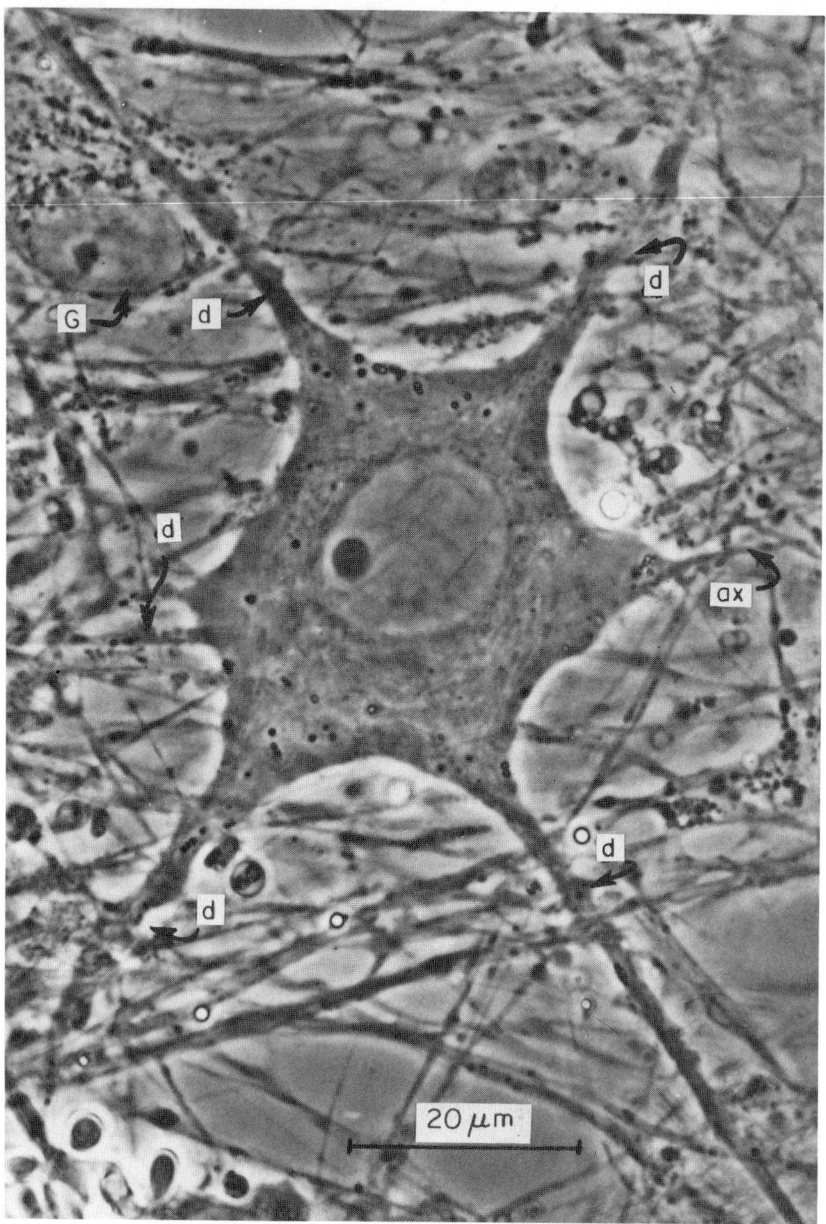

Figure 8.1a Well-isolated living neuron in a 17-day-old culture of rat brainstem. Size and shape are suggestive of the derivation of this cell from the lateral vestibular nucleus of Dieters. One small and four large dendrites are seen at d. The axon is indicated at ax. The thickness of this area is not more than one cell layer. Glial fibers form a loose network around the neuron. A neuroglia cell is seen at G. Phase contrast.

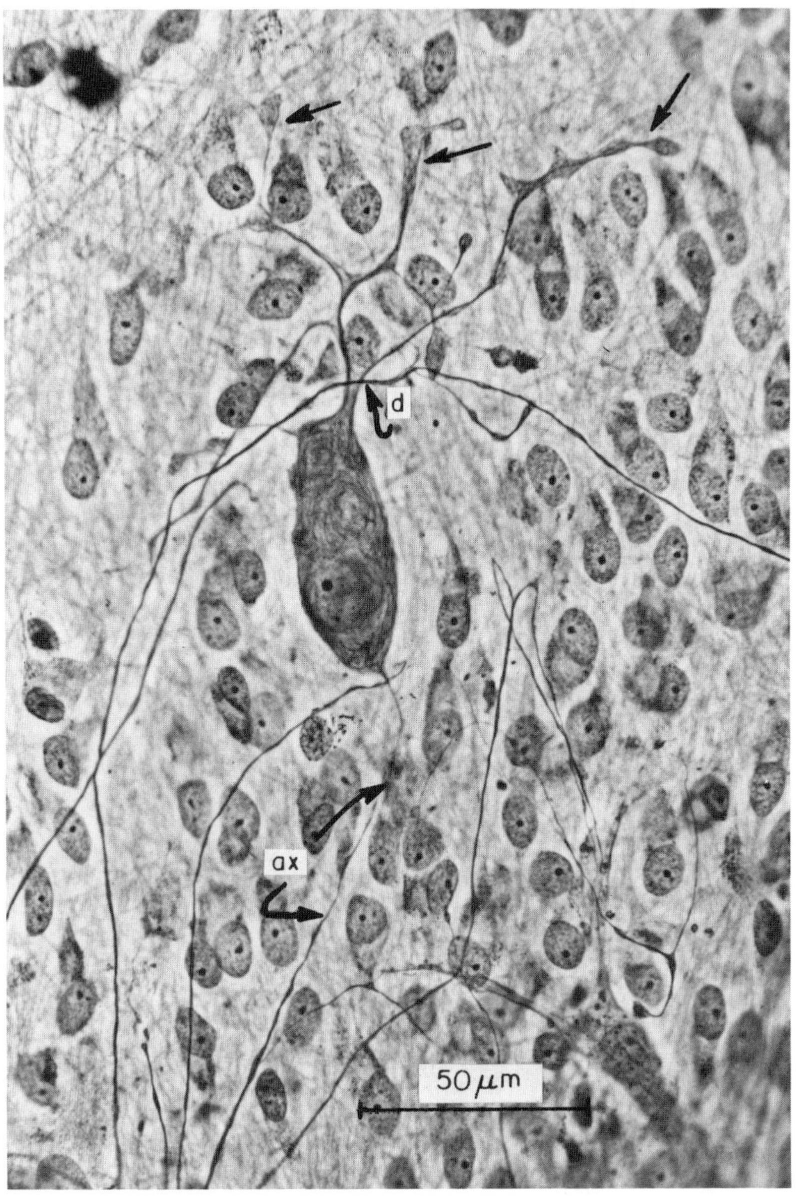

Figure 8.1b Purkinje cell in a 40-day-old culture of kitten cerebellum. The dendritic stem d of this cell extends upward. The arborization is very much reduced in complexity. Instead of numerous, finely drawn out tips there are only a few blunt club-shaped endings in which some fibrillar pattern is recognizable (arrows). The axon ax springs from the lower circumference of the cell body. Other neurites running through this field apparently are unrelated to this cell. Bodian.

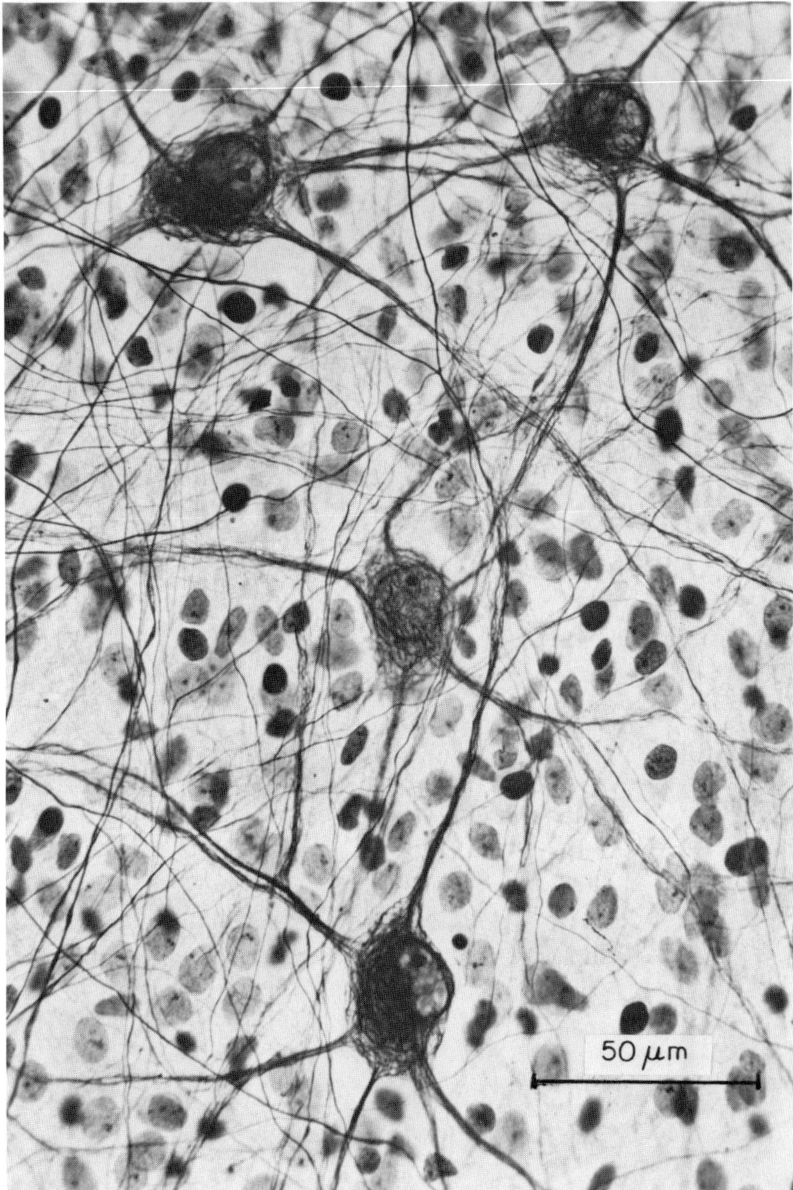

Figure 8.1c   Multipolar neurons from the deep nucleus of the cerebellum in a 12-day-old culture of rat cerebellum. Holmes.

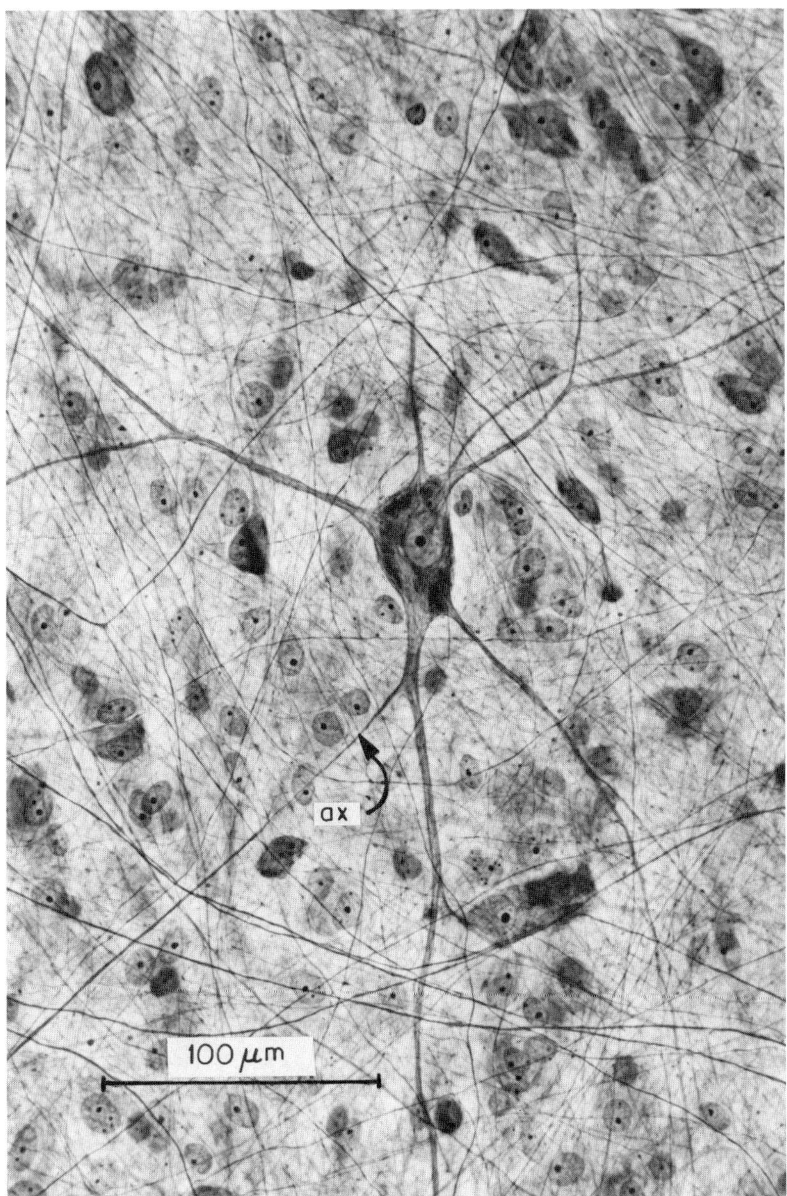

Figure 8.1d   Multipolar neuron in a 29-day-old culture of rat brain stem. The axon ax springs from one of the dendrites. Bodian. [Figures 8.1a–d courtesy of Dr. Walther Hild from W. Hild, *Z. Zellforsch. Mikrosk. Anat. 69:*155 (1966).]

Reaggregation Cultures

Before discussing the organotypic properties of brain in culture, however, it is
appropriate to discuss a few further refinements of the culture technique. First,
it should be mentioned that an important variant of the cell culture technique
involves the reaggregation culture. The reaggregation technique, first introduced
by Moscona in 1961 [5], has more recently been applied to brain. The method
involves the preparation of cells in essentially the same manner as for stationary
culture. After dissociation by trypsin, however, the cells are maintained in rotating vessels which are turned on the order of once every 1 to 3 sec. The cells can
be maintained for weeks in this condition. And, if the cells are taken from the
animal at the appropriate developmental stage, they will form aggregates in such
a rotating culture (Fig. 8.2). These aggregates are quite specific, and the properties of the aggregates reflect the tissue of origin and its developmental stage.

When reaggregation techniques are applied to brain tissue, a number of interesting things happen. Cells from different regions of embryonic chick or
mouse brain form aggregates with different patterns [6]. These aggregates vary
with the age of the embryos from which the cells were dissected, the cells from
older embryos forming smaller aggregates; that is, aggregatability declines as differentiation proceeds. Further, it has been shown that when cells from different
tissues are cultured together they tend to sort out into tissue-specific aggregates.
Brain cells from different regions of a single embryonic brain also show a tendency to sort out. If cells from the same portion of the brains of the two
species, e.g., cells from embryonic chick cerebrum and from embryonic mouse
cerebrum, are cultured together, they form aggregates of a size intermediate
between those formed by the tissues cultured separately, and the aggregates
can be seen to contain a "chimeric" fabric of both cell types. The tendency
for cells to sort out is much greater if different brain parts are used from two
species than if different brain parts are used from a single species. And the
sorting is fairly complete if neural cells from one species are mixed with nonneural cells from the other. The medium recovered after the growth of monolayer cultures of embryonic brain appears to contain a factor or factors which
specifically promote the reaggregation of similar cells in rotating culture [7].
This factor was without effect on nonneural tissues but did not seem to be
species specific, since medium from mouse cerebrum promoted the reaggregation of chick cerebrum as well. It will be interesting to know the nature
of this "neural glue" since, of course, the whole aggregation phenomenon may
be relevant to the mysterious process of cellular recognition, and this factor
may reveal something of the chemical events involved. But, as we shall see,
the process of reaggregation may have even wider developmental implications
than recognition alone.

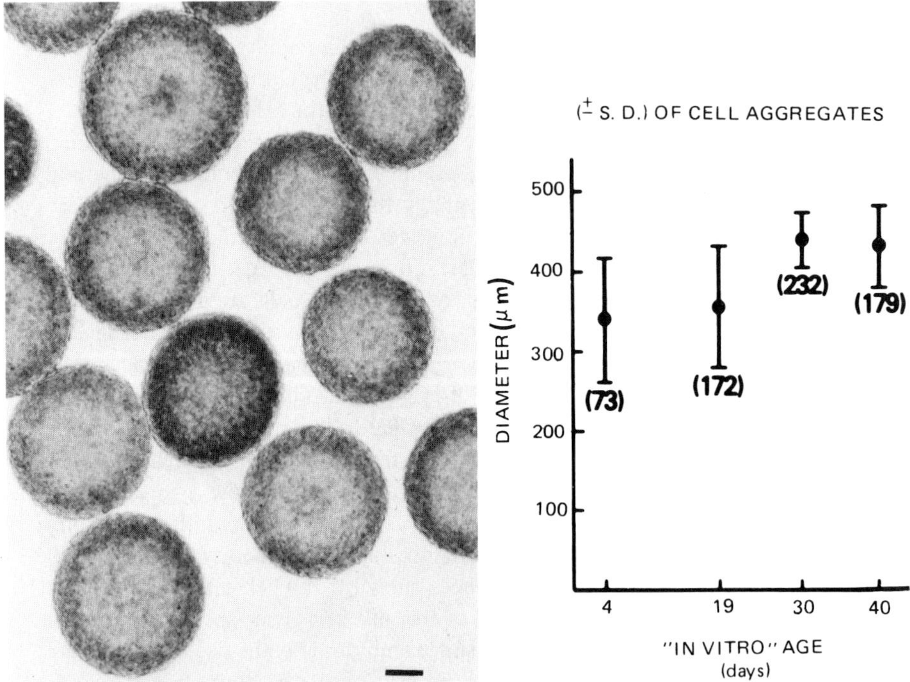

Figure 8.2 Rat brain cells in aggregating cell culture. Prior to 4 days in culture the dissociated cells coalesce to form spherical aggregates. These aggregates have been maintained in culture for 30 days. Scale bar, 100 μm. The graph indicates the mean diameter (± standard deviation) of aggregates maintained in vitro for various periods of time. (Figure courtesy of Drs. Bruce Trapp and Elliott Richelson.)

Primary Cultures, Cell Lines, and Clones

To speak again about variants in the culture technique, it is important to recognize the distinction between primary cultures, permanent cell lines, and clones. Primary cultures are merely the plating of cells removed from the tissue and the investigation of these cells through their lifetime in culture. A permanent cell line is the establishment, by a number of serial cultures, of a more or less homogeneous population of cells from a given tissue, which can be carried from one plate to another without substantial loss of the properties of the original culture. Cloned cells are those obtained originally from a single cell, and are presumably identical, one to the other. Methods for cell culture, for the establishment of permanent cell lines, and for the cloning of cells have been published in detail [8]. And information on the properties of a number of clonal cell lines from neural tissue is now available [9].

Synapse Formation In Vitro

Among the most interesting properties of neural tissue in vitro is its ability to form synapses. The formation of synaptic connections has been seen by a number of workers in a number of culture systems. The development of synapses has been observed in explants of mammalian nervous system which have been removed from the fetus before the stage of synapse formation occurs in the living brain. Synaptic connections have been seen with the electron microscope [10, 11], and have been shown by electrophysiological measurements [12,13]. Physically separate explants of nerve, or of nerve and muscle, have been shown to become linked together by synapse formation in vitro [14]. Dissociated cells also can form synapses in vitro. Fischbach [15,16] has shown that dissociated cells from spinal cord of 7-day embryonic chicks will form functional synapses after 2 to 4 weeks in vitro with muscle fibers which had developed in the same culture from myoblasts of 11-day embryonic chicks. The synapses could be seen microscopically (Fig. 8.3) and could also be shown to be functional using intracellular recording devices. The synapses were cholinergic, and the release of the

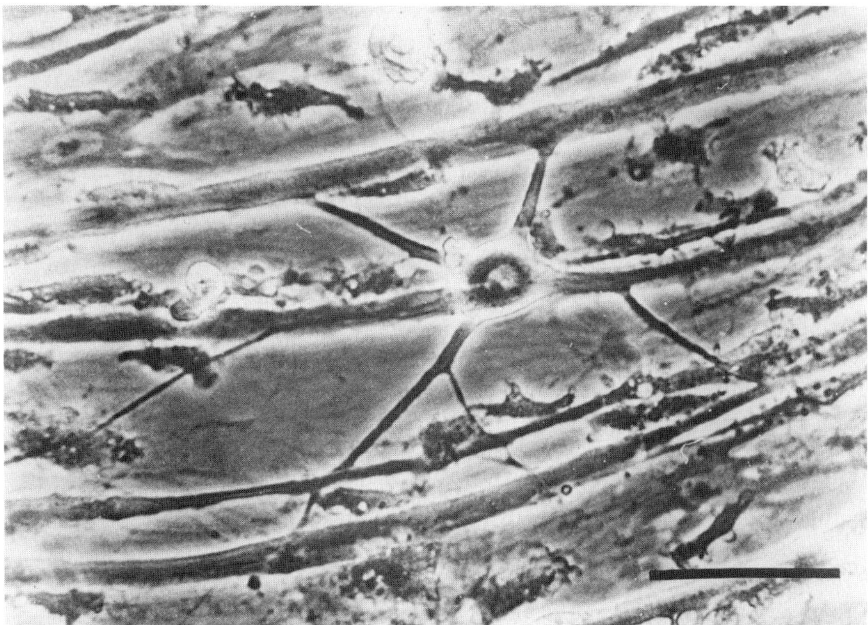

Figure 8.3 Neuron-muscle pairs in 25-day combined spinal cord-muscle cultures. Bar = 50 μm. [Figure courtesy of Dr. G. D. Fischbach, from G. D. Fischbach, *Dev. Biol. 28:*407 (1972).]

transmitter was regulated by Ca^{2+} and Mg^{2+}. Although there were some differences between the characteristics of the junctions formed in these cultures and mature neuromuscular synapses in vivo, these synapses do resemble embryonic neuromuscular junctions. Reaggregating cultures also form synapses in vitro. Studies on embryonic mouse brain in reaggregating culture have shown the formation of synapses morphologically [17] and, using extracellular recording devices, electrically [18]. Synapse formation has also been reported to occur between a hybrid cell made from mouse neuroblastoma and rat glioma, and striated muscle [19], and, more recently, between the same cloned hybrid and a clonal line from muscle [20].

Myelin Formation In Vitro

Another brain-specific property which has been observed in cell culture is the formation of myelin. Myelin formation in culture was first seen in explants from peripheral nerve [21,22]. Somewhat later myelin formation was shown in similar cultures from the central nervous system of newborn cats and rats [23, 24]. By now myelination in culture has been seen in many systems, and various aspects of the myelination process in vitro have been investigated [1]. Neuronal maintainence in the explant, and as a consequence, optimal myelin formation, seems to require a much higher glucose concentration in the medium than is required for the maintenance of other types of cells. Also, various enzymatic changes in the explant at the time of myelin formation have been elucidated. And, a number of studies have shown the incorporation into the tissue of radioactive precursors which are characteristic of myelin. The formation of myelin in explants, then, has been amply shown on a morphological and a biochemical level.

A few other properties are characteristic of brain cells in culture. Neurons exhibit axoplasmic streaming in both directions, that is, toward the cell body and away from it. This streaming, or movement, of the axoplasm, is characteristic of neurons and is related to, or responsible for, the movement of molecules and particles from the cell body to the synapse. Neural tissue in culture will extend fibers called neurites. This name is given because it is not clear whether they are dendrites or axons. In fact, these extensions may not be a characteristic property of brain at all since many other tissues will develop fiber extensions in culture. Yet another property is the ability of oligodendroglia and of Schwann cells to contract, which was first observed many years ago [25]. The contractions seen are rhythmic and resemble the contractions of smooth muscle. They exhibit a slow systole of about 90 sec, a relaxation time of 3 to 5 min, and a refractory period of varying length. The cells may contract and relax anywhere from 2 to 20 times an hour. This behavior is quite unique to nervous tissue. These regular contractions or pulsations of the cells surrounding the axon were thought to be involved in axoplasmic flow, much as the intestinal contents are moved along by peristalsis. As discussed in Chap. 4, this suggestion is no longer considered tenable and the real reason for the contractility of the glial cells is not known.

C-1300

Much of the recent work on the development of neural tissue in culture has been concerned with the biochemical properties unique to brain. A great deal of this work has been done with cells derived from brain tumors and in particular those from the mouse neuroblastoma C-1300, a tumor of sympathetic nerve. Methodology for the culture of this tissue and for the isolation of clones derived from single cells of this tissue has been developed by Nirenberg and his coworkers [9]. The major efforts have been to determine the properties of these clones with respect to the synthesis of neurotransmitters, and accordingly, methods have been developed to determine the levels of the enzymes involved in the synthesis of known neurotransmitters in the various clones isolated. Three kinds of clones have been found [26]. These are clones which have high levels of enzymes for acetylcholine synthesis, those devoted to catecholamine synthesis, and those which synthesize neither transmitter. All the clones had acetylcholinesterase activity, the enzyme necessary to inactivate acetylcholine.

In general, the specific activities of these enzymes were several-fold higher in preparations from stationary phase cells than in similar preparations from rapidly growing cells. Other properties of the clones were investigated, such as the formation of neurite processes and the excitability of the membrane. Neurite formation and the appearance of excitable rather than passive membranes, like the appearance of the transmitter-synthesizing enzymes, seems to be initiated by the cessation of growth. The data have been interpreted as follows. Differentiation is regulated and is the inverse of the rate of division. When growth of the cells stops, differentiation starts, as expressed as the formation of processes, the appearance of excitable membranes, and the increased activity of the transmitter synthesizing enzymes. The properties of the neuroblastoma clones now available have been published (Table 8.1). Also a compilation of the various other clonal cell lines and several cell types of neurobiologic interest is available (Table 8.2).

It is of interest that the development of the various neural enzymes is quite pronounced in reaggregating cultures of the type described above. It has been shown that reaggregating mouse brain cells have much higher levels of several of the enzymes involved in neurotransmitter synthesis than monolayer cultures maintained for the same period of time [27]. The rationale here is that the aggregating cultures favor cell-to-cell contacts more than the monolayer cultures and, therefore, growth is restricted and differentiation favored. It is not clear

Table 8.1   Mouse Neuroblastoma C–1300 Clones

Cell type	Clone number	Tyrosine hydroxylase	Choline acetyltransferase	Acetyl-cholinesterase	Neurites	Excitable membranes
		(pmol product formed per minute per mg protein)				
L-cells	A9		2	440	—	—
Mouse brain		5	550	69,000	+	
Tumor, in vivo		0	10	130,000		
	NS-20Y		920	47,000	+	+
	NS-20	0	490	48,000	+	+
	NS-26	1	430	47,000	+	+
	NS-25	1	125	55,000	+	+
	NS-18	1	76	80,000	+	+
	NS-21	0	52	36,000	+	+
	NS-16	0	43	67,000	+	Defective
	NIE-115	980	0.1	256,000	+	+
	NIE-124	350	0.9	175,000		
	NIE-125	330	1.3	98,000		
	NIE-116	225	0.5	109,000	+	
	NIE-122	215	0.4	40,000		
	NIE-110	150				
	NIE-126	122	0.3	73,000		

NIE-114	93	0.1	74,000	—	
NIE-112	63				
NIE-128	60				
NIE-127	60				
NIE-123	40				
NIE-111	32				
NIE-118	17			—	
N-18	2	2	105,000	+	+
N-3	9	4	174,000	+	+
N-4	4	0.1	46,000	+	+
N-7	6	0	99,000	+	
N-8	2	2	42,000	+	
N-9	11	1	53,000	+	
N-10	1	3	55,000	+	+
NIE-113	6			—	
NIE-117	1			+	
N-11	0	5	404,000	+	+
N-12	0	4	346,000	+	+
N-13	0	6	130,000	+	+
NIA-103	0	2	19,000	—	Defective
NIA-104	1	5			

Table from the work of T. Amano, E. Richelson, and M. Nirenberg, reported in B. K. Shrier, S. H. Wilson, and M. Nirenberg, in *Methods in Enzymology,* Vol. 32 (S. P. Colowick and N. O. Kaplan, eds.), Academic, New York, 1974.

Table 8.2 Glial and Other Clonal Cell Lines

Cell type	Clone designation	Species	Properties
Glioma	C6	Rat	S-100 protein; β receptor; glial fibrils
Glioma	$C2_1$	Rat	S-100 protein; β receptor; glial fibrils
Glioma	$C2_2$	Rat	S-100 protein
Glioma	CHB	Uncertain	S-100 protein; β receptor
Glioma	1181N	Human	Histamine receptor; β receptor
Glioma (transformed by polyoma virus)		Hamster	
Peripheral neurinoma	2	Rat	S-100 protein
Ependymoblastoma	5 clones	Mouse	
Mast cell tumor	P-815-x-1	Mouse	Tryptophan hydroxylase; histidine decarboxylase
Mast cell tumor	P-815-x-2	Mouse	Tryptophan hydroxylase; histidine decarboxylase
Pituitary tumor	AtT-20	Mouse	ACTH
Pituitary tumor		Rat	ACTH
Pituitary tumor	GH1	Rat	Growth hormone
Melanoma, Cloudman	M-3	Mouse	Melanin
Melanoma, Cloudman	NCTC-3960	Mouse	Melanin
Melanoma, Cloudman	NCTC-3959	Mouse	Amelanotic
Melanoma	RPM1-1846	Syrian hamster	Melanin
Melanoma	RPM1-3460-3	Syrian hamster	Melanin; 8-azaguanine resistant
Striated muscle	Many clones	Rat	Contractile

Table from the work of many authors, as reported in B. K. Schrier, S. H. Wilson, and M. Nirenberg, in *Methods in Enzymology,* Vol. 32 (S. P. Colowick and N. O. Kaplan, eds.), Academic, New York, 1974.

whether the reaggregating cultures give higher specific activities than stationary phase monolayers, but they certainly give higher levels than the logarithmic monolayer cultures of the same age.

A number of other aspects of the growth of neural cells in culture have been investigated. A series of hybrids of the mouse neuroblastoma and mouse L cells

have been made and inspected for neuronal properties. It was found that many
of the hybrids had electrically excitable membranes similar to those found in the
neuroblastoma parent [28]. This property was still present after the hybrids had
been carried through as many as 40 generations. Since the L cells have membranes
which are electrically passive, this work shows that some of the neuronal genes can
be expressed in these hybrids. As mentioned before, the hybrid formed from
mouse neuroblastoma and rat glioma will form synapses, in culture, with muscle
cells [19,20].

It now seems clear that cyclic AMP (cAMP) plays an important and unique
role in the function of normal brain. The possible role of cAMP is discussed in
Chap. 24, but it is appropriate to note here that the cAMP content of cells in culture has been studied. Assays have been done on the various clonal lines [29,30],
and in primary cultures using either the reaggregation technique [31] or the more
usual surface culture [32]. The response of cAMP concentrations to various chemical stimuli has been studied, and it has been shown that cAMP levels rise in culture, just as they do in vivo or in tissue slices, in response to the administration of
catecholamines [31,32] and to the presentation of adenosine [32].

Additionally, the following have been explored: the puzzling effects of
D_2O in stimulating growth and maturation of nervous tissue in culture [1,33,
34]; the induction of enzymes in neural tissue, including the pioneering studies
of Moscona on the precocious induction of glutamine synthetase in embryonic
neural retina by hydrocortisone [35-39]; the synthesis, in culture, of such brain-
specific proteins as the S-100 [40-43]; the use of pineal organ culture as a model
system for the study of input and output signals in a neural system and for the
elucidation of the relationship between the biosynthetic pathways for the various neurohormones [44,45]; and the dramatic effects of the nerve growth factor
on the outgrowth of processes from sensory ganglia, and so on. All these lines
of study, using cultured cells as a medium for investigation, certainly indicate a
continued and much expanded interest in brain culture as a tool. The cells
now obtainable in culture from the nervous system, on the other hand, are relatively few, and many of them, notably the neuroblastomas, are not typical. Thus,
any data obtained with such tissues must await confirmation with more typical
lines.

Clinical Applications of Culture Methods

Finally, it is important to note that the culture technique has proved of value in
the study of various neurological aberrations in at least two ways: first, in the
culture of tissues from individuals with the various conditions; and second, in the
culture of normal tissues under the aberrant conditions existing in various disease
states for possible use as model systems. In the first regard, a number of studies
have been done on tissues from the various mouse mutants described in Chap. 5.

Observations have been made on cerebellar tissue from the Jimpy mouse, and it has been found that myelination of this tissue in vitro is no more efficient than myelination in vivo [46]. In other studies it has been shown that tissues from at least one of the mouse mutants do not "sort" normally in reaggregation cultures [47], implying that the defect affects the ability of cells in the brain to recognize each other. Cells from individuals with various neurological disorders reflect, in culture, the same abnormal metabolism seen in vivo. For example, fibroblasts from galactosemics [48,49] or Refsum's patients [50] exhibit the enzyme lesions in culture that are seen in the patients. Cultured neurons from patients who had Tay-Sachs disease accumulate, in culture, the lipid characteristic of the disease state [51,52]. In other lipidoses, fibroblasts in culture show the characteristic enzyme lesion and accumulate the distinctive lipid even after a substantial period in culture.

The second approach to such studies has been to expose normal tissue in culture to conditions resembling those of a particular disease in an attempt to understand the disease mechanism. In this regard many attempts have been made to expose normal myelinating cultures to the abnormal metabolites which occur in phenylketonuria or maple syrup urine disease. It has been determined through these studies that some of the abnormal metabolites are, in fact, myelinotoxic [53] and that others cause cellular damage [54]. The inference, of course, is that the observations reflect the mechanism of the defects which occur in this condition. Similar studies have demonstrated that the heme breakdown products found in excess in patients suffering from kernicterus are damaging to myelinating cultures [55]. It has been mentioned before that the plasma of patients afflicted with various demyelinating disorders causes demyelination in vitro as well. And the demyelinating factors in the plasma of animals with experimental allergic encephalomyelitis and experimental allergic neuritis can cause the demyelination of already myelinated cultures or prevent the myelination of cultures explanted before myelination. Finally, it has been possible to show that various neurotoxins cause effects in culture similar to their effect in vitro. For example, the effects of colchicine and vinblastine upon the neurotubules and neurofibrils resemble what is seen in vivo. Diphtheria toxin causes a characteristic demyelination in vitro and in vivo. And the effects of metals and radiation on cultures have been used to understand the effects of these agents on the intact animal.

Thus, it is abundantly clear that the use of culture systems from brain not only provides a most powerful tool for the study of the normal events in the differentiation and development of neural tissue, as well as its normal neurochemistry, but also allows a detailed look at many of the diseases and toxins which impinge upon the nervous system. Certainly, one can expect that investigations of problems of neurochemical interest will be more extensively explored in vitro as tissue culture techniques become more refined and as other cell types become amenable to, and available in, cell culture.

References

1. M. R. Murray, in *Handbook of Neurochemistry*, Vol. Va (A. Lajtha, ed.), Plenum, New York, 1971.
2. C. E. Lumsden, in *The Structure and Function of Nervous Tissue*, Vol. I (G. H. Bourne, ed.), Academic, New York, 1968.
3. S. M. Crain, in *Developmental Neurobiology* (W. A. Himwich, ed.), Thomas, Springfield, Ill., 1970.
4. W. Hild, *Z. Zellforsch. Mikrosk. Anat. 69:*155 (1966).
5. A. A. Moscona, *Exp. Cell Res. 22:*455 (1961).
6. B. B. Garber and A. A. Moscona, *Dev. Biol. 27:*217 (1972).
7. B. B. Garber and A. A. Moscona, *Dev. Biol. 27:*235 (1972).
8. R. D. Cahn, H. G. Coon, and M. B. Cahn, in *Methods in Developmental Biology* (F. H. Wilt and N. Wessels, eds.), Crowell, New York, 1967.
9. B. K. Schrier, S. H. Wilson, and M. Nirenberg, in *Methods in Enzymology*, Vol. 32 (S. P. Colowick and N. O. Kaplan, eds.), Academic, New York, 1974.
10. R. P. Bunge, M. B. Bunge, and E. R. Peterson, *J. Cell Biol. 24:*163 (1965).
11. M. B. Bunge, R. P. Bunge, and E. R. Peterson, *Brain Res. 6:*728 (1967).
12. S. M. Crain, *Int. Rev. Neurobiol. 10:*1 (1966).
13. S. M. Crain and M. B. Bornstein, *Exp. Neurol. 10:*425 (1964).
14. S. M. Crain, E. R. Peterson, and M. B. Bornstein, in *Growth of the Nervous System* (G. E. W. Wolstenholme and M. O'Connor, eds.), Ciba Foundation Symposium, Churchill, London, 1968.
15. G. D. Fischbach, *Science 169:*1331 (1970).
16. G. D. Fischbach, *Dev. Biol. 28:*407 (1972).
17. M. B. Bornstein and P. G. Model, *Brain Res. 37:*287 (1972).
18. S. M. Crain and M. B. Bornstein, *Science 176:*182 (1972).
19. P. Nelson, C. Christian, and M. Nirenberg, *Proc. Natl. Acad. Sci. USA 17:* 123 (1976).
20. C. N. Christian, P. G. Nelson, J. Peacock, and M. Nirenberg, *Science 196:* 995 (1977).
21. E. R. Peterson, *Anat. Rec. 106:*232 (1950).
22. E. R. Peterson and M. R. Murray, *Am. J. Anat. 96:*319 (1955).
23. W. Hild, *Z. Zellforsch. Mikrosk. Anat. 46:*71 (1957).
24. M. B. Bornstein and M. R. Murray, *J. Biophys. Biochem. Cytol. 4:*499 (1958).
25. C. E. Lumsden and C. M. Pomerat, *Exp. Cell Res. 2:*103 (1951).
26. T. Amano, E. Richelson, and M. Nirenberg, *Proc. Natl. Acad. Sci. USA 69:* 258 (1972).
27. N. W. Seeds, *Proc. Natl. Acad. Sci. USA 68:*1858 (1971).
28. J. Minna, P. Nelson, J. Peacock, D. Glazer, and M. Nirenberg, *Proc. Natl. Acad. Sci. USA 68:*234 (1971).
29. A. G. Gilman and M. Nirenberg, *Nature (London) 234:*356 (1971).
30. A. G. Gilman and M. W. Nirenberg, *Proc. Natl. Acad. Sci. USA 68:*2165 (1971).

31. N. W. Seeds and A. G. Gilman, *Science 174:*292 (1971).
32. A. G. Gilman and B. K. Schrier, *Mol. Pharmacol. 8:*410 (1972).
33. M. R. Murray and H. H. Benetiz, *Science 155:*1021 (1967).
34. M. R. Murray and H. H. Benetiz, in *Growth of the Nervous System* (G. E. W. Wolstenholme and M. O'Connor, eds.), CIBA Foundation Symposium, Churchill, London, 1968.
35. A. A. Moscona, M. H. Moscona, and N. Saenz, *Proc. Natl. Acad. Sci. USA 61:*160 (1968).
36. R. Piddington, *Dev. Biol. 16:*168 (1967).
37. A. A. Moscona and R. Piddington, *Science 158:*496 (1967).
38. T. Alescio and A. A. Moscona, *Biochem. Biophys. Res. Commun. 34:*176 (1969).
39. P. K. Sarkar and A. A. Moscona, *Proc. Natl. Acad. Sci. USA 68:*2308 (1971).
40. P. Benda, J. Lightbody, G. Sato, L. Levine, and W. Sweet, *Science 161:* 370 (1968).
41. J. Lightbody, S. E. Pfieffer, P. L. Kornblith, and H. R. Herschman, *J. Neurobiol. 1:*411 (1970).
42. S. E. Pfeiffer, H. R. Herschman, J. Lightbody, and G. H. Sato, *J. Cell Physiol. 75:*329 (1970).
43. H. R. Herschman, *J. Biol. Chem. 246:*7569 (1971).
44. D. C. Klein and J. Weller, *In Vitro 6:*197 (1970).
45. D. C. Klein, in *The Thyroid and Biogenic Amines* (T. Rall and I. J. Kopin, eds.), North Holland, Amsterdam, 1972.
46. M. K. Wolf and A. B. Holden, *J. Neuropathol. Exp. Neurol. 28:*195 (1969).
47. G. R. DeLong and R. L. Sidman, *Dev. Biol. 22:*584 (1970).
48. R. S. Krooth and A. H. Weinberg, *Biochem. Biophys. Res. Commun. 3:* 518 (1960).
49. L. Miller, G. Gordon, and K. Bensch, *Lab. Invest. 19:*428 (1968).
50. J. H. Herndon, Jr., D. Steinberg, B. W. Uhlendorf, and H. M. Fales, *J. Clin. Invest. 48:*1017 (1969).
51. G. M. McKhann, W. Ho, C. Raiborn, and S. Varon, *Arch. Neurol. 20:*542 (1969).
52. U. Batzdorf, L. L. Sarlieve, V. A. Gold, and J. H. Menkes, *Arch. Neurol. 20:*650 (1969).
53. D. H. Silberberg, *J. Neurochem. 16:*1141 (1969).
54. L. Liss and M. D. Grumer, *J. Neurol. Neurosurg. Psychiatr. 29:*371 (1966).
55. D. Silberberg and H. Schutta, *J. Neuropathol. Exp. Neurol. 26:*572 (1967).

METABOLISM AND FUNCTION

9
AMINO ACID METABOLISM

Generally speaking, the outlines of amino acid metabolism in brain are qualitatively similar to those of amino acid metabolism in other tissues of the body. In many respects, however, the quantitative aspect of the amino acid economy is different. In a few cases the amino acid chemistry of the brain differs qualitatively. Put another way, most of the pathways found in brain are also found in liver and other organs. In many cases these pathways are working at substantially different rates in brain than elsewhere. And in a few cases, amino acid derivatives or amino acid metabolizing enzymes found in brain do not appear elsewhere in the mammalian body.

Glutamic Acid

By far the most outstanding feature of the amino acid metabolism of brain is the central role played by glutamic and aspartic acids and their derivatives [1-3]. Glutamic acid, glutamine, γ-aminobutyric acid, aspartic acid, and N-acetylaspartic acid are the predominant free amino acids in the brain of most species and contain about two-thirds of the free amino nitrogen present. Glutamate itself is in higher concentration in the brain than in any other organ of the body. Not only is the level of glutamate in brain extremely high, but its metabolism is remarkably rapid. After the injection of labeled glucose, more than 70% of the radioactivity present in the soluble fraction of rat brain is in the amino acids in 30 min primarily as glutamic acid and its derivatives [4] (Table 9.1). This is a greater conversion than occurs in other tissues and is probably due in part to the large pool of free glutamic acid in equilibrium with the α-ketoglutarate of the Krebs cycle.

Glutamate metabolism serves in brain, as in other tissues, as the link between amino acid metabolism and energy metabolism. Glutamate itself is a primary product of nitrogen incorporation from ammonia. This role of glutamic acid is possible because of its formation from α-ketoglutaric acid mediated by the enzyme glutamate dehydrogenase (Fig. 9.1). This enzyme is less active in

Table 9.1   Distribution of Radioactivity in Amino Acids of Rat Tissues 30 min After a Subcutaneous Injection of [U-^{14}C] Glucose[a]

	Distribution	
Amino acid	Brain	Liver
Aspartic acid	9	2.6
Glutamine	9	5.3
Glutamic acid	37	5.2

[a]Results expressed as a percentage of the total radioactivity found in all the acid-soluble metabolites 30 min after injection.

Table adapted from the results of M. K. Gaitonde, D. R. Dahl, and K. A. C. Elliott, *Biochem. J. 94:*345 (1965).

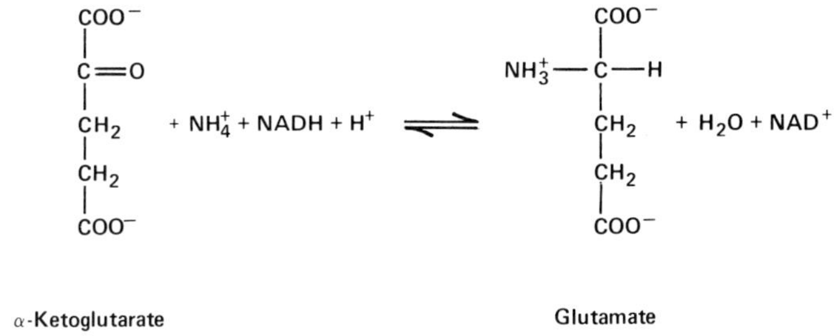

$$
\begin{array}{c}
COO^- \\
| \\
C\!=\!O \\
| \\
CH_2 \\
| \\
CH_2 \\
| \\
COO^-
\end{array}
\;+\; NH_4^+ + NADH + H^+ \;\rightleftharpoons\;
\begin{array}{c}
COO^- \\
| \\
NH_3^+\!-\!C\!-\!H \\
| \\
CH_2 \\
| \\
CH_2 \\
| \\
COO^-
\end{array}
\;+\; H_2O + NAD^+
$$

α-Ketoglutarate   Glutamate

Figure 9.1   Glutamate dehydrogenase.

brain than in liver and has been less thoroughly studied in brain. Superficially, however, it seems to have the same general characteristics as the liver enzyme; it is present in the mitochondria, requires reduced pyridine nucleotide, and is activated by ATP. The influence of various inorganic ions on the brain enzyme seems to parallel the influence of these same ions on the liver enzyme. Whether the brain enzyme is regulated by steroid hormones, as is the liver enzyme, does not seem to have been investigated. The equilibrium of this reaction in the presence of ammonium salts is strongly in favor of glutamic acid. The K_M of this enzyme for ammonia is on the order of 8 mM and the normal brain concentration is certainly below 0.5 mM, so the enzyme activity increases markedly with increasing ammonia levels.

Glutamine

Glutamate, of course, serves as the precursor of glutamine. In most species, the enzyme glutamine synthetase (Fig. 9.2) is in higher concentration in brain than

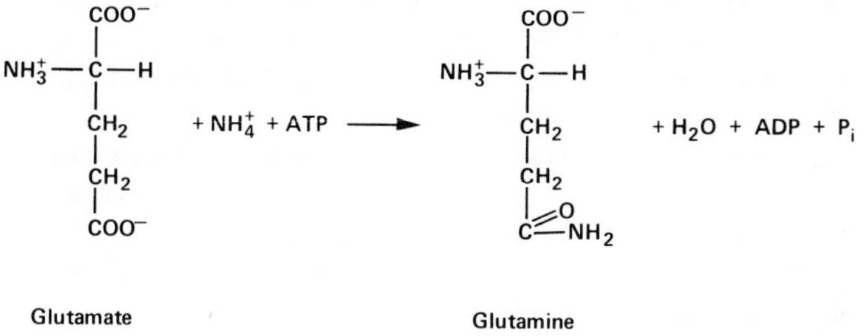

$$\text{Glutamate} + NH_4^+ + ATP \longrightarrow \text{Glutamine} + H_2O + ADP + P_i$$

Glutamate

Glutamine

Figure 9.2 Glutamine synthetase.

in any other organ of the body. When subcellular fractionation is performed the enzyme is found distributed primarily in the microsomal and the soluble compartments, but a substantial portion sediments with the crude mitochondrial fraction as well. The enzyme requires ATP and Mg^{2+} and is inhibited by some dinucleotides and trinucleotides and also by glycine and alanine. The K_M for ammonia is low, on the order of 0.39 mM, and thus the enzyme is at least half-saturated at the normal concentration of ammonia found in the brain. An extensive series of investigations by Meister and his colleagues [5] has elucidated the characteristics of purified sheep brain glutamine synthetase. It is now known to be a large enzyme (MW 450,000) with eight identical subunits [6]. The enzyme is estimated to make up about 0.2% of the total protein of the sheep brain [6]. It has an active sulfhydryl group, probably at its active site, and is inhibited by the sulfhydryl reagent, N-ethylmaleimide. This reagent is thought to react at the substrate binding site since the inhibition is prevented by the presence of ATP and Mg^{2+} [6]. The mechanism of action involves the intermediate formation of γ-glutamyl phosphate from glutamate and ATP, followed by the replacement of the phosphate by ammonia. This mechanism is consistent with more recent findings on the inhibition of the enzyme by methionine sulfoximine. This compound is known to cause convulsions and also to inhibit glutamine synthetase irreversibly in vivo and in vitro. The inhibition requires ATP and Mg^{2+} and is prevented by the simultaneous presence of glutamate and ammonia [7,8]. The mode of inhibition has been found to involve a phosphorylation of the convulsant with the ATP by the enzyme, followed by the tight binding of eight molecules of the methionine sulfoximine phosphate and eight molecules of ADP to the enzyme [7]. This elegant work has been completed with the chemical synthesis of methionine sulfoximine phosphate and the elucidation of its structure as that of an N-phosphate [9]. It is reasonable to conclude that the strong inhibition of glutamine synthetase by this drug plays some role in the convulsions which result from its administration.

It is clear from a number of lines of evidence that the formation of glutamine serves as a mechanism for the detoxification of ammonia, to which the brain is quite sensitive. Ammonium salts, when administered systemically, cause convulsions. During such infusions the brain levels of glutamine rise. The levels of glutamate remain constant, but α-ketoglutaric acid levels drop. From such studies it seems that glutamine synthesis in the brain increases in response to increases in the systemic levels of ammonia. Further, it is known that in cases of severe liver damage the systemic level of ammonia also rises, as does the cerebrospinal fluid concentration of glutamine. In such cases coma is the frequent result. It is not clear that the ammonia is the causative agent in hepatic coma. In fact, even though ammonium salts cause convulsions, there is no proof that ammonia levels mediate any other kinds of convulsions even though brain ammonia concentrations are known to double within 1 sec of an electroconvulsive shock and are also known to rise during other types of convulsions. However, although hard experimental proof is lacking, a role for ammonia in the mechanism of hepatic coma and a role for glutamine in ammonia removal are indicated by reports of the alleviation of hepatic coma by the administration of glutamate. The detailed mechanism of the coma is unknown, but it has been suggested that ammonium ion may directly antagonize one of the other cations involved in neural transmission.

Chinese Restaurant Syndrome

The administration of glutamate itself has some interesting pharmacological and clinical consequences. Topical administration to the cortex results in spreading depression. This is true even though the amounts applied are small in comparison with the normal concentration in the brain. Glutamine is much less active, if it is active at all. Activation of neuronal firing is also seen upon direct application of glutamate to the surface of the cells. Oral or intravenous administration of large amounts of glutamate has little or no central nervous system effect in most people and, indeed, glutamate salts are widely used in food preparation. In a small number of individuals the ingestion of monosodium glutamate causes sensory and motor disturbances, including a burning sensation, facial pressure, and chest pains [10]. Headache is also a frequent component of the response. This surprising and frightening condition has become known as the Chinese restaurant syndrome, for obvious reasons. Clinically there are reports that glutamate has favorable effects in promoting recovery from insulin hypoglycemia, in ameliorating some forms of petit mal seizures, and even in increasing the alertness and improving the general condition of some mentally retarded subjects. These results may be due to the reported effects of glutamate in increasing blood pressure and pulse rate and raising blood sugar levels, effects reminiscent of those caused by the administration of amphetamines. However, even in the face of these significant pharmacological effects, it is not proven that glutamate levels, glutamine levels, or the glutamate/glutamine ratio has anything to do with the mechanism of any of the many convulsive states. And as yet it is not known for sure that glutamate has any normal role in the firing or the regulation of firing of any neurons.

The removal of the amide group of glutamine by the enzyme glutaminase (Fig. 9.3) has been studied to some extent in brain. The enzyme is similar to the liver enzyme catalyzing the same reaction. The brain enzyme has been shown to be inhibited by D-glutamic acid, by L-glutamic acid, or by ammonia, and to be activated by phosphate. A number of studies indicate that the enzyme exists in two or three isozyme forms. It has been reported that the enzyme, or at least one of its forms, is further activated by N-acetylamino acids [11,12] in the presence of phosphate. In view of the unique occurrence in brain of N-acetylaspartic acid, and the lack of information about its function, these reports are intriguing. However, the concentrations needed for activation are very high compared to the physiological concentrations, and the activation is not specific for acetylaspartic acid, other acetylamino acids being even more effective. Thus, the meaning of the stimulation is not clear.

A number of transaminases occur in brain, one of the most active being the glutamate-oxalacetate transaminase (Fig. 9.4). This enzyme is as active in brain as it is in liver and, as in liver and other sources, uses pyridoxal phosphate as a

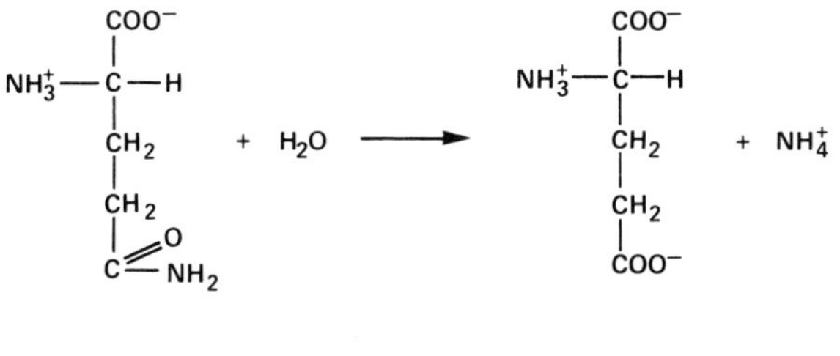

Glutamine Glutamate

Figure 9.3 Glutaminase.

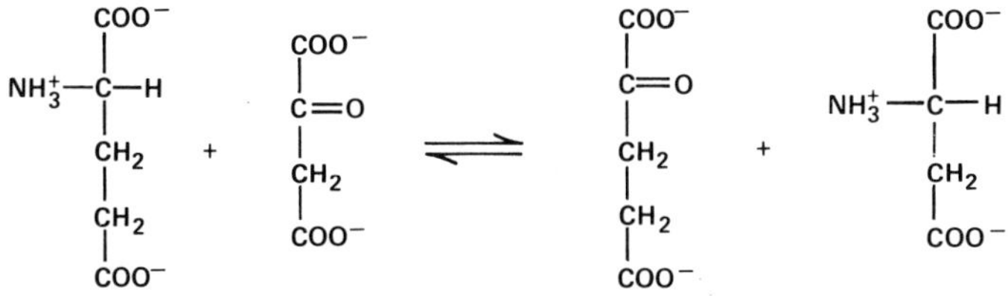

Glutamate Oxalacetate α-Ketoglutarate Aspartate

Figure 9.4 Glutamate-oxalacetate transaminase.

cofactor. It is distributed in brain in both the soluble and particulate compartments of the cell, about 20% of it occurring in the soluble and the remainder being distributed between the crude mitochondrial and the crude nuclear fractions. In brain this enzyme is the principal agent for the metabolism of glutamate. It is the most active of the many transaminases occurring, the glutamate-pyruvate transaminase being uniquely low in brain.

The other important route of glutamate metabolism in brain, namely, the decarboxylation to γ-aminobutyric acid, is discussed in Chap. 10.

Aspartic Acid

As mentioned above, the most active transaminase in the brain is the enzyme glutamate-oxalacetate transaminase. The formation of aspartic acid in brain is almost certainly due to the action of this enzyme. Probably the catabolism of brain aspartic acid occurs through the donation of the aspartic acid amino group to form argininosuccinic acid and, ultimately, urea [13]. Fumaric acid, then, is the product of the carbon chain, and this, of course, is catabolized via the Krebs cycle. Alternatively, aspartic acid amino groups can be used for the synthesis of AMP from IMP, or for the de novo synthesis of the purine or pyrimidine rings. Again, fumaric acid is the carbon chain product. These latter two routes have not been firmly established to occur in brain, but indirect evidence suggests that they are there. The formation and breakdown of asparagine have not been extensively studied in brain, but the enzyme asparaginase is known to occur in neural tissue. The metabolism of N-acetylaspartic acid is discussed in Chap. 11.

Glycine

The information about glycine metabolism in the brain is not extensive [13]. Regarding glycine formation, it is known that carbon from glucose ends up in glycine, and glycine-α-ketoglutarate transaminase has been reported [14]. Also, there is evidence for the interconversion of glycine and serine via serine hydroxymethylase, and the enzyme has been shown to occur in brain [15]. It is clear, as regards the breakdown of glycine, that the brain can oxidize glycine to CO_2. This catabolism may be a function of D-amino acid oxidase which is known to be present in neural tissue.

Conditions are known in which a high level of plasma and urinary glycine is associated with aberrant brain function and pathology. The most common of these was known until recently as idiopathic hyperglycinemia [16] and is now designated ketotic hyperglycinemia. This disease develops early in children and involves episodic vomiting, lethargy, seizures, retardation, abnormal electroencephalogram (EEG) patterns, and early death. Also, a severe ketosis caused by the excretion of a number of different ketones is involved. The high glycine in

the blood and urine is a constant feature and is not associated with any of the
episodes of seizures or vomiting. Clinically the children exhibit neutropenia,
thrombocytopenia, hypogammaglobinemia, and osteoporosis. Upon autopsy
they are found to have a spongy degeneration of the white matter of the brain
and a much delayed myelination. Within the last few years it has been shown
that this condition is in fact identical to the disease known as propionacidemia
which is a condition caused by the genetic deletion of the enzyme propionyl-
coenzyme A (CoA) carboxylase [17,18]. A complete discussion of this condition is postponed until Chap. 19, but at the moment it is appropriate to say that
it appears that the hyperglycinemia is an effect far removed from the actual defect in this condition and, in fact, is unlikely to have any direct relationship to
the observed brain dysfunction. A second condition, nonketotic hyperglycinemia, is thought to be caused by a defect in the enzyme, glycine decarboxylase.

Alanine

The metabolism of alanine in brain is probably completely dependent upon the
enzyme glutamate-pyruvate transaminase. This enzyme, as mentioned above, is
much lower in brain than in liver, for instance. It does, however, seem to catalyze both the formation and breakdown of alanine, and no other routes of metabolism for alanine in brain are known [13].

Serine

The biosynthesis of serine has been investigated specifically in brain preparations
from young adult mice [19]. The pathway for serine formation from the glycolytic intermediate, 3-phosphoglyceric acid, is outlined below (Fig. 9.5). Independently, several investigators have reported the presence of serine phosphate
in brain [13], thus providing corroborating evidence for the existence of the
serine biosynthetic pathway shown. The fate of serine in brain is not quite as
clearly known. The enzyme serine hydroxymethylase, as noted above, has been

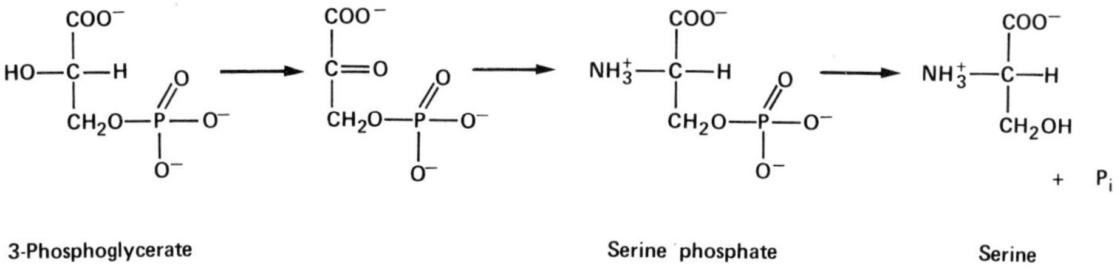

3-Phosphoglycerate   Serine phosphate   Serine

Figure 9.5 Pathway of serine biosynthesis in brain.

shown to occur. Also, both serine transaminase and serine dehydrase have been reported. All these enzymes catalyze catabolic routes for serine. Serine, in brain, serves as precursor to some of the phospholipids and to cystathionine, both of which are discussed subsequently, as well as participating in protein synthesis.

Threonine

There is little information about the metabolism of threonine in brain. Preparations of brain are known to deaminate threonine and to oxidize it to aminoacetone. The enzyme threonine dehydrase produces α-ketobutyric acid, which upon transamination would yield α-aminobutyric acid. The latter compound is known to exist in brain, but there is no direct evidence for the presence of threonine dehydrase.

Branched-Chain Aliphatics—Leucine, Isoleucine, and Valine

The branched-chain aliphatic amino acids, leucine, isoleucine, and valine, are essential for higher organisms so, of course, are not synthesized by brain. It is known that brain preparations will oxidize these amino acids [20] and, further, that brain in vitro will incorporate the carbon of leucine into cholesterol [21], so it is clear that the brain can catabolize this group of amino acids. The catabolism, studied in detail in organs other than brain, involves, first, a transamination with α-ketoglutaric acid, and, second, an oxidative decarboxylation of the α-keto acid, leading to the CoA derivative of the carboxylic acid one carbon shorter, and the formation of CO_2 from the original carboxyl carbon of the amino acid. Subsequent reactions, resembling those of long-chain fatty acid β-oxidation, lead to the complete catabolism of the carbon chain.

When the initial CoA-dependent, α-decarboxylation reactions (Fig. 9.6)

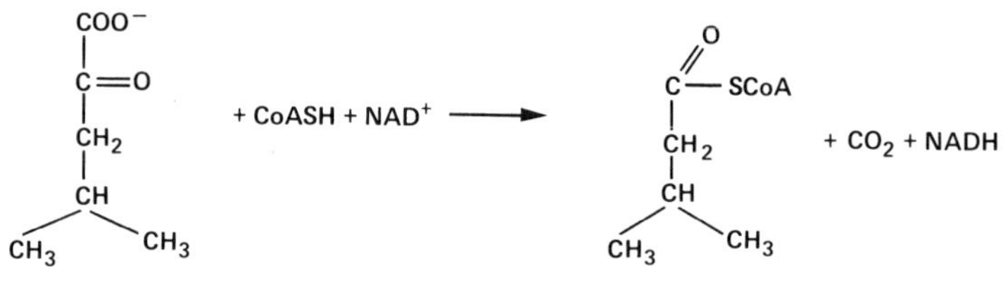

α-Ketoisocaproate

Figure 9.6 Defective reaction in maple syrup urine disease (branched-chain keto aciduria).

are impaired or deleted genetically a very severe condition known as maple syrup urine disease results. First described in 1954 [22], this syndrome has received a great deal of attention possibly because of the severity of the condition and the reasonably well-defined nature of the lesion [16,17]. The disease is transmitted as an autosomal recessive and gets its name from the first observations that the urine of afflicted children smells something like maple syrup. It is now known that the smell is caused by the presence of unmetabolized keto acids, some of which are derivatives of the aliphatic amino acids, and that these metabolites also accumulate in the blood, tissues, and cerebrospinal fluid of the victims. Thus, a more informative name, branched-chain ketoaciduria, is now used to describe this condition. Upon examination such children exhibit rigidity and seizures, among other things, and those that survive are spastic and severely retarded. Pathologically, the brains of these children at autopsy are spongy and have very marked changes in the white matter. The cortical layers are reduced in number and appear immature. There is a decreased myelin content, and areas of focal astrocytosis are apparent. Myelin breakdown products are not found so demyelination probably does not occur, but the lack of myelin formation is so pronounced that cerebrosides, sulfatides, and proteolipid protein are almost completely missing from the brain. The exact enzyme defect is not precisely known, but there is evidence that a single enzyme decarboxylates both α-ketoisocaproic acid and α-keto-β-methylvaleric acid, and that this enzyme is deleted [23]. Even if there are three separate enzymes, as now seems unlikely, the absence of even one of them could account for the accumulation of all three keto acids, since an overabundance of one keto acid could inhibit the action of the enzymes decarboxylating the other two [24]. Anyhow, it is known that the keto acids accumulate due to the deletion of the enzymes decarboxylating one or all of them. There is some evidence that the true agent of damage is α-ketoisocaproic acid, the product of the transamination of leucine. This compound has been shown to be a pronounced inhibitor of pyruvate decarboxylation and, as mentioned in an earlier context, it is the only one of the known metabolites in this disease which is toxic to in vitro myelination of cerebellar explants [25]. Whatever the exact cause of the brain damage turns out to be, there is reasonable evidence accumulating that the condition can be ameliorated by the simple expedient of lowering the dietary content of the three branched-chain aliphatic amino acids.

Sulfur Amino Acids—Methionine and Cysteine

The amino acid methionine is essential for higher animals because they cannot make the carbon chain from aspartic acid as do bacteria. Once this chain is made, and a sulfhydryl attached to give homocysteine, higher organisms can add methyl groups to form methionine. It is not known if brain per se can carry out this reaction. It is known that a series of metabolic transformations of methio-

nine can take place in brain tissue [26]. The methionine-activating enzyme, e.g.,
the one that makes S-adenosylmethionine from methionine and ATP, is present
in brain at about 10 to 20% of the activity present in liver [27]. The S-adenosylmethionine serves as methyl donor, and then the demethylated product, S-
adenosylhomocysteine, can be split to yield adenosine and homocysteine. The
homocysteine is condensed with serine to give cystathionine, a thiol ether which
occurs in high concentration in brain, and in especially high concentration in
human brain. The enzyme responsible for the condensation, cyatathionine synthetase (Fig. 9.7), is known to be present in brain [27,28], and to require pyridoxal phosphate as a cofactor. Cystathionine is split by cystathionase (Fig. 9.8)
in such a way as to transfer the sulfur from the four-carbon chain of homocysteine to the three-carbon chain of the original serine, yielding cysteine. Whether
the homoserine residue is produced and then cleaved, or cleaved immediately, is
not completely clear, but it is clear that the final products are α-ketobutyrate
and NH_3. The enzyme cystathionase also requires pyridoxal phosphate as a cofactor. Of course, the cystathionine content of the brain depends upon the relative activities of the two enzymes, cystathionine synthetase and cystathionase,
both of which require pyridoxal phosphate as a cofactor. In pyridoxine defi-

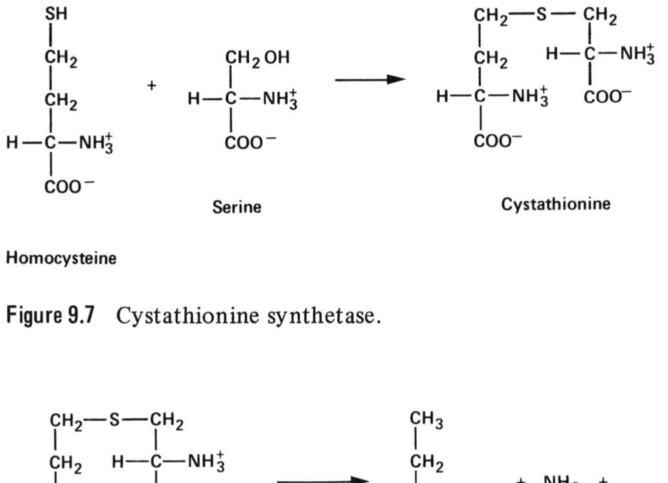

Figure 9.7   Cystathionine synthetase.

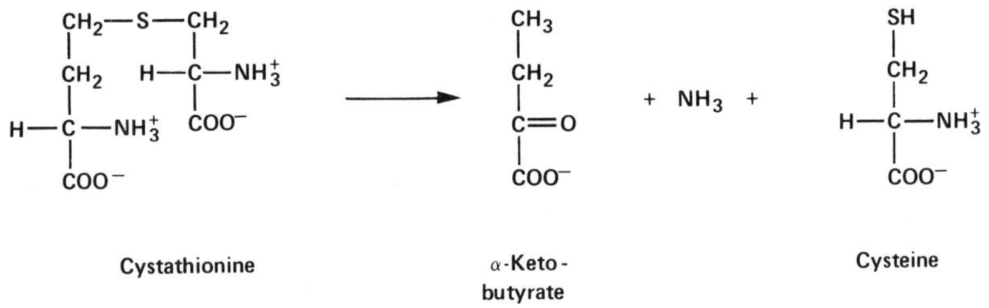

Figure 9.8   Cystathionase.

ciency the level of cystathionine in the brain rises perhaps 10-fold, as do cystathionine levels in liver, kidney, and urine, suggesting that cystathionine synthetase has a greater affinity for the cofactor than does cystathionase. This conclusion is substantiated by studies with the purified enzymes.

The deletion or malfunction of the enzyme cystathionine synthetase leads, in humans, to a condition known as homocystinuria. This condition, discovered in 1962 [29,30], is characterized by a huge excretion of homocystine in the urine, a rise in the homocystine and the methionine content of the blood, and an absence of cystathionine and cystathionine synthetase from the brain and liver upon biopsy [31]. It is the second most prevalent of the amino acidurias affecting the nervous system, less frequent only than phenylketonuria.

The disease is transmitted as an autosomal recessive. The excretion of homocystine may be lowered in some of the patients by the administration of large amounts of pyridoxine. This suggests that, in this group of patients, the cystathionine synthetase enzyme is altered in such a way that it does not bind its cofactor very efficiently. Other patients with homocystinuria do not respond to pyridoxine at all. One of the usual characteristics of the disease is a generalized fibrosis and thickening of the blood vessels. Thromboses are common, as are cerebral infarcts, and it is considered likely that the retardation seen in homocystinuria is due to random cerebral accidents, rather than to a molecular damage by the abnormal metabolism. Therapeutic approach to the disease includes the lowering of the methionine content of the diet, while providing an alternate methyl donor such as choline, as well as the inclusion of cystine which is essential to these patients since they cannot make it from methionine as do normal individuals.

Homocystinuria

A 3-year-old, white male child was examined in connection with observed dislocation of the lens of the right eye, among other complaints. A history revealed episodes of seizures and a suggestion of fairly severe mental and physical retardation. A detailed examination showed ectopia lentis, with the dislocation downward, and an irregular pupil. A malar flush was evident on the face, and the palate formation was narrow and highly arched. The child had fine, fair hair and evidence of an enlarged liver. Upon manipulation the joints of the arms and legs seemed stiff. By amino acid analysis the urine was shown to contain large amounts of homocystine; this was confirmed by the cyanide nitroprusside test.

On follow-up the child proved to be substantially retarded. The 24-hr excretion of homocystine remained approximately constant in spite of daily treatments with 500 mg of pyridoxine. At age 4 ½ a left-side hemiplegia appeared. The patient died 6 months later, as a result of thrombosis [16,32]

Another condition which results from a lesion in sulfur amino acid metabolism is cystathionuria. In this case it is the enzyme which metabolizes cystathionine, cystathionase, which is missing. Its substrate, cystathionine, spills out
in the urine, hence the name. This condition is, in general, not as severe as homocystinuria and the patients are, at most, mildly retarded [26], and some discussions of conditions which cause neurological damage fail to mention cystathionuria at all, or consider it benign [17]. As pointed out before, the level of cystathionine in brain is a function of the pyridoxal phosphate status of the animal.
This is true also of the urinary excretion, and it is known that a condition resembling cystathionuria occurs in B_6 deficiency. Finally, something like cystathionuria can be produced by the consumption of certain plant neurotoxins,
such as β-cyanoalanine, which are inhibitors of the enzyme cystathionase.

The degradation of cysteine to pyruvate, ammonia, and H_2S in mammals
is catalyzed by the enzyme cysteine desulfhydrase, but there is no evidence for
the presence of the enzyme in brain. Other routes of metabolism which do occur in brain include the oxidation of the sulfur to yield cysteine sulfinic acid,
followed by decarboxylation to taurine, or the incorporation of the cysteine
into the peptide glutathione. Both these metabolic routes are discussed in
Chap. 10.

Imino Acids—Proline and Hydroxyproline

The metabolism of the imino acids has not been extensively studied in brain
[13]. It is known that proline can be formed from glucose [33], probably
through glutamate, by brain. Further, it has been shown that proline can be
converted to other amino acids by brain in vivo [34]. The details of these conversions have not been explored. As for hydroxyproline, there is no evidence
that it is formed in brain or that it has any function in brain cells. In one specific study, it was shown not to be oxidized by brain preparations that were capable of oxidizing proline [35].

Conditions in which neurological damage is probably linked to lesions in
proline metabolism are known [16,17]. A defect in the first enzyme of proline
degradation, proline oxidase (Fig. 9.9), causes hyperprolinemia I [36], a condi-

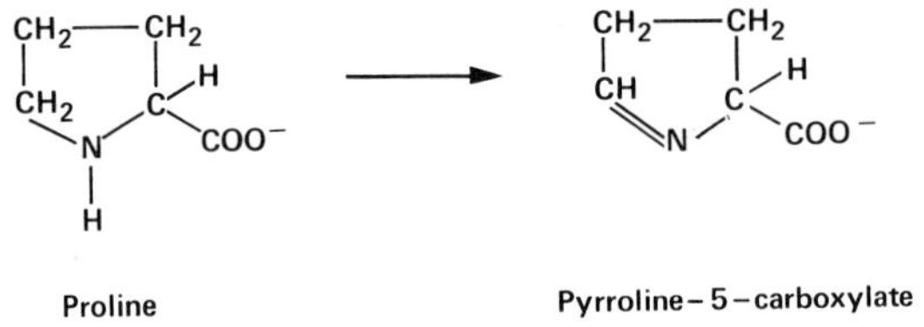

Proline   Pyrroline–5–carboxylate

Figure 9.9 Proline oxidase.

tion characterized by an increased proline concentration in the blood, and the excretion of proline, hydroxyproline, and glycine in the urine. The symptoms of the disease include mental retardation, renal disease, nerve deafness, and a photosensitive epilepsy. Neuropathological examination of such patients shows a decrease in cortical neurons, a delayed myelination, and, biochemically, a lowered cerebroside content. The connection between the biochemical lesion and the mental defect is a tenuous one because siblings of the affected individuals have, in some cases, hyperprolinemia without the associated neurological or renal problems. Hyperprolinemia II [37,38] is a condition involving a milder retardation, some seizures, but no renal disease or deafness. The proline content of the blood is frequently higher, and the individuals excrete pyrroline 5-carboxylic acid as well. By inference, the molecular lesion here is assumed to be in a later step in proline catabolism.

The literature also contains a report of hydroxyprolinemia with attendant neurological problems, even though hydroxyproline is not known to be metabolized by brain [39].

Phenylalanine and Phenylketonuria

The metabolism of the aromatic amino acids in brain has been a subject of intense interest for a number of years because of the apparent importance of some of the aromatic metabolites in neural functioning [40]. Also, a great deal of information has accumulated because of the relative ease of studying the aromatic acids in the laboratory, using as handles the intense and characteristic absorption and fluorescence exhibited by these amino acids and their derivatives. Phenylalanine is essential for higher animals, and so its biosynthesis would not be expected to occur in brain. Some early studies indicated that radioactive glucose could be converted, by minces of the brain of newborn mice, to some of the essential amino acids, including phenylalanine [41], but later work by some of the same investigators failed to confirm these observations [42]. Phenylalanine transamination and decarboxylation have been observed in brain [43-45], but these reactions probably have little or no physiological importance and are likely to be, in fact, reflections of tyrosine transaminase and dopa decarboxylase, respectively. The major route of phenylalanine metabolism in the whole animal is its hydroxylation to tyrosine by phenylalanine hydroxylase, but this enzyme, per se, is not present in brain. As is discussed in Chap. 11, certain of the brain hydroxylases, namely tyrosine hydroxylase and tryptophan hydroxylase, also hydroxylate phenylalanine to tyrosine, but the total amount of phenylalanine hydroxylated by these brain enzymes is small in comparison, and would seem to be of minimal metabolic significance.

An alteration in the overall phenylalanine metabolism of humans causes severe aberrations in brain function. The condition is known as phenylketonuria

[16,17,46]. First observed by Fölling in 1934 [47], this disease is now known to be the most prevalent of the amino acidurias which affect the nervous system. The genetics and the enzymology of this condition are now fairly well characterized. It is inherited as an autosomal recessive, and the homozygous condition is characterized by an absence of the activity of the liver enzyme phenylalanine hydroxylase [48-51]. This enzyme has been thoroughly studied [52-54], and its mechanism of action is, at least in gross outline, understood. The system for the hydroxylation of phenylalanine in mammalian liver (Fig. 9.10) consists of the hydroxylase itself, an unconjugated pteridine cofactor [55], and a pyridine nucleotide-linked reductase for recycling the pteridine cofactor [56]. The hydroxylase from rat liver is a complex, iron-containing protein. It is markedly stimulated by certain phospholipids [57,58] and by other apparently specific activator proteins [59]. It has been shown to contain covalently bound phosphate [60,61]. The cofactor, at least in rat liver, has been found to be L-erythrotetrahydrobiopterin. The enzyme is a classic monooxygenase or hydroxylase, requiring molecular oxygen as a oxidant, and the tetrahydropteridine as a reductant. The enzyme uses one atom of the molecular oxygen to oxidize the substrate, and reduces the second atom to water with the reduced pteridine as the reductant. The addition of the hydroxyl group induces the migration of the proton, or, in fact, the migration of some halogen or alkyl substituents, to the

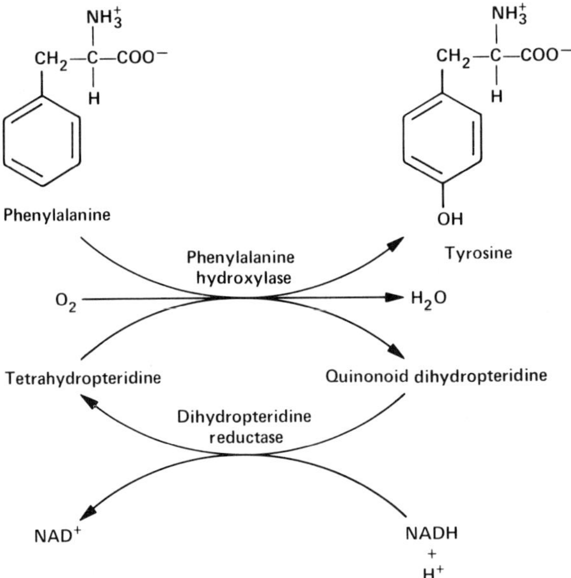

Figure 9.10   The phenylalanine hydroxylase system.

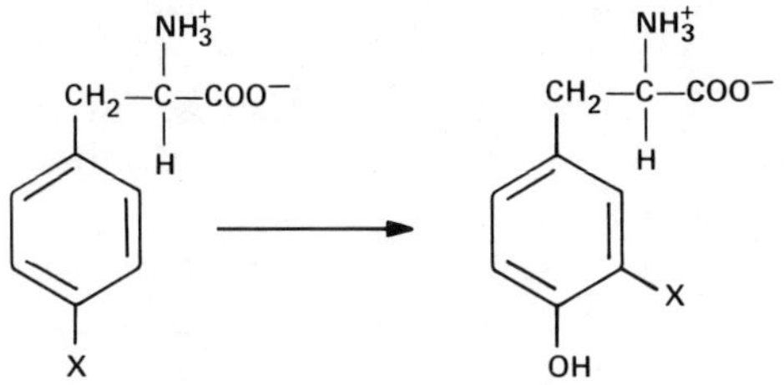

Figure 9.11   The NIH shift.

adjacent position on the aromatic ring in a reaction sequence which has been
called "the NIH shift" (Fig. 9.11) [62-66]. The tetrahydrobiopterin is truly a
cofactor, being present in the liver in small amounts. The second enzyme in the
system, dihydropterin reductase (Fig. 9.12), catalyzes the recycling of the oxidized cofactor using NADH as the ultimate source of electrons.

It is now known that the hydroxylase is the deficient part of the system in
cases of classic phenylketonuria. Further, it has been shown recently that in some
cases no significant amounts of immunologically cross-reacting material can be
found, indicating that the affected individuals cannot make the protein at all [67].
In other cases a very low level of enzymatic activity, on the order of 1% or less,
has been detected, and this activity exhibits somewhat different characteristics
than the normal human enzyme [68]. The best interpretation of these results is
that the patients studied probably made a normal level of a structurally altered
enzyme with low catalytic activity and little immunological cross-reactivity.

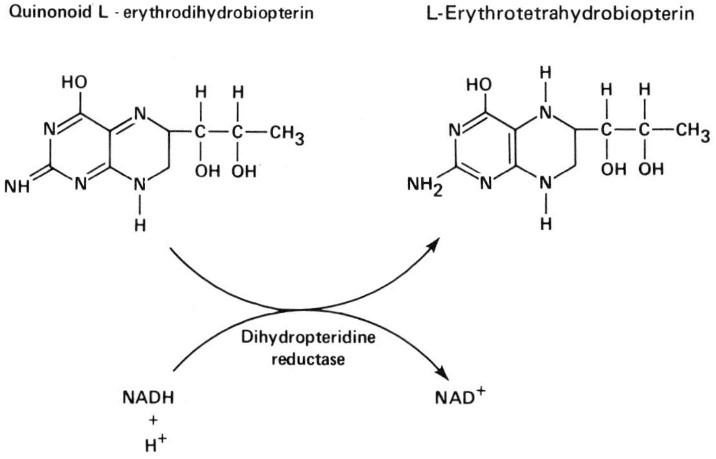

Figure 9.12   Dihydropteridine reductase.

Finally, it has been shown, at least in lower animals, that the hydroxylation system is present in kidney and pancreas, as well as in liver [69]. It is not yet established that phenylketonurics are missing the enzyme in these organs also, but the very low level of hydroxylation observed in these individuals indicates that this is likely.

It has been suggested by several workers that since the hydroxylase system is so complex, i.e., two enzymes and a reduced pteridine cofactor, variants of the disease could exist in which the reductase or the cofactor is the missing portion, but until recently no such patients have been observed. Now some cases have come to light [70], and, indeed, reductase mutants have some of the same characteristics as classic phenylketonurics. However, the lesion here has much wider consequences since the same reductase also serves the tyrosine and tryptophan hydroxylases. As a consequence, the standard therapy for phenylketonuria, lowering the phenylalanine level in the diet, is not sufficient or effective here. The full implications of this new lesion in metabolism have yet to be elucidated.

In the absence of phenylalanine hydroxylation, which is the first step in the major, if not the only, route for the catabolism of phenylalanine, a number of drastic changes take place in the whole-body chemistry of the afflicted children. First, the blood and tissue levels of phenylalanine rise, perhaps to several hundred times the normal level. Since phenylalanine hydroxylation does not occur, the level of tyrosine in tissues and plasma remains normal or even a little below normal. Some minor alterations occur in the levels of other amino acids and amino acid derivatives. Most prominent and interesting is a lowering of blood serotonin and a decreased excretion of serotonin metabolites.

The disease was first recognized as an independent entity by the presence of phenylpyruvic acid in the urine of some retarded children, hence the name. Now it is known that a large number of normally minor or undetectable metabolites appear in the urine of these children (Fig. 9.13). Apparently, the abnormally high level of phenylalanine in the body floods some pathways of metabolism which do not use phenylalanine as a substrate at all under normal conditions. Among the materials appearing are some which could be formed in brain, since enzymes are known which could act on phenylalanine in such a way. These are phenylpyruvic acid [71], phenyllactic acid [71], phenylacetic acid [72], and phenylethylamine [73]. Also found are o-hydroxyphenylacetic acid [74], phenylacetylglutamine [75], N-acetylphenylalanine [76], and hippuric acid [77]. The exact origin of this latter group of phenylalanine metabolites is not known, but it is likely that they are formed outside the brain itself. It is the o-hydroxyphenylacetic acid which, apparently, gives phenylketonuric children their characteristic "mousy" odor. Also found is an unusual Schiff's base conjugate of phenylethylamine and pyridoxal, pyridoxylidine β-phenylethylamine [78,79]. This metabolite is detectable in the urine of phenylketonurics and in the urine of rats given large doses of phenylalanine as well.

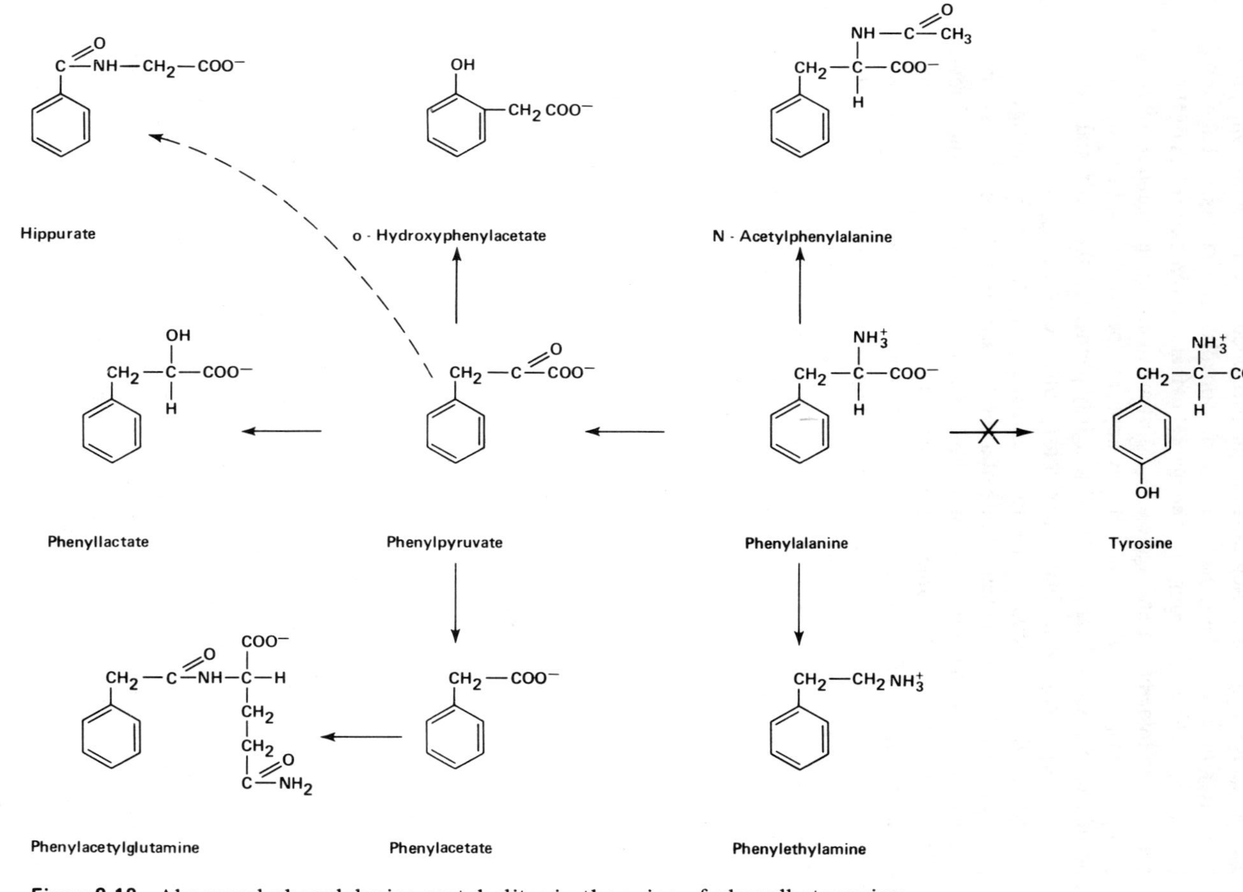

Figure 9.13 Abnormal phenylalanine metabolites in the urine of phenylketonurics.

By whatever mechanism, this combination of chemical excesses and deficiencies causes major neurological abnormalities, including irreversible and profound retardation, in the afflicted children. Convulsions, tremors, and abnormal EEG patterns usually accompany the mental defect. About two-thirds of the children are hyperactive and aggressive. Most are microcephalic, and many have skin problems. Generally, although they appear, superficially, to be relatively normal as to size and bearing, they have lightly pigmented skin, hair, and eyes, and may, upon close examination, exhibit peculiar postures or gaits.

Many different reports have appeared concerning the neuropathology of phenylketonuria, but no clear picture of a specific structural lesion in this condition has emerged. Most observers have noticed a moderate and generalized myelin deficiency [80,81]. Some studies include as findings an impairment in the cortical layering, and focal lesions in the white matter have been seen. The overall picture seems to be one of delayed or incomplete myelination instead of demyelination. And since the delay causes only a transient difference between phenylketonurics and normal children of the same age, i.e., the defect tends to become less apparent with age as the phenylketonurics "catch up," it is not difficult to see why defined structural lesions in such patients are hard to demonstrate. The structural damage observed must depend to a great extent not only on the severity of the disease, and the particular patient examined, but also on the exact age and stage of development of the case studied.

As would be expected in cases of myelin insufficiency, there are several reports in the literature of differences between the brain composition of phenylketonurics and that of normal individuals. Decreased cholesterol [81,82], cerebrosides [83,84], and a decreased content of hydroxy fatty acids in the sulfatide fraction [85] have been found in various investigations. There is also a report concerning a change in the ratio of unsaturated to saturated fatty acids [86] and one report of a lowered content of one or more fractions of the proteolipid [87].

The fool-proof diagnosis of phenylketonuria has proved to be a little difficult. The only sure proof of the disease is the absence of the enzyme phenylalanine hydroxylase. But the enzyme, so far as is known, exists only in liver, kidney, and pancreas, and liver biopsy is not a completely safe procedure, in infants anyway. Failing this direct route, initial indication of the condition can be obtained using $FeCl_3$ or $FeCl_3$-impregnated filter papers as an indicator of phenylpyruvic acid in the urine. The appearance of a green color with $FeCl_3$ was the original procedure used by Fölling in 1934. Confirmatory tests involve measurement of the phenylalanine levels in the blood by means of the Guthrie test [88], which measures the growth of bacteria in the presence of a standard amount of an inhibitory phenylalanine analog and the unknown phenylalanine-containing blood sample. Alternatively, the level can be measured by means of the fluorescence of a phenylalanine adduct [89], or the absorption of a phenylalanine derivative produced by an enzyme [90].

Prenatal diagnosis, which has proved so successful for a number of genetic errors, has not been possible here because the enzyme is not normally present in the cells obtained by amniocentesis. Detection of carrier heterozygotes has been possible since, although they suffer no metabolic disturbance and have normal blood levels of phenylalanine, they respond quite characteristically to a phenylalanine load [91].

It is clear that an accurate diagnosis is essential because the treatment is detrimental to a normal child. And since any of the above diagnostic tests yield a good number of false positives, careful and precise diagnosis is crucial. The treatment involves lowering the dietary phenylalanine by removing it from the dietary protein. This can be done by treating a protein hydrolyzate with charcoal and then replacing the amino acids, other than phenylalanine, which are removed. The product is expensive and also unpalatable. And since phenylalanine is an essential amino acid, it cannot be kept too low even in the diet of phenylketonurics, let alone normal individuals. Even in phenylketonurics, the absence of adequate phenylalanine in the diet produces growth retardation, disturbed bone metabolism, anemia, and increased susceptibility to infection [92]. So the treatment regimen requires adjustment of the dietary intake and careful monitoring by continual analysis of the blood phenylalanine of the patient.

The treatment of phenylketonurics, if done early enough, has proved successful, although there are reports of more subtle dysfunction remaining [93]. However, if the treatment is not begin within the first 6 months of life little positive value is afforded. The primary lesion, that is, the retardation, is irreversible. The damage, then, done by the altered body chemistry, occurs during some critical period in brain development. This is also suggested by the fact that continuing the treatment into adolescence provides no advantage over stopping the treatment at 6 to 8 years of age. On the other hand, such dietary treatment of mature, retarded phenylketonurics may be of some value in the reversal of some of the minor afflictions. It has been reported, for example, that lowering the phenylalanine content of the diet of adult patients relieves the eczema, the hyperactivity, the irritability, and the EEG abnormalities seen in these individuals.

A substantial amount of data is now available which shows that normal children of phenylketonuric mothers sustain some substantial damage in utero. That is, retardation of physical and mental development is apparent in children with normal body chemistry born to women with high blood phenylalanine [94-96]. Thus, dietary treatment of pregnant phenylketonurics should prevent the high incidence of abortions and malformations and produce normal children.

Phenylketonuria is a serious health problem, accounting for about 0.5% of the retarded individuals institutionalized. Figures of 1 in 20,000 to 40,000 live births have been used to describe the incidence in the United States, which indicates that individuals which are heterozygotic for the defective gene are about 1% of the population. Not only is the disease prevalent, but it is also among the

most severe of the retardations. Most untreated persons with phenylketonuria
have an IQ of less than 20. Many of them walk, but only about one-third talk.
So it is an important problem because of the severe nature of the defect and the
amount of care necessary to maintain such patients.

Of course, a great deal of effort has gone into the search for the biochemical
cause of the neural problems. Even though the chemistry of the disease is quite
well understood, the exact biochemical cause has not been identified. There are
a number of possibilities, since it has been shown that a number of crucial processes in the brain can be or are affected by the altered environment. Among the
most acceptable theories are the following:

1. Competitive inhibition, by the high level of phenylalanine, of the uptake into brain of other aromatic amino acids. It has been shown that
   the aromatic amino acids share a common transport mechanism for uptake into the brain [97-99]. Phenylalanine, specifically, has been found
   to inhibit the uptake of tyrosine and of tryptophan and others in vivo
   and in vitro and to alter their concentration in brain itself [100-102].
   Such a chronic inhibition in the afflicted children might alter, on a long-
   term basis, the levels of the other amino acids in brain and, thus, the
   rate of synthesis of the end-products of their metabolism.

2. Toxic effects of phenylalanine or its metabolites on one or more of the
   enzymes necessary for energy metabolism in the brain. Various studies
   have shown that a number of brain enzymes are susceptible to such inhibition. For example, L-phenylalanine inhibits brain pyruvate kinase,
   and phenylpyruvate inhibits brain hexokinase [103]. These brain enzymes appear to be more sensitive than corresponding liver enzymes.
   Both inhibitors, of course, inhibit glycolysis [104]. Phenylpyruvate inhibits pyruvate oxidation as well [105], and pyridoxal kinase is inhibited by the conjugate of phenylethylamine and pyridoxal found in the
   urine of phenylketonurics [106,107].

3. Interference with the production of the amines of neural importance.
   It is known that phenylalanine inhibits the hydroxylation of tryptophan
   [108, 109] and the decarboxylation of 5-hydroxytryptophan [110,111]
   on the biosynthetic pathway to serotonin, and that serotonin levels are
   lower in phenylketonurics [112]. Further, it is known that phenylpyruvic acid inhibits the decarboxylation of dopa [113], an intermediate in
   the synthesis of the catecholamines. There is also evidence that γ-amino-
   butyric acid synthesis could be disturbed since o-hydroxyphenylacetic acid
   inhibits glutamic acid decarboxylase [114].

4. A failure of myelination. Not only is there a known myelin deficit in
   young phenylketonurics, but there is now evidence that phenylalanine
   metabolites interfere specifically with the biosynthesis of the cerebrosides [115] and the sterols [116].

5. Disruption of the normal processes of protein synthesis. It has been
   shown that brain protein synthesis is uniquely susceptible to alterations
   in the free amino acid content of the brain [117], and that phenylalanine specifically inhibits the incorporation of certain amino acids into

> brain protein by limiting their uptake into brain [118]. Other studies
> suggest a more direct effect of phenylalanine on protein synthesis [119-
> 121]. Also, it is possible that phenylalanine acts on protein synthesis
> by lowering the level of tryptophan in the brain, which in turn seems to
> promote the disaggregation of brain polysomes [122].
> 6. Depletion of brain glutamine. Since one of the products of the phenyl-
> alanine overload in the urine of patients is phenylacetylglutamine, and
> since glutamine is such a key metabolite in brain, it seemed possible
> that adverse consequences could result from a lowering of brain gluta-
> mine by this route [123]. More recent studies, however, do not seem
> to support this mechanism [124,125].

Finally, it seems plausible that there is no single cause for the horrendous neural damage in this condition, but that a number of different changes occur which, together, produce the lesion. Whatever mechanism is considered, an explanation will have to be found for the fact that there are a number of cases of so-called atypical phenylketonuria or hyperphenylalaninemia known in which the phenylalanine levels of the blood are as high as in afflicted children, but no retardation is caused.

Phenylketonuria

A 28-month-old boy was admitted suffering from eczematoid dermatitis of the face, neck, and arms. He was of fair complexion and seemed quite retarded. Feeblemindedness had been present since birth and, although not precisely estimated, his intelligence quotient was considered to be less than 40 and he was classified as an idiot. The patient was not able to sit, stand, or crawl, or even raise his head. His eyes moved aimlessly although he seemed to be aware of large, bright objects. The child had frequent epileptiform seizures, as often as 4 or 5 times a day, which were characterized by upward rotation of the eyes and clutching of the hands.

Both parents were normal in mental and physical development, both dark complected, and were not of consanguinous relationship. The oldest of their three children was a 7-year-old girl, of blond type, but normal in stature and mentality. The youngest, a 16-month-old boy, was phenylketonuric.

Laboratory findings included a moderate anemia with a range of red blood cells from 3 to 4 million per cubic millimeter. The urine repeatedly and consistently yielded a transient green color with ferric chloride, characteristic of the excretion of phenylpyruvic acid. Quantitative determination gave a value of 2.5 g of phenylpyruvic acid excreted in a 24-hr period. Plasma phenylalanine was 50 mg/mg% and plasma tyrosine was 1.2 mg/mg%. Serum serotonin was 190 mg/ml, and urinary 5-hydroxyindolacetic acid in the early morning urine was 4.2 mg/g creatinine.

At age 5 ½ the patient contracted measles and died suddenly during convulsions. By this age the seizures and the eczema were less pronounced, but the patient did not walk or talk. Upon neuropathological examination 24 hr after death the brain was found to be small but well developed with normal appearing convolutions. The right optic nerve showed abnormally light staining for myelin, and neuroglia stains showed some increase in the number of astrocytes present. Myelin stain preparations disclosed light areas in the paracentral lobule in which there was a loss of myelin.

Tyrosine

The pathway of tyrosine degradation in mammals, namely through p-hydroxyphenylpyruvic acid, homogentisic acid, maleylacetoacetic acid, and fumarylacetoacetic acid, does not occur in brain. Tyrosine-α-ketoglutarate transaminase is present in brain [44,126], and tyrosine is, in fact, one of the most actively transaminated of all the amino acids in brain tissue. The tyrosine transaminase of brain is not influenced by the administration of either steroid hormones or excess substrate, as is the tyrosine transaminase of liver. Tyrosine is also a substrate for one of the nonspecific aromatic amino acid decarboxylases found in brain [45]. By far the most interesting aspect of the neural metabolism of tyrosine is its conversion to the catecholamines, which is discussed in Chap. 11.

Although a number of lesions in the overall pathway of tyrosine degradation are known, none of them produce severe neurological damage. For example, the absence of p-hydroxyphenylpyruvate oxidase results in tyrosinemia. A lack of homogentisic acid oxidase causes alkaptonuria. And a defect in the pigment-forming pathway of tyrosine metabolism, namely, the lack of tyrosinase, causes albinism. None of these diseases seem to have a prominent neurological component. The most interesting for this discussion is perhaps tyrosinemia, or tyrosinosis [127-129]. This is a hereditary condition characterized by lowered growth, renal rickets, and liver problems. It has not been extensively studied, perhaps because of the lack of neurological involvement. But it should be remembered that no mental problems are caused by an elevated blood level of tyrosine, even though the possible consequences are, in principle, the same as those incurred when the blood level of the closely related amino acid, phenylalanine, is elevated.

Tryptophan

Tryptophan, like phenylalanine, is essential for higher animals, so its biosynthesis is not found in mammalian tissues. Its complex degradative pathways are seen in liver and probably not in brain, but some more recent observations suggest that this problem might profitably be reexplored [40]. Tryptophan is transaminated in brain tissue by an enzyme using oxaloacetic acid as acceptor [43,44], and it is also decarboxylated [45], but a question remains as to whether these reactions have any significance under normal conditions. For example, the decarboxylation product, tryptamine, normally is barely detectable in the brain, and an overload of tryptophan and/or an amine oxidase inhibitor is required to produce pharmacological evidence for tryptamine formation [130,131]. The neural pathway of tryptophan metabolism which involves the production of serotonin and of melatonin is discussed in Chap. 11.

The few cases of patients with high blood and urine tryptophan which have been observed, the so-called tryptophanurias, have not been well characterized, and what information is available suggests that they do not involve neurological problems [17]. A few cases also have been reported in which a large excretion of kynurenic or of xanthurenic acids were associated with seizures and retardation [17], but little comment can be made until more information becomes available and more individuals are studied. The most prevalent and well-understood condition in which tryptophan is involved is not a lesion in metabolism per se, but involves defective tryptophan transport. This is the Hartnup disease which will be discussed in Chap. 10.

Histidine

Histidine biosynthesis does not occur in animal tissues, and so not in brain. Histidine degradation, also, seems absent from the brain [40]. The decarboxylation to histamine has been shown, and these enzymes are discussed in Chap. 11, as are the metabolism and possible role of histamine.

It should be mentioned that when histidine degradation is prevented by the genetic lesion known as histidinemia [132], which is an absence of the enzyme histidase (Fig. 9.14) from skin and liver [133,134], a characteristic set of symptoms appears [16,17]. Histidase is a sulfhydryl-containing, metal-requiring enzyme with a molecular weight of 225,000. Its metal requirement can be satisfied by Mn^{2+}, Fe^{2+}, or Cd^{2+}. Chemically, the products of the altered histidine metabolism are cognates of the aberrant products of phenylalanine metabolism in phenylketonuria. That is, the appropriate keto acid, imidazolepyruvic acid, is excreted, as are its reduced derivative, imidazolelactic acid, and its decarboxylation product,

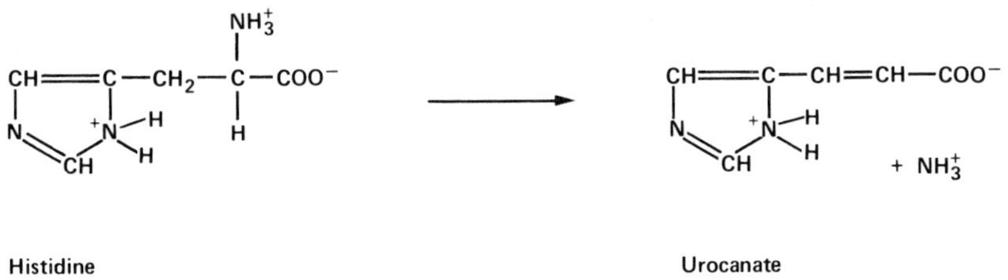

Histidine

Urocanate

Figure 9.14 Histidase.

imidazoleacetic acid. These are directly comparable to the phenylpyruvic acid, phenyllactic acid, and phenylacetic acid excreted in phenylketonuria. Of course, the histidine concentration in blood, urine, and cerebrospinal fluid is elevated, and urocanic acid, the product of the action of histidase, is missing. While the disease is interestingly similar to phenylketonuria chemically, it is startlingly different clinically. Most important, it is not a severe condition. The most common symptom is a retarded or defective speech. Sometimes there is a mild retardation, but frequently there is none. Some cases are associated with seizures and ataxia, but the connection to the metabolic lesion is far from proved. It is likely that histidinemia is a harmless, or relatively harmless, condition. Probably for this reason little or nothing has been done to understand any possible neurochemical or neuropathological changes which might occur.

Lysine

The main degradative pathway for lysine in liver and other tissues is through saccharopine, α-aminoadipic acid-γ-semialdehyde, α-aminoadipic acid, α-ketoadipic acid, glutaryl CoA, and acetoacetyl CoA. The pathway of lysine degradation in brain has not been firmly established. Some pieces of evidence suggest that it may be different than that in liver. First, lysine-ketoglutarate reductase and saccharopine reductase have been found, at most, in trace amounts in brain. Second, a recent report suggests that lysine in brain might be degraded through the pipecolic acid pathway [135]. Finally, a patient has been described with abnormally high blood and tissue concentrations of pipecolic acid and with severe destructive and demyelinating processes in the central nervous system [136].

In recent years patients with metabolic blocks at several steps in the saccharopine pathway have been identified; a major feature of each disorder is mental retardation. Hyperlysinemia is caused by a deficiency of lysine-ketoglutarate reductase which degrades lysine to saccharopine. The enzyme has been partially purified from human placenta [137] and from liver, and has been observed in trace amounts in brain [138]. Clinical findings include mental retardation, seizures, abnormal EEG patterns, and small stature [16]. Biochemically, there is a 10-fold to 20-fold increase in urinary lysine content and a 7-fold to 10-fold increase in the urinary excretion of hypusine, a conjugate of lysine and hydroxyputrescene [139]. Saccharopinuria, a defect in the conversion of saccharopine to α-aminoadipic-γ-semialdehyde, is due to a deficiency of saccharopine dehydrogenase. In the few cases reported, mental retardation was noted [140,141]. An α-aminoadipic aciduria has been described in a mentally retarded child [142]. Severe retardation has been reported to accompany α-ketoadipic aciduria [143, 144], a condition which seems to be due to a defect in the oxidative decarboxylation of α-ketoadipic acid [145]. Finally, a defect in the conversion of glutaryl CoA to acetoacetyl CoA has been shown to be accompanied by severe mental retardation [146].

Arginine and Urea Formation

The metabolism of arginine is inextricably bound up with that of urea and of
ammonia. Perhaps the link with ammonia explains the interest in arginine metabolism in brain [13,147]. The formation of arginine from proline [148] and
from ornithine [149] has been demonstrated in brain of intact animals, but the
enzymes of the complete urea cycle have not been shown to be present. Three
of the enzymes, argininosuccinate synthetase, argininosuccinase, and arginase are
clearly there [150-152]. Recently, a carbamyl phosphate synthetase, using glutamine instead of ammonia as a substrate, has been found [153,154], but there
is good evidence that this enzyme functions in pyrimidine biosynthesis and not
in the urea cycle. Ornithine transcarbamylase has been specifically sought but
not found in brain [147,152], so the presence of the urea cycle in neural tissue
is unlikely. Thus, the enzymes of the brain seem capable of making arginine and
urea from citrulline and aspartic acid (Fig. 9.15). But the synthesis of arginine
from ornithine, and thus from glutamate, does not seem to occur in brain. On
the other hand, it is quite clear from a number of measurements that the substrates for the complete urea cycle, ornithine and citrulline, are present in substantial amounts in brain. In addition to its degradation to urea and ornithine
by arginase, arginine can also serve as a source of guanido groups and, thus, as
precursor to guanidoacetic acid and γ-guanidobutyric acid, derivatives which are
discussed in Chap. 11.

There are a number of genetic defects with neurological sequelae associated with urea formation and, thus, with arginine metabolism. It is necessary to
note that these metabolic defects can not cause a total absence of the relevant
enzymes, since failure to form urea would be incompatible with life. However,
there are several conditions known which appear to be due to partial loss of the
activities associated with arginine metabolism.

A defect in the ability of the body to synthesize argininosuccinic acid from
citrulline and aspartic acid (Fig. 9.15), that is, a defect in the enzyme argininosuccinic acid synthetase, leads to an excretion of citrulline in the urine and an
elevation of the levels of citrulline in the serum and the tissues. Citrullinemia
[155,156] causes vomiting, irritability, and seizures, a postprandial hyperammonemia, and a nonspecific mental deterioration with moderate to severe retardation. The molecular defect seems to be in the ability of the enzyme to bind
its substrate, rather than the loss of the enzyme altogether [157]. These measurements indicate that the K_M for citrulline is about 25 times higher for the
defective enzyme than for the normal, so the enzyme works very inefficiently at
normal body concentrations of citrulline. There is no evidence that citrulline is
in any way toxic to the individual, and it is considered possible that the symptoms, and whatever neural damage ensues, are due to periods of ammonia toxicity during early life. Not too many patients with citrullinemia have been studied, and the enzyme defect has been shown only in a few cases, some by assays
of liver biopsy, some by culture of fibroblasts from affected children.

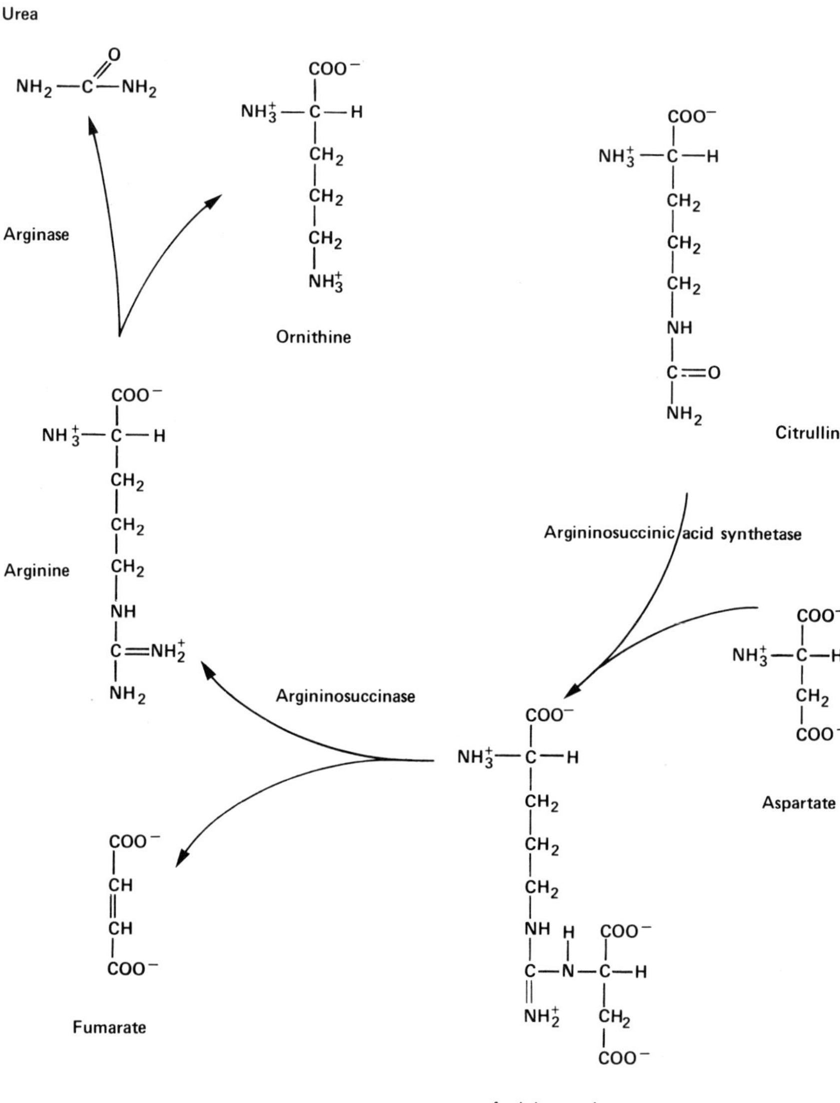

Figure 9.15   Pathway of urea synthesis and arginine biosynthesis in brain.

A somewhat more common disease, first described in 1958 [158], is argininosuccinic aciduria [16,17], which results from a defect in the enzyme argininosuccinase (Fig. 9.15), and can exist in at least two clinical forms, neonatal and late onset. A marked decrease (3% of normal in the liver at autopsy) in the activity of this enzyme [159,160] leads to the accumulation in the tissues, and the excretion in the urine, of argininosuccinic acid. In fact, the excretion of this substance in the urine can approach several grams a day, whereas normally this material does not occur in the urine at all. Urea excretion is normal so it is clear that some argininosuccinase activity persists, but postprandial hyperammonemia is a common feature, frequently to the point of coma. The disease is transmitted as an autosomal recessive, and is characterized by retardation and convulsions. Abnormal EEG patterns have been reported, and chromosomal abnormalities have been seen in some cases. Frequently there is an associated hepatomegaly. One of the most distinctive signs of this disease is abnormal hair formation, presumably because arginine is a major constituent of the protein of the hair. Although the disease is rarely fatal, the retardation can be quite severe. In one patient studied for neuropathology, edema was observed, as well as defective myelination, poorly developed cortical layers, and some degenerative changes in the white matter [161]. Of the possible causes for the neural defect the most obvious are the deficiency of arginine, the accumulation of argininosuccinic acid, and the transient episodes of hyperammonemia, but no hard information is available as to which is the most likely. Prenatal diagnosis of this disease is now possible since the enzyme can be assayed in cultured cells from amniotic fluid, as well as in red blood cells and liver [32].

Argininosuccinic Aciduria

An 8-month-old girl was admitted for evaluation in connection with several recent grand mal seizures. A history revealed the occurrence of several short periods of apparent coma. Upon examination the child was found to be of normal weight and stature with sparse, stubby, friable hair (trichorrhexis nodosa), moderate hepatomegaly, nystagmus, and some suspected difficulty in coordination. No skin lesions were evident. Laboratory studies showed a normal fasting blood ammonia concentration, a markedly elevated ammonia concentration in the blood after a moderate meal, and a 24-hr urinary excretion of 1.9 g of argininosuccinic acid. The diagnosis of argininosuccinic aciduria was confirmed by the finding of only traces of argininosuccinase activity in the red cells. Arginase and argininosuccinic acid synthetase activities were normal.

The patient was put on a low-protein diet supplemented with arginine and showed improvement. The scalp hair was restored to a near normal condition within the next several months, and the seizures did not recur. The child exhibited mild retardation at age 3, but seemed otherwise relatively normal in behavior. At age

3 ½ the child died of convulsions associated with an undiagnosed viral illness.
Upon autopsy, there was found to be a slight decrease in the myelin content, as
if caused by a moderate delay in myelination. Astrocytic nuclei were found to
be abnormal, and resembled Alzheimer's type II cells. There was a gross neuronal
loss in the thalamus [32].

A delicate and controversial situation developed around a report of the disease argininemia in two children of a single family [162]. In these children, the
concentration of arginine in the blood and cerebrospinal fluid was much increased. Lysine and cysteine were present in the urine in abnormal amounts, and
the children were severely hyperammonemic. They were also severely retarded
and suffered from spasm and seizures. A diagnosis of arginase deficiency was
made, based on the assay of arginase levels in blood cells.

The controversial part of this situation was the attempt to treat these children with a virus known to cause an increase in arginase activity in infected cells,
the Shope papilloma virus [162]. The idea, of course, was to transfer genetic information to the genes of the children using this virus as the agent. In fact, it is
stated [163] that after the infection the blood arginine content in the children
did drop. An argument can be made that the virus did transfer genetic information to the children. However, it is difficult to evaluate the success of this treatment when it is unknown whether the drop in arginine is caused by the virus, by
the activation of the arginase of the host, or by the natural evolution of the disease. In addition, some ethical questions have arisen over the treatment of the
children with massive doses of this particular virus since it is known to cause solid,
benign tumors in rabbits, and also to replicate in human cells in culture.

Hyperammonemia

There are a number of reports of hyperammonemias which seem to be associated
with defects in other enzymes. There are a few cases of a systemic lack of carbamyl phosphate synthetase activity or of ornithine transcarbamylase activity,
either of which could lead to excess ammonia in the system. In some cases, a 90
to 95% loss of ornithine transcarbamylase activity has been reported [164]. Ammonia levels are elevated, and increased glutamine excretion is observed. Associated with this condition there is intermittant stupor or coma, seizures, vomiting,
and moderate mental retardation. Again, many of the consequences of these conditions could be due to the ammonia intoxication experienced by these individuals after a meal, and not to the accumulation of the various intermediary metabolites. However, the reports of these conditions are very few, and no detailed
studies or extensive comparative surveys have as yet been possible.

Thus, a growing number of abnormalities in the enzymatic machinery for
the metabolism of the amino acids are known to have severe neurological mani-

Table 9.2 Major Abnormalities of Amino Acid Metabolism Having Neurological Consequences

Abnormality	Biochemical defect	Biochemical findings	Neurological findings	Genetics	Neuropathology
Argininemia	Arginase missing	Hyperammonemia, high blood and cerebrospinal fluid arginine	Spasm, seizures, severe retardation		
Argininosuccinic aciduria	Argininosuccinase missing or decreased	Argininosuccinic acid in the urine and cerebrospinal fluid, postprandial hyperammonemia	Retardation, seizures, abnormal EEG, coma, ataxia	Autosomal recessive	Edema, poor cortical layering, defective myelintion
Citrullinemia	Argininosuccinic acid synthetase defective	Citrulline excreted in the urine, high levels in blood and tissues, transient hyperammonemia	Seizures, hyperirritability, retardation		
Cystathionuria	Cystathionase missing	Cystathionine excreted in the urine	Moderate retardation, seizures	Autosomal recessive	
Histidinemia	Histidase	High blood and urine histidine, imidazole, keto acids excreted	Slight to moderate mental retardation, speech defects, seizures, delayed physical development	Recessive	
Homocystinuria	Cystathionine synthetase missing or defective	Homocystine excretion, high blood and urine methionine	Retardation, seizures, progressive paraplegia, dislocated eye lens	Autosomal recessive	Microgyria, spongy degeneration of the white matter, cerebral venous thrombosis

Table 9.2 (continued)

Abnormality	Biochemical defect	Biochemical findings	Neurological findings	Genetics	Neuropathology
Hyperammonemia I	Carbamyl phosphate synthetase defective	High ammonia levels in blood and cerebrospinal fluid	Spasm, seizures, coma, retardation		
Hyperammonemia II	Ornithine transcarbamylase defective	High ammonia levels in blood and cerebrospinal fluid	Retardation		
Hyperglycinemia, ketotic	Propionyl CoA carboxylase	Episodes of ketosis and acidosis, neutropenia, thrombocytopenia	Seizures, abnormal EEG		
Hyperglycinemia, nonketotic	Glycine decarboxylase	High urinary and plasma glycine	Severe retardation, seizures, microcephaly	Autosomal recessive	
Hyperlysinemia	Lysine-ketoglutarate reductase missing or defective	High urinary lysine and hypusine	Retardation, seizures, abnormal EEG		
Hyperprolinemia type I	Proline oxidase missing	Blood and urine proline increased, hydroxyproline, glycine excreted	Retardation, photosensitive seizures, nerve deafness, renal disease	Familial	Diffuse neuronal loss in one case, delayed myelination
Hyperprolinemia type II		Proline, pyrroline-5-carboxylate excreted	Convulsions, retardation		
Hypervalinemia	Valine transaminase missing	High blood and urine valine	Mental retardation, abnormal EEG, hyperactivity		

Maple syrup urine disease (branched-chain ketoaciduria)	Branched-chain keto acid oxidative decarboxylase missing	High urinary excretion of valine, isoleucine, and their keto acid analogs, hypoglycemia	Severe mental retardation, seizures, cortical blindness, progressive rigidity	Autosomal recessive	Destruction of white matter, heavy and "spongy" brain, pronounced lack of myelin
Phenylketonuria	Liver phenylalanine hydroxylase missing or defective	High blood phenylalanine, excretion of phenylketones, decreased indole excretion	Severe mental retardation, convulsions, hyperactivity, abnormal EEG	Autosomal recessive	Delayed myelination, microcephaly
Tryptophanuria	Tryptophan pyrrolase missing	Increased plasma and urine tryptophan, urinary indoles	Retardation, ataxia, dwarfism	Autosomal recessive	

festations, many of them associated with profound retardation. As with phenylketonuria, none of these manifestations, least of all the retardation, can as yet be explained at the molecular level, since we know too little about the normal functioning of the normal brain. The major known abnormalities are listed in Table 9.2, along with the metabolic lesion causing them, and the functional changes that result. More detailed tables of this nature can be found in recent reviews of the subject [17,165,166].

References

1. K. D. Neame, in *Applied Neurochemistry* (A. N. Davison and J. Dobbing, eds.), F.A. Davis, Philadelphia, 1968.
2. E. Schoffeniels, in *The Structure and Function of Nervous Tissue*, Vol. III (G. H. Bourne, ed.), Academic, New York, 1969.
3. C. F. van den Berg, in *Handbook of Neurochemistry*, Vol. III (A. Lajtha ed.), Plenum, New York, 1970.
4. M. K. Gaitonde, D. R. Dahl, and K. A. C. Elliott, *Biochem. J. 94:*345 (1965).
5. A. Meister, *Adv. Enzymol. Relat. Areas Mol. Biol. 31:*183 (1968).
6. R. A. Ronzio, W. B. Rowe, S. Wilk, and A. Meister, *Biochemistry 8:*2670 (1969).
7. R. A. Ronzio and A. Meister, *Proc. Natl. Acad. Sci. USA 59:*164 (1968).
8. A. Ronzio, W. B. Rowe, and A. Meister, *Biochemistry 8:*1066 (1969).
9. W. B. Rowe, R. A. Ronzio, and A. Meister, *Biochemistry 8:*2674 (1969).
10. H. H. Shaumberg, R. Byk, R. Gerstl, and J. H. Mashman, *Science 163:*826 (1969).
11. G. Svenneby, *Second Int. Mtg. of Int. Soc. for Neurochem.*, Milan, Sept. 1-5, 1969, Tamburini Editoire, Milan, 1969, p. 386.
12. H. Weil-Malherbe, *J. Neurochem. 16:*855 (1969).
13. H. J. Strecker, in *Handbook of Neurochemistry*, Vol. III (A. Lajtha, ed.), Plenum, New York, 1970.
14. G. A. R. Johnston and M. V. Vitoli, *Brain Res. 15:*201 (1969).
15. L. P. Davies and G. A. R. Johnston, *Brain Res. 54:*149 (1973).
16. H. E. Wiltse and J. H. Menkes, in *Handbook of Neurochemistry*, Vol. VII (A. Lajtha, ed.), Plenum, New York, 1972.
17. Y. E. Hsia, in *Basic Neurochemistry* (R. W. Albers, G. J. Siegel, R. Katzman, and B. W. Agranoff, eds.), Little, Brown, Boston, 1972.
18. Y. E. Hsia, K. J. Scully, and L. E. Rosenberg, *J. Clin. Invest. 50:*127 (1971).
19. W. F. Bridger, *J. Biol. Chem. 240:*4591 (1965).
20. K. F. Swaiman and J. M. Milstein, *J. Neurochem. 12:*981 (1965).
21. J. J. Kabara and G. T. Okita, *J. Neurochem. 7:*298 (1961).
22. J. H. Menkes, P. L. Hurst, and J. M. Craig, *Pediatrics 14:*462 (1954).
23. J. A. Bowden and J. L. Connelly, *J. Biol. Chem. 243:*3526 (1968).
24. H. J. Menkes, *Neurology 9:*826 (1959).

25. D. H. Silberberg, *J. Neurochem. 16:*1141 (1969).

26. M. K. Gaitonde, in *Handbook of Neurochemistry*, Vol. III (A. Lajtha, ed.), Plenum, New York, 1970.

27. S. H. Mudd, F. Finkelstein, F. Irreverre, and L. Laster, *J. Biol. Chem. 240:* 4382 (1965).

28. D. B. Hope, *Fed. Proc. 18:*249 (1959).

29. N. A. J. Carson and D. W. Neill, *Arch. Dis. Child. 37:*505 (1962).

30. T. Gerritsen, J. G. Vaughn, and H. A. Waisman, *Biochem. Biophys. Res. Commun. 9:*493 (1962).

31. S. H. Mudd, J. D. Finkelstein, F. Irreverre, and L. Laster, *Science 143:* 1443 (1964).

32. L. B. Holmes, H. W. Moser, S. Halldorsson, C. Mack, S. S. Pant, and B. Matzilevich, *Mental Retardation*, MacMillan, New York, 1972.

33. J. Lindsay and H. S. Bachelard, *Biochem. Pharm. 15:*1045 (1966).

34. M. B. Sporn, W. Dingman, and A. De Falco, *J. Neurochem. 4:*141 (1959).

35. J. V. Taggart and R. B. Krakaur, *J. Biol. Chem. 177:*641 (1949).

36. M. L. Efron, *N. Engl. J. Med. 273:*1243 (1965).

37. F. A. Emery, L. Goldie, and J. Stern, *J. Ment. Defic. Res. 12:*187 (1968).

38. D. J. Selkoe, *Neurology 19:*494 (1969).

39. M. L. Efron, E. M. Bixby, and C. V. Pryles, *N. Engl. J. Med. 272:*1299 (1965).

40. G. Guroff and W. Lovenberg, in *Handbook of Neurochemistry*, Vol. III (A. Lajtha, ed.), Plenum, New York, 1970.

41. R. J. Winzler, K. Moldave, M. E. Rafelson, and H. E. Pearson, *J. Biol. Chem. 199:*485 (1952).

42. H. H. Sky-Peck, C. Rosenbloom, and R. J. Winzler, *J. Neurochem. 13:*223 (1966).

43. F. Fonnum, R. Haavaldsen, and O. Tangen, *J. Neurochem. 11:*109 (1964).

44. O. Tangen, F. Fonnum, and R. Haavaldsen, *Biochim. Biophys. Acta 96:*82 (1965).

45. W. Lovenberg, H. Weissbach, and S. Udenfriend, *J. Biol. Chem. 237:*89 (1962).

46. G. Paulson and N. Allen, in *Genetic Disorders of Man* (R. M. Goodwin, ed.), Little, Brown, Boston, 1970.

47. A. Fölling, *Z. Physiol. Chem. 227:*169 (1934).

48. G. A. Jervis, *J. Biol. Chem. 169:*651 (1947).

49. G. A. Jervis, *Proc. Soc. Exp. Biol. Med. 82:*514 (1953).

50. C. Mitoma, R. M. Auld, and S. Udenfriend, *Proc. Soc. Exp. Biol. Med. 94:* 634 (1957).

51. H. W. Wallace, K. Moldave, and A. Meister, *Proc. Soc. Exp. Biol. Med. 94:* 632 (1957).

52. S. Kaufman, *Adv. Enzymol. Relat. Areas Mol. Biol. 35:*245 (1971).

53. S. Kaufman and D. B. Fisher, *J. Biol. Chem. 245:*4745 (1970).

54. D. B. Fisher, R. Kirkwood, and S. Kaufman, *J. Biol. Chem. 247:*5161 (1972).

55. S. Kaufman, *Proc. Natl. Acad. Sci. USA 50:*1085 (1963).

56. T. E. Craine, E. S. Hall, and S. Kaufman, *J. Biol. Chem. 247:*6082 (1972).

57. D. B. Fisher and S. Kaufman, *J. Biol. Chem. 247:*2250 (1972).

58. D. B. Fisher and S. Kaufman, *J. Biol. Chem. 248:*4345 (1973).

59. C. Y. Huang, E. E. Max, and S. Kaufman, *J. Biol. Chem. 248:*4235 (1973).

60. S. Milstein, J. P. Abita, N. Chang, and S. Kaufman, *Proc. Natl. Acad. Sci. USA 73:*1591 (1976).

61. J. P. Abita, S. Milstein, N. Chang, and S. Kaufman, *J. Biol. Chem. 251:* 5310 (1976).

62. G. Guroff, C. Reifsnyder, and J. W. Daly, *Biochem. Biophys. Res. Commun. 24:*720 (1966).

63. G. Guroff, M. Levitt, J. W. Daly, and S. Udenfriend, *Biochem. Biophys. Res. Commun. 25:*253 (1966).

64. G. Guroff, K. Kondo, and J. W. Daly, *Biochem. Biophys. Res. Commun. 25:*622 (1966).

65. J. Daly and G. Guroff, *Arch. Biochem. Biophys. 125:*136 (1968).

66. G. Guroff, J. W. Daly, D. M. Jerina, J. Renson, B. Witkop, and S. Udenfriend, *Science 157:*1524 (1967).

67. P. A. Friedman, S. Kaufman, and E. S. Kang, *Nature (London) 240:*157 (1972).

68. P. A. Friedman, D. B. Fisher, E. S. Kang, and S. Kaufman, *Proc. Natl. Acad. Sci. USA 70:*552 (1973).

69. A. Tourian, J. Goddard, and T. T. Puck, *J. Cell. Physiol. 73:*159 (1969).

70. S. Kaufman, N. A. Holtzman, S. Milstein, I. J. Butler, and A. Krumholz, *N. Engl. J. Med. 293:*785 (1975).

71. G. A. Jervis, *Proc. Soc. Exp. Biol. Med. 75:*83 (1950).

72. L. I. Woolf, *Biochem. J. 49:*IX (1951).

73. J. A. Oates, P. Z. Nirenberg, J. B. Jepson, A. Sjoerdsma, and S. Udenfriend, *Proc. Soc. Exp. Biol. Med. 112:*1078 (1963).

74. M. D. Armstrong, K. N. F. Shaw, and K. S. Robinson, *J. Biol. Chem. 213:* 797 (1955).

75. L. I. Woolf and D. G. Vulliamy, *Arch. Dis. Child. 26:*487 (1951).

76. F. B. Goldstein, *Biochim. Biophys. Acta 71:*204 (1963).

77. H. D. Grumer, *Nature (London) 189:*63 (1961).

78. Y. H. Loo and P. Ritman, *Nature (London) 203:*1237 (1964).

79. Y. H. Loo, *J. Neurochem. 14:*813 (1967).

80. E. C. Alvord, L. D. Stevenson, F. S. Vogel, and R. L. Engle, *J. Neuropathol. Exp. Neurol. 9:*298 (1950).

81. L. Crome, V. Tymms, and L. I. Woolf, *J. Neurol. Neurosurg. Psychiatr. 25:* 143 (1962).

82. A. L. Prensky, S. Carr, and H. W. Moser, *Arch. Neurol. 19:*552 (1968).

83. B. Gerstl, N. Malamud, L. F. Eng, and R. B. Hayman, *Neurology 17:*51 (1967).

84. J. N. Cumings, I. K. Grundt, and T. Yanagihara, *J. Neurol. Neurosurg. Psychiatr. 31:*334 (1968).

85. J. H. Menkes, *Pediatrics 37:*967 (1966).

86. J. L. Foot, R. J. Allen, and B. W. Agranoff, *J. Lipid Res. 6:*518 (1965).

87. J. H. Menkes, *Neurology 18:*1003 (1968).

88. R. Guthrie and A. Susi, *Pediatrics 32:*338 (1963).

89. M. W. McCaman and E. Robins, *J. Lab. Clin. Med. 59:*885 (1962).

90. B. N. LaDu and P. J. Michael, *J. Lab. Clin. Med. 55:*491 (1960).

91. W. E. Knox, in *The Metabolic Basis of Inherited Disease,* 2nd ed. (J. B. Stanbury, J. B. Wyngaarden, and D. S. Fredrickson, eds.), McGraw-Hill, New York, 1966.

92. B. M. Rouse, *J. Pediatr. 69:*246 (1966).

93. I. M. Hackney, W. B. Hanley, W. Davidson, and L. Lindsao, *J. Pediatr. 72:* 646 (1968).

94. C. C. Mabry, J. C. Denniston, and J. G. Coldwell, *N. Engl. J. Med. 275:* 1331 (1966).

95. W. K. Frankenburg, B. R. Duncan, R. W. Coffelt, R. Koch, J. G. Coldwell, and C. D. Son, *J. Pediatr. 73:*570 (1968).

96. R. O. Fisch, D. Doeden, L. L. Lansky, and J. A. Anderson, *Am. J. Dis. Child. 118:*847 (1969).

97. M. A. Chirigos, P. Greengard, and S. Udenfriend, *J. Biol. Chem. 235:*2075 (1960).

98. G. Guroff, W. King, and S. Udenfriend, *J. Biol. Chem. 236:*1173 (1961).

99. G. Guroff and S. Udenfriend, *J. Biol. Chem. 237:*803 (1962).

100. K. D. Neame, *Nature (London) 192:*173 (1961).

101. M. J. Carver, *J. Neurochem. 12:*45 (1965).

102. C. M. McKean, D. E. Boggs, and N. A. Peterson, *J. Neurochem. 15:*235 (1968).

103. G. Weber, *Proc. Natl. Acad. Sci. USA 63:*1365 (1969).

104. R. I. Glazer and G. Weber, *Brain Res. 33:*439 (1971).

105. B. B. Gallagher, *J. Neurochem. 16:*1071 (1969).

106. Y. H. Loo and V. P. Whittaker, *J. Neurochem. 14:*997 (1967).

107. Y. H. Loo and P. Ritman, *Nature (London) 213:*914 (1967).

108. R. C. Baldridge, L. Borofsky, H. Baird, F. Reichle, and D. Bullock, *Proc. Soc. Exp. Biol. Med. 100:*529 (1959).

109. W. Lovenberg, E. Jequier, and A. Sjoerdsma, *Adv. Pharmacol. 6A:*21 (1968).

110. A. N. Davison and M. Sandler, *Nature (London) 181:*186 (1958).

111. A. Yuwiler, E. Geller, and G. C. Slater, *J. Biol. Chem. 240:*1170 (1965).

112. C. M. B. Pare, M. Sandler, and R. S. Stacey, *Lancet 1:*551 (1957).

113. J. B. Boylen and J. H. Quastel, *Biochem. J. 80:*644 (1961).

114. R. E. Tashian, *Metabolism 10:*393 (1961).

115. L. Barbato, I. W. M. Barbato, and A. Hamanaka, *Brain Res. 7:*399 (1968).

116. S. N. Shah, N. A. Peterson, and C. M. McKean, *Biochim. Biophys. Acta 187:*236 (1969).

117. S. Roberts and C. E. Zomzely, in *Protides of the Biological Fluids,* Vol. 13 (H. Peters, ed.), Elsevier, Amsterdam, 1966.

118. H. C. Agrawal, A. H. Bone, and A. N. Davison, *Biochem. J. 117:*325 (1970).

119. N. A. Peterson and C. M. McKean, *J. Neurochem. 16:*1211 (1969).

120. K. F. Swaiman, W. B. Hosfield, and B. Lemieux, *J. Neurochem. 15:*687 (1968).

121. J. W. MacInnes and K. Schlesinger, *Brain Res. 19:*101 (1971).

122. K. Aoki and F. L. Siegel, *Science 168:*129 (1970).

123. T. L. Perry, S. Hansen, B. Tischler, R. Bunting, and S. Diamond, *N. Engl. J. Med. 282:*761 (1970).

124. C. M. McKean and N. A. Peterson, *N. Engl. J. Med. 283:*1364 (1970).

125. J. P. Columbo, *Arch. Dis. Child. 46:*720 (1971).

126. Z. N. Canellakis and P. P. Cohen, *J. Biol. Chem. 222:*53 (1956).

127. J. Gentz, R. Jagenburg, and R. Zetterstrom, *J. Pediatr. 66:*670 (1965).

128. B. N. LaDu, *Am. J. Dis. Child. 113:*54 (1967).

129. J. H. Menkes, V. Chernick, and B. Ringel, *J. Pediatr. 69:*583 (1966).

130. S. M. Hess, B. G. Redfield, and S. Udenfriend, *J. Pharmacol. Exp. Ther. 127:*178 (1959).

131. J. A. Oates and A. Sjoerdsma, *Neurology 10:*1076 (1960).

132. H. Ghadimi, M. W. Partington, and A. Hunter, *N. Engl. J. Med. 265:*221 (1961).

133. V. H. Auerbach, A. M. DiGeorge, R. C. Baldridge, C. D. Tourtellotte, and M. P. Brigham, *J. Pediatr. 60:*487 (1962).

134. B. N. LaDu, R. R. Howell, G. A. Jacoby, J. E. Seegmiller, E. K. Sober, V. G. Zannoni, J. P. Canby, and L. K. Ziegler, *Pediatrics 32:*216 (1963).

135. Y. F. Chang, *Biochem. Biophys. Res. Commun. 69:*174 (1976).

136. P. D. Gatfield, E. Taller, G. G. Honton, A. C. Wallace, G. M. Abdernour, and M. D. Haust, *Can. Med. Assoc. J. 97:*1215 (1968).

137. T. A. Fjellstedt and J. C. Robinson, *Arch. Biochem. Biophys. 168:*536 (1975).

138. J. Hutzler and J. Dancis, *Biochim. Biophys. Acta 377:*42 (1975).

139. N. C. Woody and M. B. Pupene, *Pediatr. Res. 7:*994 (1973).

140. O. Simell, J. K. Visakorpi, and M. Donner, *Arch. Dis. Child. 47:*52 (1972).

141. F. C. I. Fellows and N. A. Carson, *Pediatr. Res. 8:*42 (1974).

142. S. Lormans and A. Lowenthal, *Clin. Chim. Acta 57:*97 (1974).

143. R. W. Wilson, C. M. Wilson, S. C. Gates, and J. V. Higgens, *Pediatr. Res. 9:*522 (1976).

144. H. Przyrembel, D. Bachmann, J. Lornbeck, K. Becker, U. Wendel, S. K. Wadman, and H. J. Bremer, *Clin. Chim. Acta 58:*257 (1975).

145. U. Wendel, H. W. Rudiger, H. Przyrembel, and H. J. Bremer, *Clin. Chim. Acta 58:*271 (1975).

146. S. I. Goodman and J. G. Kohlhoff, *Biochem. Med. 13:*138 (1975).

147. H. C. Buniatian, in *Handbook of Neurochemistry,* Vol. Va (A. Lajtha, ed.), Plenum, New York, 1971.

148. M. B. Sporn, W. Dingman, and A. DeFalco, *J. Neurochem. 4:*141 (1959).

149. J. W. Kemp and D. M. Woodbury, *Biochim. Biophys. Acta 111:*23 (1965).

150. S. Ratner, H. Morrell, and E. Corvalho, *Arch. Biochem. Biophys. 91:*280 (1960).
151. S. Tomlinson and R. G. Westall, *Nature (London) 188:*235 (1960).
152. H. C. Buniatian and M. A. Davtian, *J. Neurochem. 13:*743 (1966).
153. S. E. Hager and M. E. Jones, *J. Biol. Chem. 242:*5667 (1967).
154. M. Tatibana and K. Ito, *J. Biol. Chem. 244:*5403 (1969).
155. N. C. McMurray, F. Mohyddin, R. J. Rossiter, J. C. Rathbun, and D. E. Zarfas, *Biochem. J. 84:*106P (1962).
156. G. Morrow, L. A. Barness, and M. L. Efron, *Pediatrics 40:*565 (1967).
157. T. A. Tedesco and W. J. Mellman, *Proc. Natl. Acad. Sci. USA 57:*829 (1967).
158. J. D. Allen, D. C. Cusworth, C. E. Dent, and V. K. Wilson, *Lancet 1:*182 (1958).
159. S. Tomlinson and R. G. Westall, *Clin. Sci. 26:*261 (1964).
160. J. Kint and D. Carton, *Lancet 2:*635 (1968).
161. D. Carton, F. deSchrijver, J. Kint, J. vanDurme, and C. Hooft, *Acta Paediatr. Scand. 58:*528 (1969).
162. H. G. Terheggen, A. Schwenk, A. Lowenthal, M. van Sande, and J. P. Columbo, *Lancet 2:*748 (1969).
163. S. Rogers, *Res. Commun. Chem. Pathol. Pharmacol. 2:*587 (1971).
164. B. Levin, J. M. Abraham, V. G. Oberholzer, and E. A. Burgess, *Arch. Dis. Child. 44:*152 (1969).
165. K. O. Raivio and J. E. Seegmiller, *Ann. Rev. Biochem. 41:*543 (1972).
166. D. O'Brien, in *Methods of Neurochemistry,* Vol. I (R. Fried, ed.), Dekker, New York, 1971.

10
AMINO ACID TRANSPORT

As can be seen from the previous discussion, much of the amino acid nutrition
of the brain is provided by the bloodstream. Thus, the subject of amino acid
transport into, and out of, the brain has been a subject of considerable interest
for a number of years [1-3]. The brain is generally poorly permeable to the diffusion of charged, water-soluble materials such as the amino acids. But the brain
obviously needs amino acids for adequate functioning so they must move into
the brain by some mechanism other than simple diffusion. It has been determined that the brain, like most other tissues, has specific active transport mechanisms for the capture of amino acids from the bloodstream. These transport
systems, like the systems described in other tissues, are energy requiring, concen-
trative, saturable, and have some measure of structural and steric specificity. As
in other tissues, the transport systems are less specific than most enzymes in that
the transport systems serve a family of related amino acids more or less equally.
Generally speaking, amino acid uptake by brain in vivo is slower than by other
tissues of the body, but the final concentration in brain is greater, in most cases,
than in other tissues, and the various amino acids are at least twice as concentrated in brain as in plasma. Thus, the brain has transport systems for amino
acids that fulfill the classic requirements for "active transport" and are somewhat more potent than those of other tissues.

Experimental Systems

Two types of experiments have been fruitful in the study of amino acid transport into the brain. Amino acid administration to the whole animal, of course,
represents the most direct way to obtain information. Such experiments have
been done either by injection of large amounts of amino acids into the bloodstream in order to raise the blood concentration, and observe the effects on
brain concentrations, or by injection of small amounts of radioactive amino
acid. As we shall see below, these techniques do not always give obviously con-

sistent results. Thin slices of brain, also, have been widely used to study uptake, since such a technique allows the manipulation of experimental conditions in a much more convenient and liberal manner. The slice technique has
been criticized on the grounds that it does not represent a physiological situation. Indeed, there is known to be substantial cellular damage, and marked alteration of the intracellular energy content, during the preparation and use of
slices. But, with reservations, the information provided by the slice studies has
been quite valuable in the understanding of the process. That is, although the
slice concentrates better and faster than does the brain in vivo, and although
some of the materials which are excluded from the brain in the intact animal are
taken up in the slice, the uptake in the slice most certainly can be used as a guide
to the physiological situation. Thus, a comparison of the published data shows
that the rates of uptake of the various amino acids in the slice are in the same
order as those found in vivo, the competitive relationships are the same, the
stereospecificity the same, and, so, the slice is a reasonable model.

Specificities

The work of Lajtha and of others, using both slices and the intact animal, has
led to an understanding of the number and specificities of the systems in brain
[4,5]. It is known that there are at least seven different systems for amino acid
uptake. Sites have been identified for (1) acidic amino acids, (2) γ-aminobutyric
acid and other γ-amino acids, (3) small neutral amino acids, (4) amido amino
acids, such as glutamine, (5) small basic amino acids, (6) large neutral amino acids,
including the aromatics, and (7) large basic amino acids. Perhaps there is also a
separate site for the imino acids, but this is not definitely shown as yet. All these
systems have also been found in other tissues, except the γ-aminobutyric acid
system, which is unique to brain. Thus, the picture of brain amino acid transport
is much the same in outline as the amino acid transport in other tissues, but there
are some differences in quantitative detail. Notably, brain slices are more active
in amino acid uptake than slices from other tissues.
    Since a family of amino acids is served by a given transport site, there can
be competition for uptake. Indeed, it is the evidence of competition for transport which shows that families are transported, and indicates which amino acids
belong to which families. For example, it is known that glutamate and asparate
compete for uptake [6], that the aromatics such as tryptophan and phenylalanine, as well as valine and leucine, all inhibit the uptake of tyrosine [7-9], and
that the small neutral amino acids all interact with one another [10,11]. However, it is clear that there is substantial overlap between the various sites. In many
cases a high enough concentration of an amino acid will inhibit the transport of
amino acids in other families. In other words, the groups of amino acids have reasonable affinity for their own carrier, but also have some affinity for other carriers

and, thus, at high enough concentration, will be taken in by other carriers as well.
It is clear that by careful measurement of the kinetics of uptake of an amino acid,
and of its inhibitory properties for others, an understanding of its primary and
secondary affinities can be obtained [3]. The uptake of amino acids can be described by Michaelis-Menten kinetics. Most of the carriers have about the same
K_M for the appropriate amino acids. These constants range between 0.5 and 1.5
mM, except for the γ-aminobutyric acid system which has a K_M of the order of
10^{-5} M. The competitive aspects of amino acid transport into the brain have
been well documented in vivo and in vitro. In fact, there is substantial reason to
believe that such competitive interactions play a role in the pathology of disease
states in which blood levels of individual amino acids are elevated.

Efflux

It has been shown that amino acids are actively pumped out of the brain as well
as being pumped in [12-17]. These findings account for a number of previously
puzzling observations. First, it was observed some years ago that, unlike with
some other amino acids, elevation of the blood level of glutamate did not lead to
a rise in the brain concentration of glutamate [18]. Thus, it was concluded that
the brain is not permeable to glutamate. Later experiments, however, showed
that radioactive glutamate introduced into the bloodstream crossed into the
brain quite readily. This apparent contradiction was resolved by the finding
that the active transport of amino acids takes place in the opposite direction as
well, that is, from brain to blood. The extremely rapid transport of glutamate
out of brain prevents the brain concentration from rising in response to elevated
blood concentrations. In fact, more directly, it has been shown that if radioactive amino acids are introduced into the brain, and the blood level of the same
amino acid is also raised by injection of unlabeled carrier, the amino acid will be
pumped out of the brain, even against a concentration gradient [15-17]. Similar
experiments have been done to demonstrate active transport of amino acids from
the cerebrospinal fluid into the bloodstream [19,20]. Such active efflux mechanisms also probably account for the finding that the uptake of the D isomers of
many amino acids, while slower than that of the L isomers, sometimes reaches
a higher steady-state concentration [21]. The data have been interpreted to show
that, although the uptake systems are stereopreferential, in most cases quite substantially, the efflux carriers are stereopreferential as well and to an even greater
degree, so that even though the D isomer gets in less well than the L isomer, it
gets out even more poorly in comparison. It is interesting, in passing, that all the
uptake carriers are quite a bit better with the L amino acids, except for the acidic
amino acid system, D-glutamate and L-glutamate being about equally well taken
up [6].
     So there are two systems for each group of amino acids, one for uptake and
one for efflux. The steady-state level of an amino acid achieved within the brain

substance is due mostly to a balance between these two. The characteristics of the efflux systems have not been worked out in as much detail as have the systems for influx, but it is known that the efflux is energy-requiring, and that there is competition for efflux between some amino acids. The kinetics of efflux do not seem to be straightforward Michaelis-Menten type. In sum, then, it is possible to find interactions between amino acids vis-à-vis their transport at the inside or at the outside of the membrane. And, since exchange diffusion has been shown to occur in brain [22], it is also possible to observe the so-called trans interactions i.e., where the two amino acids are on opposite sides of the membrane.

Energy Requirement

The basic requirements for amino acid transport into the brain are similar to those for amino acid transport into other mammalian cells. Energy is required for influx [3,23], and probably for active efflux as well. A number of inhibitors of energy metabolism have been used in the slice system and have been found to inhibit amino acid transport. Such substances as cyanide, iodoacetate, fluoride, malonate, and dinitrophenol have been shown to be efficient inhibitors of the transport. The question of the immediate source of energy for transport is not so well settled. Attempts to relate the exact extent of amino acid transport to the level of ATP in the slice under a variety of conditions have not been successful [24]. Specifically, in slices depleted of ATP by brief exposure to iodoacetate, or in slices whose ATP content was raised by incubation with creatine phosphate, comparable changes in amino acid transport were not seen. The interpretation of this work is that ATP is not the immediate source of energy for transport, but no information is available about what the source might be. The picture is not made any clearer by the observation that the transport of the various amino acids does not seem to be equally sensitive to the several energy variables. That is, the several inhibitors inhibit different amino acid uptakes to different extents in the same system and at the same concentration of inhibitor. Also, the different amino acids have different responses in the slice system to such variables as the presence of glucose, the omission of oxygen, the lowering of the temperature, or the alteration of the ionic composition. Ouabain in low concentrations (10^{-5} or lower) inhibits amino acid transport [6,25,26], suggesting, as does work with other tissues, that the transport of amino acids is in some as yet obscure way linked to the sodium pump. Also, sodium and, to some extent, potassium levels in the medium surrounding slices influence the uptake, but the concentration of other ions such as calcium, lithium, or ammonium have relatively little effect.

Structural Requirement

It is known that the amino acid transport systems do not accept the acid or amine analogs of amino acids. Any attempt to remove or replace either the amino group

or the carboxyl group results in a molecule which does not get into brain in vivo
and is not actively transported in vitro. For example, the brain is impermeable to
p-hydroxyphenylacetic acid, but tyrosine is taken up actively [19]. On the other
hand, the amine derivatives do not enter readily either, and the standard way to
raise the concentration of an amine in the brain is to administer the corresponding
amino acid so that it will be taken up actively and then converted to the amine.

Regional Differences

A number of studies have been done using slices from various part of brain [27].
Some quantitative differences in the amino acid transport in various regions have
been observed in this manner. In vivo studies also show regional differences in
transport [28,29]. Although the levels of amino acids in the various parts of the
brain are not the same, and the rates of transport of the several amino acids vary
also, no strong correlation between these two parameters has been shown.

Subcellular Transport

Various subcellular particles from brain also transport amino acids, presumably
across the limiting membrane. Studies have been done with nuclei, mitochondria
[30], and with synaptosomes [31]. These latter experiments are perhaps the
most interesting since they show that axonal flow mechanisms are not necessary
for the amino acid nutrition of the synaptic portions of the neurons. Very little
is known, as yet, about the possible differences in the amino acid transport of
the various cell types found in the brain.

Developmental Aspects

It is clear that the activities of the several amino acid transport systems change
during the development of the brain. What is not so clear is exactly how. Studies
in vivo almost always show that amino acids penetrate the brains of young animals more rapidly, and reach a higher concentration, than they do in adults [15,
17,32,33]. This is probably due to the fact that the blood-brain barrier is less
developed in the young, because if similar studies are carried out with brain slices,
the picture is not so clear. For many amino acids the uptake into slices increases
with age and for others the uptake rises to a maximum at some specific age after
birth and then declines as the animal becomes an adult [34–36]. Generally, the
changes seen in the slice system are thought to represent changes in the number
of transport sites for a given amino acid, since the K_M for transport usually stays
the same but the V_{max} changes. In any case, the changes in the overall transport
rate of the various amino acids as the animal develops still do not yield a consistent developmental story.

Under normal conditions the rate of transport may not directly regulate the rate of metabolism since, in general, the rates of synthesis and degradation are slow with respect to the rates of transport. Also, unlike the case of glucose transport into muscle, for instance, where metabolism is faster than transport and no free glucose accumulates in the cell, amino acids can accumulate in the brain. Transport may well be a major factor in maintaining the intracellular concentration of a given amino acid, but since the correlations between rate of uptake and intracellular concentration are not all that good, many factors other than transport are probably involved. Under pathological conditions the situation may be more crucial. That is, when one amino acid is at higher than normal levels, its influence on the rates of transport and, thus, the intracellular concentrations of other amino acids, may be much more significant.

Mechanism

Little or nothing is known about the biochemical mechanism of amino acid transport in the brain or in other tissues. One interesting recent hypothesis concerns the role of the γ-glutamyl cycle. In this postulate the metabolism of glutathione, the ubiquitous tripeptide γ-L-glutamyl-L-cysteinylglycine, is considered to be intimately involved in the transport of amino acids across cell membranes, and more specifically, into the brain [37]. The cyclic metabolism of glutathione is well understood (Fig. 10.1). Its biosynthesis is catalyzed by two sequential peptide-forming enzymes at the expense of ATP. It is catabolized via a transpeptidation which requires the transfer of the γ-glutamyl to another amino acid, the release of that amino acid, and the formation of 5-oxoproline from the γ-glutamyl group. Finally, the catabolism requires the action of a peptidase to split the remaining dipeptide, cysteinylglycine, to its constituent amino acids. Indeed, brain is a rich source of these enzymes, as well as of glutathione, and the choroid plexus is an especially rich area. It is proposed that the amino acid to be transferred reacts with glutathione on the cell surface in a transpeptidation to produce the γ-glutamyl amino acid and cysteinylglycine. The γ-glutamyl amino acid then enters the cell and is broken down by a second enzyme to yield the free amino acid, now inside the cell, and 5-oxoproline. Although this postulate is one of the most comprehensive attempts to describe the biochemical basis of amino acid transport, final proof of its validity is not yet at hand.

Clinical Implications

Aside from the suspected deleterious effects of transport competition in some of the metabolic derangements mentioned already, there are an increasing number

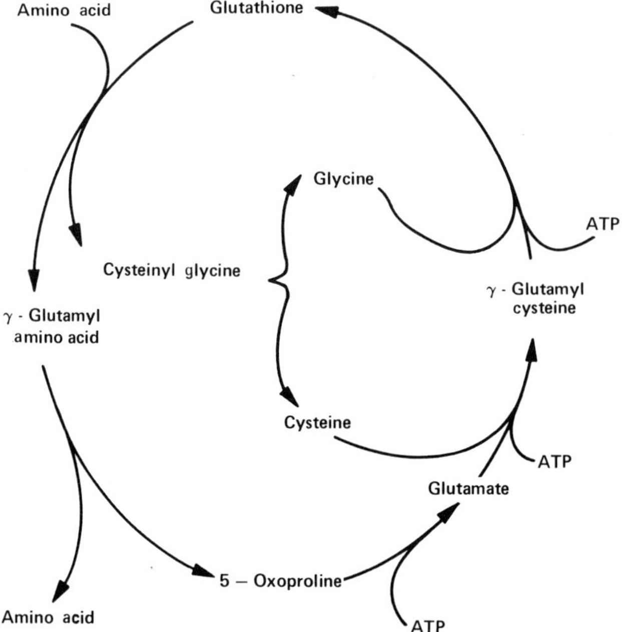

Figure 10.1   The γ-glutamyl cycle.

of diseases known which appear to be caused by direct lesions in the ability of various organs to transport amino acids. Although there are no reported cases of such defects in brain amino acid transport, perhaps because such lesions are not compatible with life, there are several which have at least moderate neurological sequelae.

The best known of these transport-related diseases, perhaps, is Hartnup disease [38,39]. First described in 1956 [40], and named after the family in which the first cases were discovered, this condition is apparently caused by a defect in the transport of tryptophan and other amino acids in the small intestine and the kidney [41]. The disease is transmitted as an autosomal recessive, and one study rates the incidence at 1 in 20,000 [42]. The symptoms are quite variable from case to case, and the retardation seen in some individuals is not always a part of the condition. There seems to be a generalized transport defect, but the amino acid which is most obviously involved is tryptophan. The fecal and urinary excretion of large amounts of indoleacetic acid and indican, usually minor excretory products, is thought to be due to the action of intestinal bacteria on the unabsorbed tryptophan in the gut. Many of the symptoms appear to be due to a niacin deficiency produced as a consequence of reduced availability of tryptophan, its metabolic precursor. So far little or no neuropathology has been done on patients suffering from Hartnup. The disease is not generally severe, and no decrease in life expectancy is involved.

Hartnup Disease

The patient, an 8-year-old girl, was admitted for treatment of an inflamed, exuding, pellagra-like rash. Physical examination showed rash and skin eruption in a symmetrical pattern about the face, neck, and limbs. The patient was of normal size and pigmentation, seemed cheerful and alert, and of normal intelligence. Vertical and horizontal nystagmus was evident, and a slight ptosis, especially of the right eyelid, could be detected. A history revealed that the rash was apparently seasonal, appearing in the spring and summer, and subsiding during the winter. Severe episodes had occurred in connection with exposure to bright sunlight. Intermittent ataxia and two fainting spells within the last year were mentioned. Reports from the child's teachers suggested a mild retardation, and periods of unexplained elation and depression, the latter sometimes verging on autism.

Laboratory studies showed a higher than normal 24-hr excretion of some of the neutral and aromatic amino acids. By amino acid analysis, alanine, serine, threonine, asparagine, glutamine, valine, leucine, isoleucine, phenylalanine, tyrosine, tryptophan, histidine, and citrulline were elevated to between 5 and 10 times the normal amounts. Taurine, glycine, cysteine, aspartic acid, glutamic acid, and lysine were within the normal range. Proline, hydroxyproline, methionine, and arginine were found in very small amounts. Blood levels, on the other hand, of all the amino acids were normal. The urine contained substantial quantities of indoleacetic acid and indican.

The diagnosis of Hartnup disease was based on the nonspecificity of the renal hyperaminoaciduria and on the light-initiated and symmetrical rash. The patient was put on an oral dose of 150 mg of nicotinamide per day. The rash improved within several weeks, but affected areas remained highly pigmented with a dirty brown color. Amelioration of the neurological symptoms was evident from periodic examination and from reports, from the child's teachers, of a cessation of the labile emotional state. The dose of nicotinamide was periodically reduced, with appropriate additions to the normal diet, until at age 12 it was discontinued altogether with no return of the symptoms [42].

There are several other conditions with more or less severe neurological consequences which probably are caused by a defect in amino acid transport. Among these are cystinosis, oasthouse urine disease (Smith-Strang disease), Lowe's syndrome (oculocerebrorenal syndrome), and possibly some of the tryptophanurias, like the blue diaper syndrome, and the lysinurias.

Cystinosis may be caused by a generalized failure to transport cystine within the cell, especially in the kidney [39,43]. Cystine accumulates as crystals in the lysosomes [44,45]. Renal tubular damage results, and a photophobia, perhaps caused by cystine accumulation and damage in the eye, is the most distinctive sign.

Oculocerebrorenal syndrome [46] is a severe retardation associated with glaucoma, cataracts, and eventual blindness, and with renal organic aciduria, aminoaciduria, and renal rickets [42,47]. The basis of the disease is not known for sure, but a defect in amino acid transport in the intestine has been demonstrated [48]. The disease is transmitted as an X-linked recessive and is generally fatal at an early age due to renal insufficiency, although some patients have reached their middle twenties. No consistent neuropathological findings have been reported.

Oasthouse urine disease is a malfunction of the absorption mechanism for methionine associated with retardation and convulsions [39]. The name stems from the unpleasant odor emanating from the patients. The absorption defect for the methionine seems in some way analogous to the intestinal defect in Hartnup disease, but the symptoms are quite different. The neural problems may originate from an intracellular deficiency of methionine which serves as a methyl donor for a number of metabolites of chiefly neural importance.

The number of diseases becoming identified with transport defects is increasing. Several of the most prominent are described in Table 10.1.

Free Amino Acid Pools

The transport of an amino acid into and out of the brain, the speed of its various metabolic conversions, and the magnitude of its incorporation into protein, together determine the concentration of that particular amino acid in the brain. In fact, the composition of the so-called "free amino acid pool" of brain has been of interest for a number of years and for a number of reasons [49-52]. The composition of the pool is fairly constant, is characteristic of brain, and does not vary much from the brain of one species to the brain of another. The total amino acid nitrogen of the brain is about 6 times higher than the amino acid nitrogen concentration of plasma, and the individual amino acids are at least twice as concentrated in the brain as in the plasma. As mentioned before, the free amino acid pool is dominated by the derivatives of glutamic and aspartic acids. Glutamic acid itself, glutamine, γ-aminobutyric acid, aspartic acid, and N-acetylaspartic acid, make up about two-thirds of the amino nitrogen of the free amino acid pool in the brains of most species. The composition of the free amino acid pool is not obviously related to the composition of the free amino acids in plasma or in cerbrospinal fluid (Table 10.2). Nor is there any relation to the composition of brain protein. There are no α-amino acids in the pool which do not occur in other tissues, and there are no amino acids in other tissues which are not found

Table 10.1 Abnormalities of Amino Acid Transport Having Neurological Consequences

Abnormality	Biochemical defect	Biochemical findings	Neurological findings	Genetics	Neuropathology
Blue diaper syndrome (tryptophan malabsorption syndrome)	Transport defect for tryptophan	Hypercalcemia, indole and kynurenine excretion	Growth retardation, irritability		
Cystinosis	Intracellular cystine transport	Cystine accumulation in tissues	Photophobia, retinopathy, renal insufficiency	Autosomal recessive	
Hartnup disease	Defect in intestinal and renal transport of tryptophan and other neutral amino acids	Specific amino aciduria; indican, indoleacetic acid in the urine	Cerebellar ataxia, tremors, mild retardation	Autosomal recessive	
Hyperlysinuria	Transport defect	Lysine excretion	Retardation		
Lowe's syndrome (oculocerebrorenal syndrome)	Intestinal transport defect shown	Aminoaciduria, organic aciduria	Retardation, seizures, deafness, cataracts	X-linked recessive	No consistent pathology noted; some focal atrophy of parietal lobes; cerebellar, cortical atrophy
Oasthouse urine disease (Smith-Strang disease; methionine malabsorption syndrome)	Transport defect for methionine	Methionine, phenylalanine, tyrosine, and α-hydroxybutyric acid excretion	Convulsions, retardation		

Table 10.2 Free Amino Acids of Human Brain, Plasma, and Cerebrospinal Fluid, and of Rat Brain

Amino acid	Rat brain	Human brain	Human plasma	Human cerebro-spinal fluid
		(μmol/g or μmol/ml)		
Glutamic acid	9.1	10.6	0.05	0.225
N-Acetylaspartic acid	5.6	5.7	–	–
Glutamine	4.2	4.3	0.7	0.03
γ-Aminobutyric acid	4.0	2.3	–	–
Aspartic acid	2.2	2.2	0.01	0.007
Cystathionine	0.2	1.7	–	–
Taurine	3.8	1.9	0.1	–
Glycine	1.8	1.3	0.4	0.013
Alanine	0.5	0.9	0.4	0.017
Glutathione	1.3	0.9	–	–
Serine	1.2	0.7	0.1	0.01
Threonine	1.0	0.2	0.15	0.025
Valine	0.1	0.2	0.25	0.013
Lysine	0.2	0.1	0.12	0.014
Leucine	0.2	0.1	0.15	0.004
Proline	0.1	0.1	0.1	–
Asparagine	–	0.1	0.07	–
Methionine	0.1	0.1	0.02	0.003
Isoleucine	0.05	0.1	0.1	0.008
Arginine	0.1	0.1	0.1	0.006
Cysteine	0.05	0.1	0.1	0.002
Phenylalanine	0.1	0.1	0.1	0.010
Tyrosine	0.1	0.1	0.1	0.006
Histidine	0.1	0.1	0.1	0.003
Tryptophan	0.02	0.05	0.05	0.010

Table adapted from T. L. Sourkes, *Biochemistry of Mental Disease,* New York, Harper and Row, 1962; H. H. Tallan, in *Free Amino Acid Pools* (J. T. Holden, ed.), Elsevier, Amsterdam, 1962; and K. Schreier, in *Free Amino Acid Pools,* (J. T. Holden, ed.), Elsevier, Amsterdam, 1962.

in the brain pool. There are, however, some amino acid derivatives, notably N-acetylaspartic acid, and γ-aminobutyric acid, which are in high concentration in the brain pool, but do not appear in other tissues.

Some difference in the composition of the free amino acid pools of different brain areas is known to exist [29,50]. The only generalization which can be drawn is that glutamate is the predominant amino acid of the whole brain, and of the forebrain. The midbrain and hindbrain portions have, on the other hand,

Table 10.3 Cerebral Amino Acid Content in Fetal, Newborn, and Adult Mice

Amino acid	15-day fetus	Newborn	Adult
	(µmol/g wet weight)		
Phosphoserine	0.23	0.25	—
Taurine	14.1	14.1	8.01
Phosphoethanolamine	1.91	2.35	0.91
Aspartic acid	2.35	2.22	3.75
Hydroxyproline	0.18	0.15	—
Threonine	4.28	0.93	0.56
Serine	2.11	1.03	0.98
Glutamic acid	7.54	4.81	11.7
Glutamine	3.73	5.69	5.59
Proline	0.89	0.66	0.15
Glycine	2.26	1.99	1.27
Alanine	5.08	4.29	0.56
Valine	0.56	0.39	0.10
Methionine	0.30	0.12	0.02
Isoleucine	0.28	0.16	0.03
Leucine	0.53	0.27	0.06
Tyrosine	0.24	0.22	0.08
Phenylalanine	0.24	0.16	0.07
γ-Aminobutyric acid	0.50	1.73	2.37
Ornithine	0.12	0.07	0.03
Lysine	0.86	0.91	0.29
Histidine	0.21	0.20	0.12
Arginine	0.45	0.16	0.11

Data from A. Lajtha and J. Toth, *Brain Res.* *55:*238 (1973).

a higher level of γ-aminobutyric acid than of glutamate. Spinal cord is much like the whole brain in free amino acid composition. Vertebrate peripheral nerve has lower concentrations of the glutamate-related amino acids, but the rest, on the whole, are much like the brain in composition. Since the functional significance of the free amino acid pool itself is not known, there is no information concerning the functional meaning of the differences in composition in the various parts of the nervous system.

A number of investigators have looked at the composition of the pool at various ages and stages of development [36,50,53,54]. Some amino acids go up with age and others go down (Table 10.3). In general, the glutamate family of amino acids tends to increase in concentration in the brain after birth. As might be expected, animals with fairly well-developed nervous systems at birth, such as the guinea pig, tend to have amino acid pools resembling the adult in composition. Those born at an earlier developmental stage exhibit more pronounced developmental changes. Again, the functional significance of these changes is obscure.

The composition of the pool is somewhat resistant to changes in the body state of the animal. Mild starvation or moderate dehydration do not alter the pool noticeably. More severe starvation and other stresses do markedly affect the amino acids in the brain. The pool is labile, and is responsive to such influences as chilling, exercise, anoxia, and hibernation. The individual amino acids sometimes show changes on the order of 50% or more [50]. It should be remembered, however, that the amino acid content of the brain, especially the concentrations of the glutamate-related amino acids, depends to a great extent on the state and rate of carbohydrate metabolism. Changes such as those mentioned above, notably the effect of hibernation, probably are due primarily to changes in the Krebs cycle activity of the brain. On the other hand, in many of these conditions, changes in the essential amino acid components are seen, which cannot be due to any changes in other metabolic systems.

A large amount of work has been done to investigate the effects of insulin on brain free amino acids, probably because of the severe effects of insulin imbalance on brain function. The administration of insulin does cause major changes in the pool. Glutamic acid, glutamine, γ-aminobutyric acid, alanine, and glycine concentrations drop. Aspartic acid and ammonia rise [55,56]. The explanation for the decrease in the glutamate-related amino acids is probably that glutamate is used as an alternate energy source in the insulin-induced absence of glucose [50]. The rise in ammonia can be explained by a lack in glutamic acid for the formation of glutamine. The rise in aspartic acid has been explained as due to a lowered acetate availability and, thus, a lowered requirement for oxaloacetate as acceptor for the Krebs cycle and a greater transamination of oxaloacetate to aspartate. It is interesting that upon treatment with insulin there is apparently not much change in the total amino acid content of the pool but only a shift in the various components. The convulsions associated with the administration of too much insulin have been ascribed to the increased ammonia content or, perhaps less likely, to the decreased content of γ-aminobutyric acid.

In liver damage the level of ammonia in the brain rises and coma results. Incidently perhaps, the level of brain glutamine rises, as would be expected. There are also changes in the histidine and methionine levels. But it is unlikely that these amino acid changes have any role in the onset of the coma. Changes in the amino acid content of brain have also been observed in various stages of anoxia. But the changes here, as those associated with coma, are unlikely to be fundamental in the condition. In fact, it is not too surprising that changes in amino acid levels, and perhaps in all other levels, accompany these very drastic physiological stresses.

A great deal of effort has gone into the study of the effects of drugs, espe-

cially the psychotropics and the convulsants, on brain amino acids. The obvious
aim is to find the root cause of the neurological action of these drugs. Unfortunately, no useful information has come out of the large body of such data
which is available. In fact, much of the data is inconsistent with much of the
other data. Chlorpromazine has been reported by one group to lower γ-amino-
butyric acid concentration, and by another to raise it. Reserpine, also, has given
conflicting data in several different reports. Many of the experiments done with
convulsants or with anticonvulsants show changes in the amino acid composition
of the brain pool. But these changes, also, have been inconsistent, and it seems
that the convulsions produce the changes and not the reverse. That is, changes in
the free amino acid pool of the brain are apparently an incidental consequence of
convulsions and are not involved in the production of the convulsion. Even the
levels of γ-aminobutyric acid, which were at one time thought to be intimately
involved in the initiation of convulsion, have been found in many cases to vary
independently with the onset of convulsion. The control of γ-aminobutyric acid
concentrations in the brain is discussed in detail in Chap. 11. But it is appropriate to mention here that some convulsants lower γ-aminobutyric acid concentration, and some raise it. And many convulsant drugs produce no effects on the
γ-aminobutyric acid concentration at all, although some produce changes in
other free amino acids. Thus, again, there are no consistent data relating the effects of psychotropic or convulsant drugs to changes in the content of any of the
free amino acids in the brain.

Large amounts of amino acids in the blood, either administered experimentally or occurring in certain disease states, do change the amino acid composition
of the pool, and do so in fairly predictable ways. Most of this work has been
done by loading experimental animals with phenylalanine or by looking at the
free amino acids of the brain and cerebrospinal fluid of phenylketonurics. In the
clinical state, the raised blood levels of phenylalanine produce lowered serotonin
levels. In animals, loading with phenylalanine lowers brain tryptophan and brain
tyrosine, as well as altering the concentrations of some of the related amino acids.
Such data can be readily explained on the basis of the current knowledge of the
competition of the amino acids for the transport sites discussed in the last few
paragraphs. The importance of these changes is obvious from the many reports
showing the severe neurological consequences which result from the ingestion of
diets containing an amino acid imbalance.

Compartmentation

Having indicated that there is a free amino acid pool in the brain, it is now necessary to point out that a large body of evidence shows that the amino acids probably do not exist in one pool but in many [2,57], and that these different pools
may very well have functional significance for the performance and separation of

various routes of metabolism. It is important to indicate that these pools must exist to explain data obtained in tracer experiments, but so far there is almost no information on the anatomical or subcellular location of the pools. The original observations were those of Berl et al. [58], which involved the measurement of the specific activities of the glutamate and glutamine of brain after the intracisternal administration of glutamate to the animal. A substantial deviation from the expected precursor-product relationship between brain glutamate and brain glutamine was observed. That is, within a very few minutes, the brain glutamine specific activity exceeded the brain glutamate specific activity by fourfold to fivefold (Fig. 10.2); the peak specific activity of the glutamine was higher than the highest specific activity of the presumed precursor, brain glutamate. Only the separation of the metabolism in compartments could account for such data. That is, the injected glutamate must have been converted to glutamine in some small active pool in the brain which was diluted by the larger inactive pool of brain glutamate when the tissue was prepared for analysis.

Most of the experiments on compartmentation in the brain have been done with the glutamate-glutamine system. Infusion experiments using labeled ammo-

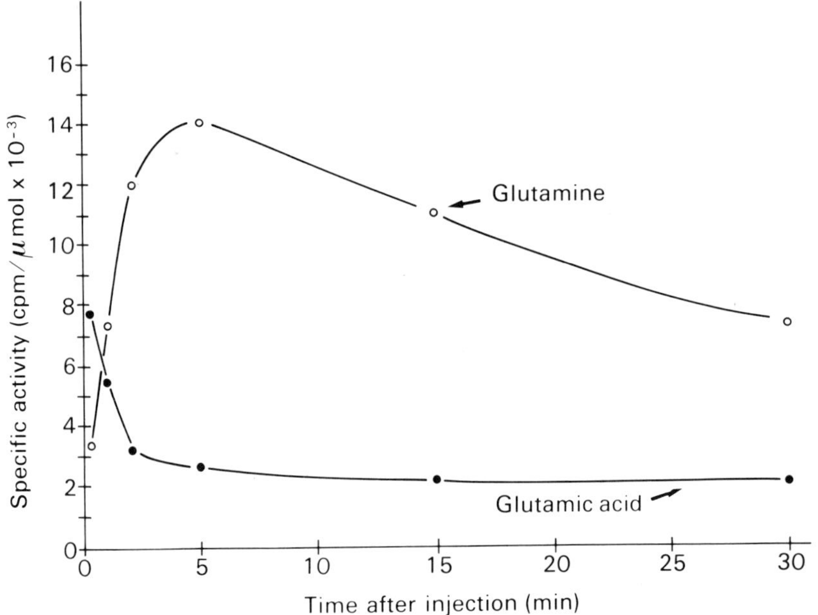

Figure 10.2   Cerebral glutamic acid metabolism after intracisternal administration of [^{14}C] glutamic acid. [Data adapted from S. Berl, A. Lajtha, and H. Waelsch, *J. Neurochem.* 7:186 (1961).]

nium acetate or bicarbonate have also indicated a compartmentation of this metabolism [59-63]. Upon infusion of ammonium acetate the α-amino group of glutamine was more highly labeled than that of glutamate in brain but not in liver. The use of the infusion technique avoided the criticism that the initial specific activity of glutamate was, in fact, very high, but dropped to lower values during the experiment.

Now a great deal of information is available on a large number of precursors. In addition to glutamate, ammonium acetate, and bicarbonate, the same relationship between glutamine and glutamate, i.e., higher brain glutamine specific activity than glutamate specific activity, is seen upon administration of acetate, citrate, butyrate, propionate, and leucine. The opposite is true if glucose, lactate, or glycerol are given. Thus, the ketogenic precursors show one pattern and the glycogenic precursors show another. The interpretation of these data is that there are large and small pools of glutamate metabolism. The small one is concerned with the conversion of glutamate to glutamine, detoxification, the synthesis of the transmitter, γ-aminobutyric acid, and perhaps with the synthesis of protein. The large pool is involved in energy metabolism. The small pool is permeable to acetate, butyrate, propionate, and the ketogenic substrates but not to the glycogenic substrates, which preferentially enter the large pool.

The anatomical site of these pools has been considered, but without much success. The endoplasmic reticulum is a prime candidate since the small pool must be at the site of glutamine synthesis and the endoplasmic reticulum has a remarkedly high content of glutamine synthetase. On the other hand, the developmental course of the glutamine synthetase activity of the brain does not correlate with the appearance of compartmentation in the animal [64]. The different metabolic roles for the compartments, glucogenic versus ketogenic, suggests that the pools might be contained in different types of mitochondria. In support of this concept is the observation that mitochondria are, in fact, heterogeneous [65] and that there are probably also two pools of α-ketoglutaric acid in brain [62,63, 66], and separate Krebs cycles. But the heterogeneity of mitochondrial population is not unique to brain, although compartmentation seems to be, and the mitochondrial location of these pools is not proved. The pools almost certainly do not reside in different cell types because all cell types must be permeable to ammonium acetate or to bicarbonate, and there is no evidence that the different cell types have disproportionate concentrations of glutamate or different glutamate-forming capacity.

A few other points with regard to compartmentation are that the compartments develop with the age of the animal but are probably not present at birth [67,68]. The compartments can be demonstrated in vitro using properly prepared slices [69] but so far not with homogenates. In such preparations the compartmentation of the metabolism of glutamate can be abolished with certain inhibitors which inhibit one or the other process more or less selectively. Thus, the compartmentation of metabolism which has been invoked by biochemists for years to explain results which did not fit expected relationships, is tractable in brain both mathematically [70,71] and experimentally.

References

1. K. D. Neame, in *Applied Neurochemistry* (A. N. Davison and J. Dobbing, eds.), F.A. Davis, Philadelphia, 1968.

2. G. Guroff, in *Basic Neurochemistry* (R. W. Albers, G. J. Siegel, R. Katzman, and B. W. Agranoff, eds.), Little, Brown, Boston, 1972.

3. S. R. Cohen and A. Lajtha, in *Handbook of Neurochemistry*, Vol. VII (A. Lajtha, ed.), Plenum, New York, 1972.

4. R. Blasberg and A. Lajtha, *Brain Res. 1:*86 (1966).

5. R. G. Blasberg, in *Progress in Brain Research,* Vol. 29 (A. Lajtha and D. H. Ford, eds.), Elsevier, New York, 1968.

6. Y. Tsukada, Y. Nagata, S. Hirano, and T. Matsutani, *J. Neurochem. 10:*241 (1963).

7. M. A. Chirigos, P. Greengard, and S. Udenfriend, *J. Biol. Chem. 235:*2075 (1960).

8. G. Guroff, W. King, and S. Udenfriend, *J. Biol. Chem. 236:*1173 (1961).

9. G. Guroff and S. Udenfriend, *J. Biol. Chem. 237:*803 (1962).

10. P. N. Abadom and P. G. Scholefield, *Can. J. Biochem. Physiol. 40:*1603 (1962).

11. R. Blasberg and A. Lajtha, *Arch. Biochem. Biophys. 112:*361 (1965).

12. G. Levi, R. Blasberg, and A. Lajtha, *Arch. Biochem. Biophys. 114:*339 (1966).

13. G. Levi, A. Cherayil, and A. Lajtha, *J. Neurochem. 12:*757 (1965).

14. A. Cherayil, J. Kandera, and A. Lajtha, *J. Neurochem. 14:*105 (1967).

15. A. Lajtha and J. Toth, *J. Neurochem. 8:*216 (1961).

16. A. Lajtha and J. Toth, *J. Neurochem. 9:*199 (1962).

17. A. Lajtha and J. Toth, *J. Neurochem. 10:*909 (1963).

18. P. Schwerin, S. P. Bessman, and H. Waelsch, *J. Biol. Chem. 184:*37 (1950).

19. R. W. P. Cutler, *J. Neurochem. 17:*1017 (1970).

20. R. W. P. Cutler and A. V. Lorenzo, *Science 161:*1363 (1968).

21. K. D. Neame and S. E. Smith, *J. Neurochem. 12:*87 (1965).

22. A. Lajtha and P. Mela, *J. Neurochem. 7:*210 (1961).

23. L. Battistin, A. Grynbaum, and A. Lajtha, *Brain Res. 16:*187 (1969).

24. M. Banay-Schwartz, L. Piro, and A. Lajtha, *Arch. Biochem. Biophys. 145:*199 (1971).

25. O. Gonada and J. H. Quastel, *Biochem. J. 84:*394 (1962).

26. H. Yoshida, K. Kanike, and J. Namba, *Nature (London) 198:*191 (1963).

27. L. Battistin, A. Grynbaum, and A. Lajtha, *J. Neurochem. 16:*1459 (1969).

28. L. Battistin and A. Lajtha, *J. Neurol. Sci. 10:*313 (1970).

29. J. Kandera, G. Levi, and A. Lajtha, *Arch. Biochem. Biophys. 126:*249 (1968).

30.   S. Navon and A. Lajtha, *Biochim. Biophys. Acta 173:*518 (1969).
31.   S. H. Appel, L. Autilio, R. W. Festoff, and A. V. Escueta, *J. Biol. Chem. 244:*3166 (1969).
32.   W. A. Himwich, J. C. Peterson, and M. L. Allen, *Neurology 7:*705 (1957).
33.   G. Guroff and S. Udenfriend, in *Progress in Brain Research,* Vol. 9 (W. A. Himwich and H. Himwich, eds.), Elsevier, New York, 1964.
34.   G. Levi, *Arch. Biochem. Biophys. 138:*348 (1970).
35.   F. Piccoli, A. Grynbaum, and A. Lajtha, *J. Neurochem. 18:*1135 (1971).
36.   A. Lajtha and F. Piccoli, in *Cellular Aspects of Neural Growth and Differentiation,* Vol. 14 (D. C. Pease, ed.), UCLA Forum Med. Sci., University of California, Berkeley, 1971.
37.   A. Meister in *Brain Dysfunction in Metabolic Disorders* (F. Plum, ed.), Raven, New York, 1974.
38.   H. E. Wiltse and J. H. Menkes, in *Handbook of Neurochemistry,* Vol. VII (A. Lajtha, ed.), Plenum, New York, 1972.
39.   Y. E. Hsia, in *Basic Neurochemistry* (R. W. Albers, G. J. Siegel, R. Katzman, and B. W. Agranoff, eds.), Little, Brown, Boston, 1972.
40.   D. N. Baron, C. E. Dent, H. Harris, E. W. Hart, and J. B. Jepson, *Lancet 2:* 421 (1956).
41.   C. R. Scriver, *N. Engl. J. Med. 273:*530 (1965).
42.   L. B. Holmes, H. W. Moser, S. Halldorsson, C. Mack, S. S. Pant, and B. Matzilevich, *Mental Retardation,* MacMillan, New York, 1972.
43.   K. O. Raivio and J. E. Seegmiller, *Ann. Rev. Biochem. 41:*543 (1972).
44.   J. A. Schneider, K. Bradley, and J. E. Seegmiller, *Science 157:*1321 (1967).
45.   J. D. Schulman, K. H. Bradley, and J. E. Seegmiller, *Science 166:*1152 (1969).
46.   C. U. Lowe, M. Terrey, and E. A. MacLachlan, *Am. J. Dis. Child. 83:*164 (1952).
47.   V. Abbassi, C. U. Lowe, and P. L. Calcagno, *Am. J. Dis. Child. 115:*145 (1968).
48.   C. S. Bartsocas, H. L. Levy, J. D. Crawford, and S. O. Their, *Am. J. Dis. Child. 117:*93 (1969).
49.   H. H. Tallan, in *Free Amino Acid Pools* (J. T. Holden, ed.), Elsevier, Amsterdam, 1962.
50.   W. A. Himwich and H. C. Agrawal, in *Handbook of Neurochemistry,* Vol. I (A. Lajtha, ed.), Plenum, New York, 1969.
51.   T. L. Sourkes, *Biochemistry of Mental Disease,* Harper and Row, New York, 1962.
52.   K. Schreier, in *Free Amino Acid Pools* (J. T. Holden, ed.), Elsevier, Amsterdam, 1962.
53.   G. Levi, J. Kandera, and A. Lajtha, *Arch. Biochem. Biophys. 119:*303 (1967).
54.   A. Lajtha and J. Toth, *Brain Res. 55:*238 (1973).
55.   R. M. C. Dawson, *Biochem. J. 47:*386 (1950).
56.   R. O. Cravioto, G. Massieu, and J. J. Izquierdo, *Proc. Soc. Exp. Biol. Med. 78:*856 (1951).

57. S. Berl and D. D. Clarke, in *Handbook of Neurochemistry,* Vol. I (A. Lajtha, ed.), Plenum, New York, 1969.
58. S. Berl, A. Lajtha, and H. Waelsch, *J. Neurochem. 7:*186 (1961).
59. G. Takagaki, S. Berl, D. D. Clarke, D. P. Purpura, and H. Waelsch, *Nature (London) 189:*326 (1961).
60. S. Berl, G. Takagaki, D. D. Clarke, and H. Waelsch, *J. Biol. Chem. 237:* 2562 (1962).
61. S. Berl, G. Takagaki, D. D. Clarke, and H. Waelsch, *J. Biol. Chem. 237:* 2570 (1962).
62. C. A. Rossi, S. Berl, D. D. Clarke, D. P. Purpura, and H. Waelsch, *Life Sci. 10:*533 (1962).
63. H. Waelsch, S. Berl, C. A. Rossi, D. D. Clarke, and D. P. Purpura, *J. Neurochem. 11:*717 (1964).
64. S. Berl, *Biochemistry 5:*916 (1966).
65. C. van den Berg, Lj Krzalic, and P. Mela, *First Int. Meeting Soc. Neurochem.,* Strasbourg, July 23-28, 1967, Pergamon, Oxford, Eng., 1969, p. 21.
66. M. K. Gaitonde, *Biochem. J. 95:*803 (1965).
67. S. Berl, *J. Biol. Chem. 240:*2047 (1965).
68. S. Berl and D. P. Purpura, *J. Neurochem. 13:*293 (1966).
69. S. Berl, W. J. Nicklas, and D. D. Clarke, *J. Neurochem. 15:*131 (1968).
70. D. Garfinkel, *J. Theoret. Biol. 3:*412 (1962).
71. D. Garfinkel, *J. Biol. Chem. 241:*3918 (1966).

11
AMINES AND OTHER
AMINO ACID DERIVATIVES

N-Acetylaspartic Acid

One of the most abundant components of the free amino acid pool of the brain
is N-acetylaspartic acid (Fig. 11.1). Its concentration in most species is 2 or 3
times higher than that of aspartic acid itself, and higher than any other amino
acid or amino acid derivative except glutamate. It is found in a concentration of
5 to 6 μmol/g in adult brain [1], and only in traces in nonneural tissue. It is
much lower in the brains of newborn animals, and rises to an adult level around
the time of myelination [2]. It is found in twofold higher concentration in grey
matter than in white [2,3], and recent studies using tumor tissue indicate a
neuronal localization for the derivative [4]. These experiments also showed that
N-acetylaspartic acid could be found in peripheral nerve and in retina, as well as
in the central nervous system.

N-acetylaspartic acid is formed biosynthetically from aspartic acid and
acetyl coenzyme A (CoA) [5,6], and the enzyme catalyzing this reaction has been
purified and studied. A number of workers have found that the formation of N-
acetylaspartic acid from its precursors in vivo is quite slow [5,7-10] and, thus,
have suggested that the turnover of the compound was also quite slow. More recent work in which labeled N-acetylaspartic acid was administered directly to
brain has indicated that it is hydrolyzed very rapidly [11]. Also it is known that
the enzyme involved in its catabolism is very active in brain [12,13]. This contradiction, i.e., slow biosynthesis and fast breakdown, has been explained on
the basis that there are separate pools of the compound [11], a large one in
which a slow biosynthesis is carried out, and a small, active one in which a rapid
catabolism of exogenously added N-acetylaspartic acid is handled.

The exact function of this compound is unknown. Suggestions have included: (1) a role as part of the intracellular "fixed" anion pool, (2) a reservoir
for acetyl groups, and (3) a potential source of N-acetylated end groups for certain brain proteins. The more recent finding of rapid catabolism has stimulated
interest in the function, and the observation that the acetyl group of exogenous

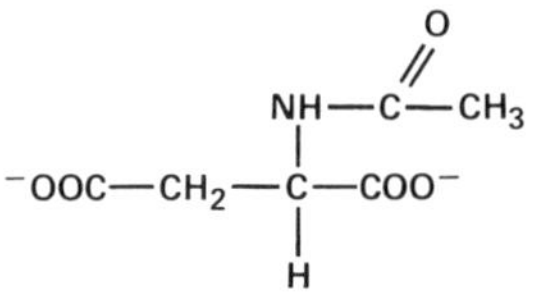

$$\begin{array}{c} \quad\quad\quad\quad O \\ \quad\quad\quad\quad \parallel \\ NH-C-CH_3 \\ \quad\quad | \\ {}^-OOC-CH_2-C-COO^- \\ \quad\quad\quad | \\ \quad\quad\quad H \end{array}$$

Figure 11.1   Structure of N-acetylaspartate.

N-acetylaspartic acid serves as an excellent and even preferential source of carbon for fatty acid synthesis in developing brain [14,15] favors an active and not a passive role for the compound.

Putreanine

Putreanine, N-(4-aminobutyl)-3-aminopropionic acid (Fig. 11.2), has been shown to be a component of the central nervous system of rats, birds, and humans. It has been isolated and characterized from bovine brain [16]. This derivative is present in small amounts in liver, but is not found in other tissues. Its concentration in brain is comparable to the concentrations of the other amino acids which are present in this tissue. Putreanine is found in highest concentration in cerebellar gray matter, but it is present in all regions of the human brain which have been examined [17]. Its concentration in rat, rabbit, and bovine brain is about equal. In rats, the amino acid appears 2 weeks after birth, and its concentration increases for several months. In humans, on the other hand, the concentration of putreanine in the parietal cortex is not substantially different between 3 days and 80 years of life. It has been suggested that putreanine is formed biosynthetically from spermidine, and experimental data are available to indicate that the conversion occurs in brain and liver catalyzed by an amine oxidase [18].

γ-Aminobutyric Acid

As mentioned before, γ-aminobutyric acid (GABA) is one of the major components of the free amino acid pool of the brain [19,20]. It is the product of the α-decarboxylation of glutamic acid and is present in the brains of various species at a level of several μmoles per gram of wet weight. It is not found in other tissues. GABA was discovered as a constituent of the brain by Awapara and by Roberts in 1950 [21,22], but no information about its function was available until many years later when it was found to be a major component of so-called

$${}^+NH_3-CH_2-CH_2-CH_2-CH_2-NH-CH_2-CH_2-COO^-$$

Figure 11.2   Structure of putreanine.

factor I preparations which inhibited impulse generation in crayfish stretch receptor [23]. Since that time a great deal of work has been done on the various aspects of the biochemistry of GABA [19,20].

It is now firmly established that GABA is the transmitter of the inhibitory fibers at the neuromuscular junction of *Crustacea*. This evidence includes the presynaptic release of GABA, in appropriate amounts, upon stimulation of inhibitory neurons [24], and the mimicking of the normal events by the iontophoretic application of GABA in the same amounts [25]. Evidence for a similar function of GABA in vertebrate systems is quite strong, but, as yet, circumstantial. Iontophoretic experiments are consistent with an inhibitory function in vertebrates, as is the localization of the compound, but the crucial studies on the presynaptic release of GABA have not yet been possible in the appropriate mammalian system. On the basis of this indirect evidence, however, it is widely assumed that GABA is the transmitter substance of the basket cells of the cerebellum which synapse with the Purkinje cells. Measurements of GABA and of the enzymes involved in its metabolism in cellular layers of the cerebellum support this assumption [26,27]. The action of GABA in this system is the best characterized of the mammalian models, but there is also strong evidence for the action of GABA in other parts of the central nervous system, notably the retina, the hippocampus, the cortex, and the spinal cord [20].

The information available about the mechanism of GABA as transmitter are discussed in detail in Chap. 24. Suffice it to say here that a number of experiments indicate a connection between GABA and Cl^- permeability. Iontophoretic application of GABA in the crustacean system leads to an increased permeability to Cl^- [28,29], and the action of GABA in vertebrate neurons can be reversed by the diffusion of Cl^- and abolished by the omission of Cl^-.

In addition to its probable function as the transmitter substance in inhibitory neurons, GABA has been implicated as an effector in several other systems [19]. Among these are a possible generalized inhibitory function, a role in osmotic regulation in frogs, and a homeostatic function. Its possible role in protein synthesis in the brain is mentioned in Chap. 15. GABA has also been reported to stimulate glucose uptake, influence the permeability of the mitochondrial membrane, enhance glycolysis, and promote evolution of ammonia from brain proteins [19].

The overall metabolism of GABA in the brain is known as the GABA shunt (Fig. 11.3). It is so-called because it circumvents a portion of the Krebs cycle, namely the oxidation of α-ketoglutarate to succinate. Various experiments have given different estimates of the amount of α-ketoglutarate which is metabolized by the shunt compared to that metabolized by the Krebs cycle enzymes. Estimates as low as 4% and as high as 44% can be found. It is likely that, overall, the amount of such metabolism is around 8% [30], but it is clear that there is a great variation in the metabolism of GABA from one brain area to another. It should

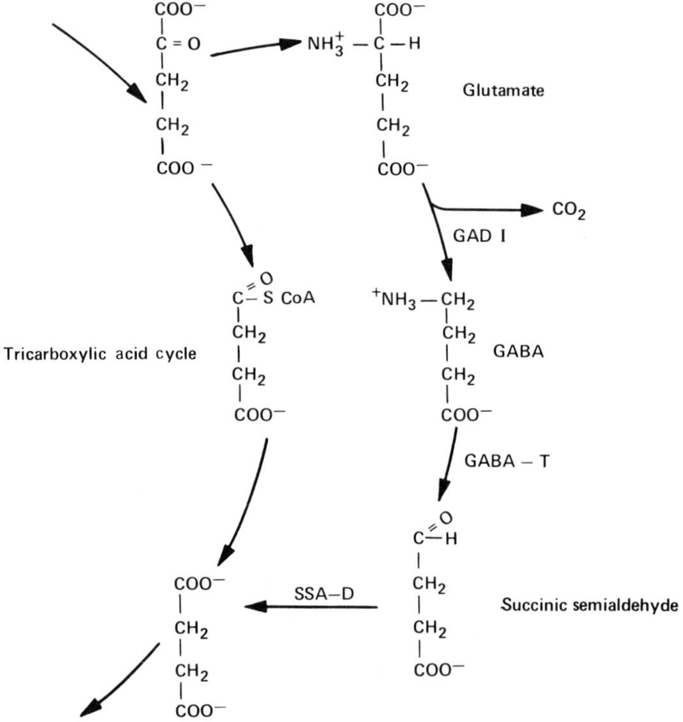

Figure 11.3   The GABA shunt.

be noted that the conversion of α-ketoglutarate to succinate via the Krebs cycle
involves the production of four molecules of ATP, one by the substrate level
phosphorylation catalyzed by succinic thiokinase, and three from the oxidation
of NADH from the α-ketoglutarate dehydrogenase reaction. The participation
of acetyl CoA is also required. The GABA shunt metabolism produces just three
ATP molecules from the oxidation of the NADH formed during the conversion
of succinic semialdehyde to succinate. But the energy yield of this pathway may
be incidental since the function is surely to produce GABA.

The individual enzymes in the shunt have been studied vigorously. The enzyme which forms GABA, glutamic acid decarboxylase (GAD I), is present only
in the central nervous system, and mostly in the gray matter. It is a pyridoxal
phosphate-requiring enzyme [21,22,31], as are most of the amino acid decarboxylases known. The cofactor is rather loosely bound to the enzyme. The enzyme is competitively inhibited by halogen ions, and, of course, by carbonyl trap-

ping agents. The enzyme resisted extensive purification for many years, becoming extremely unstable upon loss of its cofactor. Recently, however, the enzyme has been purified about 700-fold from the crude mitochondrial fraction of brain and obtained in homogeneous form [32]. It has a molecular weight of 85,000, a K_M for glutamate of about 0.7 mM, and a K_M for pyridoxal phosphate of 0.05 mM. It appears to have a subunit structure, and the smallest unit is about 15,000 and forms aggregates under physiological conditions [33]. The enzyme is almost completely specific for its substrate, glutamic acid, but it has a slight activity against aspartic acid, also. This activity is reflected in the observation that aspartic acid is a competitive inhibitor of the action of GAD I on glutamic acid [32].

It seems likely that the levels of GABA in the various parts of the central nervous system, and the changes in these levels under normal conditions, are regulated by the action of GAD I and not by the actions of the enzymes which destroy GABA. For example, exposure to high O_2 pressure produces lowered brain GABA and, eventually, convulsions in animals. Under these conditions GAD I is inhibited, but GABA-T activity is unchanged [34,35]. The regional distribution of GABA and of GAD I are quite parallel, and GABA levels and GAD I levels increase together during the early stages of the growth of the central nervous system. GAD I is localized to a large extent in the synaptosomes [36,37], but is a cytoplasmic enzyme, not membrane bound, since it is released from the synaptosomes by osmotic shock [36]. GABA, also, is enriched in the synaptosomes, and is also released by osmotic shock.

There are suggestions in the literature that the synthesis of GAD I is subject to repression and derepression under a variety of circumstances. One of the most interesting examples is the report that GAD I, and GABA as well, are lower in dark-adapted frog retina, while the levels of GABA transaminase and of glutamate are unchanged [38]. It is clear that GAD apoenzyme can be induced by the administration of glutamic acid [39]. Changes in GAD I activity can also be caused by lack of cofactor such as those which occur in B_6 deficiency in adult animals [40,41]. Recent data suggest that, in developing rat but not the adult, B_6 deficiency causes an increase in the apoenzyme as well [42], perhaps as some kind of compensatory mechanism. Also, it is now apparent that GABA itself represses the synthesis of the GAD I apoenzyme in some systems [43]. The complete story on the inductive changes in the synthesis of this important enzyme is only beginning to take shape.

Within recent years a second enzyme decarboxylating glutamate, and producing GABA, has been found (GAD II) [44-46]. This enzyme is quite different from GAD I. It is found in kidney and blood vessels, and occurs in the brain, but in the glial compartment. It is activated by carbonyl trapping agents, rather than being inhibited by them, and it requires anions for its action, again instead of being inhibited by them. The function of this second enzyme is not known, but it seems to be a real enzyme in spite of recent questions about its existence [47,48].

The location of the enzymes which catabolize GABA is quite different from that of GAD I or GABA. γ-Aminobutyric acid transaminase (GABA-T) is present in gray matter of the brain, but also occurs in other tissues. As do most transaminases, GABA-T requires pyridoxal phosphate as cofactor [49-51]. Thus, it is inhibited by inhibitors of the pyridoxal function, the carbonyl trapping agents, such as hydroxylamine, and also by aminooxyacetic acid. Although these inhibitors inhibit both GABA-T and GAD I in vitro, only GABA-T is affected in vivo, and thus GABA levels rise. As mentioned above, normally GAD I activity appears to control GABA level, but when GABA-T activity is completely or strongly inhibited, of course, GABA levels rise since it cannot be metabolized in any other way. GABA-T is found in different parts of the cell from either GAD I or GABA itself. GABA-T is a mitochondrial enzyme and is not localized in the synaptosomes. In fact, the synaptosomal mitochondria seem relatively poorer in this enzyme than do the rest of the mitochondria of the cell. GABA-T binds its pyridoxal phosphate cofactor quite tightly compared to the rather loose binding exhibited by GAD I. The enzyme is particulate and is present in many tissues. The enzyme in rats exists as isozymes, has a pH optimum of around 8.6, and is inhibited by 0.32 M sucrose. It is quite specific for the α-ketoglutaric acid acceptor, but not for the amino donor, since β-alanine can be transaminated almost as well as GABA. The K_M values for the substrates are quite high.

The enzyme which converts succinic semialdehyde to succinic acid, succinic semialdehyde dehydrogenase (SSA-D) [52-55], is distributed in the central nervous system, and in the cell, in the same pattern as is GABA-T. It is a mitochondrial enzyme which is quite specific for succinic semialdehyde and for NAD^+. It is activated by sulfhydryl compounds and exhibits substrate inhibition at substrate concentrations above 10^{-4} M. The enzyme has been purified some 250-fold and has been shown to have a sulfhydryl group essential for its action. It is inhibited by compounds which react with sulfhydryl groups. It seems to be activated by Na^+ and by K^+, and also by heat, so it is generally assayed at 44°C.

The localization of these enzymes, and of GABA itself, leads to a consistent picture of the way in which GABA acts. GAD I is synthesized in the perikaryon, presumably, of the presynaptic cell, and is transported by axonal flow to the synaptic bouton where GABA is synthesized. GABA is then released from the presynaptic cell across the synapse and acts on the postsynaptic membrane.

The action of GABA as transmitter appears, however, to be terminated, not by metabolism, but by reuptake into the presynaptic terminal. The mitochondrial localization of GABA-T, as opposed to a postsynaptic site, argues for such a mechanism. Early studies showed that a substantial portion of the GABA of the cell, and indeed of the first factor I preparations [56], was in a bound form. More formal studies of GABA binding to membrane sites [57-62] have shown that the process is highly specific, very active, and requires a high concentration of sodium ion, on the order of 0.1 M. The binding occurs, in a sodium-dependent mode, to all kinds of membranes, mitochondrial, synaptosomal, and so on. It is inhibited by compounds which react with sulfhydryl groups, as well as by certain drugs, such as chlorpromazine and imipramine. The binding is presumably to specific GABA-binding membrane sites which are involved in the transport of GABA back into the presynaptic cell. The sites are thought to be mobile within the membrane, and the binding is thought to occur outside, where there is a high concentration of Na^+. When the carrier moves in there is a dissociation due to the low concentration of Na^+ inside the cell. The energy for the movement of the carrier is not provided by ATP directly, but by the Na^+ gradient, which is itself dependent on ATP for its maintenance. The binding is seen to terminate the action of GABA by removing it from the synaptic cleft, and, presumably making it available for metabolism or for reuse.

The development of the GABA system, that is, the GABA-forming and the GABA-metabolizing enzymes, and the GABA content itself, has been the subject of several investigations [19]. Without going into the details too deeply, it can be stated that the levels of GABA and of GAD I develop in the brain together. The developmental courses of GAD I and of GABA-T are also approximately parallel. In the whole system, only the enzyme SSA-D seems to have a developmental course different from that of the other components.

The fact that the enzyme which makes GABA and the enzyme which destroys it both require pyridoxal phosphate makes the pharmacology of GABA levels somewhat complex. It has already been mentioned that the two enzymes are quite different in their affinity for the pyridoxal phosphate cofactor. GAD I binds pyridoxal much less tightly than does GABA-T. On the face of it, then, GAD I should be much more sensitive to such pyridoxal enzyme inhibitors as carbonyl trapping agents. This is, by and large, true when the enzymes are studied in vitro. In the intact animal, however, things are much more complicated. In the first place, it is likely that the mechanism of inhibition is not so simple, since other, more general enzymes such as pyridoxal kinase may be even more sensitive to such inhibitors and the general pool of pyridoxal phosphate may be affected rather than just the cofactor for these enzymes. And, in the second place, some pyridoxal trapping agents clearly raise GABA levels and inhibit the transaminase (GABA-T) more than they inhibit GAD I. The explanation here seems to be that, in many cases, the carbonyl trapping agent itself is not the true inhibitor, but the pyridoxal phosphate-carbonyl reagent complex is more active. Thus, in vivo GABA-T may be the more susceptible since it binds the cofactor, and hence the inhibitor, more tightly, while GAD I can release the cofactor-inhibitor complex more easily and be reactivated by another molecule of pyridoxal phosphate. In any case, the effect of the various pyridoxal phosphate inhibitors on the GABA enzymes and hence on GABA level is not consistent with their actions in vitro

a.

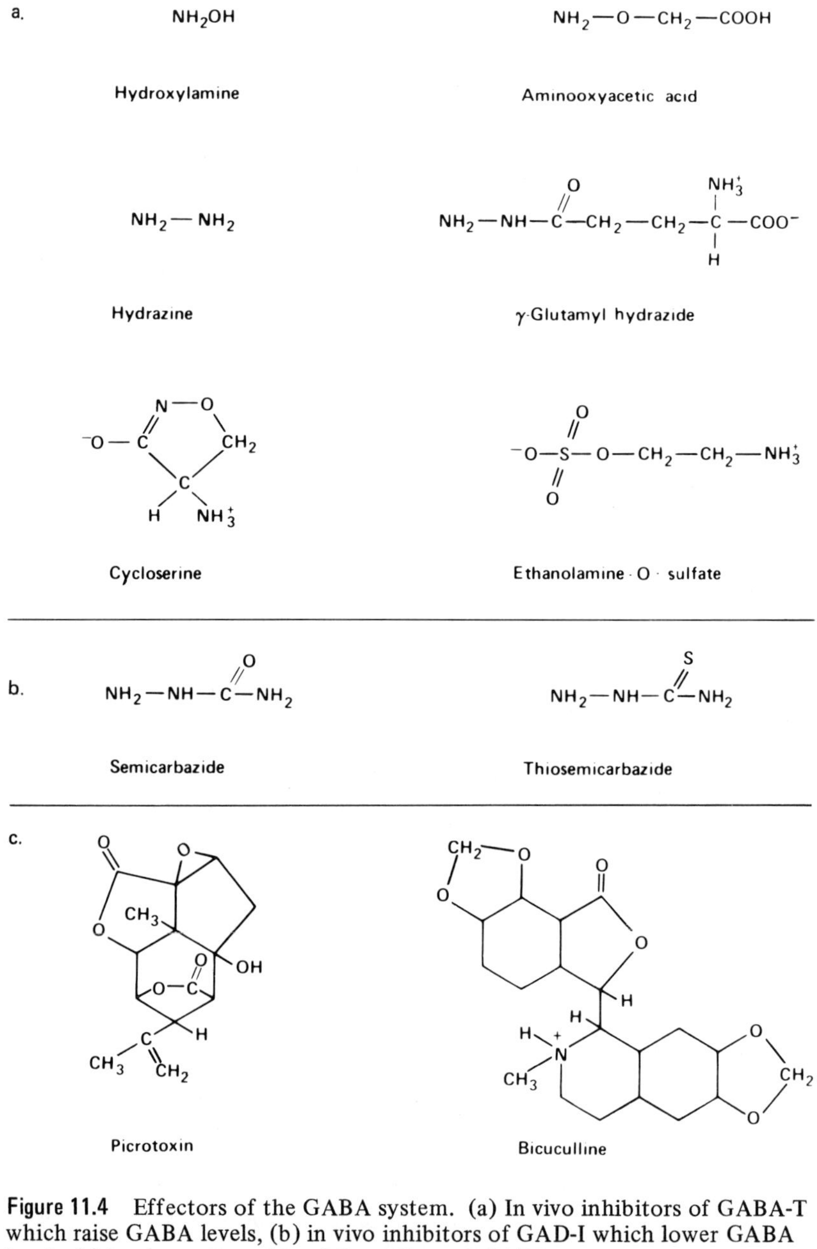

NH_2OH

$NH_2{-}O{-}CH_2{-}COOH$

Hydroxylamine

Aminooxyacetic acid

$NH_2{-}NH_2$

Hydrazine

γ-Glutamyl hydrazide

Cycloserine

Ethanolamine · O · sulfate

b.

Semicarbazide

Thiosemicarbazide

c.

Picrotoxin

Bicuculline

Figure 11.4 Effectors of the GABA system. (a) In vivo inhibitors of GABA-T which raise GABA levels, (b) in vivo inhibitors of GAD-I which lower GABA levels, (c) in vivo antagonists of the actions of GABA.

and must be evaluated in each case. A number of the most useful carbonyl reagents for the in vivo inhibition of GABA-T and, consequently, the elevation of
GABA levels are shown in Fig. 11.4. A few carbonyl reagents lower GABA levels,
presumably by inhibiting GAD I. The two most prominent are semicarbazide
and thiosemicarbazide (Fig. 11.4), the latter, incidentally, preventing the marked
elevation of GABA which occurs postmortem.

There are, in addition, a number of pharmacological tools for altering
GABA levels and actions which do not seem to be related to the pyridoxal function. Among these should be mentioned a number of GABA congeners which
have been tried as substrate analogs. Such compounds are expected to block the
action of GABA-T and thus raise GABA levels. Among the many which give substantial rise in GABA levels, one of the most interesting is ethanolamine-O-sulfate
which, for unknown reasons, causes irreversible in vivo inhibition of GABA-T
[63,64] (Fig. 11.4). Then there are some compounds which seem to block the
action of GABA directly. Picrotoxin, for example, blocks the action of GABA
on the crustacean muscle and in certain cells of the vertebrate central nervous
system. More recently the phthalide-isoquinoline alkaloid bicuculline has been
seen to block GABA actions [65,66] and appears to be considerably more specific than picrotoxin.

The observation that GABA is an inhibitory agent in nervous transmission
made experiments relating GABA levels to convulsive states very attractive. Thus,
measurements of GABA levels during both pharmacologically induced and natural convulsive states have been performed in many laboratories. In spite of early
reports to the contrary, pharmacological studies have now shown quite clearly
that there is no correlation between the production of convulsions with various
drugs and the effect of these drugs on GABA levels [19]. For example, semicarbazide lowers GABA concentrations and produces convulsions. Hydrazine
raises GABA concentrations and produces convulsions. Thiosemicarbazide produces convulsions and lowers GABA levels, but produces convulsions even if the
GABA concentration is raised above normal by the simultaneous administration
of hydroxylamine. And finally, many convulsant drugs have no effect whatsoever on the GABA concentration. So, there is no consistent data relating drug-
induced convulsions with the overall level of GABA. Indeed, there are no data
to suggest that the drug-induced alteration of GABA concentrations produce any
observable alteration in the behavior or condition of patients or animals. It has
been suggested that the concentration of GABA in some small pool in the brain
might be correlated with the appearance of the convulsive state, but there is no
evidence to support this suggestion. It seems equally possible that GABA levels
have nothing to do with any drug-induced convulsions.

There is, on the other hand, pretty good evidence to indicate that alterations in GABA levels caused by changes in normal body state or conditions can
be correlated with the level of brain excitability and with convulsions [19].

That is, changes occur in GABA which are compatible with its suspected inhibitory function. The general observation has been that in conditions associated with physiological depression or sedation, GABA levels are high. Conversely, in conditions involving hyperexcitement and convulsions, GABA levels are low. Some specific examples are the effects of exposure to high CO_2 or high O_2. In the former, GABA levels are up and the subject is refractory to convulsion. In the latter, GABA levels are down, due to the inhibition of GAD I, and the subject becomes susceptible to convulsions. In hibernation, also, the correlation seems understandable. During hibernation GABA levels rise due to a rise in the synthesis of the GAD I apoenzyme. Of course, during hibernation body temperature drops, metabolic rate drops, and the animal is in a state of torpor. The very interesting connection between GABA and normal sleep has been explored to some extent, but so far no relationship has been found. There are, however, suggestions in the literature that GABA levels can be related to certain patterns of social behavior [19].

The pharmacological actions of GABA itself have been studied, again in connection with its possible anticonvulsant properties in mind. Peripheral administration of GABA to adult animals has been shown to have little central effect, most certainly due to the low permeability of the adult central nervous system to GABA. This low degree of penetration has been demonstrated directly by administration of radioactive GABA which is excluded from all parts of the brain except the olfactory bulb and the midbrain tegmentum. It has been further shown that the permeability to GABA is increased in the adult by damage to the blood-brain barrier. In young animals, where the barrier to the penetration of GABA is not as complete, the peripheral administration of GABA causes ataxia, incoordination, and depression. Changes also occur in the electroencephalogram (EEG), and these changes are related in some way to stimulation of the retina since they are initiated by light and are accompanied by changes in the electroretinogram. The primary peripheral effect of GABA administered to adult animals is hypotension.

There are a number of metabolites of GABA known [19], as well as a number of GABA peptides which are discussed in Chap. 12. Most of these metabolites are well established, but controversy has arisen about the natural occurrence of a couple of them. Among the well-established metabolites are γ-guanidobutyric acid, which is produced by the transamidination of GABA, with arginine as donor (Fig. 11.5) [67,68]. This reaction can be observed in brain, and the derivative is found in brain, but the reaction is not specific to GABA, glycine being much the better substrate, and it is not specific to brain tissue. The γ-guanido compound can be split to GABA and urea by an enzyme which is not identical with arginase. The function of the compound is not known, but it mimics GABA at the crustacean synapse [69]. On the other hand, it is different from GABA in that it is excitatory in the mammalian cortex and causes convulsions when ad-

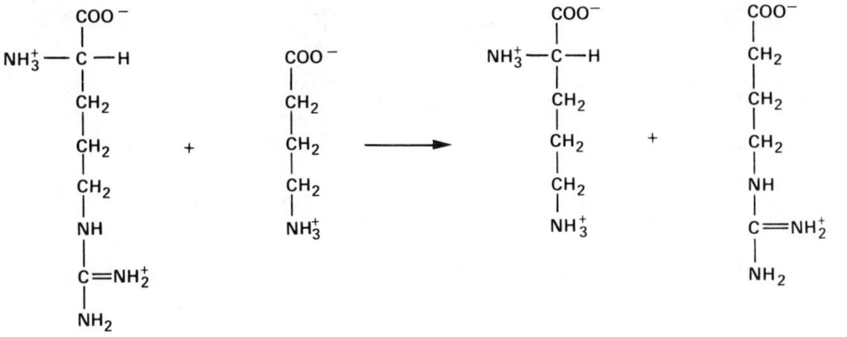

$$\text{Arginine} + \gamma\text{-Aminobutyrate} \longrightarrow \text{Ornithine} + \gamma\text{-Guanidobutyrate}$$

Figure 11.5   Formation of γ-guanidobutyrate.

ministered intracisternally to rabbits [70,71]. The compound has been found
at epileptogenic foci in the human cerebral cortex [72].

Other GABA metabolites which are known to occur include γ-butyrobe-
taine and carnitine (Fig. 11.6). These arise by the methylation and hydroxylation of GABA. They are both present in brain but are not unique to brain, and
carnitine is, in fact, in higher concentration in liver and muscle than in brain [73-
77]. Other derivatives for which some evidence exists [19] are γ-aminobutyryl-
choline, homopantothenic acid, and α,γ-diaminobutyric acid (Fig. 11.6). Not
much information is available about the possible functions of these materials.

Histamine

Histamine is another amine for which a transmitter function has been suggested
[78]. The evidence here is also quite circumstantial. Histamine has iontophoretic effects [79], and it is localized in synaptic endings [80]. Histamine and the
enzymes which are concerned with its formation and its destruction are in high
concentration in postganglionic sympathetic nerves of the hypothalamus [78].
The release of histamine has been shown to occur upon stimulation of the splenic nerve [81] and in the evoked hypotensive response of the cat [82]. But, as
in the case of GABA, the evidence has, for a long time now, been less than complete.

Histamine is formed, by decarboxylation, from histidine (Fig. 11.7) [83,
84]. This decarboxylation reaction is presumably the only source of histamine
in the central nervous system since histamine itself can not cross the blood-brain
barrier, while histidine is, as mentioned earlier, very actively transported. There

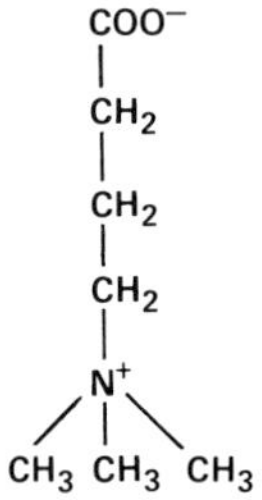

$$
\begin{array}{c}
COO^- \\
| \\
CH_2 \\
| \\
CH_2 \\
| \\
CH_2 \\
| \\
N^+ \\
\diagup \ | \ \diagdown \\
CH_3 \ CH_3 \ CH_3
\end{array}
\qquad\qquad
\begin{array}{c}
COO^- \\
| \\
CH_2 \\
| \\
CHOH \\
| \\
CH_2 \\
| \\
N^+ \\
\diagup \ | \ \diagdown \\
CH_3 \ CH_3 \ CH_3
\end{array}
$$

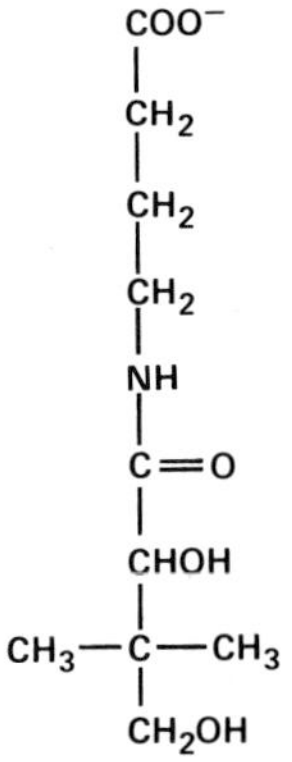

γ-Butyrobetaine $\qquad\qquad$ Carnitine

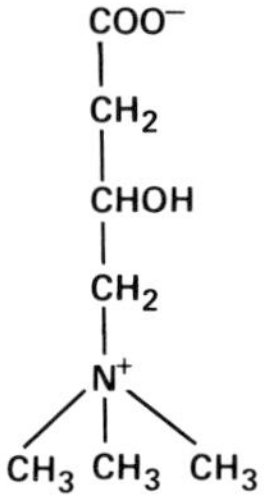

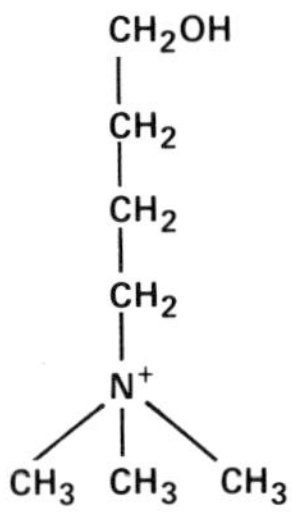

$$
\begin{array}{c}
CH_2OH \\
| \\
CH_2 \\
| \\
CH_2 \\
| \\
CH_2 \\
| \\
N^+ \\
\diagup \ | \ \diagdown \\
CH_3 \ CH_3 \ CH_3
\end{array}
\qquad\qquad
\begin{array}{c}
COO^- \\
| \\
CH_2 \\
| \\
CH_2 \\
| \\
CH_2 \\
| \\
NH \\
| \\
C=O \\
| \\
CHOH \\
| \\
CH_3-C-CH_3 \\
| \\
CH_2OH
\end{array}
$$

γ-Aminobutyrylcholine $\qquad\qquad$ Homopantothenate

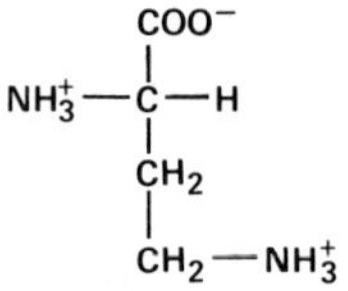

$$
\begin{array}{c}
COO^- \\
| \\
NH_3^+ - C - H \\
| \\
CH_2 \\
| \\
CH_2 - NH_3^+
\end{array}
$$

α,γ-Diaminobutyrate

Figure 11.6 Metabolites of GABA.

has been a lot of interest in the enzymes involved in the decarboxylation. There are two enzymes known to carry out the reaction [85,86]. The first is truly a histidine decarboxylase, being absolutely specific for L-histidine. It is a pyridoxal phosphate-requiring enzyme and is easily dissociated from its cofactor [87-89].

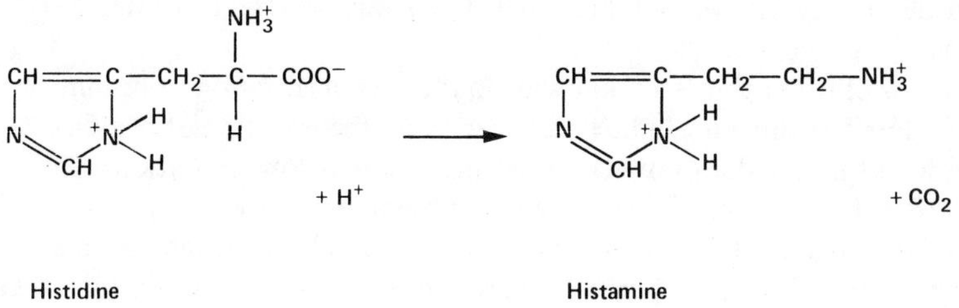

Figure 11.7   Formation of histamine.

This enzyme is found in high concentration in fetal rat tissue, in placenta, in mouse mast cells, in gastric mucosa, in bone marrow, and in tumors. In the nervous system it is found in high concentration in splenic nerve and sympathetic ganglia, lower in phrenic nerve and vagus, and quite low in the spinal cord, the brainstem, and the cortex [90]. The K_M for histidine under normal circumstances is on the order of 4×10^{-4} M [91], but the K_M and the pH optimum of the enzyme appear to depend upon the substrate concentration used [92]. Thus, with high substrate concentration, the pH optimum is between 5.5 and 6.0; at low substrate the pH optimum rises to 7.0, and the K_M drops as the pH rises. There is evidence that the enzyme is induced by stress and may be repressed by histamine [93,94].

Another enzyme which is known to decarboxylate histidine is the so-called nonspecific aromatic decarboxylase, which decarboxylates dopa, 5-hydroxytryptophan, and other amino acids, as well as histidine [84,95]. This is also a pyridoxal phosphate-requiring enzyme, but is more difficult to dissociate from its cofactor. It has quite a different distribution in the body, and quite different enzymatic characteristics. This enzyme has been studied thoroughly in connection with its action on dopa, and some reports say that even the little action it has on histidine is difficult to observe. It has a pH optimum around 9.0, requires very high concentrations of histidine to obtain substrate saturation, and, in general, has quite different properties than the specific enzyme.

Since both enzymes are found in brain, the specific enzyme probably being there primarily by virtue of the presence of mast cells, there was a question as to which enzyme was responsible for the formation of histamine in the nervous system. It seems clear now that the specific histidine decarboxylase is the responsible entity. This information is available from inhibitor studies [91,96-98]. The two enzymes have a very different spectrum of inhibitory compounds. That is, when brocresine (NSD 1055), an inhibitor of the specific enzyme, was administered, histamine concentration in the brain dropped rapidly. On the other hand when methyldopa, a classic inhibitor of the nonspecific L-aromatic amino acid decarboxylase, was administered, there was no effect on the level of histamine.

A number of other things are known about the concentration of histamine in the brain. For one thing it can be raised merely by the injection of histidine [98], suggesting that the decarboxylase is normally well below saturation. Then it is known that the turnover of histamine in the brain must be very rapid since inhibition of the histidine decarboxylase caused a drop in brain histamine characterized by a half-life of about 5 min [97], which is substantially faster than that of the other biogenic amines.

The catabolism of histamine in the whole body is mainly via histaminase, which may be identical with the enzyme diamine oxidase. However that may be, this enzyme does not seem to be present in the brain. The major route of histamine metabolism in the brain is methylation in the 1-position by a specific enzyme using S-adenosylmethionine as the methyl donor (Fig. 11.8) [99]. When the methylation is blocked, the level of histamine in the brain rises rapidly, verifying the crucial role of the methylase in the destruction of histamine, and supporting the rapid turnover of the compound in the brain. The fate of the methylhistamine in the brain seems to be similar to the fate of histamine in the rest of the body. Both compounds are oxidized to the corresponding aldehydes and then further, by aldehyde dehydrogenase, to methylimidazoleacetic acid and imidazoleacetic acid, respectively, which are then excreted. Some information is available to suggest that the methylhistamine itself is physiologically active.

Histamine is found in low concentrations in brain and spinal cord, but is present in certain postganglionic fibers at levels up to 60 μg/g. It is especially concentrated in the anterior pituitary and in such structures as the hypothalamus and the median eminence. Its concentration is higher, in general, in gray matter than in white. Histamine is found in peripheral nerve, but this content may be present in the mast cells associated with such peripheral structures. Subcellularly, histamine is found associated with the synaptosomes, and specifically with the synaptic vesicles. It is not present in isolated mitochondria or in

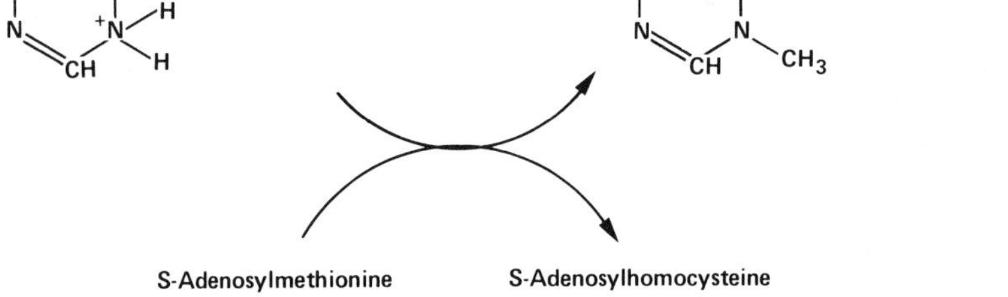

Figure 11.8   Formation of 1-methylhistamine.

myelin. The distribution of histamine is demonstrably different from that of the catecholamines or the indolamines. But one of the problems currently encountered in the study of histamine function is that, unlike the catecholamines and the indolamines, no sensitive and specific histochemical method is yet available for the in situ localization of the compound [78].

Indeed, even the direct chemical estimation of histamine in tissue extracts has proved to be a stumbling block. For a number of tissues, the measurement of histamine via its condensation with o-phthalaldehyde has provided a sensitive fluorimetric method [100]. But more recently it has been realized that in certain sources, such as brain and blood, the presence of spermidine interferes with the measurements. For these tissues it is necessary to introduce a chromatographic separation of the histamine before it can be measured [101,102]. The assay then, with chromatography in addition to the usual extraction procedures, is a formidable one. Recent micro-isotope methods give promise of easing the problems to some extent [103].

There is little or no information as yet on the nature of the receptor for histamine. It is known that particles will take up histamine from incubation media [104]. Some studies suggest that the binding in the particles may involve a sulfomucoid [78]. But the studies on histamine have not progressed to nearly the same extent as the work on other receptors, notably that for acetylcholine, to be described in Chap. 24.

Histamine levels can be altered by stress and by a number of drug treatments. Upon the imposition of stress on animals, the histamine levels in various sites drop. Interestingly, the formation of histamine, measured by the conversion of radioactive histidine to histamine, rises sharply in some sort of compensatory mechanism [105]. Some of the psychotropic drugs are known to alter histamine levels [106]. For example, reserpine lowers histamine levels in the hypothalamus and the thalamus, but not in the hypophysis. Chlorpromazine and other phenothiazines raise levels in the hypothalamus, apparently by inhibiting the enzyme catalyzing histamine methylation. Histamine concentrations in the hypophysis are again unchanged, which is consistent with the observation that the mast cells, which may be responsible for the histamine content of the hypophysis, do not have the enzyme for the methylation [107]. Iproniazide, bulbocapnine, and pentobarbital are known to raise histamine concentration in the hypothalamus and the thalamus, but morphine is without effect. Of course, the large number of decarboxylase inhibitors known lower histamine concentrations.

The actions of histamine are numerous, and no single mechanism has been suggested that will explain them all. Histamine certainly mediates gastric secretion, and probably salivary secretion as well. It is in some way involved in inflammation, and perhaps is a mediator of pain. The effects of exogenously administered histamine differ somewhat from species to species. The levels of histamine in the brain do not seem to have any behavioral associations, since when the histamine levels are raised several-fold by the injection of histidine, no overt behavioral alterations can be seen. It is of interest, however, that schizophrenics apparently have a lower incidence of allergy than does the normal population, and histamine is undoubtedly involved in the inflammation associated with allergic reaction [108]. Finally, although there are no widely accepted theories as to the molecular mechanism of histamine action, it is known that histamine, more than any of a number of other compounds of neural interest, raises the levels of cyclic AMP in brain tissue in vitro [109].

There are a number of derivatives of histamine which have been observed. Already mentioned are the aldehyde and acid derivatives of histamine itself and of 1-methylhistamine. Also of interest are reports that histamine is acetylated at the amino nitrogen, and that it can be methylated on the other ring nitrogen as well (3-methylhistamine). It is possible, but as yet unproven, that these other metabolites also have a physiological function.

Norepinephrine

Certainly among the most active areas of neurochemical study in the past 20 years has been the metabolism and function of the catecholamines [110-113]. The reasons for the interest in these materials are numerous. Norepinephrine is the most likely candidate, aside from acetylcholine itself, for transmitter function. Then, the materials involved are versatile in nature, playing perhaps multiple roles, and being involved in a number of disease processes. And, not least, the handles for the study of these compounds are reasonably good in that the materials themselves are highly absorbent, highly fluorescent, and sufficiently reactive chemically to lend themselves to a number of assay techniques. Thus, the study of the catecholamines has occupied some of the most creative and productive laboratories involved in recent neurochemistry and neuropharmacology.

The conversion of tyrosine to norepinephrine, although a minor pathway for tyrosine metabolism in the whole body, seems to be the major route of tyrosine metabolism in the brain and in the adrenal gland. The biosynthetic pathway, first postulated by Blaschko more than 30 years ago [114], and now amply verified, is shown in Fig. 11.9. This metabolism occurs in the nerve endings of the norepinephrine-containing neurons of the central nervous system, in the peripheral sympathetic nerves, and in the adrenal medulla, as well as in certain tumors originating in these tissues. Norepinephrine is thus thought to be the transmitter substance for certain neurons of the central nervous system as well as for the sympathetic neurons in peripheral nerves.

Norepinephrine, itself, is found primarily in the terminals of the neurons which make it. Evidence that it is a transmitter substance is comprised of observations on its location in neurons, and specifically in their terminals [115], on

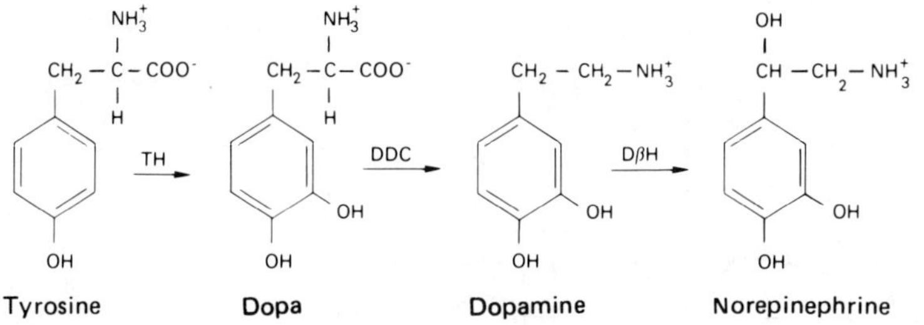

Figure 11.9   Formation of norepinephrine.

its appearance in certain vesicular ("dense core") structures [116] containing
ATP and a divalent cation and analogous to the structures containing acetylcholine, on its ability to be taken back into a neuron after it has presumably acted
postsynaptically [111], and on its release upon depolarization of sympathetic
neurons [117,118]. These pieces of evidence are admittedly indirect, but based
on the presumed characteristics of a transmitter substance, they make a strong
case.

Much of what we know about the catecholamine-containing neurons in the
central nervous system is due to the ingenious histochemical techniques devised
by the Swedish school under the direction of Hillarp [119,120]. This methodology consists of exposing thin tissue sections to formaldehyde in a humid atmos-
phere. Such exposure produces a condensation of the catecholamines with the
formaldehyde to yield highly fluorescent isoquinolines which can be readily seen
in the fluorescence microscope. Tracts containing catecholamines show a bright
green fluorescence under such conditions, where those containing serotonin, to
be described in a later section, give a yellow fluorescence. By such means it has
been shown that the norepinephrine-containing neurons of the central nervous
system are contained in the brainstem, and that some of the axons descend to
the spinal cord and some ascend to other parts of the brain. The norepinephrine-
containing cell bodies are found in 10 or more clusters in the medulla oblongata,
the pons, and the midbrain, with the most densely packed region being in the
locus coeruleus. A schematic representation of the central norepinephrine-containing tracts revealed by this method is shown in Fig. 11.10 [121]. In addition
to the basic information about the distribution of the various tracts, this method
has been widely applied in the study of such problems as axonal transport [122,
123], and the effects of the various drugs impinging on the metabolism of catecholamines [120,122]. Although it is a semiquantitative tool, it would be hard
to overstate the impact of the histochemical fluorescence method on this area of
neurochemistry.

The first enzyme in the biosynthetic pathway to norepinephrine, tyrosine

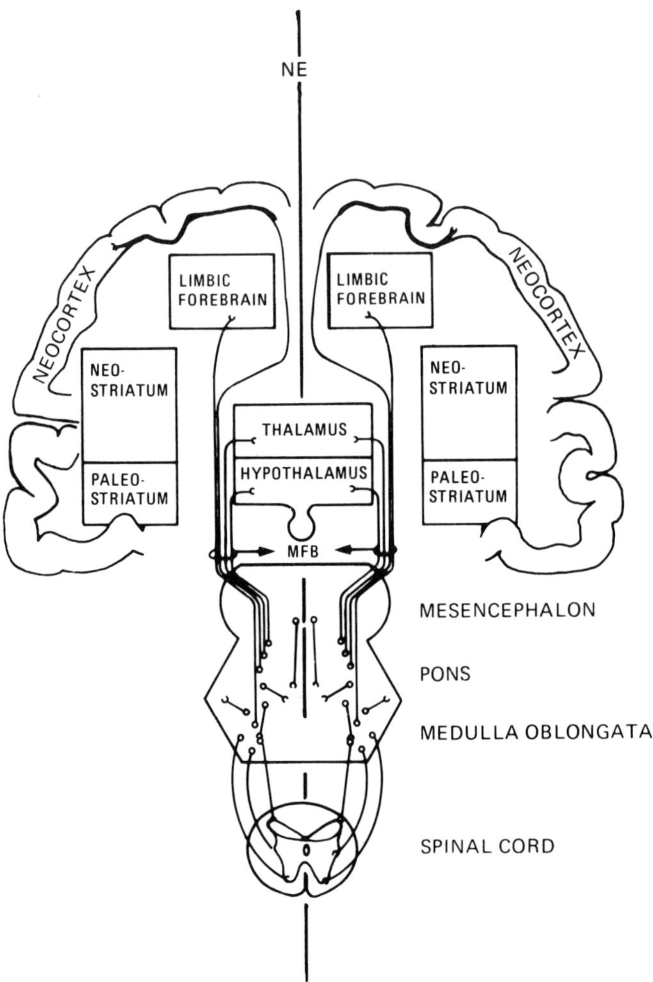

Figure 11.10   Norepinephrine-containing tracts in the central nervous system. [Redrawn from Anden et al., *Acta Physiol. Scand. 67:*313 (1967).]

hydroxylase, was the last one to be actually observed in the laboratory in cell-free form. For a number of years it was considered possible that the enzyme tyrosinase, which also makes dopa from tyrosine, was responsible for the first step in the biosynthesis of norepinephrine. However, the distribution of this enzyme in the animal body was quite different from the distribution of norepinephrine. The clean observation of tyrosine hydroxylase by Nagatsu and co-

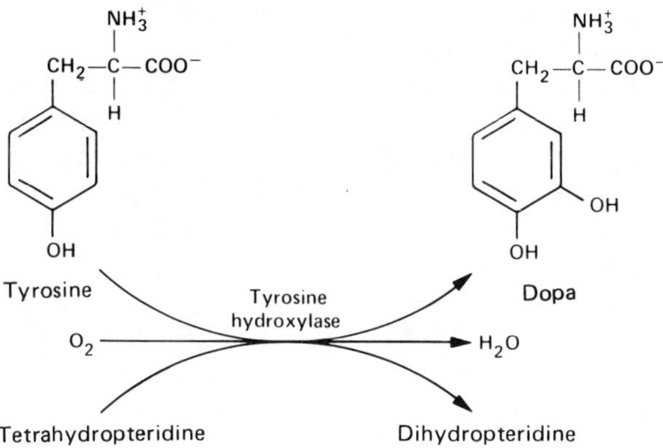

Figure 11.11   Tyrosine hydroxylase.

workers in 1964 [124], resolved most of the questions. This enzyme (Fig. 11.11) is now known to be a mixed-function oxidase, superficially similar to phenylalanine hydroxylase in its mode of action. It occurs in adrenal gland, midbrain, and the sympathetic nervous system. It uses a reduced pteridine as cofactor, and the cofactor, at least in rats, has been shown to be tetrahydrobiopterin [125], which has also been shown to be the specific pteridine serving the phenylalanine hydroxylase reaction. The enzyme also requires molecular oxygen and probably contains [126,127] ferrous iron. It has been shown to incorporate oxygen into the 3-position of the tyrosine ring [128].

It is now clear that the tyrosine hydroxylase reaction is the rate-limiting step for the overall pathway [129]. Further, and as a consequence, it is at this step that the pathway is regulated. The inhibition of the enzyme by catecholamines, including norepinephrine, indicates that the enzyme is under feedback control, and that the catecholamines control the rate of their own synthesis [130, 131]. The mode of inhibition of the enzyme by catecholamines is not clear, but the best data available indicate that the catecholamines compete with the reduced pteridine cofactor and thus cause inhibition [130]. In any case, the in vitro studies on the inhibition of the enzyme by norepinephrine are consistent with any number of studies in vivo in which the synthesis of norepinephrine from tyrosine, but not from dopa, is lowered when the tissue level of norepinephrine is raised, for example, by the administration of inhibitors of enzymes involved in the degradation of the catecholamines. A number of other inhibitors of the enzyme are known, including various catechols and tyrosine analogs, the most useful of which in a pharmacological sense has been α-methyltyrosine [130].

It has become apparent that some of the characteristics of the action of

tyrosine hydroxylase depend upon the nature of the cofactor used to support
the reaction. Because of the instability and expense of tetrahydrobiopterin, it
has been common practice to carry out tyrosine hydroxylation in the laboratory
using the synthetic cofactor 6,7-dimethyltetrahydropteridine. With this cofactor
it was shown that the enzyme is inhibited by phenylalanine because it uses phenylalanine as substrate [132]. The hydroxylation of phenylalanine appeared to
be quite weak, but a sequential p-hydroxylation and m-hydroxylation could be
shown. More recent studies, using the natural cofactor, tetrahydrobiopterin,
have shown that phenylalanine is indeed a rather good substrate for the adrenal
enzyme [126]; it appears to be as good a substrate as tyrosine itself under these
conditions. These studies are consistent with comparable studies on phenylalanine hydroxylase, which also show alterations in the characteristics of the action
of the enzyme upon changing from the synthetic to the natural cofactor.

It has been shown by several groups that the incubation of tyrosine hydroxylase under phosphorylating conditions alters the enzymatic characteristics
of its action [133,134]. These studies have indicated that such alterations include a decreased K_M for the pteridine cofactor, an increased V_{max}, and a shift
in the pH optimum of the enzyme activity. In these studies it was found that
cyclic AMP (cAMP), ATP, and Mg^{2+} were all required and that cAMP or ATP
alone were ineffective. These experiments strongly suggest that tyrosine hydroxylase can be phosphorylated by a cAMP-dependent protein kinase and that such
phosphorylation might regulate its action. Indeed, incorporation of radioactive
phosphate into tyrosine hydroxylase has been demonstrated in organ cultures of
adrenal medulla and superior cervical ganglia [135].

The enzyme has not yet been prepared in active, homogeneous form,
and so a number of questions remain. An active chymotryptic fragment of
the enzyme has been obtained pure and used to produce a specific antibody to
the enzyme [136]. Such antibody is found to inhibit tyrosine hydroxylases from
other species, but not to cross-react with related hydroxylases from the same
species.

Dopa decarboxylase (DOPA-D), the second enzyme in the pathway [84],
seems to be a typical amino acid decarboxylase (Fig. 11.12), acting on a large
number of related substrates, and requiring pyridoxal phosphate for its action.
It is not restricted to the nervous system in its distribution, and appears to be in
substantial excess of tyrosine hydroxylase in the appropriate sites, and is, thus,
not a factor in the regulation of the synthesis of norepinephrine. Correspondingly, inhibitors of the enzyme are not too important pharmacologically since
95% inhibition does not change amine levels. The enzyme has been studied
in a number of tissues, most exhaustively by Lovenberg and coworkers [95],
who purified the decarboxylase 50-fold to 100-fold from guinea pig kidney.
The guinea pig kidney enzyme decarboxylates a wide variety of substrates, including dopa, 5-hydroxytryptophan, and, as mentioned before, histidine. In fact,

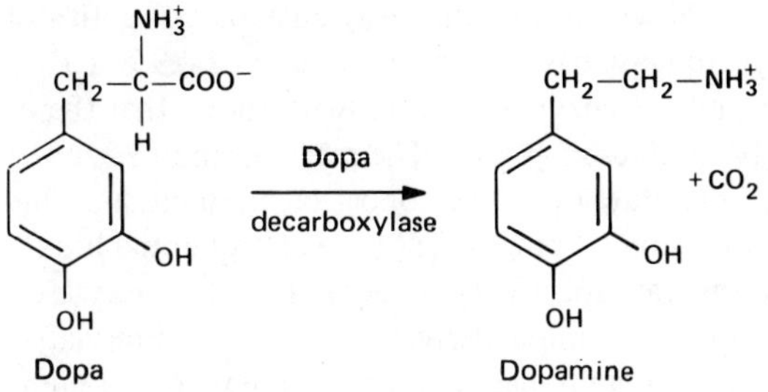

NH₃⁺ | CH₂—C—COO⁻ | H — (Dopa) → Dopa decarboxylase → CH₂—CH₂—NH₃⁺ + CO₂ (Dopamine)

Dopa

Dopamine

Figure 11.12 Dopa decarboxylase.

the enzyme is so nonspecific that it has been generally termed "aromatic amino
acid decarboxylase." The concept that one enzyme served both the pathway to
norepinephrine and the pathway to serotonin has some substantial implications
as far as the regulation and separation of these two important pathways are concerned. Recently, however, some dispute has arisen about the nature of the two
activities in brain, and whether, as in kidney, there is truly only one enzyme. Immunological evidence favors both activities being on one protein [137], but some
other data involving different pH optima, solubility, and subcellular distribution
suggest that the activities, in brain at least, reside on separate molecules [138].
A resolution of this debate perhaps must await the complete purification of these
activities from the brain.

The final step in this pathway is catalyzed by dopamine β-hydroxylase
(DβH). This enzyme (Fig. 11.13) is, of course, present in brain but has not been
thoroughly studied in this tissue because of difficulties in the assay, and because
of the presence of potent natural inhibitors. The corresponding enzyme from
adrenal medulla has been extensively purified by Friedman and Kaufman [139].
It requires, for its action, molecular oxygen and ascorbic acid. The activity is

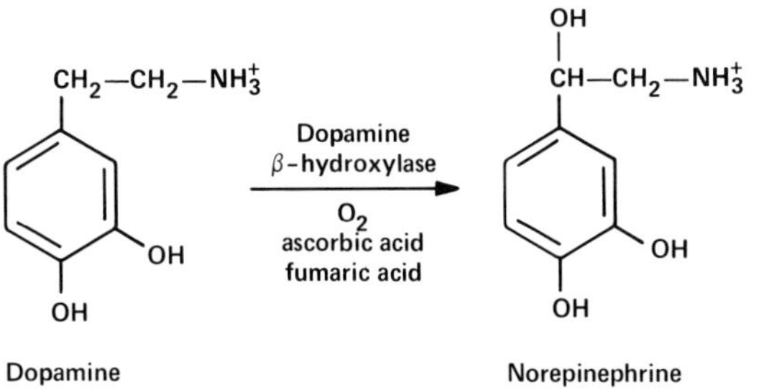

CH₂—CH₂—NH₃⁺ (Dopamine) → Dopamine β-hydroxylase, O₂, ascorbic acid, fumaric acid → OH | CH—CH₂—NH₃⁺ (Norepinephrine)

Dopamine

Norepinephrine

Figure 11.13 Dopamine β-hydroxylase.

markedly stimulated by fumaric acid which is in some way an allosteric activator of the enzyme. Dopamine β-hydroxylase has a molecular weight of 2.9 X 10^5 and contains 2 mol of Cu^{2+} per mole of enzyme. Recent work shows that there are three identical subunits of 100,000 each [140]. The mechanism of action seems to involve a cyclic oxidation-reduction of the copper component, and the catalysis can apparently be described by a "ping-pong" type of sequential binding to the enzyme [141]. The substrate specificity of dopamine β-hydroxylase is not rigid since a number of phenylethylamine derivatives will be hydroxylated. In fact, tyramine is a better substrate than dopamine. As a consequence, a number of materials act as inhibitors of the enzyme, as do a number of copper chelators. One interesting, naturally occurring inhibitor is fusaric acid [142]. Since the enzyme is not normally rate limiting for the synthesis of catecholamines, these inhibitors have limited pharmacological significance.

In the adrenal, catecholamine metabolism continues with the formation of epinephrine from norepinephrine. The enzyme phenylethanolamine N-methyltransferase (PNMT; Fig. 11.14) is responsible for this conversion [143]. This is a soluble enzyme using S-adenosylmethionine as the methyl donor. It is specific for phenylethanolamine derivatives since the corresponding phenylethylamines are not methylated. The enzyme is found in the adrenal medulla and its action is regulated by adrenal cortical steroids [144] providing cortical control over the functions of the medulla. Although the enzyme has been reported in brain and sympathetic ganglia, there is little information about the possible significance of its occurrence.

Shortly after the early work on tyrosine hydroxylase, it was suggested that all the enzymes in the pathway to norepinephrine synthesis must be contained in a unique particle [145], the function of which would be the coordinated synthe-

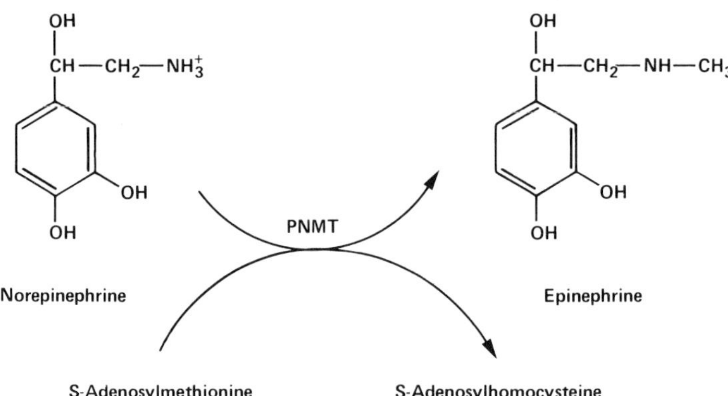

Figure 11.14   Phenylethanolamine-N-methyltransferase.

sis, and movement through the neuron, of the transmitter substance. This economical and attractive picture seems not to be supported by the facts. The evidence against such a concept takes two forms. First, studies on the subcellular distribution of the three enzymes have shown that, although all three can be found in the terminals, only dopamine β-hydroxylase is truly particle bound. Tyrosine hydroxylase is found in a bimodal distribution, some being particle bound and some soluble, the proportion of each depending somewhat on the method of preparation of the tissue. Dopa decarboxylase can be found completely in the soluble portion after reasonably vigorous homogenization. Thus, dopamine β hydroxylase, alone of the enzymes, seems to be present in the particle which contains the transmitter, and this is consistent with its particle bound localization in the adrenal gland as well. A second and confirming area of evidence has been the more recent studies on the axonal transport of the three enzymes. Clearly, if all three were in a single particle they should all move down the axon at the same speed. Just as clearly this is not the case. Work from several laboratories [146-148], although differing a bit in the quantitative aspects of the results, consistently shows that dopamine β-hydroxylase and the norepinephrine storage particles travel at a rapid rate, tyrosine hydroxylase at a somewhat slower rate, and dopa decarboxylase at an even slower rate. Thus, even though the proposal that all the enzymes were contained in the same particle proved not to be the case, the information generated while it was being investigated has added much to our knowledge of these enzymes.

There are several levels of regulation of the enzymes which synthesize norepinephrine, and so any discussion must involve several different physiological conditions. First, in most species there is a developmental increase in the concentration of catecholamines and of the enzymes which synthesize them. This increase parallels the development of the sympathetic nervous system and has its morphological counterpart in the appearance of the catecholamine-containing nerve terminals, which, in most species, are not fully developed at birth [149]. After the pathway is fully developed the regulation of the synthesis is, as mentioned before, at the first and rate-limiting step, tyrosine hydroxylase. The kinetic characteristics of the enzymes in the pathway are such [112] that the K_M of the overall pathway is the K_M of tyrosine hydroxylase, since the K_M values of the other two are substantially higher. For example, the K_M of tyrosine hydroxylase for its substrate is on the order of 50 μM, the K_M of dopa decarboxylase is about 400 μM, and the K_M of dopamine β-hydroxylase is about 6,000 μM. Since the plasma and tissue concentration of tyrosine is about 100 μM, tyrosine hydroxylase is generally saturated and the others are not. Thus, the rate of tyrosine conversion to norepinephrine is virtually independent of the blood concentration of tyrosine and depends on the activity of tyrosine hydroxylase alone, since as the rate of dopa production increases, the rate of action of the unsaturated dopa decarboxylase and dopamine β-hydroxylase also increases.

There are two ways, then, in which the needs of the body can result in an increase in the activity of tyrosine hydroxylase, and thus in the synthesis of norepinephrine. First, as mentioned before, the enzyme is subject to feedback inhibition by the end-product, norepinephrine. Stimuli which lower the level of norepinephrine in the nerve terminal instantly produce a rise in the synthesis of norepinephrine. Thus, it is found during sympathetic activity, when norepinephrine is liberated at the synapse, more norepinephrine is made from tyrosine. Stimulation of the sympathetic nervous system results very quickly in a markedly increased norepinephrine synthesis from radioactive tyrosine. Conversely, inhibition of monoamine oxidase, which raises norepinephrine levels, decreases norepinephrine synthesis. When the level of tyrosine hydroxylase is measured in vitro, however, no increase is found, indicating that what has happened is probably a relief of the end-product inhibition normally present. Since there is no marked change in the overall level of norepinephrine observed, it must be that as sympathetic output increases during stimulation the concentration of transmitter is maintained by increased synthesis. Thus, a persuasive argument can be made for the concept that the true state of the sympathetic nervous system cannot be understood by static measurements of catecholamine levels, but only by more complex studies involving the turnover of the sympathetic transmitters [110].

More recently it has become clear that prolonged stimulation of sympathetic nerves can lead to an increase in the synthesis of the enzyme proteins in addition to an increase in the level of their activities. Specifically, it has been shown that the synthesis of the enzymes tyrosine hydroxylase and dopamine β-hydroxylase, but not dopa decarboxylase, is increased by prolonged demand on the sympathetic system [149,150]. For example, a prolonged cold stress can increase the levels of these enzymes in the superior cervical ganglia. And since this induction can be prevented by decentralizing the ganglia, that is, by cutting the preganglionic cholinergic nerves, this increase in the synthesis of enzyme in the postganglionic neurons clearly depends on the activity of the preganglionic nerves. Since this process goes on in some way across the synapse it has been termed "transsynaptic induction." The exact mechanism of this induction is not known, but it is known that it plays a role in the maintenance of the normal levels of enzyme as well as in the increased levels during sympathetic nerve overactivity, since decentralization produces a fall in the basal levels of these enzymes in adult animals under normal conditions [151].

There are two enzymes involved in the degradation of the catecholamines and it is not completely clear in which order they act under various conditions. One is catechol-O-methyltransferase (COMT; Fig. 11.15). This enzyme catalyzes a methylation of the hydroxyl in the 3-position using S-adenosylmethionine as the methyl donor, and a divalent cation [152]. It acts on a wide variety of substrates, including dopa, dopamine, norepinephrine, and the acid or alcohol derivatives of these materials. It is a soluble enzyme not confined to the nervous sys-

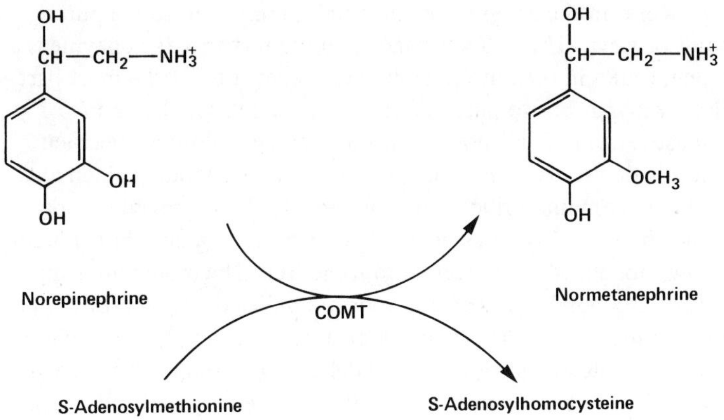

Figure 11.15   Catechol-O-methyltransferase.

tem, nor even particularly high in brain [153] , and is thought, for that reason, to
be involved primarily in the inactivation of circulating catechol compounds [154] .
Since substantial levels of O-methylated amines are found in the circulation, it is
thought that catechol-O-methyltransferase acts on the amines after they have
acted synaptically and then escaped into the circulation without being taken up
into any cells and, thereby, becoming exposed to the other degradative enzyme,
monoamine oxidase (MAO; Fig. 11.16).

Much more information is available about monoamine oxidase than about
catechol-O-methyltransferase, primarily because of the intense pharmacological
interest in the inhibition of this enzyme in the treatment of angina, hypertension,
high blood pressure, and some of the emotional disorders [155,156] . In spite of

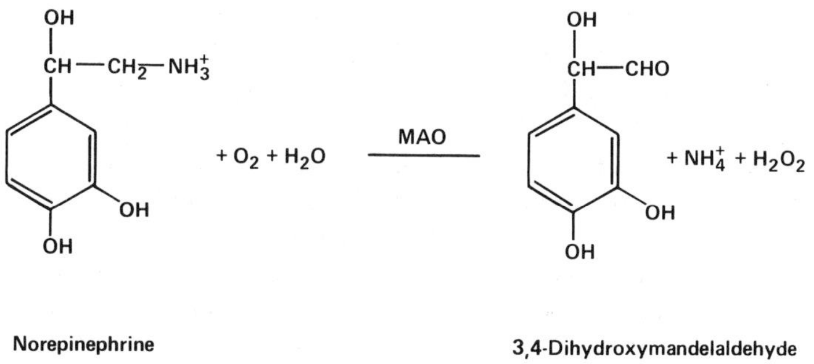

Figure 11.16   Monoamine oxidase.

the large body of work on the enzyme, much remains unknown from a purely biochemical point of view. This is due primarily to the fact that the enzyme is a mitochondrial one, found in the outer membranes. Since it is for the most part difficult to solubilize, and thus to purify, many questions about the nature of monoamine oxidase remain. It is known that the enzyme has broad specificity, acting to oxidize a wide variety of monoamines to the corresponding aldehydes. The enzyme probably contains a riboflavin component. Some previous work also indicated that the enzyme contained a copper prosthetic group, but this copper content is now thought to be incidental to its action. The monoamine oxidase from rat liver mitochondria has an approximate molecular weight of 15,000 [157]. The enzyme uses molecular oxygen as an acceptor, and produces ammonia and hydrogen peroxide in addition to the aldehyde corresponding to the amine substrate oxidized. In addition to the intractability of the enzyme itself, it has been somewhat difficult to distinguish the action of monoamine oxidase from the actions of the several other enzymes in the body which oxidize amines. Now it is clear that the true monoamine oxidase can be observed by its characteristic insensitivity to certain carbonyl reagents such as semicarbazide since other amine oxidases such as the diamine and the polyamine oxidases are quite sensitive to such reagents [155]. Monoamine oxidase is a very nonspecific enzyme and is widely distributed in the body. One of the other major problems in the study of this enzyme has been the fact that there are apparently several forms of the enzyme or perhaps isozymes, which may be tissue specific [158,159]. Although it is clear that this allows a great deal of flexibility, perhaps permitting "tailor-made" enzymes for various brain parts to act selectively on the amine constituents of those specific parts, such variability complicates the study of the enzyme. But, in spite of these problems, some progress is now being made toward the solubilization and eventual study of this obviously complex enzyme [160,161].

In any case, it seems that the function of monoamine oxidase in the metabolism of catecholamines is to inactivate them at their site of action, perhaps even in the terminal in case they escape from the vesicle. But since catecholaldehydes are known to occur without the 3-O-methyl, MAO does act on these transmitters at some stage before they are released into the circulation and thus become susceptible to the action of catechol-O-methyltransferase. The further metabolism of the catecholamines involves, then, 3-O-methylation and oxidation by monoamine oxidase, either independently or sequentially in either order. Then the aldehyde products are either oxidized to the corresponding carboxylic acids or reduced to the alcohols. Quantitatively, vanillyl mandelic acid is the major metabolite of norepinephrine in the periphery and is measured clinically as an index of sympathetic function or to diagnose the presence of tumors which produce norepinephrine. The catabolism of norepinephrine is outlined in Fig. 11.17.

The function of these two degradative enzymes is certainly to regulate amine levels in the body, and to destroy circulating amines. But it is clear that

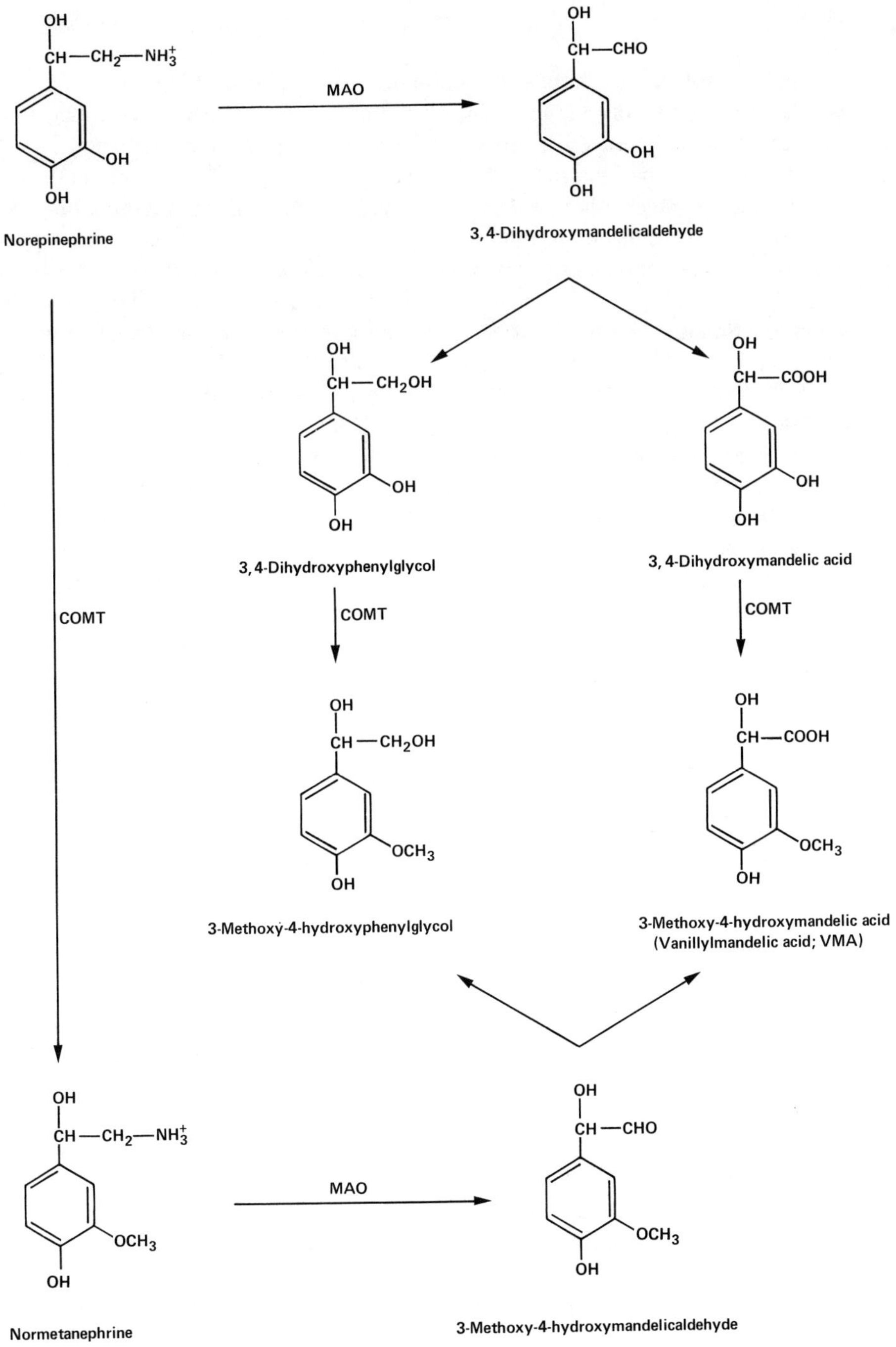

Figure 11.17   The catabolism of norepinephrine.

neither has a role in terminating the action of norepinephrine at the synapse. This has been shown in a number of studies, among them the many experiments demonstrating that none of the many inhibitors of monoamine oxidase, nor the few of catechol-O-methyltransferase, have any potentiating effect on the action of the catecholamines. If either of these enzymes were involved in terminating the actions of the transmitter, inhibition of the enzymes should enhance transmission. It is now abundantly clear that the process that terminates the action of norepinephrine is one of reuptake into the neural terminal from which it was liberated. Similar observations have already been discussed in regard to termination of the action of GABA, but the situation in the case of the catecholamines is even more thoroughly understood. The observations of Axelrod [162,163] on the fate of administered catecholamines pointed the way for future thought on the termination of action of these materials. These experiments showed that the injected materials accumulated specifically in tissues with dense sympathetic innervation and that the uptake into these tissues could be prevented by cutting the nerves leading to these tissues. The amines accumulated in the synaptic boutons of the nerves, and drugs which prevented this accumulation also potentiated nerve stimulation. Thus, the concept that reuptake normally terminates the action of the transmitter and that inhibitors of reuptake increase the effectiveness of the nerve action was established. It is thought that certain of the tricyclic antidepressants work by blocking this reuptake mechanism. Studies of this system in isolated brain tissue, or in synaptosomes, have shown that the reuptake is due to an amine "pump," presumably in the membrane, which is energy dependent, saturable, and inhibited by ouabain. Thus, the amine pump, although it has very high efficiency by comparison, is somewhat like the amino acid transport systems discussed in Chap. 10. Also, the amine pump probably depends in some way, on the Na^+ pump as a driving force, as do the amino acid transport systems [164].

Thus, norepinephrine is most certainly the transmitter substance in the sympathetic nervous system and in certain neurons of the central nervous system. As such it is made in the neuron, stored in "packets" in the synaptic bouton, and released upon stimulation of the nerve cell. It is released into the synapse and acts on the postsynaptic cell by interacting with a surface receptor, and then its action is terminated by an active and efficient "reuptake" into the presynaptic nerve cell ending.

Most of the current information about the action of norepinephrine has come from pharmacological studies in which the concentration of the amine has been altered by interfering with one of the steps in its biosynthesis, its storage, its reuptake, or its degradation. Enormous numbers of compounds have been found to be more or less effective in altering these processes and thus changing the levels of the catecholamines in various parts of the nervous system. Although many drugs have been found to impinge on norepinephrine concentrations they can be divided, after the classification of Pletscher et al. [165], into five categories, as follows.

1. Drugs which decrease storage of the amine. Of these the most prominent is reserpine, one of the *Rauwolfia* alkaloids, which blocks the storage of the
amine. Reserpine depletes the overall level of the amine in the system by preventing the binding to the storage sites and thereby making the amine more available to the destructive effects of the normal metabolizing enzymes. Reserpine is
a potent tranquilizer.

2. Drugs which decrease the synthesis of the amine. Since the tyrosine
hydroxylase step is rate limiting, it is not surprising that inhibitors of this enzyme
are the most useful. At the moment the most used of these is the inhibitor α-
methyltyrosine. Its action is, in fact, so well documented that inhibition of a
biochemical process, or a physiological function, by α-methyltyrosine is diagnostic for the dependence of that process or function on the action of tyrosine hydroxylase or the synthesis of norepinephrine. For example, by using α-methyl-
tyrosine it has been shown that the weak hydroxylation of tyrosine found in
brain is, in fact, due to the action of tyrosine hydroxylase and not to the presence in brain of phenylalanine hydroxylase, an enzyme upon which this inhibitor
has no effect. Pharmacologically, α-methyltyrosine produces sedation. Inhibitors
of dopa decarboxylase are available, α-methyldopa being the best known, and so
are inhibitors of dopamine β-hydroxylase, which are nearly always copper
chelators. But, as mentioned above, these have received relatively little attention.
More recently, a number of antibiotics with more or less specific action on these
enzymes have been found.

3. Drugs which decrease the destruction of the amine. Of these, the inhibitors of monoamine oxidase have received wide attention [155]. The most used
include the potent competitive inhibitors, harmine and harmaline; amphetamine,
which is also reversible; and the irreversible inhibitors, iproniazide, pargyline, and
nialamide. These materials have a variety of effects in the whole animal in addition to, and as a consequence of, their inhibition of monoamine oxidase. Clinically, they have found use in the treatment of angina, hypertension, and some
kinds of depression. Iproniazide, which raises amine levels in brain, is a potent
excitant. It is known that monoamine oxidase inhibition, as would be expected,
leads to whole-body increases in the catecholamine content, and it is further
known that, since monoamine oxidase is in excess, no such increase occurs unless the enzyme is 85 to 90% inhibited. A few inhibitors of catechol-O-methyltransferase are known, but the pharmacological properties of these compounds
have not been widely explored.

4. Drugs which inhibit reuptake. These materials are active because they
selectively block the amine pump which removes norepinephrine from the synaptic cleft. Thus, they effectively prolong and enhance the action of norepinephrine at its receptors. The most potent of these is imipramine, which markedly
potentiates sympathetic nerve stimulation in the peripheral nervous system. It
is a potent antidepressive agent and seems to act in the central nervous system
not only on the norepinephrine pump, but also, and even more strongly, on the

α-Methylnorepinephrine

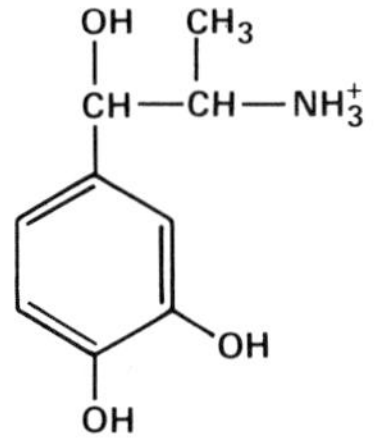

α-Methyl-m-tyramine
(metaraminol)

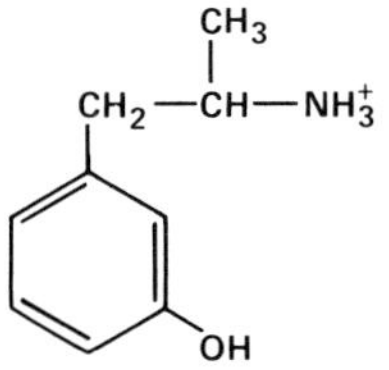

Chlorpromazine

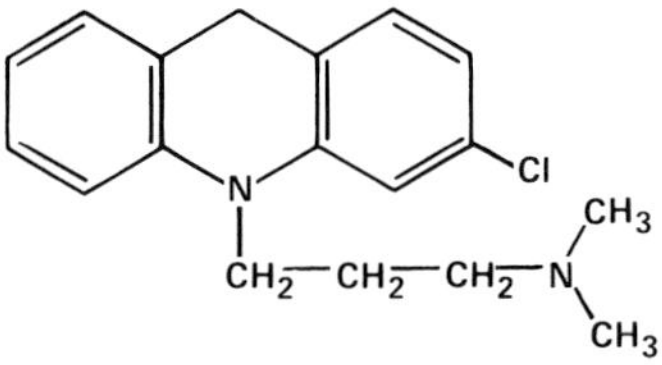

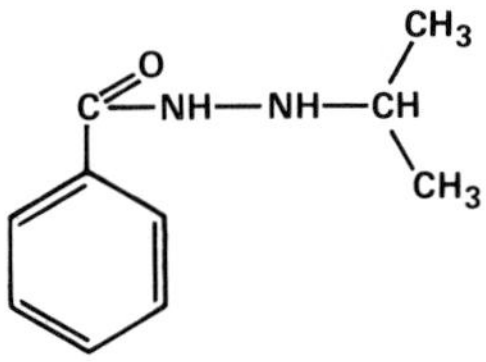

Iproniazide

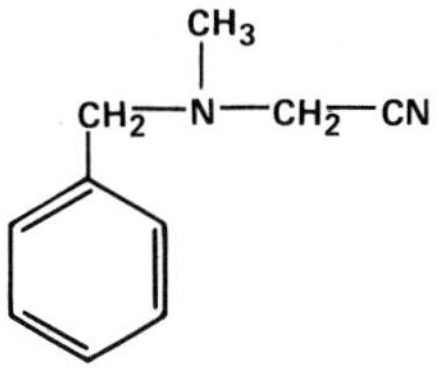

Pargyline

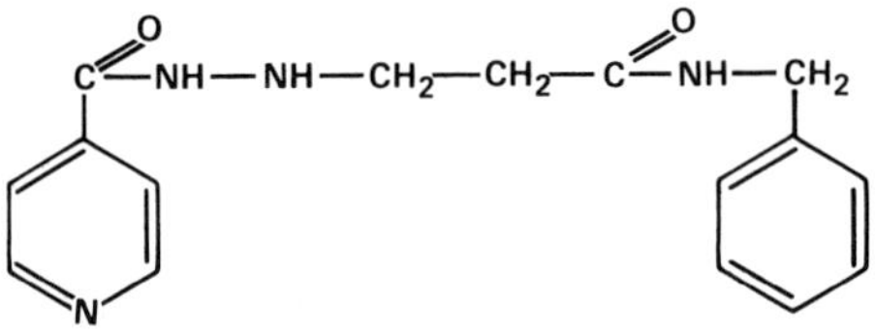

Nialamide

Figure 11.18 Effectors of the norepinephrine system.

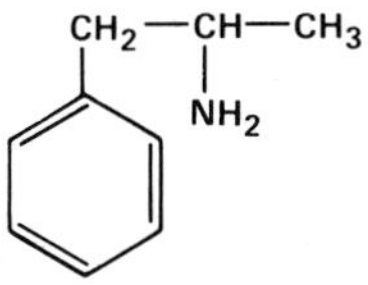

Figure 11.18 (continued)

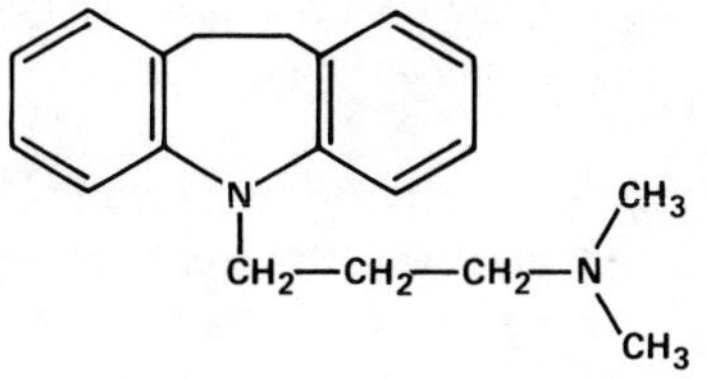

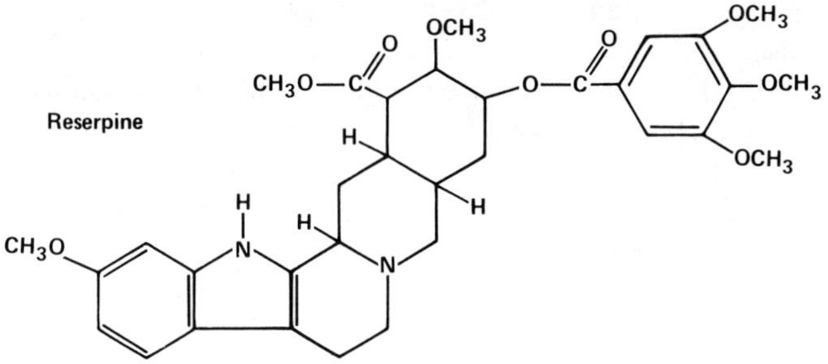

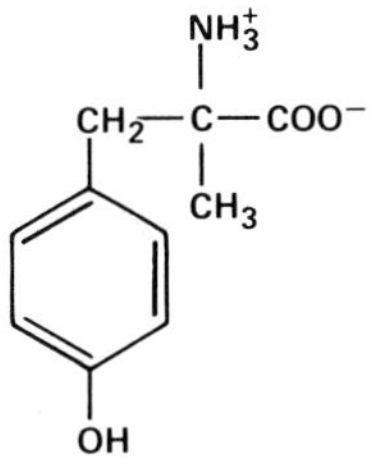

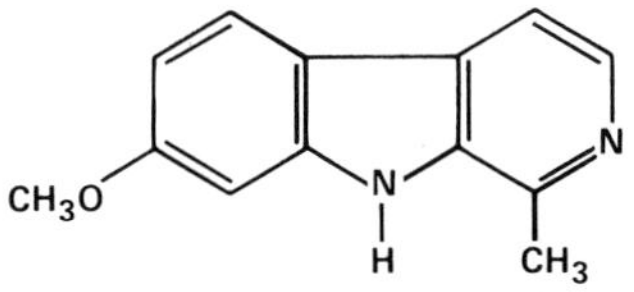

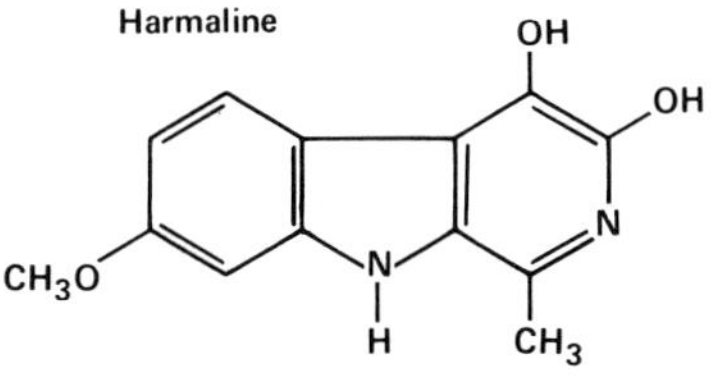

pump for serotonin. Related compounds which also produce behavioral activation, but are more selective for the norepinephrine mechanism, include protriptyline and desipramine.

5. Drugs which inhibit the binding of norepinephrine to its receptors by competition. These are the so-called "false transmitters" which prevent the action of norepinephrine [166]. They are structural analogs of norepinephrine precursors and are converted to amine antagonists by the normal pathway, but then are apparently unable to function as transmitters. The most notable of these are α-methylnorepinephrine and α-methyl-m-tyramine (metaraminol), which come from α-methyldopa and α-methyl-m-tyrosine, respectively, by decarboxylation and β-hydroxylation. These drugs decrease norepinephrine concentrations in brain and in peripheral nerves and cause hypotension.

Chlorpromazine is also known to have effects on the catecholamine system. But its actions are complex, and some of them may be a secondary consequence of its known ability to produce hypothermia or its demonstrated effects on membrane structures. Thus, its actions on the catecholamine system do not lend themselves to classification and, in fact, appear to depend on the amount administered and the method of administration, among other things. The structures of some of the compounds mentioned above are presented in Fig. 11.18.

Very generally speaking, then, drugs which raise the effective levels of catecholamines are, clinically, antidepressants, and those which lower the effective levels are tranquilizers. But since, as will be discussed below, most of these drugs have effects on indolamines also, it is still not known what would be the effects of raising or lowering the norepinephrine concentration completely independently. Further, it is as yet not possible to change the norepinephrine concentrations in any single portion of the brain and to inspect the effect on the animal. The long-term changes, if any, brought about by chronic alterations in amine levels or by genetically determined differences in levels from one individual to another are still unexplored.

Dopamine

Dopamine was originally thought to function merely as an intermediate in the biosynthesis of norepinephrine. Now it is known that dopamine is present and has independent functions, probably as a transmitter, in certain distinct portions of the brain [167,168], and in certain interneurons of the sympathetic ganglia. The dopaminergic neurons in the brain have been mapped using the histofluorescence technique described before (Fig. 11.19). With this method the fluorescence of dopamine is indistinguishable from that of norepinephrine, but the tracts can be distinguished through the use of certain specific drugs. The dopamine-containing neurons have no dopamine β-hydroxylase and thus cannot complete the conversion of tyrosine to norepinephrine. The breakdown of dopamine is similar to that of norepinephrine (Fig. 11.19).

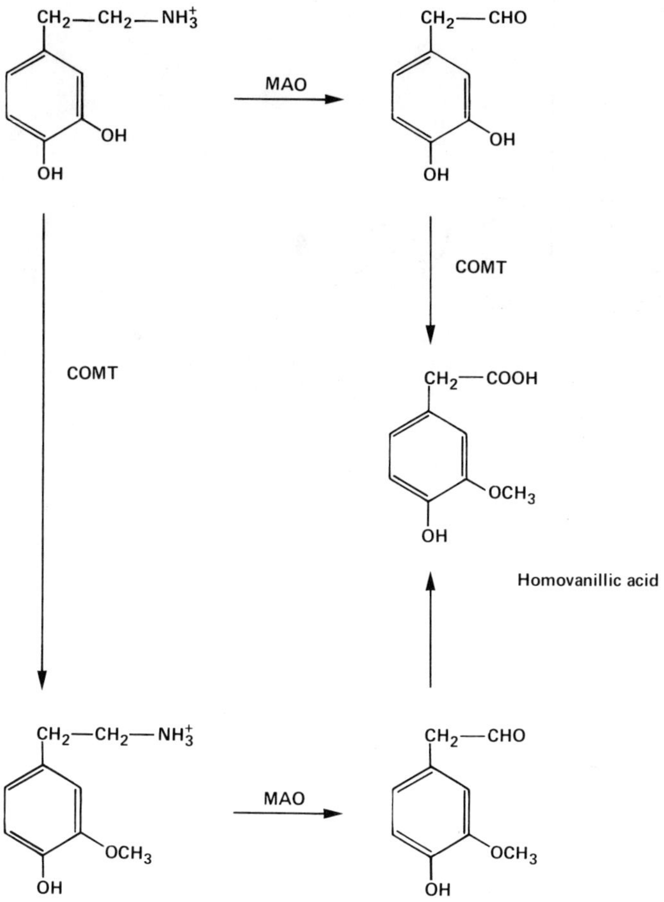

Figure 11.19   The catabolism of dopamine.

The dopaminergic tracts originate in the nigroneostriate and in the diencephalon and the telencephalon (Fig. 11.20). In fact, there are three discrete dopaminergic systems known [113]. The one of most contemporary interest originates in the zona compacta of the substantia nigra, and the axons terminate in the caudate nucleus and the putamen of the corpus striatum. When this tract malfunctions or degenerates, the condition known as parkinsonism results. Parkinsonism is the best understood of the several afflictions of the basal ganglia.

The disease is characterized by shaking, tremor, muscular rigidity, abnormal facial movements, and difficulties in walking, speaking, and writing. It is known from studies on parkinsonism, and from other information, that the basal

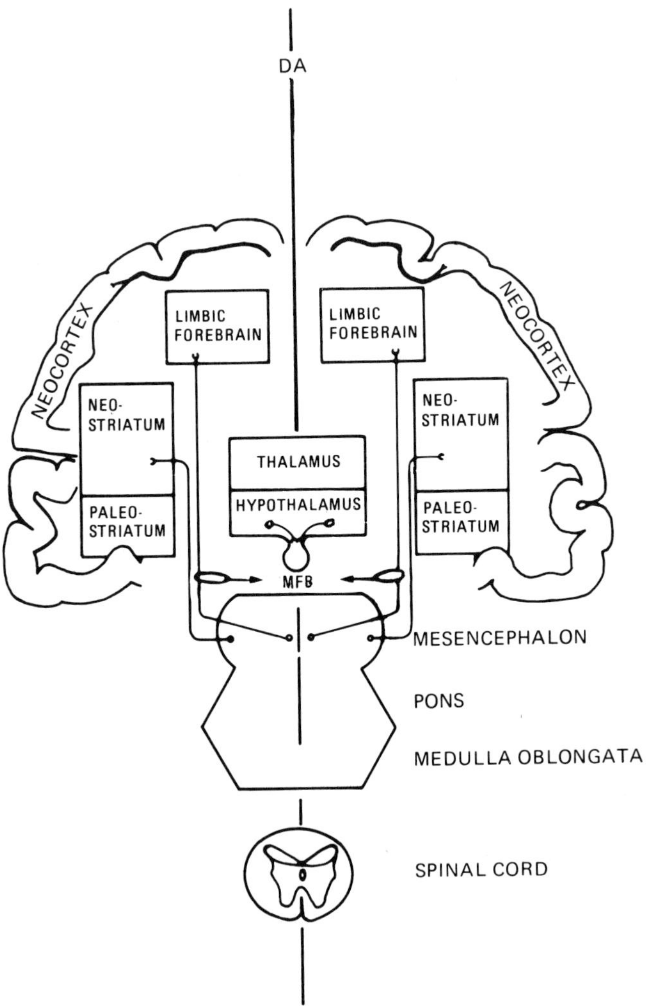

Figure 11.20 Dopamine-containing tracts in the central nervous system. [Redrawn from Anden et al., *Acta Physiol. Scand. 67:*313 (1967).]

ganglia, that portion of the brain which encompasses most of the dopaminergic tracts, regulates, among other things, the tone and the posture of the limbs.

An important and dramatic advance in chemotherapy has been the successful treatment of patients with Parkinson's disease with large amounts of L-dopa [169]. The treatment was suggested by the observation that dopamine, and its specific catabolite homovanillic acid (Fig. 11.19), were lower in Parkinson patients. Further, it was known that inhibitors of monoamine oxidase were not effective in raising dopamine levels in Parkinson patients, but that such inhibitors did raise norepinephrine levels in these individuals. These findings led to the conclusion that the dopamine neurons were affected. Replacement of the transmitter substance by systemic administration of exogenous dopamine is not feasible because the amines, including dopamine, do not cross from blood into brain. But administration of dopa had substantial ameliorative action on the rigidity and the shaking. Large doses were necessary presumably due to the presence of dopa decarboxylase in the endothelial cells of the brain capillaries. Alternatively, smaller amounts of dopa could be given if a decarboxylase inhibitor was given at the same time. The effectiveness of the treatment is dramatic; more than half of all Parkinson patients respond with substantial and visible improvement in their condition.

While at first glance the treatment seems to make good sense, there are still certain questions remaining about why it works at all [168] . It is reasonable to assume that the success of the treatment is due to the increase in dopamine in affected portions of the brain. However, if the dopaminergic neurons are, in fact, destroyed, one can wonder how the administered dopa gets converted to dopamine. It has been suggested that dopa itself, in high doses, is effective, or that dopa is decarboxylated by dopa decarboxylase in some other neurons in the same portion of the brain, or even that dopa undergoes condensation with some other molecule to form the actual active substance. Even if one of these mechanisms is the correct one, we are left with the question as to how flooding the system with massive doses of the wrong material can substitute for the delicate and detailed neural transmission machinery. Thus, the mechanism of the treatment is not yet clear. But whatever the mechanism, it has proved to be one of the most widely publicized and dramatic treatments in recent years, perhaps because the condition of parkinsonism in itself is such a vivid one.

Parkinsonism

The patient, a 64-year-old man, was admitted showing tremor and rigidity of the extremities. The patient required assistance in most daily activities and was able to walk only with difficulty. He had violent grade 4 tremor of the right extremities and grade 1 tremor of the left extremities, with grade 4 cogwheelism of the right side and minimal rigidity on the left side. He also had generalized bradykinesis. His gait was propulsive and festinating. His right upper extremity was useless, and his facies were masked. Tremor and rigidity of the right extremities had appeared 15 years before admission and had become progressively more severe; within the last few years they had also spread to the left extremities.

L-Dopa was administered orally beginning at a dose level of 100 mg, 3 times a day, and increased by 200 to 300 mg every 2 to 4 days with appropriate monitoring for undesirable side effects. An optimum dosage of 4 g/day was established. Tremor and rigidity decreased to barely detectable levels. Nausea and vomiting were encountered at dose levels above 4 g/day. The patient was discharged after 8 weeks of such treatment and continued to function well on dopa therapy with outpatient monitoring.

Serotonin

The metabolism and function of serotonin has also been an area of intense interest [170], and a substantial body of indirect data suggests that serotonin is a transmitter substance in certain neurons. It is synthesized in the neuron, concentrated in the terminals, bound into storage granules or vesicles, and a reuptake mechanism similar to the one for norepinephrine has been described in slices of brain. Serotonin is localized in cell bodies in the nuclei of the lower midbrain and the upper pons called the raphe nucleus [171]. All the cells in the raphe are probably serotonergic. The axons go to all parts of the brain through the median forebrain bundle. Most of the terminals lie in the hypothalamus and very few in the cerebellum and the cortex (Fig. 11.21). These tracts have been revealed by the fluorescence histology described above for norepinephrine [171]. Serotonin gives a bright yellow fluorescence under these conditions. There are no serotonergic neurons known outside the brain. In brain, the serotonin-containing neurons serve the primordial or "older" structures, that is, those which appeared early on in the evolution of the species. Serotonin is the major transmitter in the brains of invertebrates.

The conversion of tryptophan to serotonin is a minor pathway in the whole animal, representing perhaps 1% of the total tryptophan metabolism of the body, but it is the major if not the only pathway of tryptophan metabolism in brain. It is catalyzed by two enzymes, tryptophan hydroxylase and 5-hydroxytryptophan decarboxylase. The first of these, tryptophan-5-hydroxylase, is present in brainstem, pineal, mouse mast cells, and the carcinoid tumor [83,172,173] (Fig. 11.22). The enzyme is a mixed-function oxidase requiring tetrahydropteridine and molecular oxygen, as are the tyrosine and the phenylalaninc hydroxylases. This enzyme is stimulated as well by high levels of mercaptoethanol and by ferrous iron. In the case of the tryptophan hydroxylase the exact pteridine cofactor present in tissues has not been identified formally, but one can anticipate that it will also be tetrahydrobiopterin as in the case of the other two enzymes. The enzyme is rate limiting for the synthesis of serotonin, but unlike tyrosine hydroxylase, which is rate limiting for the production of norepinephrine, tryptophan hydroxylase appears to have a high K_M and is not saturated under normal body conditions. The evidence for this is that elevations in the blood level of tryptophan lead to increases in the tissue content of serotonin [174,175]. As with the other pteridine-dependent hydroxylases, the kinetics of the enzyme vary depend-

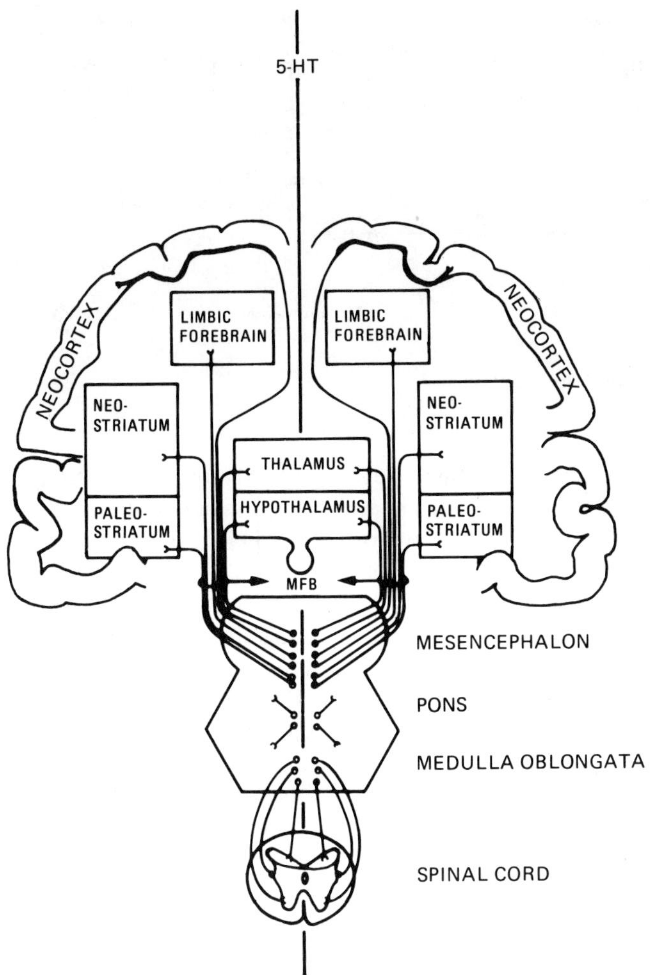

Figure 11.21   Serotonin-containing tracts in the central nervous system.   [Redrawn from Anden et al., *Acta Physiol. Scand.* 67:313 (1967).]

ing on the pteridine used as cofactor [176]. When assayed under optimal conditions the enzyme appears to be primarily in the soluble fraction of the cell [172]. It is known to be localized to some extent in the synaptic portion of the neuron [173]. Phenylalanine and catecholamines inhibit the activity of tryptophan hydroxylase. The latter observation indicates that there is a close relationship be-

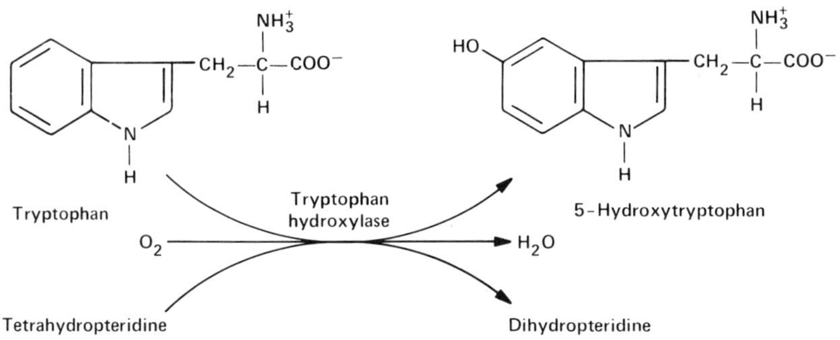

Figure 11.22   Tryptophan hydroxylase.

tween the indole and the catechol pathways of amine formation. Serotonin has
been found to repress the synthesis of the hydroxylase, thus indicating control
at the level of synthesis as well as at the level of activity [177]. Further, prolonged cold stress raises the levels of tryptophan hydroxylase in the brain and
also raises serotonin concentrations, confirming that enzyme levels can be influenced by environmental alterations. Some initial comparative work aimed at a
study of the properties of the brain and the pineal tryptophan hydroxylases
showed that the two enzymes are not identical [178].

The second enzyme, 5-hydroxytryptophan decarboxylase (Fig. 11.23), is
the typical, abundant, and ubiquitous aromatic amino acid decarboxylase discussed previously in the sections relating to the decarboxylation of histidine and
of dopa. As discussed earlier, there has been general agreement for the last several years that the enzyme in kidney, at least, was the same for all these substrates. Just now, however, there has been some suggestion that, in the brain,
which has not been inspected too closely in the past, there are two separate enzymes, one for 5-hydroxytryptophan and one for dopa [137,138].

The major pathway by which serotonin is metabolized to 5-hydroxyindole-
acetic acid (Fig. 11.24), a compound which is a clinical indicator of the state of
the hydroxyindole metabolism of the body, includes the enzymes monoamine
oxidase and aldehyde dehydrogenase. In cases of carcinoid tumor, where the
synthesis of serotonin goes up markedly due to the presence of this pathway in
the tumor, the excretion of 5-hydroxyindoleacetic acid is markedly raised. Inhibition of tryptophan hydroxylase in the tumor gives relief of some of the symptoms, but does not destroy the tumor [179]. Some small amounts of serotonin
are metabolized by methylation either on the amine group or at the 1-position.

It is said that serotonin levels have something to do with control of mood
[170]. Certainly the effects of serotonin depletion are not dramatically obvious.

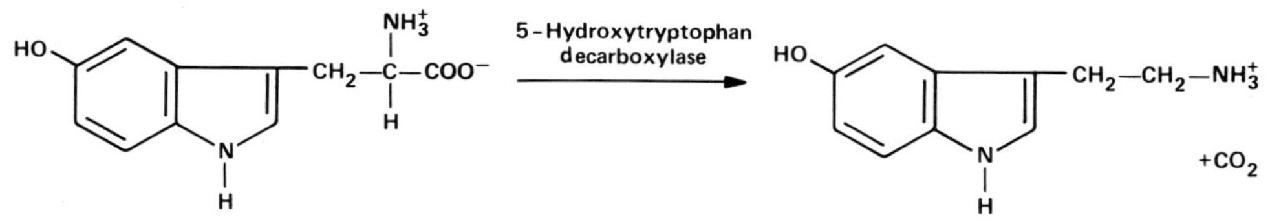

5-Hydroxytryptophan

Serotonin

Figure 11.23 5-Hydroxytryptophan decarboxylase.

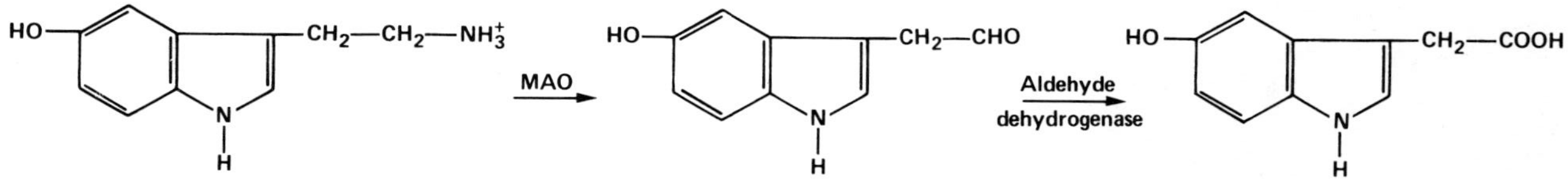

Serotonin

5-Hydroxyindoleacetic acid

Figure 11.24 Catabolism of serotonin.

Drugs which reduce serotonin levels in the brain by 90 or 95% do not produce overt effects in experimental animals. It is only when the behavior of animals is inspected by rather subtle means that the possible actions of serotonin become discernible. Among the most interesting lines of study are the experiments done by Jouvet [180,181], and now by others, relating serotonin levels and sleep patterns. Insomnia in experimental animals can be produced by either a physical destruction of the raphe nucleus, where the serotonin neurons are, or by inhibiting the tryptophan hydroxylase. Sleep, then, can be restored by giving 5-hydroxytryptophan along with the inhibitor. Stimulated by this work, a number of investigators have looked at the connection between the indoleamines and sleep patterns. It seems that alterations in the level of the amines or of the amine-forming enzymes can induce changes in sleep patterns, and vice versa [182]. Although the effects on sleep are the best documented and the most "respectable" of the effects of serotonin manipulation, the use of specific inhibitors of tryptophan hydroxylase have suggested that levels of serotonin influence a number of different types of behavior. Specifically, serotonin levels have been reported to be involved in aggression, sexual behavior, alcoholism, learning and retardation, and drug addiction, as well as in sleep patterns.

The inhibitor which has been of most use in these studies is p-chlorophenylalanine. First observed by Koe and Weissman to deplete brain serotonin [183], it was later shown to be a potent and irreversible inhibitor of tryptophan hydroxylase [184]. Use of p-chlorophenylalanine as a specific inhibitor of the synthesis of serotonin has sparked most of the reports as to its functions. However, a number of other inhibitors are also of use. Of course, most of the drugs discussed in the last section which impinge on the norepinephrine pathway, also are involved in altering serotonin levels. Inhibitors of synthesis lower the levels, and inhibitors of the breakdown, specifically the monoamine oxidase inhibitors, raise the levels. Reserpine blocks the storage of serotonin [185] as it does the storage of the catecholamines. There seems to be some specificity among the various inhibitors of the amine pumps. Imipramine, for example, blocks serotonin reuptake, and it is more active against serotonin reuptake than against the reuptake of norepinephrine, as is its derivative, amitryptyline. Other derivatives, such as protriptyline and desipramine, are more active against norepinephrine. The differences in effect are subtle, the norepinephrine blockers producing behavioral activation, while the serotonin blockers are known for elevation of mood. The structural relationship between serotonin and some of the hallucinogenic agents has caused speculation about the role of serotonin in the mechanism of these drugs. Such speculations have been reinforced by the observation that lysergic acid diethylamide (LSD) antagonizes the action of serotonin in certain nonneural test systems [186,187], such as the isolated uterus, in which serotonin causes contraction. It is widely assumed that the hallucinogens interact with serotonin receptors, but there is very little data to support this contention [110,170].

The pineal, while strictly not a part of the brain of higher animals, has received the attention of neurochemists because of its high concentration and unique metabolic pathways of serotonin [188]. The pineal in amphibia is a photoreceptor and sends impulses directly to the brain. In mammals, the pineal is joined to the brain by connective tissue only. It is not influenced directly by light. It responds indirectly to visual events through a neural pathway from the eyes, through the inferior accessory optic system, and, finally, the pineal sympathetic system. The gland is innervated sympathetically from the superior cervical ganglia. The pineal cells themselves are a unique type called pinealocytes which have long processes terminating near the capillaries which run through the gland.

The indoleamine hormone of the pineal, melatonin, was discovered in 1958 by Lerner [189,190] who was studying the action of melanocyte-stimulating hormone (MSH). The pineal principal which antagonized the action of MSH was eventually characterized as melatonin. Melatonin acts on the brain and elsewhere and influences the development of the various other glands of the body, including the thyroid, the pituitary, and the gonads.

The pineal contains the highest concentration of serotonin in the body. It also contains tryptophan-5-hydroxylase and 5-hydroxytryptophan decarboxylase for the production of serotonin. The further metabolism of serotonin to melatonin (Fig. 11.25) occurs in mammals only in the pineal, but in amphibia this pathway is present in brain and in the eye as well. The work of Axelrod and Weissbach elucidated the enzymatic steps leading to melatonin [191-193]. The first step, an N-acetylation of serotonin, is catalyzed by an acetylating enzyme using acetyl CoA as a source of acetyl groups. The second enzyme, hydroxyindole-O-methyltransferase, produces melatonin from N-acetylserotonin using S-adenosylmethionine as the methylating agent. An alternate pathway for the metabolism of serotonin in the pineal is known. This pathway is also unique to pineal and involves the conversion of serotonin to the corresponding aldehyde and then to 5-hydroxytryptophol, the corresponding alcohol. 5-Hydroxytryptophol can also be methylated, presumably by hydroxyindole-O-methyltransferase, but neither the function of this compound nor the significance of this alternate pathway have received much attention.

The amine levels and the levels of the relevant enzymes fluctuate with a daily rhythm. Serotonin content itself shows a 10-fold fluctuation between 12 noon and 12 midnight. This rhythm does not seem to depend on the exposure of the animal to light or dark since it is also found in blinded animals or in animals kept in constant illumination [113,194]. The rise and fall in the serotonin concentration seems to be due to a periodic release of serotonin from its storage sites, thereby making it available to degradative enzymes [195]. The activity of hydroxyindole-O-methyltransferase, the second enzyme in the synthesis of melatonin, also exhibits a diurnal rhythm. The activity is at least 10-fold higher at night than in the daytime. This rhythm, however, seems to be directly influenced by illumination since it no longer persists in blinded animals [188].

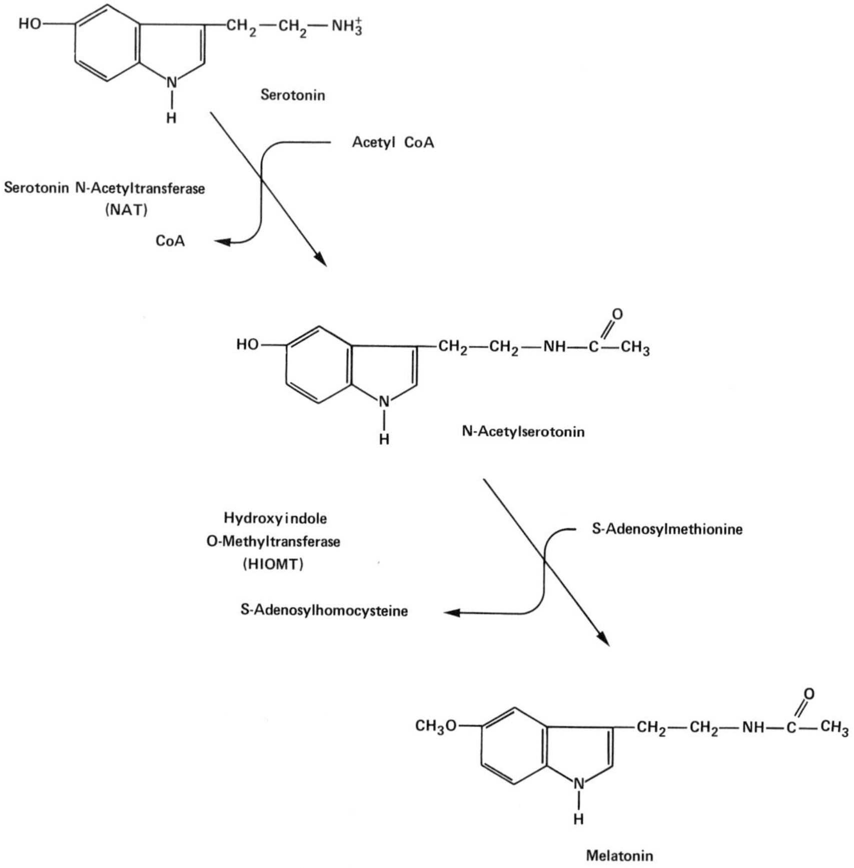

Figure 11.25   Formation of melatonin.

The whole process of melatonin metabolism and release and, in fact, pineal function in general, is mediated by the sympathetic nervous system. The pineal, as mentioned before, is innervated by norepinephrine-containing neurons from the superior cervical ganglion. The gland has a high content of sympathetic nerve endings, and, thus, a high content of norepinephrine and the enzymes which are involved in its synthesis. Norepinephrine content and the activity of tyrosine hydroxylase in these nerve endings also exhibit a diurnal rhythm. The terminals are unique in that they store serotonin as well as norepinephrine, but this may be due only to the fact that a uniquely high concentration of serotonin exists around them. The synthesis of melatonin has been shown to be under direct control by the catecholamine levels. Studies using the pineal in organ culture, where it will make serotonin and melatonin from tryptophan, have shown that serotonin N-acetyltransferase, the first biosynthetic enzyme, can be markedly stimulated by the addition of norepinephrine, the normal product of the noradrenergic nerve terminals [196,197]. These studies have further implicated cAMP as the "second messenger" in this system [198,199]. That is, the data presented indicate that norepinephrine administration, and presumably the normal release of norepinephrine from the terminals, raises the level of cAMP, which, in turn, increases the level of the acetylase enzyme.

In any case, the metabolism of indoleamines in the pineal leads to the release of melatonin, which serves in the body as some kind of regulating influence on hormonal development. This release is controlled by the catecholamines released, in turn, from the nerve terminals under the influence of the light impinging on the eye. The apparent transfer of responsibility between the catecholamines and the indoleamines via the pineal gland has led to the concept of this gland as a neurohumoral transducer.

Taurine

Taurine is one of the major constituents of the "free amino acid pool" of brain. There are at least three possible routes by which it could be formed [200], but the major route appears to be by oxidation from cysteine, through cysteine sulfinic acid and hypotaurine (Fig. 11.26). The presence of taurine in brain was originally observed by Stein and Moore [201]. In addition to its abundance in brain, it is now known to be in very high concentration in the pineal gland, the pituitary, and the retina. In support of the biosynthetic pathway outlined above, the presence in brain of low amounts of hypotaurine has now been reported.

The function of taurine in brain is not known, but several lines of evidence support its action as a neurotransmitter. First, it has depressant action when applied iontophoretically to neurons of the rat spinal cord or the cortex [202]. Second, it can be shown that taurine is released from the cortex upon electrical

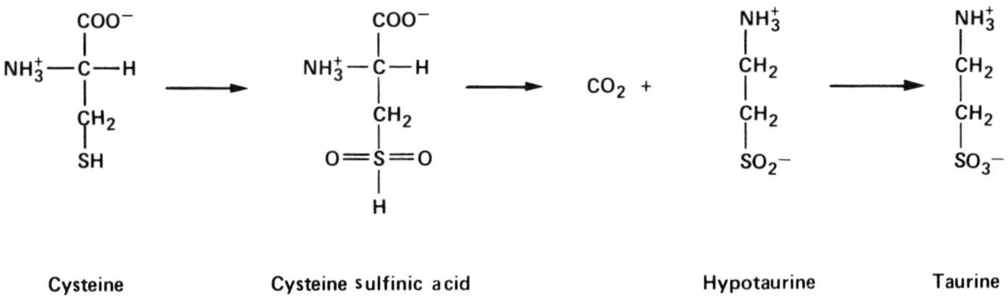

Figure 11.26   Formation of taurine.

stimulation and that calcium is required for the release [203]. Third, it has been found that taurine is localized in the nerve endings through work on the synaptosome fraction, and the decarboxylase functioning in its synthesis from cysteine has also been found in this site [204]. And finally, a specific reuptake system with high affinity for taurine has been observed using brain slices [203] suggesting that a mechanism exists for the termination of the synaptic action of taurine. Thus, the criteria [205] generally accepted for the assignment of transmitter function to a molecule are fulfilled for taurine. As yet, however, no specific tracts for which taurine might serve have been identified. Unfortunately, the several studies on the regional distribution of taurine in brain do not give a clue to this problem.

The Polyamines

The polyamines have attracted an increasing amount of attention in recent years because of their apparent involvement in the processes of rapid cell growth and development [206]. In brain, as well, these agents have been a subject of some interest, especially in studies on brain development. The biosynthesis of the polyamines involves the decarboxylation of ornithine by ornithine decarboxylase (Fig. 11.27) to give putrescine, which is thought to be the rate-limiting step for the further production of the other amines, spermidine and spermine. The developmental course of this enzyme in brain supports the suggestion that it is the controlling step in the biosynthesis [207,208]. It was found in these studies that the levels of ornithine decarboxylase in various brain regions increased at about the same point in time as the most rapid development of that particular portion. And, further, the increases in the enzyme paralleled increases in the spermidine content, thus highlighting the controlling importance of the activity

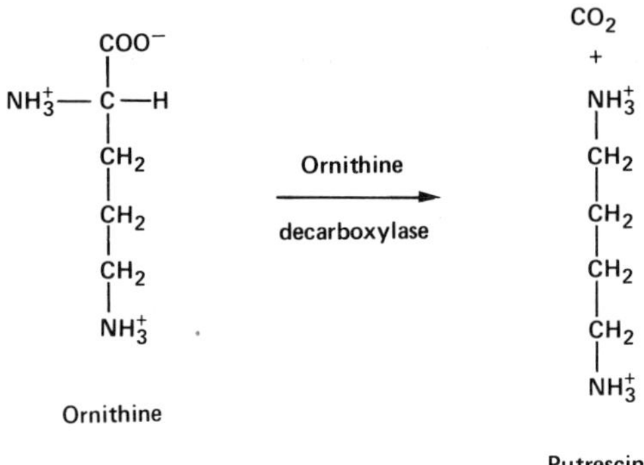

Figure 11.27 Ornithine decarboxylase.

of the decarboxylase. The characteristics of this enzyme in brain do not appear
to have been studied in detail, and, in fairness, it should be mentioned that some
workers have been unable to assay the enzyme in brain at all [209].

The formation of spermidine from putrescine in brain requires the participation of two enzymes [210]. One enzyme is involved in the decarboxylation
of S-adenosylmethionine and the other in the transfer of the propylamine portion to the putrescine moiety (Fig. 11.28). The nature of the decarboxylase in
brain has recently been investigated. It was found [211] that the brain enzyme
resembled the homologous enzyme from liver in its kinetics and pH optimum
and that it had a definite developmental course which exhibited a peak at 10 days
of age in the rat, a point of rapid brain growth, but one which precedes the formation of myelin. The conversion of spermidine to spermine has not been studied
in brain. As is mentioned in an earlier section, there is evidence that putre-

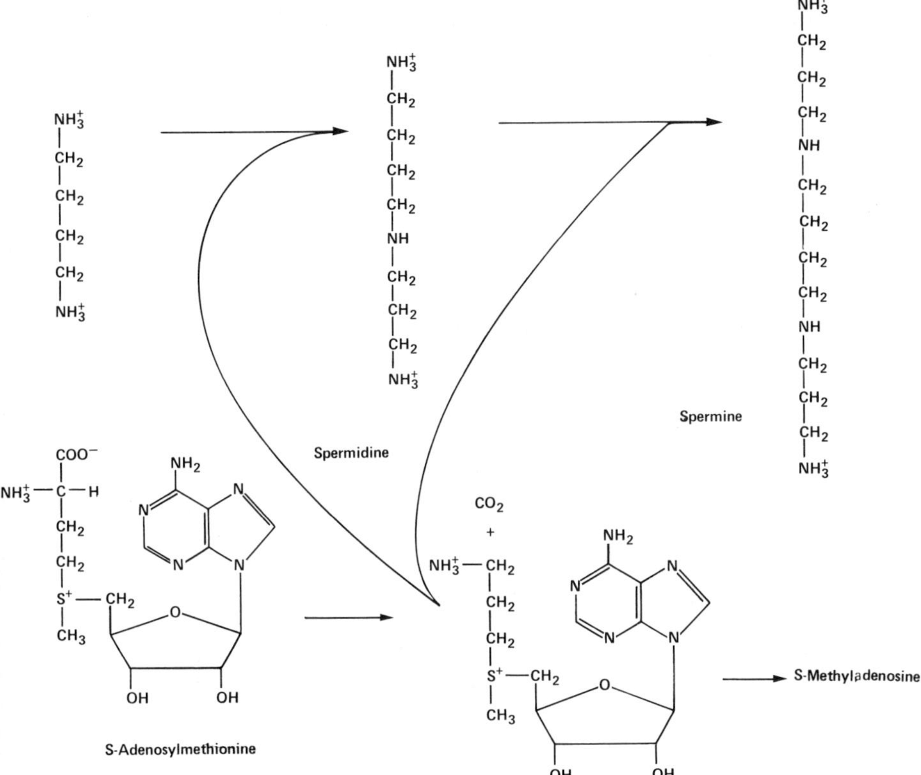

Figure 11.28   Formation of the polyamines.

anine, a recently discovered brain constituent, is derived metabolically from spermidine.

The function of the polyamines has been a subject of wide interest and speculation, but as yet no conclusive experimental work. Suggestions that they are involved in processes of rapid growth, perhaps by some effects on nucleic acid structure or on membranes, have been discussed. In brain, the information which is available could lend itself to the acceptance of such a role. As mentioned before, the levels of spermidine and of the spermidine-synthesizing enzymes increase markedly at the time of the most rapid development of the relevant brain portion. The amines overall are widely and evenly distributed in the various brain regions. Some studies suggest that the two amines, spermidine and spermine, are not distributed in a parallel manner in the brain. These studies indicate that spermine predominates in gray matter while spermidine predominates in white matter and in peripheral nerves [212,213]. But these observations have not led to any cogent suggestions about function. There have been reports that one or other of the amines will enhance RNA synthesis in brain nuclei [214], stimulate RNA polymerase in brain [215], stabilize brain ribosomes [216], and stimulate protein synthesis [217] in brain. More recent has been the observation that the amine-forming enzyme, ornithine decarboxylase, can interact directly with RNA polymerase I and enhance its activity [218]. However, the experiments presented so far do not specify which, if any, of these functions are served in the normal course of brain metabolism.

References

1.   H. H. Tallan, S. Moore, and W. H. Stein, *J. Biol. Chem. 219:*257 (1956).
2.   H. H. Tallan, *J. Biol. Chem. 224:*41 (1957).
3.   M. C. Fleming and O. H. Lowry, *J. Neurochem. 13:*779 (1966).
4.   J. V. Nadler and J. R. Cooper, *J. Neurochem. 19:*313 (1972).
5.   F. B. Goldstein, *J. Biol. Chem. 234:*2702 (1959).
6.   F. B. Goldstein, *J. Biol. Chem. 244:*4357 (1969).
7.   S. Berl, G. Takagaki, D. D. Clarke, and H. Waelsch, *J. Biol. Chem. 237:* 2562 (1962).
8.   R. U. Margolis, S. S. Bankulis, and A. Geiger, *J. Neurochem. 5:*379 (1960).
9.   R. M. O'Neal and R. E. Koeppe, *J. Neurochem. 13:*385 (1966).
10. K. L. Reichelt and E. Kvamme, *J. Neurochem. 14:*987 (1967).
11. J. V. Nadler and J. R. Cooper, *J. Neurochem. 19:*2091 (1972).
12. S. M. Birnbaum, L. Levintow, R. B. Kingsley, and J. P. Greenstein, *J. Biol. Chem. 194:*455 (1952).
13. A. F. D'Adamo, J. Crispin Smith, and C. Woiler, *J. Neurochem. 20:*1275 (1973).
14. A. F. D'Adamo and F. M. Yatsu, *J. Neurochem. 13:*961 (1966).
15. A. F. D'Adamo, L. I. Gidez, and F. M. Yatsu, *Exp. Brain Res. 5:*267 (1968).

16. T. Kakimoto, T. Nakajima, A. Kumon, Y. Matsuoka, N. Imaoka, A. Kanazawa, and I. Sano, *J. Biol. Chem.* *244:*6003 (1969).
17. T. L. Perry, S. Hansen, and M. Kloster, *J. Neurochem.* *19:*1395 (1972).
18. T. Nakajima, *J. Neurochem.* *20:*735 (1973).
19. C. F. Baxter, in *Handbook of Neurochemistry,* Vol. III (A. Lajtha, ed.), Plenum, New York, 1970.
20. E. Roberts and R. Hammerschlag, in *Basic Neurochemistry* (R. W. Albers, G. J. Siegel, R. Katzman, and B. W. Agranoff, eds.), Little, Brown, Boston, 1972.
21. J. Awapara, A. J. Landau, R. Fuerst, and B. Seale, *J. Biol. Chem.* *187:*35 (1950).
22. E. Roberts and S. Frankel, *J. Biol. Chem.* *187:*55 (1950).
23. A. Bazemore, K. A. C. Elliott, and E. Florey, *Nature (London)* *178:*1052 (1956).
24. M. Otsuka, L. L. Iversen, Z. W. Hall, and E. A. Kravitz, *Proc. Natl. Acad. Sci. USA* *56:*1110 (1966).
25. A. Takeuchi and N. Takeuchi, *J. Physiol. (London)* *177:*225 (1965).
26. K. Kuriyama, B. Haber, B. Sisken, and E. Roberts, *Proc. Natl. Acad. Sci. USA* *55:*846 (1966).
27. K. Obata, M. Otsuka, and Y. Tanaka, *J. Neurochem.* *17:*697 (1970).
28. J. Dudel and S. W. Kuffler, *J. Physiol. (London)* *155:*543 (1961).
29. A. Takeuchi and N. Takeuchi, *J. Physiol. (London)* *183:*433 (1966).
30. R. Balazs, Y. Machiyama, B. J. Hammond, T. Julian, and D. Richter, *Biochem. J.* *116:*445 (1970).
31. E. Roberts and S. Frankel, *J. Biol. Chem.* *188:*789 (1951).
32. J. Y. Wu, T. Matsuda, and E. Roberts, *J. Biol. Chem.* *248:*3029 (1973).
33. T. Matsuda, J. Y. Wu, and E. Roberts, *J. Neurochem.* *21:*167 (1973).
34. J. D. Wood and W. J. Watson, *Can. J. Biochem. Physiol.* *41:*1907 (1963).
35. J. D. Wood and W. J. Watson, *Can. J. Physiol. Pharmacol.* *42:*277 (1964).
36. L. Salganicoff and E. deRobertis, *J. Neurochem.* *12:*287 (1965).
37. E. Roberts and K. Kuriyama, *Brain Res.* *8:*1 (1968).
38. L. T. Graham, Jr., C. F. Baxter, and R. N. Lolly, *Brain Res.* *20:*379 (1970).
39. P. Krause, *Hoppe-Seylers Z. Physiol. Chem.* *349:*1425 (1968).
40. J. K. Tews and R. A. Lovell, *J. Neurochem.* *14:*1 (1967).
41. R. A. Bayoumi, J. A. Kirwan, and W. R. D. Smith, *J. Neurochem.* *19:*569 (1972).
42. R. A. Bayoumi and W. R. D. Smith, *J. Neurochem.* *19:*1883 (1972).
43. B. Haber, P. Sze, K. Kuriyama, and E. Roberts, *Brain Res.* *18:*545 (1970).
44. K. Kuriyama, B. Haber, and E. Roberts, *Brain Res.* *23:*121 (1970).
45. B. Haber, K. Kuriyama, and E. Roberts, *Biochem. Pharmacol.* *19:*1119 (1970).
46. B. Haber, K. Kuriyama, and E. Roberts, *Science* *168:*598 (1970).
47. D. L. Martin and L. P. Miller, in *GABA in Nervous System Function* (E. Roberts, T. N. Chase, and D. B. Tower, eds.), Raven, New York, 1976.
48. J. Y. Wu, in *GABA in Nervous System Function* (E. Roberts, T. N. Chase, and D. B. Tower, eds.), Raven, New York, 1976.

49. S. P. Bessman, J. Rossen, and E. C. Layne, *J. Biol. Chem. 201:*385 (1953).
50. C. F. Baxter and E. Roberts, *J. Biol. Chem. 233:*1135 (1958).
51. Y. Nishigawa, T. Kodama, and S. Konishi, *J. Vitaminol. 5:*117 (1959).
52. R. W. Albers and G. J. Koval, *Biochim. Biophys. Acta 52:*29 (1961).
53. F. N. Pitts, Jr., and C. Quick, *J. Neurochem. 12:*893 (1965).
54. C. Kammeratt and H. Veldstra, *Biochim. Biophys. Acta 151:*1 (1968).
55. Z. W. Hall and E. A. Kravitz, *J. Neurochem. 14:*55 (1967).
56. K. A. C. Elliott and N. M. van Gelder, *J. Physiol. (London) 153:*423 (1960).
57. K. Sano and E. Roberts, *Biochem. Pharmacol. 12:*489 (1963).
58. S. Varon, H. Weinstein, T. Kakefuda, and E. Roberts, *Biochem. Pharmacol. 14:*1213 (1965).
59. H. Weinstein, S. Varon, D. R. Muhleman, and E. Roberts, *Biochem. Pharmacol. 14:*273 (1965).
60. S. Varon, H. Weinstein, C. F. Baxter, and E. Roberts, *Biochem. Pharmacol. 14:*1755 (1965).
61. P. Strasberg and K. A. C. Elliott, *Can. J. Biochem. 14:*1795 (1967).
62. K. Kuriyama, E. Roberts, and T. Kakefuda, *Brain Res. 8:*132 (1968).
63. L. J. Fowler and R. A. John, *Biochem. J. 130:*569 (1972).
64. L. J. Fowler, *J. Neurochem. 21:*437 (1973).
65. H. McLennan, *Nature (London) 228:*674 (1970).
66. A. W. Duggan and H. McLennan, *Brain Res. 25:*188 (1971).
67. F. Irrevere, R. I. Evans, A. R. Hayden, and R. Silber, *Nature (London) 180:*704 (1957).
68. J. J. Pisano, D. Abraham, and S. Udenfriend, *Arch. Biochem. Biophys. 100:*323 (1963).
69. C. Edwards and S. W. Kuffler, *J. Neurochem. 4:*19 (1959).
70. H. Takahashi, B. Arai, and C. Koshino, *Jpn. J. Physiol. 11:*403 (1961).
71. D. P. Purpura, M. Girado, and H. Grundfest, *Science 127:*1179 (1958).
72. D. Jinnai, A. Sawai, and A. Mori, *Nature (London) 216:* 17 (1966).
73. E. A. Hosein, *Arch. Biochem. Biophys. 100:*32 (1963).
74. E. A. Hosein, M. Smart, K. Hawkins, S. Rochon, and Z. Strasberg, *Arch. Biochem. Biophys. 96:*246 (1962).
75. M. A. Mehlman and G. A. Wolf, *Arch. Biochem. Biophys. 98:*146 (1962).
76. G. Lindstedt and S. Lindstedt, *J. Biol. Chem. 240:*316 (1965).
77. G. Lindstedt and S. Lindstedt, *Biochem. Biophys. Res. Commun. 7:*394 (1962).
78. J. P. Green, in *Handbook of Neurochemistry,* Vol. IV (A. Lajtha, ed.), Plenum, New York, 1970.
79. J. W. Phillis, A. K. Tebecis, and D. H. York, *Br. J. Pharmacol. 33:*426 (1968).
80. E. A. Carlini and J. P. Green, *Br. J. Pharmacol. 20:*264 (1963).
81. U. S. von Euler, in *Histamine* (G. E. W. Wolstenholme, ed.), CIBA Foundation Symposium, Little, Brown, Boston, 1956.
82. R. S. Tuttle, *Am. J. Physiol. 213:*620 (1967).

83. G. Guroff and W. Lovenberg, in *Handbook of Neurochemistry*, Vol. III (A. Lajtha, ed.), Plenum, New York, 1970.

84. D. Aures, R. Hakanson, and W. G. Clark, in *Handbook of Neurochemistry*, Vol. IV (A. Lajtha, ed.), Plenum, New York, 1970.

85. P. A. Ganrot, A. M. Rosengren, and E. Rosengren, *Experientia 17:*263 (1961).

86. H. Weissbach, W. Lovenberg, and S. Udenfriend, *Biochim. Biophys. Acta 50:*177 (1961).

87. A. M. Rothschild and R. W. Schayer, *Biochim. Biophys. Acta 34:*392 (1959).

88. P. Ono and P. Hagen, *Nature (London) 184:*1143 (1959).

89. R. Hakanson, *Eur. J. Pharmacol. 1:*383 (1967).

90. P. Holtz and E. Westermann, *Arch. Exp. Pathol. Pharmacol. 227:*538 (1956).

91. J. C. Schwartz, C. Lampart, and C. Rose, *J. Neurochem. 17:*1527 (1970).

92. R. Hakanson, *Eur. J. Pharmacol. 1:*42 (1967).

93. R. W. Schayer and O H. Ganley, *Am. J. Physiol. 197:*721 (1959).

94. G. Kahlson, E. Rosengren, D. Swann, and R. Thunberg, *J. Physiol. (London) 174:*400 (1964).

95. W. Lovenberg, H. Weissbach, and S. Udenfriend, *J. Biol. Chem. 237:*89 (1962).

96. R. W. Schayer and M. A. Reilly, *J. Neurochem. 17:*1649 (1970).

97. K. M. Taylor and S. H. Snyder, *J. Neurochem. 19:*341 (1972).

98. J. C. Schwartz, C. Lampart, and C. Rose, *J. Neurochem. 19:*801 (1972).

99. D. D. Brown, R. Tomchick, and J. Axelrod, *J. Biol. Chem. 234:*2948 (1959).

100. P. A. Shore, A. Burkhalter, and V. H. Cohn, *J. Pharm. Exp. Ther. 127:* 182 (1959).

101. L. T. Kremzer and C. C. Pfeiffer, *Biochem. Pharmacol. 14:*1189 (1965).

102. I. A. Michaelson and P. Z. Coffman, *Anal. Biochem. 27:*257 (1969).

103. K. M. Taylor and S. H. Snyder, *J. Neurochem. 19:*1343 (1972).

104. J. D. Robinson, J. H. Anderson, and J. P. Green, *J. Pharmacol. Exp. Ther. 147:*236 (1965).

105. K. M. Taylor and S. H. Snyder, *Science 172:*1037 (1971).

106. H. M. Adam and H. K. A. Hye, *Br. J. Pharmacol. 28:*137 (1966).

107. A. V. Furano and J. P. Green, *J. Physiol. (London) 170:*263 (1964).

108. J. P. Green, *Fed. Proc. 23:*1095 (1964).

109. S. Kakiuchi and T. W. Rall, *Mol. Pharmacol. 4:*367 (1968).

110. E. Costa and N. H. Neff, in *Handbook of Neurochemistry*, Vol. IV (A. Lajtha, ed.), Plenum, New York, 1970.

111. J. Glowinski, in *Handbook of Neurochemistry*, Vol. IV (A. Lajtha, ed.), Plenum, New York, 1970.

112. L. L. Iversen, in *Handbook of Neurochemistry*, Vol. IV (A. Lajtha, ed.), Plenum, New York, 1970.

113. S. Snyder, in *Basic Neurochemistry* (R. W. Albers, G. J. Siegel, R. Katzman, and B. W. Agranoff, eds.), Little, Brown, Boston, 1972.

114. H. Blaschko, *J. Physiol. (London)* 96:5-51P (1939).

115. J. Glowinski and R. J. Baldessarini, *Pharmacol. Rev.* 18:1201 (1966).

116. T. Hokfelt, *Z. Zellforsch. Mikrosk. Anat.* 79:110 (1967).

117. G. Burnstock and M. E. Holman, *Pharmacol. Rev.* 18:481 (1966).

118. W. W. Douglas, in *Mechanisms of Release of Biogenic Amines* (U. S. von Euler, S. Rosell, and B. Uvnas, eds.), Wenner Gren Center International Symposium Series, Pergamon, New York, 1966.

119. N. A. Hillarp, K. Fuxe, and A. Dahlstrom, *Pharmacol. Rev.* 18:727 (1966).

120. N. A. Hillarp, K. Fuxe, and A. Dahlstrom, in *Mechanisms of Release of Biogenic Amines* (U. S. von Euler, S. Rosell, and B. Uvnas, eds.), Wenner Gren Center International Symposium Series, Pergamon, New York, 1966.

121. N. E. Anden, A. Dahlstrom, K. Fuxe, K. Larsson, L. Olson, and U. Ungerstedt, *Acta Physiol. Scand.* 67: 13 (1966).

122. A. Dahlstrom and J. Haeggendal, *Acta Physiol. Scand.* 67:278 (1966).

123. J. Haeggendal and A. Dahlstrom, *J. Pharm. Pharmacol.* 21:55 (1969).

124. T. Nagatsu, M. Levitt, and S. Udenfriend, *J. Biol. Chem.* 239:2910 (1964).

125. T. Lloyd and N. Weiner, *Mol. Pharmacol.* 7:569 (1971).

126. R. Shiman, M. Akino, and S. Kaufman, *J. Biol. Chem.* 246:1330 (1971).

127. B. Petrack, F. Sheppy, V. Fetzer, T. Manning, H. Chertock, and T. Ma, *J. Biol. Chem.* 247:4872 (1972).

128. M. Levitt, J. Daly, G. Guroff, and S. Udenfriend, *Arch. Biochem. Biophys.* 126:593 (1968).

129. S. Udenfriend, *Pharmacol. Rev.* 18:43 (1966).

130. S. Udenfriend, P. Zaltzman-Nirenberg, and T. Nagatsu, *Biochem. Pharmacol.* 14:837 (1965).

131. S. Spector, R. Gordon, A. Sjoerdsma, and S. Udenfriend, *Mol. Pharmacol.* 3:549 (1967).

132. M. Ikeda, M. Levitt, and S. Udenfriend, *Biochem. Biophys. Res. Commun.* 18:482 (1965).

133. W. Lovenberg, E. A. Bruckwick, and I. Hanbauer, *Proc. Natl. Acad. Sci. USA* 72:2955 (1975).

134. T. Lloyd and S. Kaufman, *Biochem. Biophys. Res. Commun.* 66:907 (1975).

135. C. H. Letendre, P. MacDonnell, and G. Guroff, *Biochem. Biophys. Res. Commun.* 74:891 (1977).

136. T. Lloyd and S. Kaufman, *Mol. Pharmacol.* 9:438 (1973).

137. J. G. Christenson, W. Dairman, and S. Udenfriend, *Proc. Natl. Acad. Sci. USA* 69:343 (1972).

138. K. L. Sims, G. A. Davis, and F. E. Bloom, *J. Neurochem.* 20:449 (1973).

139. S. Friedman and S. Kaufman, *J. Biol. Chem.* 240:4763 (1965).

140. A. Foldes, P. L. Jeffrey, B. N. Preston, and L. Austin, *J. Neurochem.* 20:1431 (1973).

141. M. Goldstein, T. H. Joh, and T. Q. Garvey, *Biochemistry* 7:2724 (1968).

142.   T. Nagatsu, H. Hidaka, H. Kuzuya, and K. Takeya, *Biochem. Pharmacol.*
       *19:*35 (1970).
143.   J. Axelrod, *J. Biol. Chem. 237:*1657 (1962).
144.   R. J. Wurtman and J. Axelrod, *J. Biol. Chem. 241:*2301 (1966).
145.   S. Udenfriend, *Harvey Lect. 60:*57 (1966).
146.   F. Oesch, U. Otten, and H. Thoenen, *J. Neurochem. 20:*1691 (1973).
147.   G. F. Wooten and J. T. Coyle, *J. Neurochem. 20:*1361 (1973).
148.   W. Dairman, L. Geffen, and M. Marchelle, *J. Neurochem. 20:*1617 (1973).
149.   H. Thoenen, *Pharmacol. Rev. 24:*255 (1972).
150.   P. B. Molinoff and J. Axelrod, *Ann. Rev. Biochem. 40:*465 (1971).
151.   I. A. Hendry, L. L. Iversen, and I. B. Black, *J. Neurochem. 20:*1683
       (1973).
152.   J. Axelrod and R. Tomchick, *J. Biol. Chem. 233:*702 (1958).
153.   J. Axelrod, *Physiol. Rev. 39:*751 (1959).
154.   I. J. Kopin, *Pharmacol. Rev. 16:*179 (1964).
155.   J. H. Quastel, in *Handbook of Neurochemistry,* Vol. IV (A. Lajtha, ed.),
       Plenum, New York, 1970.
156.   T. L. Sourkes, in *Basic Neurochemistry* (R. W. Albers, G. J. Siegel, R. Katz-
       man, and B. W. Agranoff, eds.), Little, Brown, Boston, 1972.
157.   T. L. Sourkes, *Adv. Pharmacol. 6A:*61 (1968).
158.   V. Z. Gorkin, *Pharmacol. Rev. 18:*115 (1966).
159.   M. Sandler and M. B. H. Youdim, *Pharmacol. Rev. 24:*331 (1972).
160.   K. T. Yasunobu, I. Igaue, and B. Gomes, *Adv. Pharmacol. 6A:*43 (1968).
161.   B. Gomes, I. Igaue, H. G. Kloepter, and K. T. Yasunobu, *Arch. Biochem.*
       *Biophys. 132:*16 (1969).
162.   G. Hertting and J. Axelrod, *Nature (London) 197:*172 (1961).
163.   J. Axelrod, *Recent Prog. Horm. Res. 21:*597 (1965).
164.   L. L. Iversen, *The Uptake and Storage of Noradrenaline in Sympathetic*
       Nerves, Cambridge University Press, London, 1967.
165.   A. Pletscher, P. A. Shore, and B. B. Brodie, *J. Pharmacol. Exp. Ther. 116:*
       84 (1956).
166.   I. J. Kopin, *Ann. Rev. Pharmacol. 8:*377 (1968).
167.   O. Hornykiewicz, *Pharmacol. Rev. 18:*925 (1966).
168.   T. L. Sourkes, in *Basic Neurochemistry* (R. W. Albers, G. J. Siegel, R. Katz-
       man, and B. W. Agranoff, eds.), Little, Brown, Boston, 1972.
169.   G. D. Cotzias, M. H. VanWoert, and L. M. Schiffer, *N. Engl. J. Med. 276:*
       374 (1967).
170.   I. H. Page and A. Carlsson, in *Handbook of Neurochemistry,* Vol. IV (A.
       Lajtha, ed.), Plenum, New York, 1970.
171.   A. Dahlstrom and K. Fuxe, *Acta Physiol. Scand.,* Suppl. 232, *62:*1 (1965).
172.   W. Lovenberg, E. Jequier, and A. Sjoerdsma, *Adv. Pharmacol. 6A:*21
       (1968).
173.   A. Ichiyama, S. Nakamura, Y. Nishizuka, and O. Hayaishi, *Adv. Pharma-*
       *col. 6A:*5 (1968).
174.   J. D. Fernstrom and R. J. Wurtman, *Science 173:*149 (1971).

175. J. D. Fernstrom and R. J. Wurtman, *Science 174:*1023 (1971).

176. P. A. Friedman, A. H. Kappelman, and S. Kaufman, *J. Biol. Chem. 247:* 4165 (1972).

177. S. Eiduson, *J. Neurochem. 13:*923 (1966).

178. W. Lovenberg, E. Jequier, and A. Sjoerdsma, *Science 155:*217 (1967).

179. K. Engelman, W. Lovenberg, and A. Sjoerdsma, *N. Engl. J. Med. 277:* 1103 (1967).

180. M. Jouvet, *Science 163:*32 (1969).

181. M. Jouvet, *Adv. Pharmacol. 613:*265 (1968).

182. A. K. Sinha, R. D. Ciaranello, W. C. Dement, and J. D. Barchas, *J. Neurochem. 20:*1289 (1973).

183. K. Koe and A. Weissman, *J. Pharmacol. Exp. Ther. 154:*499 (1966).

184. E. Jequier, W. Lovenberg, and A. Sjoerdsma, *Mol. Pharmacol. 3:*274 (1967).

185. A. Pletscher, P. A. Shore, and B. B. Brodie, *J. Pharmacol. Exp. Ther. 116:* 84 (1956).

186. J. H. Gaddum and K. A. Hameed, *Br. J. Pharmacol. 9:*240 (1954).

187. E. Costa, *Proc. Soc. Exp. Biol. Med. 91:*39 (1956).

188. R. J. Wurtman, in *Handbook of Neurochemistry,* Vol. IV (A. Lajtha, ed.), Plenum, New York, 1970.

189. A. B. Lerner, J. D. Case, Y. Takahashi, T. H. Lee, and W. Mori, *J. Am. Chem. Soc. 80:*2587 (1958).

190. A. B. Lerner, J. D. Case, and Y. Takahashi, *J. Biol. Chem. 235:*1992 (1960).

191. J. Axelrod and H. Weissbach, *J. Biol. Chem. 236:*211 (1961).

192. H. Weissbach, B. G. Redfield, and J. Axelrod, *Biochim. Biophys. Acta 43:* 352 (1960).

193. H. Weissbach, B. G. Redfield, and J. Axelrod, *Biochim. Biophys. Acta 54:* 190 (1961).

194. S. H. Snyder, M. Zweig, and J. Axelrod, *Life Sci. 3:*1175 (1964).

195. S. H. Snyder and J. Axelrod, *Science 149:*542 (1965).

196. J. Axelrod, H. M. Shein, and R. J. Wurtman, *Proc. Natl. Acad. Sci. USA 62:*554 (1969).

197. D. C. Klein, J. L. Weller, and R. Y. Moore, *Proc. Natl. Acad. Sci. USA 68:*3107 (1971).

198. D. C. Klein, G. R. Berg, and J. Weller, *Science 168:*979 (1970).

199. D. C. Klein and J. L. Weller, *J. Pharmacol. Exp. Ther. 186:*516 (1973).

200. M. K. Gaitonde, in *Handbook of Neurochemistry,* Vol. III (A. Lajtha, ed.), Plenum, New York, 1970.

201. W. H. Stein and S. Moore, *J. Biol. Chem. 211:*915 (1954).

202. D. R. Curtis and J. C. Watkins, *Pharmacol. Rev. 17:*347 (1965).

203. L. K. Kaczmarek and A. N. Davison, *J. Neurochem. 19:*2355 (1972).

204. H. C. Agrawal, A. N. Davison, and L. K. Kaczmarek, *Biochem. J. 122:* 759 (1971).

205. R. Werman, *Comp. Biochem. Physiol. 18:*745 (1966).

206.   H. Tabor and C. W. Tabor, *Pharmacol. Rev. 16:*245 (1964).
207.   L. Pearce and S. M. Schanberg, *Science 166:*1301 (1969).
208.   T. R. Anderson and S. M. Schanberg, *J. Neurochem. 19:*1471 (1972).
209.   E. C. Shaskan, J. H. Haraszti, and S. H. Snyder, *J. Neurochem. 20:*1443 (1973).
210.   A. Raina and P. Hannonen, *FEBS Lett. 16:*1 (1971).
211.   G. L. Schmidt and G. L. Cantoni, *J. Neurochem. 20:*1373 (1973).
212.   H. Shimizu, Y. Kakimoto, and I. Sano, *J. Pharmacol. Exp. Ther. 143:*199 (1964).
213.   N. Seiler and J. M. Schröder, *Brain Res. 22:*81 (1970).
214.   G. R. Dutton and H. R. Mahler, *J. Neurochem. 15:*765 (1968).
215.   V. K. Singh and S. C. Sung, *J. Neurochem. 19:*2885 (1972).
216.   R. K. Datta, W. Antopol, and J. J. Ghosh, *Biochem. J. 125:*213 (1971).
217.   P. P. Giorgi, *Biochem. J. 120:*643 (1970).
218.   C. A. Manen and D. H. Russell, *Science 195:*505 (1977).

12
PEPTIDES

The brain produces peptides in a bewildering array of sizes and functions [1-3].
The literature on the peptides, especially the smaller peptides, is confusing and
vague in many cases due to the identification of trace amounts of many peptides
which later turned out to be present artifactually by virtue of the breakdown of
brain protein. Even the study of those which are now well-established brain constituents has proved quite difficult because of the low levels of these materials
present, because of the problems involved in separating one closely related substance from another, and because of the lack, in most cases, of specific and sensitive methods for the assay of these materials. In spite of these problems, a
number of peptides are known to exist in brain, and some are, in fact, unique to
brain.

Glutathione and Other γ-Glutamyl Peptides

One which is clearly not unique to brain, but which is present in large amounts, is
glutathione. This peptide, which is found in all cells, is the tripeptide γ-glutamyl
cysteinylglycine (Fig. 12.1). It is present in brain to the extent of more than
100 mg/100 g of fresh brain and is almost completely in the reduced form. Within a few minutes after death, however, about half of the glutathione is oxidized
[4]. The biosynthesis of glutathione has been accomplished in cell-free systems
[5], and the enzymes catalyzing the two steps of the synthesis have been studied,
but have not yet been extensively purified from brain. The biosynthesis involves
the formation of γ-glutamylcysteine at the expense of ATP and the further condensation of this intermediate with glycine by a second enzyme and at the expense of another molecule of ATP. The biosynthesis is probably carried out by
enzymes in brain, since the precursor amino acids are rapidly incorporated into
brain glutathione [6,7]. The turnover of glutathione in brain is rapid, $t_{1/2}$ being
about 70 hr, but not as rapid as in liver [8]. Glutathione will serve as a cofactor
in vitro for a number of enzymatic reactions, some of them carried out by en-

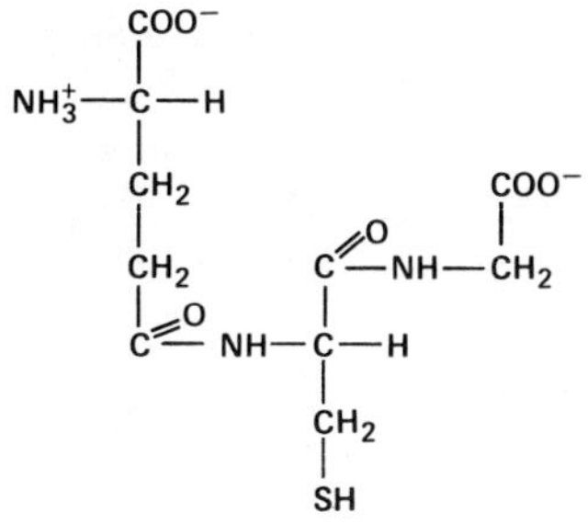

$$\begin{array}{l} \text{COO}^- \\ | \\ \text{NH}_3^+ - \text{C} - \text{H} \\ | \\ \text{CH}_2 \qquad\qquad\qquad \text{COO}^- \\ | \qquad\qquad\qquad\qquad | \\ \text{CH}_2 \qquad \text{C}{=}\text{O}, \text{--NH} - \text{CH}_2 \\ | \qquad\qquad | \\ \text{C}{=}\text{O}, \text{NH} - \text{C} - \text{H} \\ \qquad\qquad\qquad | \\ \qquad\qquad\qquad \text{CH}_2 \\ \qquad\qquad\qquad | \\ \qquad\qquad\qquad \text{SH} \end{array}$$

Figure 12.1 Structure of glutathione.

zymes found in brain, but the true function of glutathione in the cell is not known. Recent studies, described in Chap. 10, have suggested that glutathione is a constituent of the amino acid transport system in brain [9], as well as in other mammalian cells.

Large numbers of γ-glutamyl peptides, other than glutathione, have been observed in brain. Among these are the γ-glutamyl derivatives of glutamine, glutamate, glycine, α-aminoisobutyric acid, serine, alanine, and valine [1,10-14]. More recent reports have shown the presence of many other dipeptides, including glutamylisoleucine, α-aspartylserine, alanylphenylalanine, and isoleucylleucine. In addition, the tripeptides glutamylalanylglycine and S-methylglutathione are known to occur. Many of the γ-glutamyl peptides are present in reasonable amounts, approaching, in some cases, 1 mg/100 g of fresh brain. They are not unique to the brain, however, and there is some suggestion that they arise by transpeptidation reactions from glutathione [1]. That is, there are enzymes in brain and in other tissues which will remove the glutamate from glutathione and transfer it in a γ linkage to other amino acids. Whether these, and the other nonglutamate dipeptides and tripeptides have any function in brain is not known.

One dipeptide which has been well characterized and for which a function is suspected is N-acetyl-α-aspartylglutamate (Fig. 12.2). The structure of this material has been established, and the levels of the peptide in various brain parts and in the brains of various species have been determined [15,16]. This peptide is found only in the central nervous system. It occurs in large amounts, approximating 50 mg/100 g of fresh brain in the horse and more than 100 mg/100 g in most species. It is fairly evenly distributed in the various areas of the brain, but

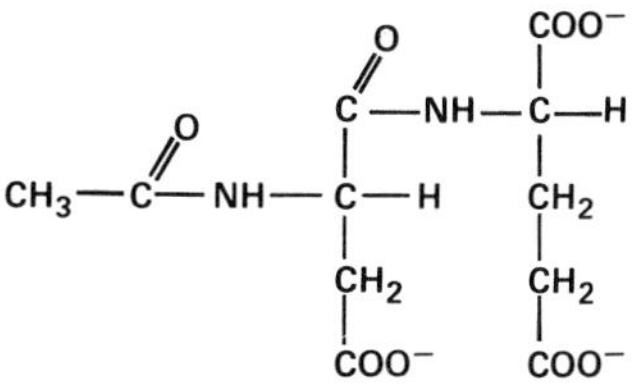

$$\begin{array}{l} \qquad\qquad\qquad\qquad \text{O} \qquad \text{COO}^- \\ \qquad\qquad\qquad\qquad \| \qquad | \\ \qquad\qquad\qquad\quad \text{C} - \text{NH} - \text{C} - \text{H} \\ \qquad\qquad \text{O} \qquad\qquad | \qquad | \\ \qquad\qquad \| \qquad\qquad | \qquad \text{CH}_2 \\ \text{CH}_3 - \text{C} - \text{NH} - \text{C} - \text{H} \qquad | \\ \qquad\qquad\qquad\quad | \qquad\quad \text{CH}_2 \\ \qquad\qquad\qquad\quad \text{CH}_2 \qquad | \\ \qquad\qquad\qquad\quad | \qquad\quad \text{COO}^- \\ \qquad\qquad\qquad\quad \text{COO}^- \end{array}$$

Figure 12.2 Structure of N-acetylaspartylglutamate.

its concentration can be shown to decrease gradually toward the caudal regions. Although it is attractive to try to construct a relationship between this peptide and the large amounts of N-acetylaspartic acid found in brain, no such connection has been found. It is known, however, that the sequence of this acetylated dipeptide is found at the N terminus in the contractile protein of muscle. It has been suggested that it may play a similar role in the contractile protein more recently discovered in brain, which is discussed in Chap. 15.

Homocarnosine

Another peptide which is uniquely present in the brain and which has been quite well characterized is homocarnosine. Discovered by Pisano in 1961 [1,17] during experiments on possible bound forms of GABA, this peptide has been fairly extensively studied [18-20]. The structure was readily established as γ-aminobutyrylhistidine, and its relation to the well-known muscle constituent, carnosine (β-alanylhistidine; Fig. 12.3), gives rise to its name. The peptide is present in large amounts in the central nervous system. In fact, there is more histidine present in the central nervous system as homocarnosine than as histidine itself. Homocarnosine is highest in human brain, reaching levels of 8 mg/100 g of wet weight. It is also high in the brains of monkeys, guinea pigs, rabbits, and cows. It is lower in rats and chickens, on the order of 1 mg/100 g, and much lower in the brains of pigs, dogs, cats, ducks, and fish. There is no obvious pattern to its distribution in the brain; it is pretty evenly distributed in all portions.

The biosynthesis of homocarnosine is most probably catalyzed by the same enzyme which makes carnosine, carnosine synthetase. This reaction which has been studied in muscle preparations, is quite nonspecific in that a number of amino acids can substitute for the precursors of carnosine, β-alanine and histidine. In fact, the substitution of γ-aminobutyric acid for β-alanine leads to the in vitro synthesis of homocarnosine [21]. In confirmation of this concept is the observation that if γ-aminobutyric acid is fed in large enough amounts that it begins to

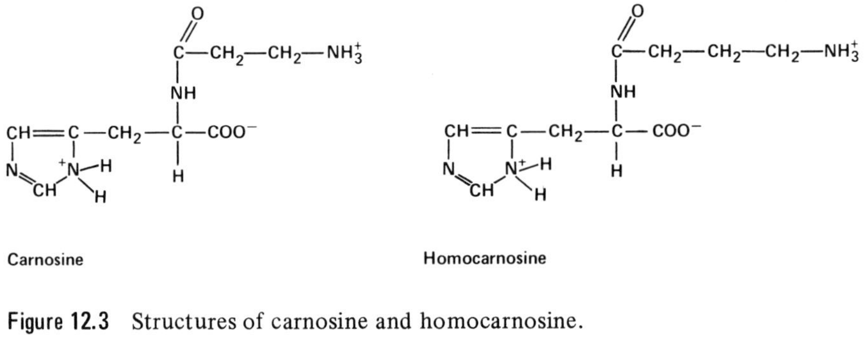

Carnosine   Homocarnosine

Figure 12.3   Structures of carnosine and homocarnosine.

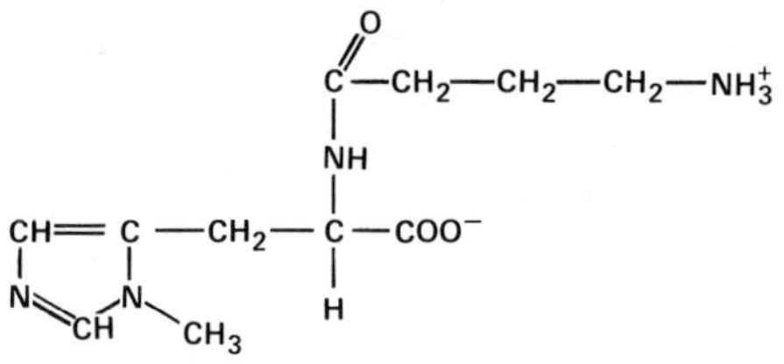

$$\begin{array}{c}
\mathrm{O} \\
\parallel \\
\mathrm{C\!-\!CH_2\!-\!CH_2\!-\!CH_2\!-\!NH_3^+} \\
\mid \\
\mathrm{NH} \\
\mid \\
\mathrm{CH\!=\!C\!-\!CH_2\!-\!C\!-\!COO^-} \\
\end{array}$$

Figure 12.4 Structure of homo-anserine.

appear in muscle, then homocarnosine appears as well [22]. Thus, it is thought that homocarnosine is present in brain, in higher amounts than carnosine, because γ-aminobutyric acid, rather than β-alanine, predominates in this tissue. Homocarnosine is normally absent from muscle because of the absence of the precursor, γ-aminobutyric acid.

Somewhat later, another derivative of γ-aminobutyric acid, homoanserine (Fig. 12.4), was found in brain [23]. This peptide, γ-aminobutyryl-1-methyl-histidine, is the homolog of another muscle peptide, anserine (β-alanyl-1-methyl-histidine). Thus, both these γ-aminobutyryl peptides are uniquely present in brain, homoanserine in lower amounts. On the other hand, both carnosine and anserine are also present in brain, but in much lower amounts than in muscle, and in much lower amounts than the γ-aminobutyryl homologs. Although no work has been done on the biosynthesis of homoanserine, it is thought that it, like homocarnosine, is present in brain by virtue of the presence of its precursor, γ-aminobutyric acid, and not because of the action of some unique brain enzyme.

The function of homocarnosine is not known, which is not surprising, because the function of the much longer recognized and much more abundant carnosine is not known either. There are many suggestions, one of the most often heard being that carnosine in muscle serves as a buffer. The observation that homocarnosine is present in the urine and, in low amounts, in the cerebrospinal fluid [18] has led to the suggestion that its measurement could serve as some kind of an index of nervous system function. In fact, there are now several reports that homocarnosine is elevated in the cerebrospinal fluid of patients suffering from neurological disorders [24-27]. Such observations lead to questions about the levels of homocarnosine and homoanserine in the cerebrospinal fluid and the urine in other neurological diseases, especially those involving carnosine or γ-aminobutyric acid. The somewhat difficult and time-consuming assay for these compounds has discouraged such measurements in the past. Newer methods of assay may make such studies feasible in the future [28].

Substance P

Substance P [29] was discovered many years ago by von Euler and Gaddum [30], who showed that alcoholic extracts of brain or intestine caused the contraction of isolated intestinal preparations or a drop in the blood pressure of intact rats. The peptide is present in inordinately small amounts in brain or intestine. Estimates of substance-P activity are difficult, however, since the bioassays used, such as in vitro intestinal contraction or in vivo blood pressure lowering, are time consuming, imprecise, and nonspecific. In fact, to be able to measure any activity at all, the animals must be pretreated with antagonists of serotonin, histamine, and acetylcholine. These compounds have similar actions under some circumstances and are present in crude extracts. So, the purification of such a molecule by these techniques assumes heroic proportions. The situation is made even more difficult by the fact that the active principal seems not to be a single entity; the activity may reside in a family of peptides.

Following an entirely different experimental line [3], Leeman and Hammerschlag [31] obtained a peptide from the hypothalamus which stimulated the secretion of saliva. Its resemblance to substance P prompted a comparison. Using this simplified detection system, a pure protein was isolated relatively rapidly [32,33], and its structure duplicated synthetically [34]. It is currently felt that this sialogogic peptide is identical to substance P, although, apparently, no material has been prepared pure, using classic substance-P assay methods, which can be used for comparison. The information which is now available indicates that substance P is a small molecule with 11 amino acids. It is stable in acid but not in base. Distribution studies have indicated that the material is present in the neurons and not in the glia, and, in fact, that it is localized in the synaptic bouton. Only very low concentrations have been found in the peripheral system. Substance P, then, is a very potent peptide. It is present in minute amounts, e.g., 1 mg/100 kg of goat hypothalami. It resembles in structure other peptides with potent actions, such as physalaemin and eledoisin [33]. Its localization in neurons, more specifically in the synaptic bouton, has prompted the suggestion that it is a transmitter in central or peripheral neurons.

Pituitary Hormones

It is well known that the pituitary makes and releases a large number of hormonal peptides [35,36]. These include growth hormone, thyrotropic hormone, corticotropic hormone, the gonadotropins, prolactin, and melanocyte-stimulating hormone. The structures of many of these peptides are completely known, and their actions are recognized, if not understood. Since there is little known about their biosynthesis, and since most do not seem to affect the brain itself in a major way, these will not be discussed in detail.

Hypothalamic Peptides and Releasing Factors

More to the point here are the peptide products of the hypothalamus. First, the hypothalamus makes the two important hormones oxytocin and vasopressin.

Oxytocin is involved in uterine contraction and lactation, and vasopressin in the retention of fluids. The structures of these two have been known for a number of years and will not be reviewed here. The biosynthesis has been studied by Sachs et al. [3,37] in organ culture of the anterior hypothalamus and presents several interesting aspects. Specifically, these long-term cultures were able to incorporate labeled amino acids into vasopressin, and this incorporation was blocked by puromycin. The synthesis of the hormone-binding proteins, the neurophysins, was also seen.

In addition to the two small peptide hormones, the hypothalamus also produces a number of small peptide-releasing factors which are now known to act on the pituitary and regulate the release of the many large peptide hormones mentioned above. It is now generally accepted that the hypothalamus controls the action of the pituitary via these releasing factors carried by the portal circulation [36] (Fig. 12.5). Evidence for the existence of these factors came from several lines of experiments. It is known that removal or damage to the hypothalamus produces, as one consequence, changes much like the removal of the pituitary

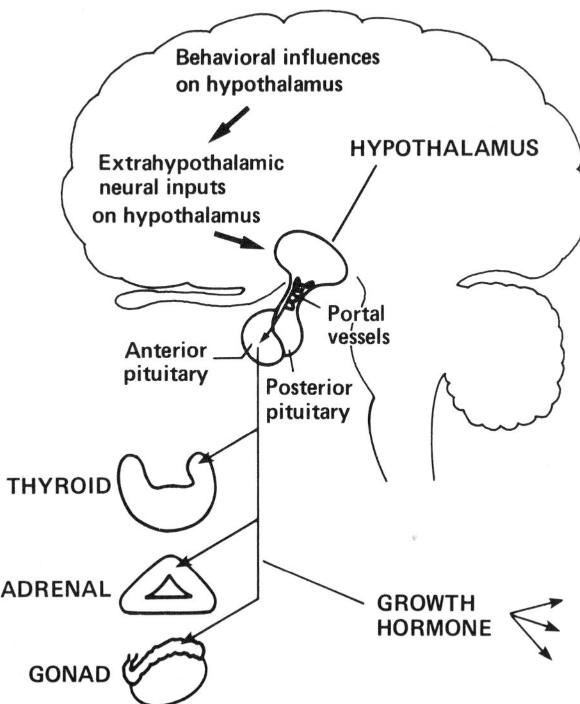

Figure 12.5   Schematic representation of the interactions among hypothalamic, pituitary, thyroid, adrenal, and gonadal hormones.

Growth Hormone Releasing Factor	Unknown
Growth Hormone Release Inhibiting Factor (Somatostatin)	Ala-Gly-Cys-Lys-Asn-Phe-Phe-Trp-Lys-Thr-Phe-Thr-Ser-Cys
Thyrotropin Releasing Factor	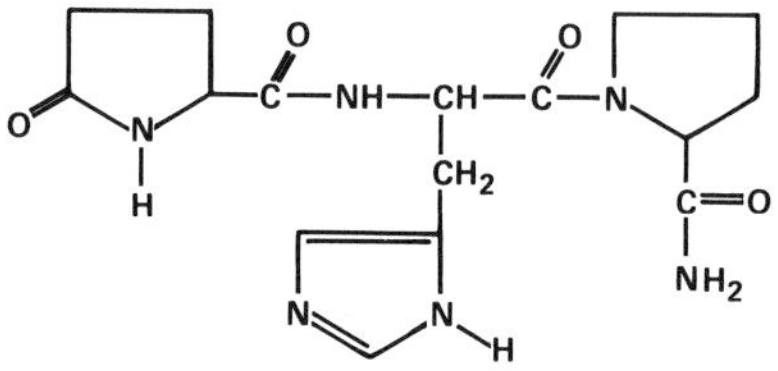 L-pyroglutamyl-L-histidyl-L-prolinamide
Luteinizing Hormone Releasing Factor	(pyro) Glu-His-Trp-Ser-Tyr-Gly-Leu-Arg-Pro-Gly(NH_2)
Follicle Stimulating Hormone Releasing Factor	Possibly identical to luteinizing hormone releasing factor
Corticotropin Releasing Factor	Acetyl-Ser-Tyr-Cys-Phe-His-Gln-Asn-Cys-[Pro-Val]-Lys-GlyNH_2
Prolactin Release Inhibiting Factor	Unknown
Melanocyte Stimulating Hormone Release Inhibiting Factor	L-leucyl-L-prolylglycinamide (probably released from the C-terminal portion of oxytocin)

Figure 12.6 Structures of the hypothalamic releasing factors.

itself. Then, it was shown that the removal of the hypothalamus leads directly to the lowering of levels of many of the pituitary hormones, and that these levels could be raised by hypothalamic extracts not containing the hormone itself. Hypothalamic extracts increased the amounts of hormones in their various target tissues, but this did not occur in hypophysectomized animals. Finally, it has been shown in vitro that extracts of the hypothalamus will stimulate the release of the various pituitary hormones from slices of the pituitary in physiological media.

Now it is known that the hypothalamus elaborates releasing factors for growth hormone (somatotropin release factor) [38,39], thyrotropin (TRF) [40,41], gonadotropins (LRF and FSHRF) [42,43], corticotropin (CRF)

[44] , and factors which inhibit the release of prolactin (PIF) and melanocyte-stimulating hormone (MSH-R-IF) [45,46] . These are very small, heat-stable proteins, some of which have been completely characterized (Fig. 12.6). They are present in exquisitely small amounts. Thus, Guillemin used 55 kg of sheep hypothalami from 300,000 sheep, to isolate 0.5 mg of TRF. They are elaborated in the hypothalamus and transported to the pituitary. There, in as yet unknown ways, they release the appropriate stored hormone.

So, the brain makes a bewildering variety of peptides, from dipeptides of unknown function to large peptide hormones of well-defined mode of action. Several studies in the neurochemical literature indicate that there are large numbers of peptides of all sizes present in the brain which are yet to be isolated and characterized.

References

1. J. J. Pisano, in *Handbook of Neurochemistry*, Vol. I (A. Lajtha, ed.), Plenum, New York, 1969.
2. I. Sano, *Int. Rev. Neurobiol. 12:*235 (1970).
3. F. E. Bloom (ed.), *Neurosci. Res. Program Bull. 10*(2) (1972).
4. H. Martin and H. McIlwain, *Biochem. J. 71:*275 (1959).
5. K. Bloch, *J. Biol. Chem. 179:*1245 (1949).
6. A. Lajtha, S. Berl, and H. Waelsch, *J. Neurochem. 3:*322 (1959).
7. T. Takahashi and Y. Akebane, *J. Neurochem. 7:*89 (1961).
8. G. W. Douglas and R. A. Mortensen, *J. Biol. Chem. 222:*581 (1956).
9. A. Meister, in *Brain Dysfunction in Metabolic Diseases* (F. Plum, ed.), Raven, New York, 1974.
10. Y. Kakimoto, T. Nakajima, A. Kanazawa, M. Takesada, and I. Sano, *Biochim. Biophys. Acta 93:*333 (1964).
11. Y. Kakimoto, A. Kanazawa, T. Nakajima, and I. Sano, *Biochim. Biophys. Acta 100:*426 (1965).
12. A. Kanazawa, Y. Kakimoto, T. Nakajima, and I. Sano, *Biochim. Biophys. Acta 111:*90 (1965).
13. A. Kanizawa, Y. Kakimoto, T. Nakajima, H. Shimizu, M. Takesada, and I. Sano, *Biochim. Biophys. Acta 97:*460 (1965).
14. T. Nakajima, Y. Kakimoto, A. Kumon, M. Matsuoka, and I. Sano, *J. Neurochem. 16:*417 (1969).
15. J. V. Auditore, E. J. Olson, and L. Wade, *Arch. Biochem. Biophys. 114:*452 (1966).
16. E. Miyamoto, Y. Kakimoto, and I. Sano, *J. Neurochem. 13:*999 (1966).
17. J. J. Pisano, J. D. Wilson, L. Cohen, D. Abraham, and S. Udenfriend, *J. Biol. Chem. 236:*499 (1961).
18. D. Abraham, J. J. Pisano, and S. Udenfriend, *Arch. Biochem. Biophys. 99:*210 (1962).
19. A. Kanazawa, Y. Kakimoto, E. Miyamoto, and I. Sano, *J. Neurochem. 12:*957 (1965).

20. A. Kanizawa and I. Sano, *J. Neurochem. 14:*211 (1967).

21. G. D. Kalyankar and A. Meister, *J. Biol. Chem. 234:*3210 (1959).

22. D. Abraham, J. J. Pisano, and S. Udenfriend, *Biochim. Biophys. Acta 50:*
     570 (1961).

23. T. Nakajima, F. Wolfgram, and W. G. Clark, *J. Neurochem. 14:*1107 (1967).

24. S. P. Bessman and R. Baldwin, *Science 135:*789 (1962).

25. J. Levenson, K. Lindahl-Kiessling, and S. Rayner, *Lancet 2:*756 (1964).

26. C. R. Scriver, S. Pueschel, and E. Davies, *N. Engl. J. Med.* 274:635 (1966).

27. T. L. Perry, S. Hansen, B. Tischler, R. Bunting, and K. Berry, *N. Engl. J.
     Med. 23:*1219 (1967).

28. A. B. Young and S. H. Snyder, *J. Neurochem. 21:*387 (1973).

29. G. Zetler, in *Handbook of Neurochemistry,* Vol. IV (A. Lajtha, ed.), Ple-
     num, New York, 1970.

30. U. S. von Euler and J. H. Gaddum, *J. Physiol. (London) 72:*74 (1931).

31. S. E. Leeman and R. Hammerschlag, *Endocrinology 81:*803 (1967).

32. M. M. Chang and S. E. Leeman, *J. Biol. Chem. 245:*4784 (1970).

33. M. M. Chang, S. E. Leeman, and H. D. Niall, *Nature New Biol. (London)
     232:*86 (1971).

34. G. W. Tregear, H. D. Niall, J. T. Potts, S. E. Leeman, and M. M. Chang,
     *Nature New Biol. (London) 232:*87 (1971).

35. M. Reiss, in *Handbook of Neurochemistry,* Vol. IV (A. Lajtha, ed.), Plenum,
     New York, 1970.

36. G. J. Siegel and J. S. Eisenman, in *Basic Neurochemistry* (R. W. Albers, G. J.
     Siegel, R. Katzman, and B. W. Agranoff, eds.), Little, Brown, Boston, 1973.

37. H. Sachs, R. Goodman, J. Osinchak, and J. McKelvy, *Proc. Natl. Acad. Aci.
     USA 68:*2782 (1971).

38. A. V. Schally, A. Arimura, and A. J. Kastin, *Science 119:*341 (1973).

39. A. V. Schally, T. W. Redding, J. Takahara, D. H. Coy, and A. Arimura,
     *Biochem. Biophys. Res. Commun. 55:*556 (1973).

40. A. V. Schally, T. W. Redding, C. Y. Bowers, and J. F. Barrett, *J. Biol.
     Chem. 244:*4077 (1969).

41. R. Burgus, T. Dunn, D. Desiderio, D. Ward, W. Vale, and R. Guillemin,
     *Nature (London) 226:*321 (1970).

42. H. Matsuo, Y. Baba, R. M. G. Nair, A. Arimura, and A. V. Schally, *Biochem.
     Biophys. Res. Commun. 43:*1334 (1971).

43. R. Burgus, M. Butcher, M. Amoss, N. Ling, M. Monahen, J. Rivier, R. Fel-
     lows, R. Blackwell, W. Vale, and R. Guillemin, *Proc. Natl. Acad. Sci. USA
     69:*278 (1972).

44. Y. Koch, P. Chobsieng, U. Zor, M. Fridkin, and H. R. Lindner, *Biochem.
     Biophys. Res. Commun. 55:*623 (1973).

45. M. E. Celis, S. Taleisnik, and R. Walter, *Proc. Natl. Acad. Sci. USA 68:*
     1428 (1971).

46. R. M. G. Nair, A. J. Kastin, and A. V. Schally, *Biochem. Biophys. Res.
     Commun. 43:*1376 (1971).

13
NUCLEOTIDE METABOLISM

Composition and Distribution

All the nucleotides found in other cells are found in cells of the brain [1] , and
even the relative amounts and the distribution are in no way unique (Table 13.1).
There is reasonable quantitative similarity between the nucleotide distributions
in the brains of the several species. In most brains, the adenine nucleotides predominate. In fact, the adenine nucleotides make up more than half of the total
of all the nucleotides in the brain and most of the nucleotide is as the triphosphate, the ATP/ADP ratio ranging from 10 to 20 for most brains studied. Very
small amounts of the monophosphate, AMP, are present. Guanine nucleotides
make up a larger portion of the pool of nucleotides in nervous tissue than in
other tissues of the body, and about 90% of the guanine nucleotide is also in the
form of the triphosphate. Uridine nucleotides and guanine nucleotides are, quantitatively, about equivalent, but the content of cytidine nucleotides is very low,
as in most tissues, leading to speculation that this component is limiting for RNA
biosynthesis. In general, the monophosphates are in very small amounts. There
are a number of other nucleotides found in brain which are involved in various
routes of metabolism, such as the functioning of the cytidine nucleotides in the
biosynthesis of phospholipids, or the uridine nucleotides in polysaccharide biosynthesis, but these are discussed in other chapters.

Studies on the distribution of nucleotides have shown that, generally, they
are low in white matter and higher in gray. Such a distribution is not surprising
in view of the high lipid content of the white matter. The total nucleotide content of the brain rises with the age of the animal when expressed in terms of the
DNA content of the brain [1] . The individual nucleotides, on the other hand,
do not all rise, the cytidine nucleotides decreasing with age as the adenine nucleotides increase.

The literature contains many different reports of the nucleotide content of
neural tissues, not all of them in agreement. This is now known to be due in
some cases to the remarkably rapid changes in the content of the nucleotides,

Table 13.1 Free Nucleotides in the Cerebral Hemispheres of
Different Species (μmol/100 g wet weight)

Nucleotide	Mouse	Rat	Guinea pig	Rabbit
CMP	traces	—	—	—
AMP	—	2.1	5.0	5.4
ADP	10.9	12.2	20.0	18.4
ATP	194.0	208.2	201.4	213.0
GMP	—	—	traces	traces
GDP	2.2	2.4	5.5	9.8
GTP	21.8	29.9	32.2	35.7
UMP	1.2	1.7	traces	2.3
UDP	1.7	2.2	3.1	0.8
UTP	17.9	22.0	26.0	25.6
IMP	4.8	5.1	2.6	4.7

Table from the work of P. Mandel, in *Handbook of Neurochemistry,*
Vol. VA (A. Lajtha, ed.), Plenum, New York, 1971.

especially of the ATP, during procedures used in killing animals and obtaining
tissue. For example, it is known that the ATP content of rat brain decreases by
70% within 20 sec after the death of the animal by decapitation [2]. Thus, for
studies on brain nucleotide composition, it is important to be aware of the
marked alterations which can occur merely as a result of the methods employed
in obtaining the data [1].

On the other hand, there are a number of rather drastic treatments which
are known not to influence the ATP content of the brain of experimental animals [1,3]. Anesthetics of all kinds have been investigated and do not seem to
have either a marked or a consistent effect on the nucleotide content of brain.
Nor do the nucleotide levels change when an animal sleeps or is deprived of
sleep. The literature contains contradictory data on the effects of various tranquilizers. However, some stimulant drugs which produce psychomotor excitation increase ATP levels in brain, as do certain antidepressants. Amphetamines,
while not raising ATP levels, appear to increase the turnover of the phosphates
of ATP. The effect of ischemia is generally found to be a lowering of ATP levels;
there is some evidence, however, that the drop in the nucleotide is secondary to
a drop in the levels of other high-energy phosphates in the brain, notably that of
phosphocreatine. Electric shock lowers high-energy phosphates in the brain
rapidly, ATP dropping to 50% of its normal level in a few seconds. Experimentally induced seizures have, in some cases, been reported to reduce levels of high-
energy phosphates, for example, in the case of seizures induced with the convulsant gas Indoklon; in other types, such as audiogenic seizures, no lowering of

ATP levels was found. The preparation and utilization of brain slices has generally been found to cause a marked depletion of ATP in the tissue, and a substantial decrease in the ATP/ADP ratio.

Biosynthesis

The biosynthesis of the nucleotides for use in nucleic acid synthesis in brain, as in other tissues, can proceed by two quite different pathways. The first of these are the de novo pathways. De novo pathways for both purines and pyrimidines were elucidated several years ago in mammalian and bacterial systems. In brain, each of these pathways can be shown to operate to some extent. It has been observed that brain preparations will incorporate glycine into adenine nucleotides [4] , as well as into the earlier intermediates such as formylglycinamide ribotide [5] . Studies in vivo are consistent with these data in that radioactive precursors, such as formate or glycine injected intracerebrally, can be found incorporated into purines of the brain [4,6] . There is no reason to suspect that this pathway differs qualitatively from that found in other mammalian tissues.

The entire pathway for the de novo synthesis of pyrimidines has been a bit more difficult to demonstrate in brain, and the apparent absence of carbamyl phosphate synthetase in brain led to the suggestion that the de novo pathway was not present intact. However, it is not difficult to demonstrate the presence and functioning of the later stages in the de novo pathway for pyrimidines. Orotic acid can be incorporated into brain RNA in vitro, and the enzymes orotic acid: PRPP pyrophosphorylase and OMP decarboxylase are readily demonstrable. These enzymes have, in fact, been studied in brain where they copurify as a single protein [7] . More recent data on the enzyme carbamyl phosphate synthetase show that this activity also is present in brain, and the difficulty in finding it in the earlier studies was caused by the fact that the activity in brain is quite low compared to that in liver, and is present as the labile, soluble glutamine-dependent enzyme now known to function in pyrimidine biosynthesis in many tissues [8,9] , instead of the mitochondrial, ammonia-dependent enzyme for urea formation known to predominate in the liver. Again, there is no evidence to indicate that the steps of the de novo pathway for pyrimidine synthesis in brain are different from in other tissues.

Salvage and the Lesch-Nyhan Syndrome

The other route of formation of precursors for nucleic acid synthesis are the so-called salvage pathways which utilize preformed purine and pyrimidine components. In brain, these components could arise from the breakdown of existing nucleic acid or from the blood, the ultimate source of the latter being the liver. Not only are such pathways present in brain, but they are known to be extremely active [1,10]. For pyrimidines, uracil is not a good precursor for RNA, at least in the rat. But the nucleosides uridine and cytidine are readily incorporated into RNA by brain preparations [11]. The situation with the purines is even more striking since the enzymes hypoxanthine-guanine phosphoribosyltransferase (HGPRTase) and adenine phosphoribosyltransferase (APRTase) have been shown to be present, and, in fact, the former is most active in neural tissues [12,13]. The pathways of nucleotide metabolism observed in brain are summarized in Fig. 13.1.

In addition to the salvage and the de novo pathways, a number of other reactions have been observed in brain preparations. Among them are the conversion of UMP to CMP [14], and the deamination of AMP to IMP by adenylate deaminase as prelude to its conversion to GMP [15,16]. The kinases for most of the nucleosides and all the nucleotides involved in RNA metabolism, as well as for the thymine nucleotides, have been demonstrated [1].

A question which has attracted some experimental interest is the following. Does the brain make its own RNA precursors via the de novo pathway, or does it

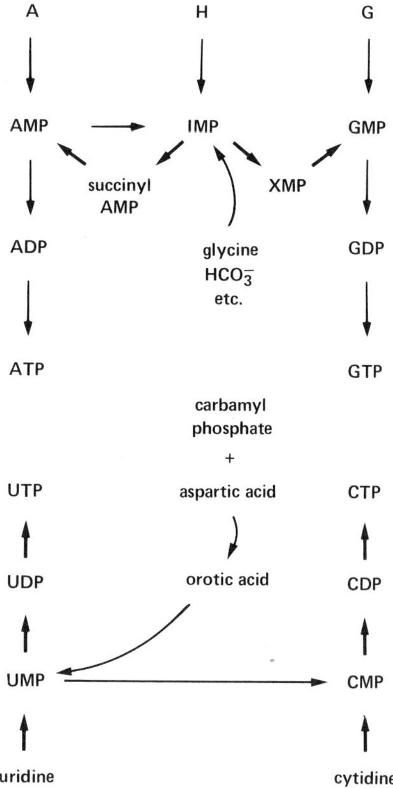

Figure 13.1 Pathways of nucleotide metabolism in brain. [Adapted from J. N. Santos, K. W. Hamstead, L. E. Kapp, and R. P. Miech, *J. Neurochem.* *15*:367 (1968).]

rely on the liver for preformed precursors and utilize them by way of its salvage
pathways? The answer seems to be that the brain, of rats at least, has some de
novo capacity for both purines and pyrimidines, but uses the salvage pathways
to a much greater extent, relying probably on preformed, blood-borne precursors for its source. The evidence includes several different experimental observations. First are the many reports of high salvage pathway activity and the com-
paratively lower levels of the activities of the de novo route [1,10,11,17,18].
Second are experiments in which the de novo pathway for pyrimidines has been
blocked with little effect on the brain content of RNA [19]. Third, there are
the experiments in which the brain has been isolated surgically from the viscera
and the relative incorporation of the precursor, orotic acid, and the preformed
pyrimidine, uridine, were compared [18]. Fourth are the data showing that preformed bases can be transported from the liver to the brain [20], and further, as
shown by Geiger and Yamasaki if they are not, the brain cannot function in a
normal way [21]. In this regard, however, it should be pointed out that if the
brain were totally dependent on the circulation, it might be expected to have
very active mechanisms for the capture of these compounds from the blood. Although brain is permeable to nucleosides [1], active uptake mechanisms have
been sought but not found in brain [11,22]. Perhaps the phosphorylation steps
serve the same purpose by trapping the nucleosides in a form to which the cells
are not permeable. Nonetheless, the brain does not seem to have active transport mechanisms for the bases or the nucleosides. Finally, the drastic consequences, in humans, of the lack of one of the salvage enzymes, HGPRTase, discussed in detail below, perhaps attests to the importance of this pathway. All
in all, then, the general feeling is that the brain, while not totally dependent on
the presence of preformed precursors, relies more heavily on the salvage pathways than it does on de novo synthesis [1,17,20].

     Some degradation of the nucleotides, the nucleosides, and the free purine
and pyrimidine bases occurs in brain. The nucleotides can be hydrolyzed to
nucleosides in brain by 5'-nucleotidase. The further catabolsim of the purines
has been shown also, namely, deamination by adenosine or guanosine deamin-
ases [23], phosphorolysis by purine nucleoside phosphorylase [24], and breakdown via xanthine oxidase [25], although others have failed to find this latter
enzyme [26]. The enzyme guanase has also been shown [23], as has the presence in brain of hypoxanthine [27]. Not much work has been done on the
catabolism of the pyrimidines in brain, but it is known that pyrimidine nucleoside phosphorylase is present in brain and that it follows a definite developmental course, as it does in liver [28]. In fact, the activities of both the purine and
the pyrimidine nucleoside phosphorylases are much higher in the brains of adult
animals than in the brains of young. It is interesting, perhaps, that the levels of
the enzymes involved in the synthesis of the nucleotides and of RNA itself are
relatively constant in the brain during the growth of the animal. The levels of

the degradative enzymes, on the other hand, increase markedly [24], perhaps explaining, at least in part, the many observations that the synthesis of RNA, as measured by the incorporation of radioactive precursors into the product, is much faster in young animals than in adults [11,29,30].

Interest in the neurochemical aspects of nucleotide metabolism has increased sharply within the last few years, stimulated to a great extent by the discovery of a bizarre behavioral characteristic associated with the absence of one of the enzymes of nucleotide metabolism. The original observation of this condition in two young brothers [31] has been followed by a great volume of work on what is now known as the Lesch-Nyhan syndrome. Perhaps this has been so because of the extreme and terrible nature of the behavioral lesion. Also, the disease has provoked interest because it is perhaps the clearest link between a defined and unique behavioral pattern and a distinct and discrete biochemical lesion. The condition occurs only in males and is thus transmitted as an X-linked recessive [32].

The clinical signs of the disease [33], now observed in about 150 children, include spastic cerebral palsy (choreoathetosis), motor and mental retardation, and, most striking of all, a compulsive self-mutilation, primarily of the fingers and lips. The motor defect is generally more severe than the mental retardation in that the children cannot usually sit up alone or even hold their head up unaided. They have difficulty speaking and swallowing, and also vomit frequently. However, they do have lowered mental capacity, usually rated as IQ of less than 50. Generally, they are short of stature and underweight, probably as a result of the difficulty in eating which can lead to death from inanition. Usually these children develop normally up to the age of 6 or 8 months before the symptoms begin to appear.

The self-mutilation associated with Lesch-Nyhan has attracted the most interest, since it seems the most unique. The children must be continuously restrained and protected by mouth guard. If not so protected they will bite both fingers and lips to the point of amputation. The behavior is compulsive in the extreme, since the children appear fearful of being unrestrained, and scream in pain during the mutilative episodes. The destructive behavior is not limited to themselves, as they are outwardly destructive as well. But it is generally observed that, when restrained, the children are rather pleasant and good-natured.

The children excrete large amounts of uric acid and suffer the consequences of hyperuricemia. They are found to have urinary stones, kidney dysfunction, arthritis, and, generally, the problems of gout. Aside from the always present excess of uric acid excretion, and some reports of increased excretion of xanthine, hypoxanthine, and aminoimidazole carboxamide [34] and decreased excretion of adenine and adenosine, little else of a biochemical or pathological nature has been reported in these children.

The exact biochemical lesion causing this condition was shown by Seeg-

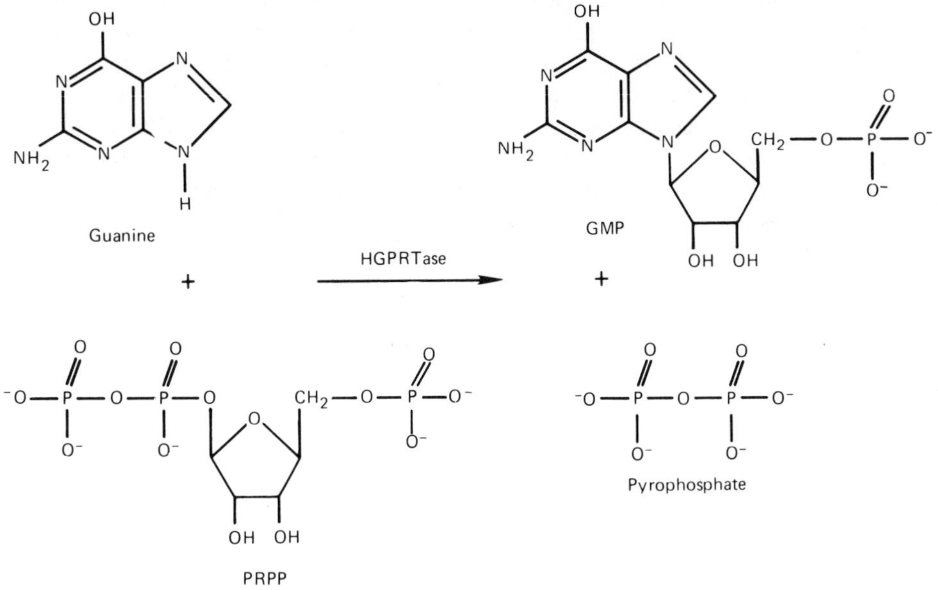

Guanine

HGPRTase

GMP

+

+

PRPP

Pyrophosphate

Figure 13.2   HGPRTase reaction.

miller and his colleagues [35] to be an absence of the enzyme hypoxanthine-guanine phosphoribosyl transferase (HGPRTase) (Fig. 13.2). This enzyme, originally considered merely a part of the salvage pathway for purines, has now assumed a greater importance as has salvage metabolism itself [36]. The children suffering from the Lesch-Nyhan syndrome have been shown to lack the enzyme in all tissues, including brain. The normal enzyme has been studied by a number of workers, using human red blood cells as the primary source. The normal enzyme requires Mg^{2+}, and is inhibited strongly by nucleotides, primarily the products of its action, GMP and IMP, which compete with PRPP for the active site [37-39]. It occurs in many tissues in the body; the highest activity occurs in the basal ganglia [40]. It is of interest, therefore, that the neurological symptoms are indicative of basal ganglia dysfunction. The purified erythrocyte enzyme has been studied as to its kinetics and mechanism of action [41,42]. It has a molecular weight of 68,000 and is made up of two identical subunits [43] The enzyme from erythrocytes appears to have isozyme forms which have been observed by several workers with a number of techniques [43-47]. Depending on the report and the technique, two, three, or four isozymes have been seen. There is no information as yet as to the structural basis of these isozyme forms although they are apparently due to some "post-transcriptional" alteration in the molecule [43]. The enzyme has also been purified from rat brain [48]. The distribution of the enzyme in rat brain is apparently more uniform than in the human, but otherwise the rat enzyme seems quite similar to the enzyme from other sources, even down to the presence of isozymes.

It is now clear that one enzyme catalyzes the conversion of both guanine and hypoxanthine to their respective nucleotides. This is different from the situation in bacteria where separate enzymes are involved. The evidence in mammalian systems includes the following [41]. There is a deletion of both activities in the Lesch-Nyhan syndrome. The two activities have identical heats of inactivation and cannot be physically separated. They are equally inhibited in the presence of either of the products of the reactions, and the two substrates, guanine and hypoxanthine, inhibit each other. The K_M for PRPP is identical no matter which substrate is used. It has been shown that the enzyme will also use xanthine as a substrate at about 1% of the rate for guanine and for hypoxanthine [49], but a separate enzyme, APRTase, is involved in the conversion of adenine to AMP.

This second enzyme, APRTase (Fig. 13.3), is related in function but seems completely different in many ways. First, it is different immunologically [44]. Also, it is on a different chromosome in the human [50]. It is not particularly high in the brain [39]. Most striking of all, neither a partial [51], nor a com-

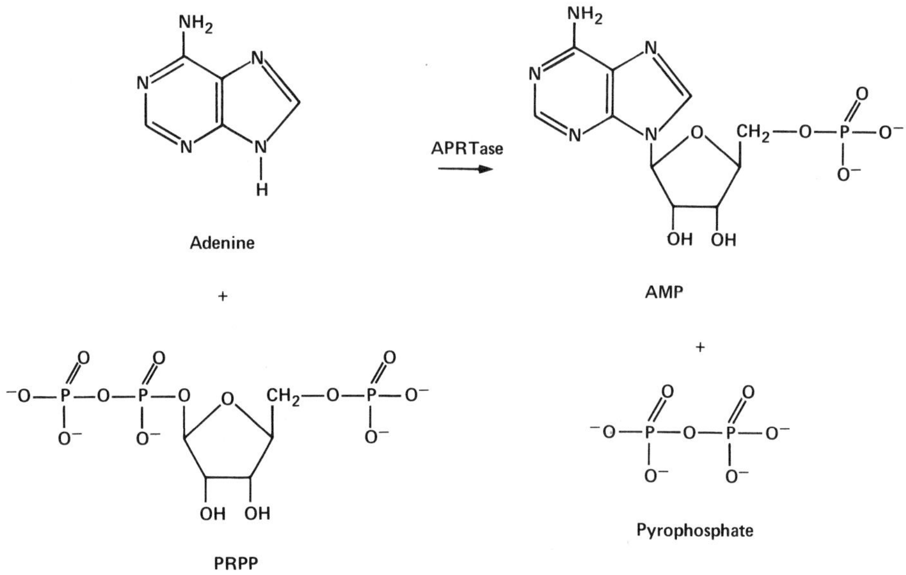

Figure 13.3 APRTase reaction.

plete absence [52] of this enzyme in humans seems to convey any detectable neurological consequences. A routine finding in children with Lesch-Nyhan syndrome, however, has been that APRTase is higher in their erythrocytes than in those of normal children. The explanation for this observation seems to be that, in cells missing HGPRTase, the concentration of PRPP is elevated perhaps 20-fold because it is not used for nucleotide formation. This elevated PRPP concentration probably protects APRTase against the normal inactivation as the cells age. This has been shown to be the case with purified APRTase, which is protected against heat inactivation by PRPP [53]. An elegant further proof involves experiments in which erythrocytes from normal humans and from Lesch-Nyhan patients have been treated by hypotonic lysis. In such an experiment it was shown that the oldest fraction of red blood cells from the patients had a higher relative specific activity of APRTase than the comparable fraction in normals, again arguing that the enzyme is more stable in the Lesch-Nyhan individuals [53]. Thus, the increased stability of the enzyme in erythrocytes is pretty clear. The cause, however, is still open to some doubt, since APRTase levels in Lesch-Nyhan fibroblasts are normal even though the PRPP levels are elevated in these cells as well as in erythrocytes [54].

Several studies have shown that the HGPRTase lesion in Lesch-Nyhan is a structural mutation in the enzyme, and not a complete deletion of the enzyme protein. Further, several different types of lesion have been shown, indicating a marked degree of genetic heterogeneity in this syndrome [33,55]. Some patients have very low activity of the enzyme, on the order of 1 to 4% [56-58], and what enzyme there is has altered properties, as demonstrated by disk gel electrophoresis [59]. Other patients have been shown to have lots of enzyme, but enzyme with kinetic properties altered so as to make it inactive at the normal concentrations of substrate found in the body [60]. Also, some Lesch-Nyhan patients have been found to have material in their cells which cross-reacts immunologically with normal HGPRTase, although little enzyme activity could be shown [44]. In addition to these many cases of altered and poorly functioning HGPRTase in Lesch-Nyhan patients, a number of individuals have been found who have low levels of activity, on the order of 5% or thereabouts, and who may have altered metabolism, and even gout [61], but are normal in behavior. It is clear, then, that there is a large degree of molecular heterogeneity in the Lesch-Nyhan syndrome, i.e., a large number of alterations in the enzyme protein all leading to deleted activity. Further, it is known that some individuals have markedly lowered enzyme activity, and may share some of the biochemical consequences, but that in the presence of even small amounts of activity the people are spared the drastic neurological abnormality.

The abnormal excretion of uric acid in these individuals stimulated investigations on the level of purine synthesis in Lesch-Nyhan cells, and it is now well established that such cells synthesize abnormally large amounts of purines by the de novo pathway. Evidence for this comes from experiments in which the amount of various precursors incorporated into the purine nucleus was measured in normal and in Lesch-Nyhan individuals. In such experiments it has been found that radioactive glycine is incorporated 20 times as rapidly into the uric acid of Lesch-Nyhan patients than into uric acid in control subjects [33]. Corroborating data have been obtained in cell culture where the amount of formate incorporated into purine intermediates in fibroblasts has been measured. Again, purine synthesis in the Lesch-Nyhan cells was several-fold more rapid than in the controls [35]. So, clearly, the de novo synthesis of purines is markedly increased in cells lacking the HGPRTase. The reason for this overproduction was, at first, thought to be a relief of feedback inhibition. That is, the first enzyme in the de novo pathway is subject to inhibition by nucleotides, and the lack of HGPRTase could cause a lowering of nucleotide levels in the cell and thus a heightened activity of this enzyme. It is known now, though, that cultured cells from Lesch-Nyhan patients have normal nucleotide levels, but still overproduce purines. It is now thought that the reason for the overproduction is the raised level of PRPP caused by the lack of HGPRTase. Since PRPP is the substrate for the first enzyme, and its concentration may be rate limiting for de novo synthesis, i.e., the K_M for the first enzyme is higher than the normal cellular PRPP concentration, the raised level in the defective cells may cause overproduction of purines [33,55].

Thus, the HGPRTase has assumed a much more interesting place in biochemical contemplation than previously. Its true function in the cell probably is not yet known. Certainly, this enzyme provides nucleotides for RNA synthesis via the salvage pathway. Also, however, it may be involved in the normal regulatory mechanism for control of purine biosynthesis. Indeed it has been stated that the HGPRTase maintains the balance between de novo and salvage pathways [36]. There is evidence accumulating that the HGPRTase functions in the transport of purines into cells. The first evidence was that such enzymes do function in purine transport in bacteria [62,63]. More recently it has been found that loss of HGPRTase activity by Lesch-Nyhan fibroblasts is accompanied by loss of hypoxanthine transport [64] and that HGPRTase is indeed a component of membrane vesicles from mouse cells [65]. Also, it has been shown that in mutant cell lines missing HGPRTase, a specific, saturable component for the transport of hypoxanthine was also missing [66]. Although others have been unable to find such a connection in other systems, it now seems possible that the function of the HGPRTase is as a component of purine transport. In an even larger sense, the HGPRTase may be important for the normal functioning of certain neurons and perhaps even have "a role in preventing a stereotyped form of compulsive behavior" [55]. In this regard it would be of remarkable interest if it were found that appropriate parts of the mammalian brain contained purinergic neurons. Although such neurons have not been definitely identified in the central nervous system, there is some evidence for their presence [67].

The disease itself has become more tractable in one sense. It is now possible to diagnose the disease prenatally, and there are methods available to detect the carrier state [53]. Prenatal diagnosis is possible through standard techniques of amniocentesis. Detection of the carrier state is more difficult. Males are normal. Females, according to the Lyon hypothesis, should be, and are observed to be, mosaics of normal and deficient cells. But, simply measuring enzyme levels in blood cells is not useful, since some unknown selection process during the production or maturation of red blood cells leads to selection out of the defective cells and a normal blood level of the enzyme. A number of simple manipulations are available, however, to reveal the carrier state in cell culture. These include cloning of cells and some pharmacological methods of cell selection by which the defective cells may be detected if present and, in fact, selectively grown. The most widely used is the addition of analogs of the purine bases which are toxic only after conversion to their nucleotide form. In defective cells, no such conversion is possible due to the lack of HGPRTase and so these cells will grow even though normal cells are killed. Such diagnosis is possible in cell culture, and, more recently, techniques have appeared for the diagnostic use of hair follicles.

There is no known treatment for the condition, although many have been tried. Allopurinol has some ameliorative effect on the gout associated with Lesch-Nyhan, but has no effect on the behavior. There is no information whatsoever on the cause of the bizarre behavior. In attempts to relate behavior to chemistry, some self-mutilative response has been produced in rats and rabbits by the injection of large amounts of methylated purines into the cerebrospinal fluid of these animals [68,69]. Possibly these data suggest that an aberrant purine metabolite, even a methylated derivative, is responsible for the behavior. But the relationship of the model to the disease has not been explored in sufficient detail to allow such a conclusion.

Lesch-Nyhan Syndrome

The patient, a 4 ½-year-old boy, was admitted for the treatment of hematuria. He had advanced cerebral palsy and was markedly retarded. Muscle tone was increased, and legs were continually in the scissor position with feet and toes plantar-flexed. He could not walk, sit, or stand without assistance. He spoke little but appeared to comprehend. His most striking behavioral characteristic was destructive biting of the fingers and lips. The patient had partially amputated the distal phalanx of two fingers and completely chewed away his lower lip.

A history indicated reasonably normal development until 12 months of age although the child was small. Self-mutilation was noted at 1 ½ years of age. The biting of lips occurred initially after a severe fall and an injury to the mouth. Finger biting was observed 6 to 8 months later. An older brother was normal, but a cousin was known to be afflicted with Lesch-Nyhan.

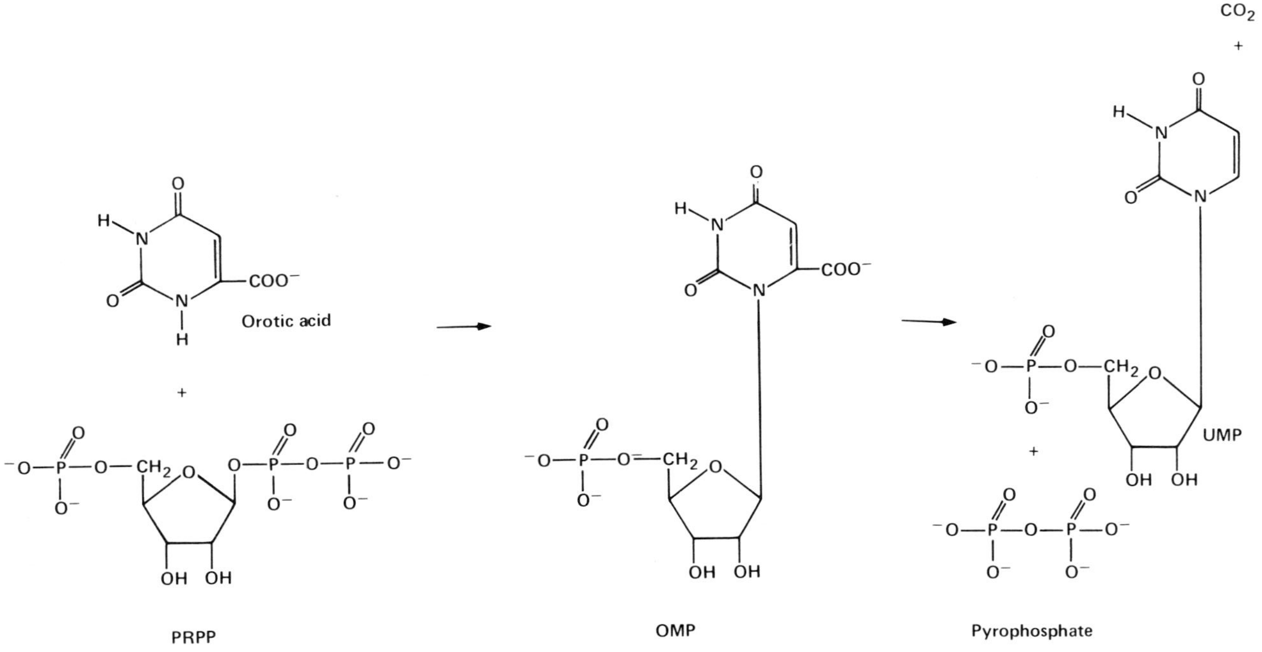

Figure 13.4   Orotic acid:PRPP transphosphorylase and OMP decarboxylase reactions.

Upon hospitalization the patient was found to have a blood uric acid concentration of 11.5 mg/100 ml and a urinary uric acid to creatinine ratio of 6.0. No HGPRTase was detectable in his erythrocytes. Continual vomiting caused difficulty in maintaining proper nutrition. Episodes of self-mutilation, accompanied by screaming and other evidence of extreme fright, necessitated continuous restraint. Under these conditions the patient appeared cheerful and happy.

During hospitalization the condition of the child worsened due to apparent kidney failure, and he died 7 weeks after hospitalization of gouty nephritis. Allopurinol was given in dosages of 10 mg kg^{-1} day^{-1} for the last 2 weeks of the child's life but appeared to have little or no effect on his condition.

Another disease in which an aberration in nucleotide metabolism is connected with neurological malfunction is orotic aciduria [17,33]. In this condition both OMP pyrophosphorylase and OMP decarboxylase are markedly lowered in activity or missing altogether (Fig. 13.4). The absence of both enzymes provides further evidence that the two are located, as mentioned before, on the same protein molecule. Genetically, the condition is transmitted as an autosomal recessive. The disease is characterized by a B_{12}, ascorbic acid, and folic acid-resistant megaloblastic anemia. In addition, the children have retarded growth, sparse hair, and are subject to frequent infection. The neurological involvement is variable, although severe intellectual retardation is often seen. Whether the neurological syndrome is primary in the condition is unknown.

References

1. P. Mandel, in *Handbook of Neurochemistry,* Vol. Va (A. Lajtha, ed.), Plenum, New York, 1971.
2. P. Mandel and S. Harth, *J. Neurochem. 8:*116 (1961).
3. P. J. Heald, *Phosphorus Metabolism of the Brain,* Pergamon, Oxford, 1960.
4. J. F. Henderson and G. A. LePage, *J. Biol. Chem. 234:*2364 (1959).
5. W. J. Howard, L. A. Kevson, and S. H. Appel, *J. Neurochem. 17:*121 (1970).
6. W. A. Mannel and R. J. Rossiter, *Biochem. J. 61:*418 (1955).
7. S. H. Appel, *J. Biol. Chem. 243:*3924 (1968).
8. S. E. Hager and M. E. Jones, *J. Biol. Chem. 242:*5667 (1967).
9. M. Tatibana and K. Ito, *J. Biol. Chem. 244:*5403 (1969).
10. J. N. Santos, K. W. Hamstead, L. E. Kopp, and R. P. Miech, *J. Neurochem. 15:*367 (1968).
11. G. Guroff, A. F. Hogans, and S. Udenfriend, *J. Neurochem. 15:*489 (1968).
12. F. M. Rosenbloom, W. N. Kelley, J. M. Miller, J. F. Henderson, and J. E. Seegmiller, *J. Am. Med. Assoc. 202:*175 (1967).
13. A. W. Murray, *J. Biochem. 100:*664 (1966).
14. D. M. Dawson, *J. Neurochem. 15:*31 (1968).

15. B. Setlow, R. Burger, and J. M. Lowenstein, *J. Biol. Chem. 241:*1244 (1966).
16. B. Setlow and J. M. Lowenstein, *J. Biol. Chem. 242:*607 (1967).
17. R. L. Levine, N. J. Hoogenraad, and N. Kretchmer, *Pediatr. Res. 8:*742 (1974).
18. A. F. Hogans, G. Guroff, and S. Udenfriend, *J. Neurochem. 18:*1699 (1971).
19. W. Wells, D. Gaines, and H. Koenig, *J. Biol. Chem. 245:*2199 (1970).
20. A. W. Murray, *Ann. Rev. Biochem. 40:*811 (1971).
21. A. Geiger and S. Yamasaki, *J. Neurochem. 1:*93 (1956).
22. S. Nakagawa and G. Guroff, *J. Neurochem. 20:*1143 (1973).
23. W. K. Jordan, R. March, O. Boyd-Houchin, and E. Popp, *J. Neurochem. 4:*170 (1959).
24. G. Guroff and M. Brodsky, *J. Neurochem. 18:*2077 (1971).
25. G. G. Villela, *Experientia 24:*1101 (1968).
26. U. A. S. Al-Khalidi and T. H. Chaglassian, *Biochem. J. 97:*318 (1965).
27. A. Mori, H. Ohkusu, M. Kohsaka, and M. Kurono, *J. Neurochem. 20:*1291 (1973).
28. G. Guroff and C. A. Rhoads, *J. Neurochem. 16:*1543 (1969).
29. T. C. Johnson, *J. Neurochem. 14:*1075 (1967).
30. F. Orrego, *J. Neurochem. 14:*851 (1967).
31. M. Lesch and W. L. Nyhan, *Am. J. Med. 37:*561 (1964).
32. W. L. Nyhan, J. Pesek, L. Sweetman, D. G. Carpenter, and C. H. Carter, *Pediatr. Res. 1:*5 (1967).
33. W. L. Nyhan, in *Biology of Brain Dysfunction,* Vol. 1 (G. E. Gaull, ed.), Plenum, New York, 1973.
34. L. Sweetman and W. L. Nyhan, *Biochem. Med. 4:*121 (1970).
35. J. E. Seegmiller, F. M. Rosenbloom, and W. N. Kelley, *Science 155:*1682 (1967).
36. A. W. Murray, *Ann. Rev. Biochem. 40:*811 (1971).
37. A. W. Murray, *Biochem. J. 103:*271 (1967).
38. A. W. Murray, *Biochem. J. 100:*664 (1966).
39. T. A. Krenitsky, *Biochim. Biophys. Acta 179:*506 (1967).
40. F. M. Rosenbloom, W. N. Kelley, J. Miller, J. F. Henderson, and J. E. Seegmiller, *J. Am. Med. Assoc. 202:*175 (1967).
41. J. F. Henderson, L. W. Brox, W. N. Kelley, F. M. Rosenbloom, and J. E. Seegmiller, *J. Biol. Chem. 243:*2514 (1968).
42. T. A. Krenitsky and L. Papaioannou, *J. Biol. Chem. 244:*1271 (1969).
43. W. J. Arnold and W. N. Kelley, *J. Biol. Chem. 246:*7398 (1971).
44. C. S. Rubin, J. Dancis, L. C. Yip, R. C. Nowinsky, and M. E. Balis, *Proc. Natl. Acad. Sci. USA 68:*1461 (1971).
45. B. Bakay and W. L. Nyhan, *Biochem. Genet. 5:*81 (1971).
46. M. R. Davies and B. M. Dean, *FEBS Lett. 18:*283 (1971).
47. M. M. Muller and H. Dobrovits, *Prep. Biochem. 2:*375 (1972).
48. W. Gutensohn and G. Guroff, *J. Neurochem. 19:*2139 (1972).

49. W. N. Kelley, F. M. Rosenbloom, J. F. Henderson, and J. E. Seegmiller, *Biochem. Biophys. Res. Commun. 28:*340 (1967).

50. J. F. Henderson, W. N. Kelley, F. M. Rosenbloom, and J. E. Seegmiller, *Am. J. Hum. Genet. 21:*61 (1969).

51. W. N. Kelley, R. I. Levy, F. M. Rosenbloom, J. F. Henderson, and J. E. Seegmiller, *J. Clin. Invest. 47:*2281 (1968).

52. H. Debray, P. Cartier, A. Temstet, and J. Cendrin, *Pediatr. Res. 10:*762 (1976).

53. M. L. Greene, J. R. Boyles, and J. E. Seegmiller, *Science 167:*887 (1970).

54. W. N. Kelley, *J. Lab. Clin. Med. 77:*33 (1971).

55. J. E. Seegmiller, *Biochemie 54:*703 (1972).

56. W. N. Kelley, M. L. Greene, F. M. Rosenbloom, J. F. Henderson, and J. E. Seegmiller, *Ann. Intern. Med. 70:*155 (1969).

57. W. N. Fujimoto and J. E. Seegmiller, *Proc. Natl. Acad. Sci. USA 65:*577 (1970).

58. W. N. Kelley and J. C. Meade, *J. Biol. Chem. 246:*2953 (1971).

59. B. Bakay and W. L. Nyhan, *Biochem. Genet. 6:*139 (1972).

60. J. A. McDonald and W. N. Kelley, *Science 171:*689 (1971).

61. W. N. Kelley, F. M. Rosenbloom, J. F. Henderson, and J. E. Seegmiller, *Proc. Natl. Acad. Sci. USA 57:*1735 (1967).

62. J. Hochstadt-Ozer and E. R. Stadtman, *J. Biol. Chem. 246:*5304 (1971).

63. J. Hochstadt-Ozer, *J. Biol. Chem. 247:*2419 (1972).

64. P. J. Benke, N. Herrick, and A. Herbert, *Biochem. Med. 8:*309 (1973).

65. C. Li and J. Hochstadt, *J. Biol. Chem. 251:*1181 (1976).

66. J. Epstein and J. W. Littlefield, *Exp. Cell Res. 106:*247 (1977).

67. G. Burnstock, *Pharmacol. Rev. 24:*509 (1972).

68. W. L. Nyhan, *Fed. Proc. 27:*1044 (1968).

69. L. L. Morgan, N. Schneiderman, and W. L. Nyhan, *Psychem. Sci. 19:*37 (1970).

14
NUCLEIC ACID METABOLISM

DNA

The DNA content of adult rat brain tissue is about 0.7 to 1.0 mg/g of fresh
weight. Of this, 98% or more is located in the nucleus, with the remaining 1 to
2% being mitochondrial. Since a DNA content of about 6.5 pg per cell is constant from one diploid cell to the next, the DNA content can be, and frequently
is, used to estimate the number of cells in a given tissue under specific conditions,
or at a specific developmental stage [1]. For example, the DNA content of early
fetal guinea pig [2] or embryonic chick cortex per gram of tissue is much higher
than that of newborn or adult individuals. These data are consistent with the observation that the cells are much smaller and more closely packed in fetal tissue,
and during postnatal growth the cytoplasm increases and the space between the
nuclei becomes much greater [3]. Thus, the DNA content drops even though
the cell number may remain constant or even increase. Caution must be observed in such studies, since in a tissue with a high content of tetraploid cells
the DNA values will be misleading as to cell number and give, obviously, a large
overestimate. Such is the case for the DNA content of cerebellum. In this tissue
the cells are densely packed, but the high DNA content observed is primarily due
to the large proportion of tetraploid Purkinje cells present [4].

DNA polymerase has been studied only cursorily in brain tissue. Activity
has been detected in rat brain cerebellum [5], in developing chick brain [6], and
in the brains of octopi [7]. The characteristics of the brain enzyme are unremarkable and comparable to similar preparations from other mammalian tissues.
DNase activity has also been found in brain, and at least two separate DNase enzymes, with different pH optima and substrate specificity, have been reported [8].

Autoradiography of DNA Synthesis

Since DNA is known to be stable in brain, and since not even DNA repair has
been detected in this tissue, any DNA synthesis found can reasonably be assumed

to represent cell proliferation. A great deal of very useful information has been obtained through studies of DNA synthesis in brain at various stages of development. Early investigations on DNA synthesis, done by measuring the DNA content of various areas directly, were not too revealing. The development of the autoradiographic method for measuring the incorporation of tritiated thymidine as an index of DNA synthesis [9] has opened wide this area of investigation. Basically, the method involves systemic administration of tritiated thymidine followed by a visualization of the site at which this isotope was incorporated through autoradiography. Suitably fixed and processed tissue is brought into contact with a radiosensitive emulsion, specifically responsive to the tritium emissions. After an appropriate exposure the emulsion is developed. The blackened grains give a location and a quantitation of the amount of thymidine incorporated in various cells. Thymidine is the precursor of choice because of its selectivity for labeling DNA. Also, it is metabolized by the animal in a short time so that systemic administration is equivalent to a pulse labeling, i.e., the label is present in the system for only about 30 min and labels the cells for that period and that period only. The fate of cells tagged during such a period can be followed by sacrificing injected animals at intervals. Since DNA does not turn over, loss of DNA from a site can be due only to cell death, cell migration, or cell division. Tritiated material is preferred to other isotopes because its half-life is long compared to the experiment itself, and especially because its low energy limits its penetration into the emulsion and increases the resolving power of the technique. "The application of fine-resolution thymidine-^{3}H autoradiography to the study of DNA metabolism and cellular proliferation in the central nervous system has revolutionized this area of investigation. Thus, the vicissitudes of a class of tagged cells, their migrations, mode of differentiation, and life span can be traced...." [9].

Primary among the questions approached by this methodology has been the problem of cell division in the mature central nervous system. Early on it had been assumed that little or no cell division takes place in the brain postnatally. Now, through the use of autoradiography, it has been thoroughly demonstrated that DNA synthesis, and thus cell division, does occur in mammals postnatally, but that only certain types of cells divide. Almost all the DNA synthesis which takes place does so in glial components. Most of the long-axoned or macroneurons are formed prenatally, except for some in the hippocampus [10,11]. Many of the microneurons or interneurons are formed postnatally, and account for some of the incorporation into DNA [12-14]. But the formation of all types of glia is, in mammals, essentially a postnatal phenomenon [15,16], and accounts for most of the incorporation of precursors into DNA, as well as the increasing glia/neuron ratio which accompanies maturation.

Another application of the autoradiographic tool has been to investigate the cellular embryogenesis of the brain. The classic description [17] of neural tube differentiation in mammals suggested that two kinds of cells existed in the ependymal or neuroepithelial layer, the germinal cells and the spongioblasts. These were differentiated on the basis of their morphology and were thought to give rise to the neurons and the glia elements, respectively. This view was challenged in 1935 by the observation [18] that the germinal cells and the spongioblasts were a single class of cells and differed in morphology due to the fact that they represented cells in different parts of the cell cycle. Thus, glia and neurons were different differentiated forms of the same cell line. This latter concept was fully confirmed by the autoradiographic tool [19,20].

It is of interest that there are a number of experimental and environmental conditions which alter the proliferation of cells, and thus the synthesis patterns of DNA, in the central nervous system [9]. Injury to the brain causes a proliferation of glia elements [21-23], and also of endothelial cells, indicating an increased vascularization of the injured area. There are certain viruses which specifically affect rapidly dividing cells, such as glia, and when they attack the brain they generally cause locomotor ataxia [24]. X-ray treatment is also rather specific for dividing cells and causes ataxia in some experimental situations [9,25]. Malnutrition during infancy causes a reduced brain weight in the resulting adult, and does so by causing a reduced number of cells to be formed, rather than by interference with the formation of any specific biochemical constituent [26]. Conversely, an increased caloric intake during infancy is reported to lead to an increased number of brain cells in the adult [27]. Finally, environmental influences are said to lead to differences in cell number in the brain. Animals reared in a so-called "enriched environment," which have been shown to have heavier brains, thickened cortices, and increased levels of some of the enzymes leading to transmitter synthesis, also show increased proliferation of glial cells compared to controls reared in isolation [28].

RNA Content

Recent interest in the RNA metabolism of the brain has been intense because of the obvious informational possibilities inherent in the RNA structure. The efforts to tie RNA directly to the engram will be discussed in a later section. But studies on the metabolism if RNA in the brain have proceeded, to some extent, independently of this effort [29-31]. Such studies, both biochemical and behavioral, have received enormous methodological support from the elegant techniques for the microanalysis of RNA, even from single cells and parts of cells, devised by Hydén and by Edstrom and their colleagues [32-35]. Using such methods, levels of RNA in single cells, and even in single nuclei, have been measured. Micromethods for analyzing the base composition of these RNAs have been described, and data have been presented relating changes in content and composition to changes in the environment, training, and chemical treatment of experimental animals.

The RNA content of fresh cortical tissue of rats is about 0.7 to 0.8 mg/g. There is some histological evidence that the concentration of RNA in oligodendrocytes is higher than that in neurons, and that astrocytes, on the other hand, have a very low RNA concentration [36]. Upon direct assay, neuronal content of RNA is found to range from about 1500 pg per Dieters' cell to about 70 pg per neuron in the supraoptic nucleus [30]. The glia have generally lower content, on the order of 50 to 150 pg per cell. The different classes of glia differ in their RNA content, the oligodendroglia having the highest level. The preponderance of the RNA in neurons is of the ribosomal type. Some estimates set the ribosomal component in whole brain as accounting for 80% of the total brain RNA [37]. With the exception of a single report [38], there are no unique RNAs in brain. The ribosomal RNA (rRNA), transfer RNA (tRNA), and, as far as is known messenger RNA (mRNA) of brain are much like the corresponding RNAs from other tissues [29-31].

Ribosomal RNA

Brain ribosomes, as discussed in an earlier section, are about 30% RNA. The composition of this RNA is typical of mammalian ribosomal RNA from any tissue, and closely similar from the brain of one species to that of the next. These RNAs characteristically have a high content of guanine and cytidine, the G + C: A + T ratio from all species ranging around 1.5 to 1.7. There is some evidence that the composition of brain ribosomal RNA changes with the age of the animal [39,40]. On the other hand, it has been found that no difference in the functional aspects of ribosome action can be seen in the ribosomes from young or old rats [41]. The ribosomal RNA from brain, as from other tissues, is made up of three classes, an 18S species from the small ribosomal subunits, and a 5S and a 28S species from the large ribosomal subunit [37]. Ribosomal precursor RNAs of 45S and 32S have been observed in brain [42-45]. These appear to be made in the nucleolus and methylated during processing, as are ribosomal precursors in other tissues. Although estimates vary, the whole-brain ribosomal RNA component has been found to have a turnover time of 12 days, which is the same as that for ribosomal protein, and suggests that the ribosomes turn over as a unit [46]. It is known that the fraction of the genome coding for the ribosomal RNA of brain is on the order of 0.15% [47].

Transfer RNA

Transfer RNA has been prepared from brain [48,49]. The properties of these RNAs are not known to be different from those from other tissues in the same species. The proportion of the various tRNAs appears to change with age, but there is no relationship between the changes in the various tRNAs and concomi-

tant changes in the proportions of the amino acids in the free amino acid pool
[50].

Messenger RNA

The attempts to obtain messenger RNA from brain have been somewhat more intense than for other tissues, again because of the possible informational implications of such a messenger. Early attempts to investigate brain messenger involved
assessment of its characteristics without actual isolation. Thus, the presence of
mRNA was suggested by the detection, in brain, of RNA fractions which had
rapid turnover [42,51-54], DNA-like composition [42,47,52,54], and polydis-
persity [42,52], and which stimulated the incorporation of radioactive amino
acids into TCA-insoluble precipitates when added to cell-free, protein-synthesizing systems [55-58]. It was shown that the brain messenger fraction had an
average turnover time of between 30 min and 2 hr [52,58,59], although a longer-
lived fraction was also detected [51]. The G + C : A + T ratio was found to approximate 0.9 [52,53], which is close to the 0.7 found for DNA, and certain
fractions from brain, especially those from the nucleus, were active as templates
for the incorporation of amino acids into TCA-insoluble material in cell-free, protein-synthesizing systems from various sources [55-58].

More recent studies, using newer methods of isolation, have begun to give
a detailed picture of the messenger fraction from the brain, and to allow its trans-
lation into specific brain proteins. Experiments of Roberts and his colleagues
using EDTA to free cytoplasmic mRNA from its polyribosome attachment, led
to the isolation of a messenger fraction from brain which sedimented with discrete peaks at 8S and 16S (Fig. 14.1a) [60]. The properties of this fraction were
consistent with its identity as messenger. Thus, it had a "DNA-like" composition (Table 14.1), hybridized to DNA, was rapidly labeled, was active as a template, and comprised on the order of 2% of the total polysomal RNA. The subsequent work of Zomzely-Neurath, using the EDTA procedure, has shown that
it is possible to prepare a messenger fraction which will direct the cell-free synthesis, by brain preparations, of two brain-specific proteins, the S-100 [61] and
the 14-3-2 [62].

It has been found that many, if not all, eukaryotic mRNAs have a 3' terminal sequence of somewhere between 50 and 200 adenylate residues. The function of this poly(A) sequence is not yet known, but it has been quite useful in
the preparation of mammalian messengers for various purposes. Such a fraction
has been prepared from rat brain and its properties examined [63,64]. The isolation included phenol extraction at high temperature and pH, followed by chromatography of the purified RNA on cellulose esterified with short chains of poly-
thymidylic acid. Such a procedure yielded an RNA preparation which was heterodisperse and essentially free of ribosomal RNA contamination. Total brain RNA

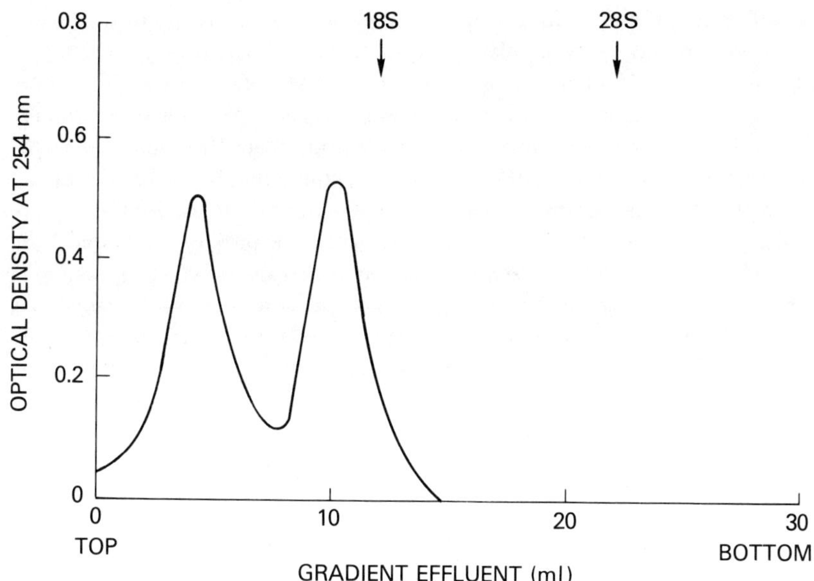

Figure 14.1a Sedimentation profile of the polysomal mRNA preparation from cerebral cortex of the adult rat. The sucrose gradient was centrifuged for 18 hr at 25,000 rpm. Arrows at 18S and 28S mark the positions of the two major peaks of cerebral rRNA analyzed under similar conditions. [From C. E. Zomzely, S. Roberts, and S. Peache, *Proc. Natl. Acad. Sci. USA 67*:644 (1970).]

Table 14.1   Base Composition of Cerebral RNA Fractions from Adult Rats

RNA Fraction	C	A	G	U (or T)	G + C:A + U, (or T)
Polyribosomal mRNA	19.1	23.2	27.3	30.4	0.87
Ribosomal RNA	27.6	17.0	32.4	21.5	1.57
Rat liver DNA	21.6	28.8	21.5	28.5	0.75

Data taken from: C. E. Zomzely, S. Roberts, and S. Peache, *Proc. Natl. Acad. Sci. USA 67*: 644 (1970); S. C. Bondy and S. Roberts, *Biochem. J. 105*:1111 (1967); F. Labrie, *Nature (London) 221*:1217 (1969).

from adult rats contains about 5% of such RNA, while brain RNA from young rats has somewhat more. The poly(A) sequence on the RNA from adult brain was on the order of 188 nucleotides long, while the sequence of the RNA from the brains of young animals contained 208 nucleotides. The poly(A)-containing RNA fraction was rapidly labeled in vivo and in brain slices [63], and can also be synthesized by isolated nuclei [65]. Poly(A)-containing RNAs have been found in the nuclei from brain, as well as attached to polysomes. In both fractions the RNA was heterodisperse (Fig. 14.1b), with the peak being around 16S. Using a wheat-germ, cell-free system, it was found that the poly(A)-containing RNAs were better as templates than the RNAs which did not contain the poly(A) sequence, and hence did not bind to the oligo(dT)cellulose. This relatively facile preparation of what appears to be the messenger fraction from brain should prove useful for future studies.

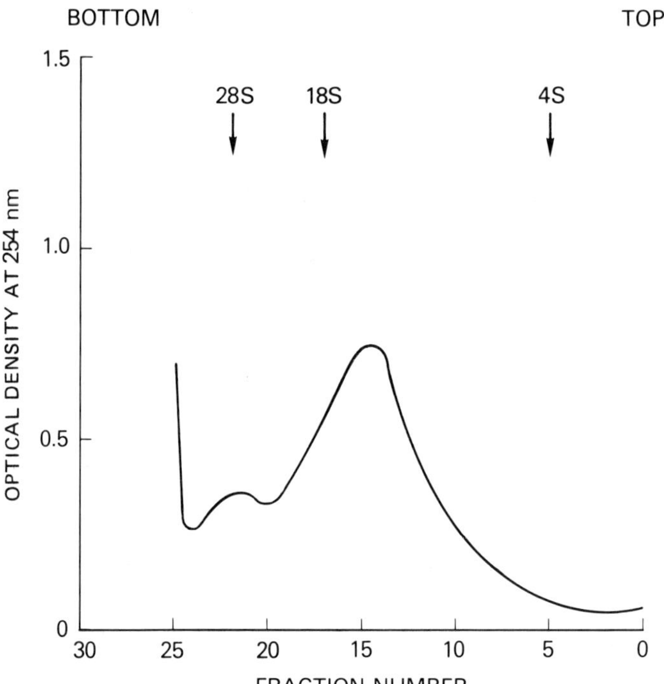

Figure 14.1b Glycerol density gradient analysis of rat brain polysomal RNA which bound to oligo(dT)cellulose. Centrifugation was carried out for 18 hr at 112,500 g. [Reprinted with permission from J. DeLarco, A. Abramowitz, K. Bromwell, and G. Guroff, "Polyadenylic Acid-Containing RNA from Rat Brain," *J. Neurochem. 24:*215 (1975). Copyright 1975, Pergamon Press, Ltd.]

Hybridization

Estimating the fraction of the genome in brain which is devoted to the synthesis of mRNA has been accomplished by hybridization methodology. Early studies showed that rapidly labeled RNA from whole brain hybridized with 1.2% of the DNA, and a similar fraction from polysomes alone hybridized with 0.8% [47,66]. Thus, less than 1% of the DNA from brain was thought to be devoted to the synthesis of cytoplasmic brain message. However, it is now known that the early hybridization studies were done in such a way as to lead to hybridization of the RNA with only the highly repetitive sequences now known to be present in mammalian DNA. Since it is unlikely that these very redundant sequences are coding for mRNA, newer experiments have been designed to measure the hybridization of the mRNA to unique sequences in DNA only. This has been done by allowing the DNA to hybridize to itself first, and then separating those portions which do not hybridize rapidly. The rapidly reassociating DNA fragments are discarded, and the brain RNA is then hybridized to the remainder. Such experiments have given quite different data from the early ones.

The work of McCarthy and his colleagues, and of others, has shown that the hybridization to these unique sequences is much higher in mouse brain RNA than with RNA from other organs, being 4 to 5%, for liver, kidney, and so on, but on the order of 11% for brain (Fig. 14.2 left) [67-69]. This percentage increased as the animal matured (Fig. 14.2 right). Furthermore, when the RNA

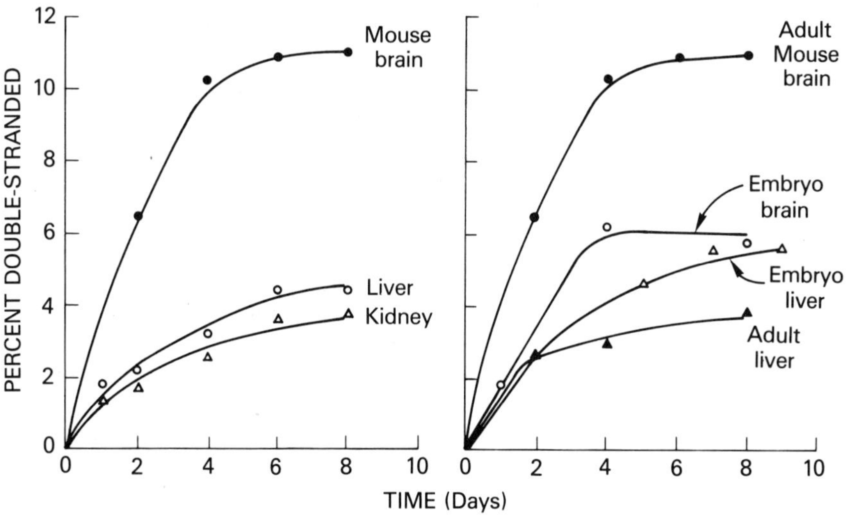

Figure 14.2   Hybridization of RNA from various mouse tissues to mouse unique-sequence DNA. [Data taken from L. Grouse, M. D. Chilton, and B. J. McCarthy, *Biochemistry 11:*798 (1972).]

from human brain was hybridized back to human DNA a greater percentage of the DNA was used [70]. For example, 12% of the DNA was hybridized to RNA from fetal brain and up to 24% from adult human. That is, assuming asymmetric transcription, almost one-half of the overall genetic information present in unique DNA may be expressed in brain. In a startling experiment the RNA from the left frontal cortex was compared with that of the right. It was shown that 2 to 3 times more hybridization occurred with the RNA from the left side, leading the authors to suggest that the usual left-sided dominance for language might be related to such data.

The studies described above concerned total tissue or nuclear RNA, and the RNA diversity observed reflected transcriptional RNA diversity. Subsequent work has shown that the diversity in the RNA which is actually translated, i.e., polysomal mRNA, is less than that found in the total cell RNA. Mouse brain polysomal RNA diversity has been found to be 5.5% [71]; the poly(A)-containing mRNA from mouse brain had a sequence diversity of 3.7% [72]. The overall significance of the greater diversity of RNA in the brain compared to other tissues is not entirely clear, although recent data on RNA frequency distributions [73] in brain suggest that greater heterogeneity in cell type and cell function is responsible. In any case, the data show that more of the genetic information available in the DNA is used by brain than by any other tissue, and that human brain is using more than the brain of the other species tested.

It has been known for some time that brain has a much greater mRNA content than, for example, liver [30]. The work of Grouse and McCarthy indicates that this higher messenger content represents, as well, a much greater informational diversity. It is of interest that the diversity present in whole-cell brain RNA increases with age. On the other hand, it is known that the brains of older animals make less poly(A)-containing RNA per unit time than do younger animals. Thus, these observations suggest that in adults there is less brain message, but that what is there represents greater RNA diversity.

There is specific information on how the mRNA is transferred to the polysomes from the nucleus in brain. It seems that, as in other tissues, it is conveyed as a ribonucleoprotein particle or complex. The evidence that the RNA portion of the complex is mRNA is that it is more rapidly labeled than the polysome-bound mRNA, and that it has the same size as mRNA, namely, 8S to 15S [74,75].

There is little specific information about the RNA component of brain mitochondria, but there is no reason to suspect that it is any different from that from other mitochondria. It is known that, in common with other mitochondria, those from the brain contain RNA with poly(A) sequences. The poly(A) segments, however, are much shorter than those found attached to either the nuclear or the messenger RNA [76].

One of the few reports of an unusual RNA associated with a specific disease state involves biopsy material from the globus pallidus of Parkinson's disease patients. In the early stages of this disease, a highly aberrant RNA has been reported in certain glial cells. This RNA has a high adenine content and a low amount of guanine and uracil. As the disease progresses, similar but not identical changes appear in the RNA of neurons. The RNA of both types of cells increases. What role this unusual RNA plays in the disease state is not known [77].

RNA Polymerase

The RNA polymerase of brain has been studied off and on for several years. The first measurements were done on purified nuclei [78] or on an "aggregate" insoluble enzyme preparation [79]. These studies showed that the activity was present in brain and, in fact, that it is at least as high in activity as it is in liver. The characteristics of the brain enzyme were unremarkable when compared to polymerases from other tissues. The pH optimum was found to be 8.0 and the activity was inhibited by actinomycin D, stimulated by spermidine, and higher in gray matter than in white.

More recently it has been shown that there are at least two RNA polymerases in brain. As in other tissues, the nucleolar polymerase (polymerase I), is stimulated by Mg^{2+} and by conditions of low ionic strength, resistant to α-amanitin, and makes a ribosomal-type RNA with a high content of guanine and cytidine. A second enzyme, polymerase II, is found in the nucleoplasm, is stimulated by Mn^{2+} and high ionic strength, is sensitive to α-amanitin, and is responsible for the synthesis of a DNA-like RNA. The RNA polymerases of ox brain have now been solubilized and purified [80,81]. The procedure involves sonication of the nuclei, and then an ammonium sulfate fractionation, followed by the separation of the two activities on a DEAE-cellulose column. The preparations are DNA dependent and rather unstable. They are stimulated, but not equally, by spermidine [82]. Fractionation of brain nuclei into populations from neuronal and glial cells showed that both polymerase I and polymerase II were present in the nuclei of both types of cells, and only quantitative differences occurred in the level of activity of each in the various nuclei isolated [83]. No report has appeared on the presence of an enzyme comparable to the RNA polymerase III now found in other cells.

Enzymes in brain nuclei will synthesize and degrade the so-called homopolymers as well. Homopolymers of adenine and cytidine have been made from the appropriate mononucleotide using brain nuclei as an enzyme source. The rapid degradation of polyadenylic acid has been demonstrated as well [84,85]. The possible function of these homopolymers could be as storage sites for nucleotides and their further utilization for the synthesis of RNA, but so far this is only speculation and even the occurrence of these materials as normal nuclear constituents has yet to be demonstrated. The newer information on the poly(A)-sequences in mRNA from eukaryotic cells may serve to implicate these enzymes in mRNA processing rather than in the metabolism of homopolymers.

Other enzymes of RNA metabolism have been shown in brain. These include RNA methylases, three of which are known, differing in their pH optima [86]. Also present are acid and alkaline RNase [87], and an endogenous RNase inhibitor, which is acidic and has a molecular weight of about 60,000, and which seems similar to the inhibitor which has been purified and characterized from liver [88].

The rate of synthesis of RNA in the brain is quite substantial when compared to other organs. Many neurons produce RNA as fast or faster than any other somatic cells in the body. RNA precursors are incorporated rapidly into brain RNA in vivo or in vitro. There is some evidence that RNA synthesis is more rapid in glia than in neurons [89]. There is also some suggestion, to be discussed later, that the RNA made in the glia can be transferred to the neurons.

Developmental Changes

The RNA content of brain is known to vary with the age of the animal. Generally speaking, the RNA content of brain rises rapidly during the postnatal period, and this rise can be as much as 25-fold, for example, in the human [90]. Then, in senescent animals the RNA content drops from its level in the mature animal. These changes are reflected in the RNA content of specific cells. In the rat the RNA content of the hippocampal pyramidal neurons rises from 24 pg per cell in the newborn to 110 pg in the adult, and then decreases to 53 pg per cell in the senescent rat [91]. Correspondingly, in the human, RNA measurements in certain motor neurons of the spinal cord show that RNA goes from 402 pg per cell between 0 and 20 years of age, to 553 pg between ages 21 and 40, to 640 pg between ages 41 and 60, and then falls to 420 pg in individuals over 80 years of age [92].

Consistent with the finding that the RNA content of brain rises during young adulthood are the many different studies which show that RNA synthesis is much more active in the brains of young rats than in old [93-95]. These data have been obtained in vivo and in vitro with various precursors and techniques. They are consistent with other data which show that the rates of protein synthesis in the brain also decrease as the animal matures. Inquiries into the biochemical basis of the increased RNA synthesis have indicated a number of things about the developmental course of brain RNA synthesis. The RNA polymerase enzyme decreases in activity with age, perhaps twofold in the rat [96]. On the other hand, levels of total tRNA, free or aminoacylated, do not change with maturation [50]. Ribosomes from the brains of young animals are the same as those from adults [41], but the proportion of the ribosomes in the brain which are in the form of polysomes does seem to be age dependent. While the proportion of polysomes in liver seems to go up as the animal ages, the amount in brain drops [41,97,98]. Since the polysomes are held together by mRNA, the number of polysomes should reflect the amount of mRNA present. In fact, data is available to show that the turnover of mRNA in the newborn is greater than in the adult [54], suggesting that mRNA should be more stable in the adult. Also, it has been shown that there is more poly(A)-containing RNA in the brains of neonatal rats than in adults, and more is synthesized per unit time [63]. Thus, it seems that the greater RNA synthesis in the brains of young animals is due to greater synthesis and turnover of the mRNA component, perhaps due to a higher activity of the RNA polymerase of brain.

Physiological Changes

There are numerous reports of changes in RNA content or rate of synthesis of brain RNA by physiological, motor, or sensory stimuli [29-31]. For example, the content of RNA in appropriate neurons can be increased by violent exercise [99,100], mechanical rotation [101], olfactory [102] or visual stimulation [103], or thirst [104]. The synthesis of RNA, as measured by the incorporation of precursors, can be increased in certain cells by visual stimulation [105] and, apparently, by certain kinds of training [106,107]. Electrical stimulation raises the content [108] and the synthesis of RNA [109] in neurons when applied in vivo. On the other hand, experiments with brain slices [94] show that synthesis goes down in vitro upon stimulation, perhaps due to a depletion of ATP which lowers the conversion of the nucleotides to the triphosphate level, a conversion which must precede their incorporation. The various stimuli mentioned above can, in some cases, raise the levels of RNA in "target" neurons by as much as 80%. Generally speaking, the conditions which raise the RNA of neurons lower the RNA content of the glial elements associated with the affected neurons. The increased neuronal RNA is, in some cases, of the ribosomal type, as is the RNA lost from the associated glial elements.

Chemical Influences

There are a number of chemical influences which also alter RNA synthesis in the brain (Fig. 14.3). Of course, the classic RNA inhibitors, such as the actinomycins, act on brain RNA synthesis as they do on RNA synthesis in other tissues. The nucleotide analogs, 8-azaguanine, 6-azauracil, 5-fluorouracil, and 5-fluoroorotic acid, inhibit the synthesis of RNA [31]. In addition to these, a few other compounds have influence on brain RNA. Tricyanoaminopropene (TCAP), a dimer of malononitrile, raises the RNA content of neurons in the Deiters' nucleus [110]. A drop in glial RNA also occurs, and this increase and loss are in terms of ribosomal RNA. Pemoline has been reported to enhance RNA synthesis [111] and

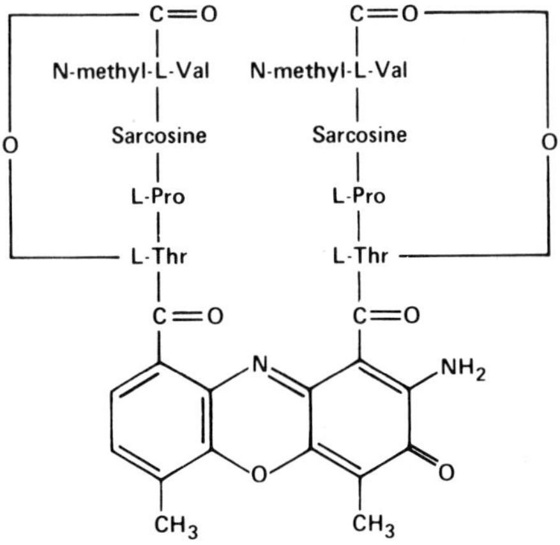

Actinomycin D

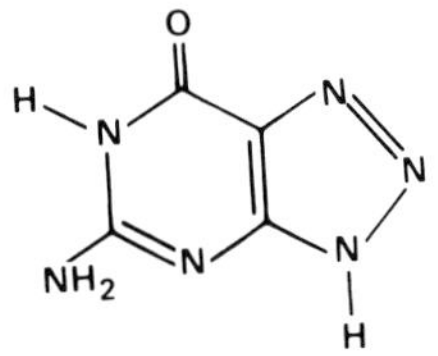

8-Azaguanine

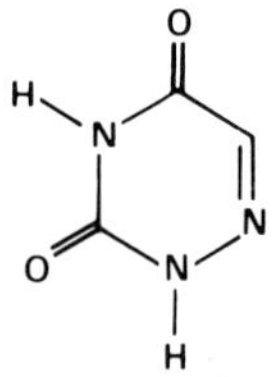

6-Azauracil

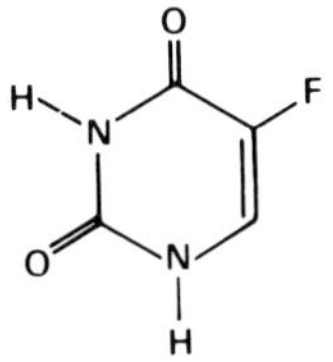

5-Fluorouracil

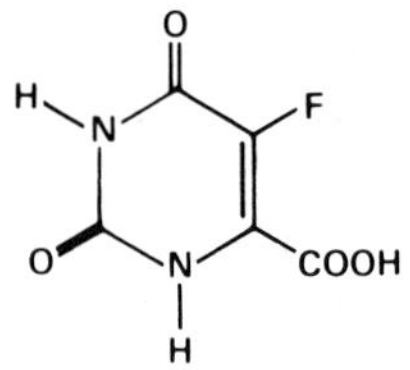

5-Fluoroorotic acid

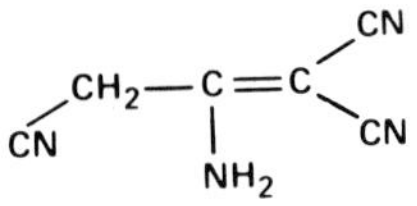

1, 1, 3-Tricyano-
2-Aminopropene
(TCAP)

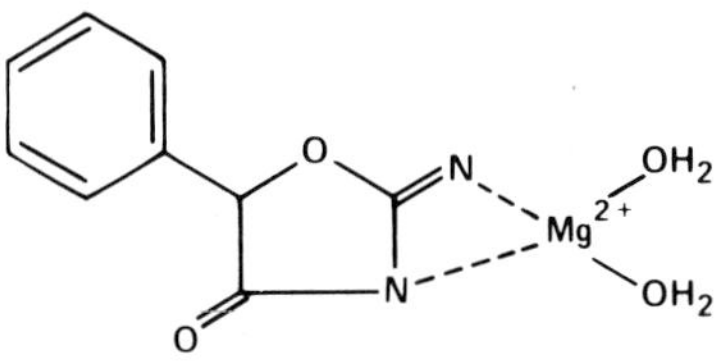

Magnesium pemoline

Figure 14.3 Reported effectors of brain RNA synthesis.

RNA polymerase [112], but these reports have been difficult to reproduce [113, 114]. The administration of certain hormones has also been shown to increase neuronal RNA.

Reciprocal Changes in Neurons and Glia

It should be mentioned that the many instances of reciprocal changes in neurons and their associated glia, i.e., increases in RNA of neurons and losses of the same type of RNA from the supporting glia, have been interpreted by Hydén [115-117] as a transfer of RNA between these two cell types. This transfer would be one expression of the close relationship which appears to exist between a given neuron and the glial cells which surround it. However, a transfer of RNA molecules, if it exists, suggests that the glial cells in this way influence at least quantitatively the protein synthesis in a given neuron directly, as well as perhaps by indirect nutritive processes.

Axonal RNA

Neurons, then, contain all the normal classes of RNA found in other cells. If there is anything unique about neuronal RNA it is almost certainly not the presence of any unique class of RNA. One problem which is unique to neurons, however, and one which has stimulated a good deal of study and controversy, is the question of the nature, or even the existence, of axonal RNA [118]. Clearly there is RNA in axons [119-121], although the concentration is quite low. Nevertheless, because of the great volume of the total axonal network, compared to the volume of the cell body, there is more RNA in axons than in cell bodies of neurons. What kind of RNA is present is yet another question. The base composition of this RNA suggests that it, like the RNA of neurons, is primarily ribosomal in nature [120,121], but ribosomes, as such, have never been seen in differentiated axons. Although axonal preparations as conventionally made could be contaminated with mitochondria, glial fragments, or synaptosomes, the RNA present does not seem to be totally mitochondrial in nature [122].

The origin of this axonal RNA is also not known. As with other materials, there is good evidence that RNA is transported down the axon by axonal flow [123-125]. There is also evidence, although it is not strong, that there is local, extramitochondrial synthesis of RNA [126,127], but most of it probably comes from the cell body. The possible function of this axonal RNA is still a matter of speculation. If it is truly nonmitochondrial, even in part, then it might be there to support some specialized protein synthesis. There is evidence for the local axonal synthesis of certain enzymes [128-130]. The problem of the mechanism of such synthesis in the apparent absence of ribosomes is a difficult one, although it has been suggested that either a low concentration of ribosomes has gone undetected [31] , or that membrane sites serve the function of ribosomes in the axon [118] . If there is local protein synthesis directed by the axonal RNA, then the RNA must have messenger function. Perhaps only a limited complement of messengers are present, including only those of relevance to the maintenance of the axon or the synapse.

Another approach to this problem has been the inspection of synaptosomal RNA, especially with regard to its poly(A) content and the length of its poly(A) segment. The results of such experiments [76] showed that there is poly(A)-containing RNA in the synaptosomes, but that the RNA has a poly(A) content and length characteristic of that of the mitochondria. The conclusion reached through these studies is that all the RNA in the synaptosomes is due to the presence of synaptosomal mitochondria. This conclusion has been supported by more recent studies [131] . In contrast, other workers have presented data which indicate that although the preponderance of synaptosomal RNA is mitochondrial, there is a very small portion of synaptosomal RNA which is not found in the RNA of purified mitochondria [132] . Thus, the question remains open.

References

1.   I. H. Heller and K. A. C. Elliott, *Can. J. Biochem. Physiol. 32:*584 (1954).
2.   J. B. Flexner and L. B. Flexner, *J. Cell. Comp. Physiol. 38:*1 (1951).
3.   V. B. Peters and L. B. Flexner, *Am. J. Anat. 86:*133 (1950).
4.   L. W. Lapham, *Science 159:*310 (1968).
5.   J. F. Chiu and S. C. Sung, *Biochim. Biophys. Acta 209:*34 (1970).
6.   F. L. Margolis, *J. Neurochem. 16:*447 (1960).
7.   M. Libonati, G. Luguori, and A. Giudetta, *J. Neurochem. 19:*1959 (1972).
8.   S. C. Sung, *J. Neurochem. 15:*477 (1968).
9.   J. Altman, in *Handbook of Neurochemistry,* Vol. II (A. Lajtha, ed.), Plenum, New York, 1969.
10.   J. Altman, *Anat. Rec. 145:*573 (1963).
11.   J. Altman and G. D. Das, *J. Comp. Neurol. 124:*319 (1965).
12.   J. Altman and G. D. Das, *Nature (London) 207:*953 (1965).
13.   J. Altman and G. D. Das, *J. Comp. Neurol. 126:*337 (1966).
14.   J. Altman, *J. Comp. Neurol. 128:*431 (1966).
15.   I. Smart and C. P. LeBlond, *J. Comp. Neurol. 116:*349 (1961).
16.   O. R. Hommes and C. P. LeBlond, *J. Comp. Neurol. 129:*269 (1967).
17.   W. His, *Sächs. Wiss. 15:*313 (1889).
18.   F. C. Sauer, *J. Comp. Neurol. 62:*377 (1935).
19.   R. L. Sidman, I. L. Miale, and N. Feder, *Exp. Neurol. 1:*322 (1959).
20.   A. Martin and J. Langman, *J. Embryol. Exp. Morphol. 14:*25 (1965).
21.   H. Koenig, M. B. Bunge, and R. P. Bunge, *Arch. Neurol. 6:*177 (1962).
22.   J. Altman, *Exp. Neurol. 5:*302 (1962).

23.   E. K. Adrian and B. E. Walker, *J. Neuropathol. Exp. Neurol. 21:*597 (1962).
24.   J. Kilham and G. Margolis, *Am. J. Pathol. 48:*991 (1966).
25.   S. P. Hicks, C. J. C'Amato, and M. J. Lowe, *J. Comp. Neurol. 113:*435 (1959).
26.   M. Winick and A. Noble, *J. Nutr. 89:*300 (1966).
27.   M. Winick and A. Noble, *J. Nutr. 91:*179 (1967).
28.   J. Altman and G. D. Das, *Nature (London) 204:*1161 (1964).
29.   P. Mandel and M. Jacob, in *Handbook of Neurochemistry,* Vol. Va (A. Lajtha, ed.), Plenum, New York, 1971.
30.   E. Koenig, in *The Structure and Function of Nervous Tissue,* Vol. IV (G. H. Bourne, ed.), Academic, New York, 1972.
31.   H. R. Mahler, in *Basic Neurochemistry* (R. W. Albers, G. J. Siegel, R. Katzman, and B. W. Agranoff, eds.), Little, Brown, Boston, 1972.
32.   H. Hydén, *Acta Physiol. Scand.,* Suppl. 17, 6 (1943).
33.   J. E. Edström, *Biochim. Biophys. Acta 12:*361 (1953).
34.   J. E. Edström, *J. Neurochem. 3:*100 (1958).
35.   J. E. Edström, *J. Biophys. Biochem. Cytol. 8:*39 (1960).
36.   H. Koenig, in *Morphological and Biochemical Correlates of Neural Activity* (M. M. Cohen and R. S. Snider, eds.), Harper, New York, 1964.
37.   H. R. Mahler, W. J. Moore, and R. J. Thompson, *J. Biol. Chem. 241:*1283 (1966).
38.   W. Oesterle, K. Kanig, W. Buchel, and A. K. Nickel, *J. Neurochem. 17:*1403 (1970).
39.   D. H. Adams, *Biochem. J. 98:*636 (1966).
40.   J. Bernsohn and H. Norgello, *Proc. Soc. Exp. Biol. Med. 122:*22 (1966).
41.   M. R. V. Murthy, in *Protein Metabolism of the Nervous System* (A. Lajtha, ed.), Plenum, New York, 1970.
42.   C. Vesco and A. Giuditta, *Biochim. Biophys. Acta 142:*385 (1967).
43.   Z. S. Tencheva and A. A. Hadjiolov, *J. Neurochem. 16:*769 (1969).
44.   J. L. Saborio and V. Aleman, *J. Neurochem. 17:*91 (1970).
45.   H. Løvtrup-Rein, *J. Neurochem. 17:*853 (1970).
46.   K. von Hungen, H. R. Mahler, and W. J. Moore, *J. Biol. Chem. 243:*1415 (1968).
47.   J. Stevenin, J. Samec, M. Jacob, and P. Mandel, *J. Mol. Biol. 33:*777 (1968).
48.   T. C. Johnson, *J. Neurochem. 16:*1125 (1969).
49.   L. Chou, M. P. Lerner, and T. C. Johnson, *J. Neurochem. 18:*2535 (1971).
50.   T. C. Johnson and L. Chou, *J. Neurochem. 20:*405 (1973).
51.   S. H. Appel, *Nature (London) 213:*1253 (1967).
52.   M. Jacob, J. Stevenin, R. Jund, C. Judes, and P. Mandel, *J. Neurochem. 13:*619 (1966).
53.   R. H. Kimberlin, *J. Neurochem. 14:*123 (1967).
54.   S. C. Bondy and S. Roberts, *Biochem. J. 109:*533 (1968).
55.   M. H. Samli and S. Roberts, *J. Neurochem. 16:*1565 (1969).

56. I. D. Herriman and G. D. Hunter, *J. Neurochem. 12*:937 (1965).

57. S. C. Bondy and S. Roberts, *Biochem. J. 105*:1111 (1967).

58. S. Yamagami, R. R. Fritz, and D. A. Rappoport, *Biochim. Biophys. Acta 129*:532 (1966).

59. F. Orrego and F. Lipmann, *J. Biol. Chem. 242*:665 (1966).

60. C. E. Zomzely, S. Roberts, and S. Peache, *Proc. Natl. Acad. Sci. USA 67*:644 (1970).

61. C. Zomzely-Neurath, C. York, and B. W. Moore, *Proc. Natl. Acad. Sci. USA 69*:2326 (1972).

62. C. Zomzely-Neurath, C. York, and B. W. Moore, *Arch. Biochem. Biophys. 155*:58 (1973).

63. J. DeLarco and G. Guroff, *Biochem. Biophys. Res. Commun. 49*:1233 (1972).

64. J. DeLarco, A. Abramowitz, K. Bromwell, and G. Guroff, *J. Neurochem. 24*:215 (1975).

65. S. P. Banks and T. C. Johnson, *Science 181*:1064 (1973).

66. J. Stevenin, P. Mandel, and M. Jacob, *Proc. Natl. Acad. Sci. USA 62*:490 (1969).

67. L. Grouse, M. D. Chilton, and B. J. McCarthy, *Biochemistry 11*:798 (1972).

68. V. Hahn and C. D. Laird, *Science 173*:158 (1971).

69. I. R. Brown and R. B. Church, *Biochem. Biophys. Res. Commun. 42*:850 (1971).

70. L. Grouse, G. A. Omenn, and B. J. McCarthy, *J. Neurochem. 20*:1063 (1973).

71. L. Grouse, Ph.D. Dissertation, Department of Biochemistry, University of Washington, Seattle, 1973.

72. J. G. Bantle and W. E. Hahn, *Cell 8*:139 (1976).

73. L. D. Grouse, B. K. Schrier, E. L. Bennett, M. R. Rosenzweig, and P. G. Nelson, *J. Neurochem. 30*:191 (1978).

74. J. Samec, P. Mandel, and M. Jacob, *J. Neurochem. 14*:887 (1967).

75. J. Samec, M. Jacob, and P. Mandel, *Biochim. Biophys. Acta 161*:377 (1968).

76. J. DeLarco, S. Nakagawa, A. Abramowitz, K. Bromwell, and G. Guroff, *J. Neurochem. 25*:131 (1975).

77. G. Gomirato and H. Hydén, *Brain 86*:773 (1963).

78. S. C. Bondy and H. Waelsch, *J. Neurochem. 12*:751 (1965).

79. S. H. Barondes, *J. Neurochem. 11*:663 (1964).

80. V. K. Singh and S. C. Sung, *Brain Res. 25*:677 (1971).

81. V. K. Singh and S. C. Sung, *Can. J. Biochem. 50*:299 (1972).

82. V. K. Singh and S. C. Sung, *J. Neurochem. 19*:2885 (1972).

83. R. J. Thompson, *J. Neurochem. 21*:19 (1973).

84. P. Mandel, A. R. Dravid, and N. Pete, *J. Neurochem. 14*:301 (1967).

85. A. R. Dravid, N. Pete, and P. Mandel, *J. Neurochem. 18*:299 (1971).

86. L. N. Simon, A. J. Glasky, and T. H. Rejal, *Biochim. Biophys. Acta 142*:99 (1967).

87. R. K. Datta, D. Bhattacharyya, and J. J. Ghosh, *J. Neurochem. 11*:87 (1964).

88. Y. Takahashi, K. Mase, and Y. Suzuki, *J. Neurochem. 17:*1433 (1970).
89. B. Daneholt and S. O. Brattgard, *J. Neurochem. 13:*913 (1966).
90. P. Mandel, H. Rein, S. Harth-Edel, and R. Mandell, in *Comparative Neuro-chemistry* (D. Richter, ed.), MacMillan, New York, 1964.
91. U. Ringborg, *Brain Res. 2:*296 (1966).
92. H. Hydén, in *The Neurosciences* (G. C. Quarton, T. Melnechuk, and F. O. Schmitt, eds.), Rockefeller University Press, New York, 1967.
93. G. Guroff, A. F. Hogans, and S. Udenfriend, *J. Neurochem. 15:*489 (1968).
94. F. Orrego, *J. Neurochem. 14:*851 (1967).
95. T. Itoh and J. H. Quastel, *Science 164:*79 (1969).
96. G. Guroff and M. Brodsky, *J. Neurochem. 18:*2077 (1971).
97. M. R. V. Murthy, *Biochim. Biophys. Acta 119:*586 (1966).
98. M. R. V. Murthy, *Biochim. Biophys. Acta 119:*599 (1966).
99. H. Hydén, *Recent Adv. Biol. Psychiatry 6:*31 (1964).
100. W. J. Watson, *J. Physiol. (London) 180:*741 (1965).
101. H. Hydén and A. Pigon, *J. Neurochem. 6:*57 (1960).
102. D. A. Rappoport and H. F. Daginawala, *J. Neurochem. 15:*991 (1968).
103. J. Jarlstedt, *Acta Physiol. Scand. 67:*243 (1966).
104. J. E. Edström, D. Eichner, and N. Schor, in *Regional Neurochemistry* (S. S. Kety and J. Elkes, eds.), Pergamon, Oxford, 1961.
105. A. J. Dewar and H. W. Reading, *Nature (London) 225:*869 (1970).
106. J. W. Zemp, J. E. Wilson, K. Schlesinger, W. O. Boggan, and E. Glassman, *Proc. Natl. Acad. Sci. USA 55:*1423 (1966).
107. B. E. Kahan, M. C. Krigman, J. E. Wilson, and E. Glassman, *Proc. Natl. Acad. Sci. USA 65:*300 (1970).
108. L. Z. Pevzner, *J. Neurochem. 12:*993 (1965).
109. R. P. Peterson and D. Kernell, *J. Neurochem. 17:*1075 (1970).
110. E. Egyhazi and H. Hydén, *J. Biophys. Biochem. Cytol. 10:*403 (1961).
111. L. N. Simon and A. J. Glasky, *Life Sci. 7:*197 (1968).
112. A. J. Glasky and L. N. Simon, *Science 151:*702 (1966).
113. D. F. Cain, *Life Sci. 6:*1919 (1967).
114. A. Yuwiler, W. Greenouich, and E. Geller, *Psychopharmacologia 13:*174 (1968).
115. H. Hydén, *Nature (London) 184:*433 (1959).
116. H. Hydén, *Endeavor 21:*144 (1962).
117. H. Hydén, in *The Neurosciences* (G. C. Quarton, T. Melnechuk, and F. O. Schmitt, eds.), Rockefeller University Press, New York, 1967.
118. E. Koenig, in *Handbook of Neurochemistry,* Vol. II (A. Lajtha, ed.), Plenum, New York, 1969.
119. J. E. Edström, D. Eichner, and A. Edström, *Biochim. Biophys. Acta 61:*178 (1962).
120. A. Edström, *J. Neurochem. 11:*309 (1964).
121. E. Koenig, *J. Neurochem. 12:*357 (1965).
122. L. Austin, I. G. Morgan, and J. J. Bray, in *Protein Metabolism of the Nervous System* (A. Lajtha, ed.), Plenum, New York, 1970.

123. J. J. Bray and L. Austin, *J. Neurochem. 15:*731 (1968).
124. L. Casola, G. A. Davis, and R. E. Davis, *J. Neurochem. 16:*1037 (1969).
125. S. C. Bondy, *J. Neurochem. 19:*1769 (1972).
126. L. Austin, J. J. Bray, and R. J. Young, *J. Neurochem. 13:*1267 (1966).
127. E. Koenig, *J. Neurochem. 14:*437 (1967).
128. E. Koenig and G. B. Koelle, *J. Neurochem. 8:*169 (1961).
129. D. H. Clouet and H. Waelsch, *J. Neurochem. 8:*201 (1961).
130. E. Koenig, *J. Neurochem. 12:*343 (1965).
131. J. M. England and G. Attardi, *J. Neurochem. 27:*895 (1976).
132. A. Cupello and H. Hydén, *J. Neurochem. 25:*399 (1975).

15
PROTEIN METABOLISM

Although, as is noted in Chap. 14, active cell division in the mature brain is limited to certain of the glial elements and only a few of the neurons, brain cells have a very active RNA metabolism. Correspondingly, the level of protein synthesis and protein breakdown is comparable to that found in other tissues of the body, and greater than that of most. Early studies in which radioactive precursors, given systemically, were incorporated very poorly into brain protein were taken to indicate the opposite, namely, that brain protein was relatively inert metabolically. These data are now considered to reflect the limited uptake of precursors into the brain due to the blood-brain barrier, and not the true activity of the brain protein-synthesizing machinery. More recent work has shown that appropriate precursors, properly administered, are readily incorporated, and that in whole-animal experiments and studies with cell-free systems, the brain exhibits an active and interesting protein synthetic and protein degradative machinery [1,2].

Comparison of Neurons and Glia

Attempts have been made to compare the rates and characteristics of protein synthesis in neurons and in glia. These studies suffer from the general problems of the separations of cell types from the brain, which are discussed in Chap. 1. Nevertheless, certain general conclusions may be warranted. Primarily, it has been found that neurons have a much faster rate of protein synthesis than do glia. This conclusion is based on autoradiographic experiments [3,4], on in vivo labeling followed by isolation of the various cell types and their subcellular constituents [5], and on labeling studies done in vitro with isolated fractions specifically enriched in one cell type or the other [5,6]. The properties of protein synthesis in the two cell types may differ in their sensitivity to ions; specifically, there are data to suggest that protein synthesis in glia cells has a greater sensitivity to the K^+ levels in the medium [7].

Protein Turnover

Studies on the turnover of brain proteins have been done by pulse labeling in order to get information on whether unique fractions occur in brain which might be related in some way to brain function. These studies, of course, give data pertaining to a very heterogeneous population of protein species. The studies, taken together, indicate that brain proteins of the rat, in toto, have an average half-life of about 14 days, and that 90% of the proteins in the brain turn over, with a $t_{1/2}$ of between 10 and 20 days [8]. Proteins from the brains of young animals turn over somewhat more rapidly. Very little, if any, of the protein of the brain is inert, or even metabolically very stable [9]. Even the proteins in the myelin sheath are in dynamic equilibrium. For example, the myelin basic protein in the brains of adult rats has been shown to have a $t_{1/2}$ on the order of 21 days [10], which is somewhat faster than the 35-day turnover reported for total myelin [11]. Specific subcellular particles, and specific proteins, have turnover times quite different from the average (Table 15.1). It should be understood that there are many methods for the study of turnover. The pulse labeling used in most studies is not necessarily the best. Obviously, pulse labeling does not detect those proteins whose rates of synthesis are very slow with respect to the length of the pulse. Consequently, as can be seen in Table 15.5, the $t_{1/2}$ determined in such studies can vary as a function of the length of the pulse, and also with respect to the time of sampling after the pulse. Continuous infusion techniques are perhaps less subject to such errors. Using continuous infusion, a somewhat shorter $t_{1/2}$ is found [8,12]. In any case, such studies have shown that brain proteins are in dynamic equilibrium, as are most proteins in the body, that very little of the brain protein is metabolically inert, and that the brain has an average protein turnover slower than that of liver, and faster than that of muscle. Consequently, very few, if any, of the neural proteins are stable for any substantial portion of the lifetime of the organism.

Cell-Free Systems from Brain

Fundamentally, cell-free protein synthesis systems prepared from brain homogenates are like cell-free systems from other tissues. This indicates that the mechanism of protein synthesis in brain is much like that in other tissues. Two exceptions should be noted, however. One is that the systems from brain seem to be more sensitive than others to changes in the environment, such as in the concentrations of ions and amino acids. This is discussed further below. Second, there is at least a possibility that protein synthesis occurs in the axon and the bouton of neuronal processes by a mechanism not involving polyribosomes. This also is discussed more fully later in this chapter.

The requirements for cell-free protein synthesis in the brain are the stan-

Table 15.1 Half-Life Values for Total Brain Proteins and for Proteins in Selected Brain Fractions and Components

Protein source or protein	$t_{1/2}$ (days)	Method
Area		
Rat brain	14	–
Rat brain	10–22	Feeding
Rat cerebral cortex	2–3	1-hr pulse
	11–26	6-hr pulse
	31–37	12-hr pulse
Subcellular Fraction		
Nuclei	24	^3H_2O in diet
Mitochondria	16	^3H_2O in diet
Microsomes	13	^3H_2O in diet
Ribosomes	12	^{14}C
Synaptic components	20–24	
Component		
S-100	16	^3H
Histones	54–117	^{14}Leu
Mouse neurotubule	4	5-day-old
Myelin		
Basic protein	21	^3H
Basic protein	14–21	$[^{14}C]$ Lys
Wolfgram	30	$[^{14}C]$ Lys
Folch-Lees	10–25	$[^{14}C]$ Lys
Glycoprotein	1–4	Glucosamine, newborn
	12.5	Glucosamine, adults

Data taken from Table IA and IB of A. Lajtha and N. Marks, in *Handbook of Neurochemistry,* Vol. Vb (A. Lajtha, ed.), Plenum, New York, 1971.

dard ones. A source of ribosomes, transfer RNA (tRNA), the pH 5 fraction, ATP, GTP, Mg^{2+}, and some source of messenger RNA (mRNA) are needed [2] . Such systems do not have either the amount of activity, or the stability of the more commonly used cell-free systems from *Escherichia coli,* Krebs II cells, or wheat germ. They are, however, much cleaner than the original brain preparations of a few years ago. There are now methods for the isolation of brain polysomes free of RNase activity [2,13,14] , and for the preparation of a cell-free system from rat brain which is completely dependent on the addition of exogenous mRNA, and so on, and which is composed of only the ribosomal subunits from brain ribosomes [15] . This system contains 50 μmol of Tris-HCl (pH 7.6), 5 μmol of $MgCl_2$, 100 μmol of KCl, 1 μmol of dithiothreitol (Cleland's reagent), 200 μg of microsomal wash factor protein, 500 μg of tRNA, 2 μmol of GTP, 100 μg of creatine phosphokinase, 20 μmol of phosphocreatine, 4 mg of a 70% ammonium sulfate fraction from brain, 10 A_{260} units of 40S ribosomal subunits, 20 A_{260} units of 60S ribosomal subunits, 40 μg of RNase inhibitor, and 500 μg of mRNA derived from brain polysomes, all in 1 ml final volume. Incubation is at 37°C for 60 min.

Several points should be mentioned about this system. The microsomal wash factors are prepared by suspending microsomes from brain in 0.5 M KCl buffer containing $MgCl_2$, glycerol, and Cleland's reagent for 1.5 hr. This preparation contains the various initiation factors. The tRNA used is a commercial preparation from rat liver. The ammonium sulfate preparation is made from rat brain supernatant fraction in a standard way and contains the activating enzymes generally provided as a pH 5 precipitable fraction. The ribosomal subunits were obtained from brain polysomes by preincubating under standard protein synthesis conditions, but in the presence of puromycin to dissociate polysomes into subunits which can then be isolated by sucrose gradient methodology. The RNase inhibitor is prepared from rat liver. Finally, the mRNA is isolated from brain polysomes by treatment with EDTA, and large amounts are needed. Such systems, as is mentioned below, have been used for the in vitro biosynthesis of a number of brain-specific proteins [15,16]. Details of the preparation of less elaborate polysomal and ribosomal cell-free systems from brain are also available.

The special sensitivities of the brain systems have been noted by several workers. Polysomes from brain seem inordinately sensitive to the action of RNase [17]. In the presence of 10 mM Mg^{2+}, the polysomes are stable. In lower Mg^{2+} concentrations or with the addition of RNase, the polysomes fall apart. Also, brain polysomes seem sensitive to other influences [18,19] that do not affect comparable preparations from liver. Further, protein synthesis by brain exhibits sensitivity to changes in the ionic composition of the medium [20], the method of ATP presentation [20], the availability of amino acids [21], and the composition of the amino acid mixture [22], which do not seem to influence the activity of synthetic systems from other tissues. Whether these seeming sensitivities play any role in some special regulation of brain protein synthesis is not known at the present.

Synthesis of Brain-Specific Proteins

The cell-free system from brain has been used to carry out the synthesis of certain brain-specific proteins. Earlier work using a less purified brain system indicated that one of the major products synthesized using the endogenous brain messenger RNA contained on the polysomes was the brain-specific protein known as S-100 [23]. In fact, 5 to 10% of the soluble proteins made during these experiments was thought to be S-100. Now it is known that these very high percentages were due to nonspecific coprecipitation of other proteins, with the immunoprecipitate used to determine the level S-100 of synthesis. Nevertheless, S-100 and 14-3-2, another brain-specific protein, have been made in a cell-free system from brain, using mRNA isolated from cerebral polysomes. Much lower amounts are found when the products are assayed by a procedure starting with gentle heating to remove large proteins, followed by concentration, immunoprecipitation, and, finally, SDS-acrylamide gel electrophoresis [16,24]. The percentage of the total amino acid incorporation which is due to the synthesis of either S-100 or 14-3-2 is substantially less than 1%, but this percentage can be raised by an initial selection of the polysome population used to prepare the mRNA [15]. Using only the free polysomes from white matter of rat brain, about 2% of the product appeared to be S-100. When mRNA from the free polysomes of cortex was used, almost 5% of the product was 14-3-2 (Table 15.2). In addition to indicating the possible level of synthesis of these proteins in the whole brain, these data suggest that both proteins function in the interior of the cells which make them, glia in the case of the S-100 and neurons in the case of the 14-3-2, since proteins synthesized on the "free" polysomes are thought to remain in the cell, while those made on the membrane-bound polysomes are thought to be exported from the cell.

In less detailed work, the synthesis of tubulin has also been reported in a cell-free system derived from a clonal line of neuroblastoma [25]. It should also be noted that the in vitro synthesis of the encephalitogenic basic protein of myelin has been reported [26] in cell-free systems primed with cerebral nuclear RNA. A most interesting observation indicates that proteolipids, some of which are major proteins of the myelin sheath, can be made by cerebral mitochondria [27,28]. In any case, protein synthesis systems from brain are available, and they can be programmed, using brain RNA fractions, to make specific brain proteins. There is certainly no reason to believe that brain systems are uniquely able to make brain proteins, or that they cannot make any proteins for which they are properly programmed.

The synthesis of protein in cell-free systems from brain and in brain in vivo is inhibited by the same inhibitors which are effective in other mammalian systems. Cycloheximide and puromycin are potent inhibitors, while chloramphenicol, which is known to inhibit bacterial protein synthesis, has no effect on brain polyribosomal synthesis. It is of interest that the inhibition in vivo by cycloheximide has no apparent neurological disturbance associated with it. On the other hand, the inhibition with puromycin produces disturbances in the electrical activity of the brains of experimental animals as well as substantial protein synthesis inhibition [1].

An interesting system, which seems to be quite different than the classic polysomal cell-free system described above, is found in the high-speed supernatant fraction of brain [29,30]. This system incorporates tyrosine or phenylalanine, but only these amino acids, into the carboxyl terminal position of some

Table 15.2   In Vitro Synthesis of S-100 and 14-3-2 Proteins

Type of polysomes used for isolation of mRNA	cpm, total soluble protein	cpm, S-100 protein	cpm, 14-3-2 protein	% of cpm in S-100	% of cpm in 14-3-2
Whole brain, free	873,450	11,355		1.30	
Whole brain, membrane bound	827,806	414		0.05	
White matter, free	800,250	18,486		2.26	
White matter, membrane bound	762,300	726		0.095	
Liver total	794,822	69		0.009	
Cortex, free	720,130		34,566		4.8
Cortex, membrane bound	785,178		786		0.10

Table taken from the work of C. Zomzely-Neurath, C. York, and B. W. Moore, *Arch. Biochem. Biophys. 155:*58 (1973).

as yet unidentified acceptor protein. The incorporation is RNase insensitive, so
it does not appear to require a template. The incorporation requires ATP, Mg^{2+},
and K^+. Whatever enzymes are involved seem to be very unstable, and they are
in by far the highest concentration in the brain. The function of this incorporation is unknown, but its specificity for either tyrosine or phenylalanine, its inability to use other amino acids, and its presence only in the brain make it an
interesting and intriguing one.

There are a number of observations on the protein synthetic capacity of
brain nuclei both in vivo and in vitro [2,8]. Since other mammalian nuclei have
been shown to contain ribosomes and to synthesize protein, the observations on
the brain are not surprising. Because it is known that the histones residing in the
nucleus are actually made in the cytoplasm and then transferred to the chromatin, the nucleus itself is thought to be the site of the synthesis of the other major
nuclear protein species, namely, the nonhistone chromatin proteins [31,32].
Using carefully controlled and sterile conditions it has been shown the brain nuclei incorporate labeled amino acids into protein linkage, and that this incorporation is stimulated by the presence of brain high-speed supernatant fraction [33].
The incorporation is not inhibited by puromycin or by cycloheximide, but this
failure may be due to the limited permeability of the nucleus to these inhibitors.

Brain mitochondria will also synthesize proteins [2]. The incorporation by
these particles is much like the incorporation by other mitochondria in that it is
insensitive to the inhibitors of polysomal synthesis, such as puromycin and cycloheximide, but is at least partially inhibited by chloramphenicol, normally an inhibitor of bacterial systems. Characteristically, the synthesis of protein by mitochondria is self-contained in that it does not require the addition of ATP or the
pH 5 fraction. Also, the synthesis is not inhibited by RNase, but this might be
due to the impermeability of the mitochondrion to the RNase molecule. The
synthesis is inhibited by such things as dinitrophenol and antimycin A, classic
inhibitors of energy production by the mitochondrion. As mentioned before,
there are reports that brain mitochondria incorporate amino acids into proteolipids [27,28].

Synthesis in Axons and Synapses

A continuing controversy surrounds the problem of the synthesis of protein by
the axonal processes and the terminal bouton of the neuron. It has been clear
for a number of years that some portion of the protein found in the axon and
the synaptic bouton of neurons comes from the cell body by the process of axonal flow. On the other hand, some convincing evidence is available to suggest
that the axon also is the site of local protein synthesis. This evidence takes the
form of studies in which the enzyme acetylcholinesterase has been irreversibly
inhibited with organophosphate inhibitors, and the return of activity, presum-

ably by new enzyme synthesis, is observed. When this is done it is found that
the activity of the enzyme reappears fairly uniformly along the axon, and not as
a wave traveling down the axon, as would be the case if the enzyme was made in
the cell body [34-36]. Supporting data have been obtained in a few in vivo
studies [37,38]. Conceptually, it seems reasonable that local synthesis mechanisms are present in the axon since the transport of membrane proteins, for instance, from the cell body seems to be an inefficient way in which to satisfy the
continuing needs of the dynamic neuronal surface. Mechanistically, though, the
outlines of such a protein synthesis system have been difficult to visualize, mainly because of the lack of identifiable polysomes in the axon.

In response, then, to the suggestions from the acetylcholinesterase studies
cited above, several groups have attempted to investigate the local synthesis in axonal cell-free systems similar to those prepared from cell bodies. The difficulties
which attend such studies are many and certainly are the reason for the conflicting literature on the subject. Some of the problems include the following. First,
preparations of axons or synaptosomes are notably impure, so it is difficult to
know if any synthesis observed is truly due to the axons or if it is due to contaminating materials. Second, since the axons and the synaptosomes have, mitochondria inside them, it is necessary to clearly differentiate between the synthesis
which might be thought to be nonmitochondrial. Finally, what synthesis is there
is quite weak.

In spite of these difficulties most of the relevant studies indicate that, indeed, there is local synthesis of protein in the terminal portion of the axon, and
perhaps throughout its length. First, it was demonstrated that radioactive amino
acids administered in vivo rapidly label the proteins of the axons or synaptosomes
in preparations made soon after administration [39]. Then, it has been shown
that isolated axons incorporate amino acids in vitro [40,41]. Finally, a number
of studies show that isolated preparations enriched in synaptosomes incorporate
amino acids, and that the incorporation is, in part at least, nonmitochondrial
[42,43]. In fairness it should be pointed out that other studies are not consistent with this view [44]. But the bulk of the work shows that some of the incorporation is not inhibited by chloramphenicol, but is by cycloheximide, and
that the characteristics, in general, of some of the incorporations are different
than those of pure mitochondrial incorporation. The most recent work may
provide an answer to the troubling question of the lack of ribosomes, since it has
been shown that the nonmitochondrial protein synthesis of the synaptosome is
found in the membrane fraction of the lysed preparation. This may indicate that
the template for the specialized synthesis is in the membrane itself which is known
to contain RNA. In very careful and complete studies [45], it has been shown
that the membrane incorporation is not N terminal and requires Mg^{2+}, monovalent ions, ATP, GTP, synaptosomal soluble enzymes, and amino acids. It is
inhibited by RNase, cycloheximide, and puromycin, but not by chloramphenicol.

In this work the possible interference by contaminating microorganisms, or by
mitochondria or mitochondrial membranes, was convincingly ruled out, as was
contamination with intact synaptosomes or microsomes. Even further, the products of the system were analyzed by acrylamide gel chromatography and shown
to be quite different than those produced by the microsomal system of the same
forebrain source. So, it seems quite likely that there is local synthesis of proteins
in the axon and the synaptosome, in addition to appearance of protein by axonal
flow. Just what proteins are made by this system remains to be determined. In
passing, it should be mentioned that the synaptosomes have the capacity to incorporate glucosamine into glycoproteins, but the protein acceptor seems to arise
in the cell body and be transported to the synaptic portion of the cell.

Regulation of Synthesis

We do not know as much about the regulation of protein synthesis as we do about
its mechanism. Nor do we know much about the special regulation, if any, of protein synthesis in the nervous system. There are a few suggestions about possible
sites of regulation. But whether any or all are physiologically important is not
clear. First, it has been suggested, as noted before, on the basis of studies of cell-
free synthesis, that the brain system is unusually sensitive to changes in the amino
acid composition of the free amino acid pool [21]. Indeed, there is information
available on the competition between amino acids for incorporation into brain
proteins in vitro [22]. On another level, studies on amino acid concentrations
in blood and brain, discussed before in a different context, indicate that the free
amino acid pool of brain in vivo can be altered by altering the relationship between the concentrations of various related amino acids in the blood [46,47].
Since the level of tryptophan, specifically, appears to influence the level of polysome formation in brain [48] and other tissues, and since polysomes from brain
seem unusually unstable, and, finally, since polysome formation is crucial for
synthesis of protein, these observations provide yet another level at which the
concentration of certain amino acids can regulate the synthesis of protein in the
brain. Changes in ion concentration which alter the rate of brain protein synthesis have also been referred to earlier [20]. Thus, alterations in the protein synthetic milieu are certainly of significance in determining the rate of protein synthesis in the brain.

Physiological changes in the state of the animal also seem to impinge on
the synthetic machinery. Electrical stimulation in vivo increases the incorporation of amino acids into protein [49,50], while in vitro in a slice system the opposite is observed [51,52]. Light stimuli are reported to increase protein synthesis
in the appropriate parts of the brain [53], and increases in incorporation have
also been reported due to exercise [54], changes in circulating Na^+ concentration
[55], and alterations in sleep patterns [56].

The literature on the influence of transmitter substances on brain protein synthesis is confusing and incomplete. Isolated studies have implicated serotonin, acetylcholine, dopa, epinephrine, and norepinephrine in the regulation of protein synthesis in various brain systems [2,57]. None of these reports, however, have provided detailed follow-up studies on the possible mechanisms of these effects, nor upon their suspected physiological significance. The studies on γ-aminobutyric acid (GABA) are by far the most complete but, at the moment, perhaps the most puzzling of all. Several years ago it was reported that GABA stimulated brain protein synthesis [58]. The increases were not gigantic, but were substantial, and in line with what might be a modulatory effect of the transmitter. These reports were confirmed and extended in Baxter's laboratory [59,60], and the mechanism sought in a systematic and careful way. The resulting data showed that the effect was fairly specific for GABA, quite specific for brain, and required Na^+ for expression. A number of mechanisms were ruled out and, although some questions remained about the high concentration of GABA required, the effect was localized to a stimulation of the brain aminoacyl tRNA synthetases, specifically to the step in which the amino acid is transferred from the acyladenylate to the tRNA. At that point, apparently, the effect disappeared. Numerous detailed attempts to recapture the exact methodology failed to reproduce the effect. So, in spite of searching review of the status of the experimental animals, the purity of various chemical sources, and the smallest details of procedure [57], no success has been reported in reproducing the earlier results, even though those results were obtained repeatedly over several years, and have been reported from more than one laboratory. What the eventual outcome of the studies on the GABA stimulation will be remains unknown.

In passing it should be mentioned that certain hormones, growth factors, and drugs have been shown to impinge on the protein synthetic systems from neural tissue. The work of Sokoloff on the stimulatory effects of thyroxine on protein synthesis in brain mitochondria should be mentioned [61-63]. There are also reports on the inhibition of mitochondrial protein synthesis by chlorpromazine [58], and the effects of certain narcotics, including morphine [64,65], and of the synthetic morphine congener, levorphanol [66-68], the last, however, in nonneural systems.

Developmental Aspects

A consistent finding in many different laboratories has been that protein synthesis in the brain declines as the animal develops [2,69]. This has been found in vivo [70,71] in tissue slices [51], in cell suspensions [72], and in various cell-free preparations [73-75]. It is not at all clear, however, what the biochemical basis is for this developmental decline. One of the most convincing suggestions

about the molecular mechanism of this effect is the observation that the polysome profiles of brain change during development. That is, it has been found
that there are fewer polysomes, and more monosomes, in the brains of adult rats
than in the brains of infant animals [76-79]. Such a decline in the relative proportion of polysomes is not observed in the livers of the same animals. But, there
are several other experiments which point in other directions. For example, it is
reported that mRNA turnover is faster in young than in adult brain [80]. Again,
it is suggested that the synthesis of protein in vitro is limited by the available ATP,
and that this amount is less in adults than in young [61,81]. Then, there are recent data, contradicting the polysome work cited above, that suggest that polysomes from young and adult, properly prepared so as to remove inactive components, are identical, and that the difference lies in the pH 5 enzymes which are 30
to 40% less active in the brain of adult animals [82]. Finally, some data suggest
that the change is located in the ribosomes [83], and some other data suggest that
the K_M of the system for certain amino acids increases with age [84]. The point is
that nothing is known for sure except that the level of protein synthesis in the
brains of young animals is greater than in the brains of adults. Where the change
or changes in the system occur during development is quite unsure. Most likely
the changes are many. All that can be said is that the decrease "... reflects both
quantitative and qualitative changes in the components of the protein synthesizing system such as mRNA, pH 5 fraction, and polysomes " [69].

Proteolysis

Obviously, since brain protein turns over, there must be proteolytic activity in
the brain. The enzymes responsible have been studied to some extent [85], but,
like such enzymes from other mammalian tissues, they are weak, labile, and complex. As a result, knowledge about these enzymes is incomplete. Basically, there
are two classes of proteolytic enzymes found in tissues and both are present in
brain. The cathepsins, or acid proteinases, were observed in brain by a number
of workers several years ago [86-88]. Before the multiplicity of these enzymes
was recognized, partial purification of the acid-proteinase fraction was achieved.
Now it is known that there are several cathepsins in most tissues and that the specificity and characteristics of these enzymes vary, but that they are all active at
low pH and most likely reside, under normal conditions in the cell, in an inactive
form in the lysosomes [89]. Cathepsin A, a sulfhydryl enzyme with pepsin-like
specificity, has not been convincingly shown to occur in brain. Cathepsin B, also
sulfhydryl in nature with a specificity similar to that of trypsin, has been seen in
brain but in extremely low levels [90]. Cathepsin C, which is also a potent di-
peptidyl transferase of wide specificity, has been shown in brain, but in levels
approximating 5% of those in, for example, spleen [90,91]. It is possible that

in the case of cathepsin C, and in others as well, that the true activity is masked
by the presence of potent natural inhibitors. Such an inhibitor is definitely present in brain for cathepsin C [92,93]. Cathepsin D, active at pH 3 to 4, acts on
proteins such as denatured hemoglobin, but not on synthetic substrates such as
the dipeptides normally used to study the specificity of such enzymes. This is
perhaps the major acid proteinase found in brain [94], and accounts for about
two-thirds of the hemoglobin-splitting activity of this tissue [95]. It has a specificity somewhat like that of pepsin, but not as wide. The enzyme does not require sulfhydryl or metal for its action. Generally speaking, the overall level of
acid proteinase in brain is greater than that of muscle and not so great as that of
liver or spleen.

Also present in brain are a number of the even weaker and, if anything,
more mysterious neutral proteinases. A number of these enzymes apparently
occur in brain [89,94-100]. They are nonlysosomal, many are soluble, and they
probably have specific functions and consequently specific substrates. As yet
neither the function nor the substrate of any of these are known. One such enzyme has been studied to a small extent. A Ca^{2+}-activated proteinase, at first
found only in brain, was purified slightly and shown to act on a specific bond
in the β chain of insulin which was used as a model [98]. Later, this enzyme
was shown to be present in many other tissues, and further, to be the enzyme involved in one of the in vitro activation routes for the enzyme phosphorylase kinase [101]. It seems possible that this enzyme serves a wider role in cellular
metabolism since it has been shown to convert a proenzyme of protein kinase to
the active form [102]. But, since this conversion is irreversible it is questionable
if it could serve a regulatory role in the cell. Thus, the function of this enzyme,
and of the other neutral proteinases in brain tissue, remains unknown.

Proteinase and peptidase activities have been found widely distributed in
the nervous system. A number of peptidases are known to be present. As noted
earlier, a proteinase has been widely thought to be one of the few enzymatic
components of the myelin sheath. Proteolytic activity of peripheral nerve increases during Wallerian degeneration, as mentioned before, and may be responsible in part, for the degeneration of the myelin matrix. There are some indications that the proteolytic action is the initial trigger for the entire degenerative
chain of events. Proteinase activity is found in the spinal cord [89] and in the
pituitary [85,103]. Both acid and neutral proteinases are found in the synaptosomes isolated from various parts of the central nervous system [85,97].

Little is known of the function of these enzymes. Most of the proteinase
activities that have been so studied have been found to increase with the age of
the animal [104-106]. One of the best substrates for the acid proteinases of
brain is the myelin basic protein which, when degraded, gives rise to peptides,
producing experimental allergic encephalomyelitis [107]. It is possible that one
of the functions of these enzymes is the activation or inactivation of neurally

active peptide hormones [85] , such as substance P, but not much supporting data
are available. Other functions, such as those of tissue remodeling or adaptation,
are equally likely.

Brain-Specific Proteins

Although the mechanism of protein synthesis in the brain does not seem to be in
any way unique, the protein composition of the nervous system is known to be
unique in at least two respects. First, there are a number of proteins which are
specific to the nervous system and are not found in other tissues. Second, there
are a few proteins which, while not specific to brain, are found in much higher
amounts in the nervous system than in other organs.

S-100

The best characterized of the proteins in the first class, that is, of those specific
to the brain, is the S-100 protein discovered in 1965 by Moore and his coworkers
[108,109] . This protein, so-called because it is soluble in 100% ammonium sulfate at pH 7.0, was observed first by the straightforward experimental route of
chromatographing the soluble extracts from a number of tissues in two dimensions and looking for protein spots that occurred only in extracts of brain. The
S-100 protein was found in the region corresponding to the small acidic proteins,
and since that time a great deal of work has gone into its characterization and
into a search for its function [110] . It is now known that the S-100 is indeed
brain specific but, on the other hand, appears to be species nonspecific in that
it can be detected antigenically in the brains of all species tested [111] . The
molecular weight of the S-100 is around 24,000, and the molecule is composed
of three nonidentical subunits each of a molecular weight of about 7,000 [112] .
There is no lipid, carbohydrate, or nucleic acid attached to the protein, nor is
any metal found associated with it, but calcium, as will be mentioned below,
does have a profound effect on the molecule. The acidic nature of the S-100 is
due to its high content of glutamic and aspartic acids. By very sensitive antibody
techniques the localization to brain has been confirmed, and it is known that the
S-100 is at least 1,000 times higher in concentration in the brain than in any
other organ [110] .

The exact localization of the S-100 in the brain has been a subject of some
mild controversy. The protein is clearly present in the glia. This has been shown
by a number of techniques, including microdissection and immunofluorescence
[110] . Information consistent with this conclusion has been provided by some
experiments in which the concentration of the S-100 in degenerating nerve tracts
was measured [113,114] . During such degeneration the content of S-100 remains fairly constant while that of clearly neuronal constituents decreases. Fur-

ther confirmation is found in the observation that the S-100 is made in cultures of rat astrocytoma [115] or of human glia [24], but not in cultures of neuro-blastoma, and that the mRNA which codes for the S-100 is in higher concentrations in the white matter than in the gray [15]. The protein has also been found to be present in certain neurons [116] and, by immunofluorescence, in the area of the neuronal nuclei [117]. Thus, there is some evidence that the S-100 is present in both glia and neurons, but it is clearly present in glia.

Although the function of the S-100 is not known, some interesting leads have become available. Perhaps the most complete studies concern the interaction of the S-100 with calcium ions [118,119]. In the presence of calcium ions the conformation of the S-100 changes. This change can be detected by the presence of multiple bands on acrylamide gel electrophoresis and by a change in the fluorescent properties of the molecule. The physical basis of the multiple bands on acrylamide is a bit elusive. The multiple bands are not due to aggregation of the protein or to a reversible equilibrium among several forms. It has been ascribed to "multiple combinations of subunits with different affinities for calcium ion. . . ." [119]. The changes in fluorescence seem more easily understood. The fluorescence of the molecule is due in large measure to the excitation of the single tryptophan residue in one of the subunits. The addition of calcium ion increases the fluorescence twofold, indicating that a conformational alteration has taken place. Since no corresponding change in the optical rotatory dispersion takes place, the change in shape caused by calcium ion is considered to be a relatively minor one. In addition to the change in the fluorescent properties, there are other alterations in the molecule observed upon the addition of calcium. There is a spectral shift seen, and titratable sulfhydryl groups appear. Also, the molecule becomes much more susceptible to attack by chymotrypsin, so it is thought that the effect of calcium is to unfold the structure a bit. Whatever the intimate nature of the structural changes, they are quite specifically caused by calcium. Strontium and magnesium are not effective. Potassium ions, on the other hand, do have a pronounced effect. The action of calcium is antagonized by potassium, and it has been shown that the potassium competes with calcium for certain high-affinity sites. That is, in the absence of potassium, calcium binds to the S-100 in a bimodal fashion at one set of sites with an affinity constant on the order of 10^{-5} M and to another set with a much lower affinity, on the order of 10^{-3} M. In the presence of potassium ions these high-affinity sites for calcium disappear, as if now occupied by potassium. Studies involving calcium ion and the S-100 protein have been done with model membranes. Although the studies are complicated and open to several interpretations, the findings are that the permeability of artificial lipid membrane spheres to rubidium, as a test ion, increases when calcium ion and the S-100 protein are present [120]. One possible interpretation of this data is that the S-100 is in some way involved with, or even identical to, the carrier for monovalent ions in the neural cell membrane.

In addition to the suggestion that S-100 is involved in ion movement across cell membranes, a number of other functions have been postulated. Some data indicate that the S-100 may have a role in behavior or in learning [121]. Also, there are some indications that the S-100 is involved in the visual response [122]. Finally, data are available to indicate a nuclear localization for at least a small part of the S-100 [123], which, together with reports on its ability to stimulate the activity of RNA polymerase [124], would suggest that it is involved in the regulation of gene expression. None of these suggestions, including the one on the role of the S-100 in ion transport, have received wide experimental support.

14-3-2; Neuron-Specific Enolase

Several other proteins unique to the brain have been reported within the last several years. Among these is a small acidic protein known as 14-3-2, so called because of its position on the chromatograms. This protein was also discovered by Moore and his associates [125], using much the same methodology as that used to find the S-100. The 14-3-2 has a molecular weight of about 50,000 daltons and is quite acidic. Antibody measurements show that it is present in brain in at least 100 times higher levels than in other tissues [110]. Unlike the S-100, which is antigenically the same in all species, the 14-3-2 exhibits some antigenic variation in various species, being different in birds than in mammals, and undetectable in fish or lower species. The localization of the 14-3-2 is also unlike that of S-100. There are several firm lines of evidence that indicate that the 14-3-2 is neuronal and a few reports which suggest that it is present in glia as well. The protein is made by neuroblastoma lines in culture [126]. The 14-3-2 protein has recently been shown to be a brain-specific, or even a neuron-specific, form of the enzyme enolase [127,128].

There are reports of a number of other proteins unique to the central nervous system. Some of the better-characterized proteins are listed in Table 15.3. Several of these have more or less general distribution throughout the brain. The α_2 glycoprotein, found in human brain, is an acidic protein localized in the oligodendroglia [129,130]. The protein is made up of two subunits, one of 50,000 and one of 12,000 molecular weight [131]. It is found in the microsomal fraction of the cell, indicating that it is a membrane-bound species, and it appears, developmentally, around the time of myelination [130].

Another brain-specific protein found at least primarily in the glia is the glia fibrillary acidic protein (GFAP) [131-134]. This protein is found in many species, has a molecular weight of 43,000, a tendency to aggregate, and contains no sugar or lipid. It seems to be localized to the astrocytic elements. Antibody studies show that while the protein is not at all species specific, it is present only in the central nervous system and is not found in the peripheral nerves or in other tissues.

Table 15.3 A Summary of Proteins of Interest Found in the Nervous System and Some of Their Characteristics

Protein	Characteristics
S-100	Primarily glial; acidic; MW 24,000; conformational change upon binding of Ca^{2+}
14-3-2	Primarily neuronal; acidic; MW 50,000; also called antigen α or, neuron-specific protein (NSP), a neuron-specific isozyme of enolase
α_2–Glycoprotein	Oligodendroglia membranes, acidic MW 62,000
Glia fibrillary acidic protein	Astrocytes, acidic MW 43,000
GP 350	Glycoprotein; acidic; MW 11,600; high concentration in ganglia, low in myelin
Vesiculin	Cholinergic vesicles; acidic; MW 10,000
Olfactory bulb protein	Neuronal; MW 20,000
Visual system specific protein	—
Calcium-binding protein	MW 13,000
β-Trace protein*	MW 31,000
Basic protein	Histone; MW 18,000
Nucleoprotein	MW 19,000
Neurofilament protein	—
Tubulin*	Subunit of microtubules; MW 110,000; binds GTP, colchicine, vinblastine
Neurostenin*	Contractile protein; ATPase activity
Al	Myelin protein; basic; "unfolded" conformation" MW 18,400
Proteolipid*	Myelin proteins; soluble in $CHCl_3-CH_3OH$; MW 12,000

*Indicates identical or closely related proteins found in other tissues.

Between 4 and 15% of brain protein is, in fact, glycoprotein [135]. The glycoprotein in the brain has between 10 and 33% of its weight as carbohydrate. The glycoproteins are probably intercellular in location, and seem to occur in the synaptic cleft. They are constructed, seemingly, in the Golgi region and the carbohydrate portion is added while the protein is still on the ribosomes, although, as mentioned before, there are studies suggesting that carbohydrate is also added to protein in the synaptic bouton. Enzymes which synthesize or degrade the carbohydrate portion of the glycoproteins are known. Glycoproteins are probably secreted from the cell to form a cell coat or a superficial protein layer on the plasma membrane.

A number of glycoproteins apparently specific to brain are known. The best characterized of these is the GP-350 [136,137], which derives its name from the fact that it is eluted from DEAE-cellulose at about 0.35 M NaCl. The protein has a molecular weight of around 11,600, and contains one sialic acid residue per molecule. It is extremely acidic, with an isoelectric point of about 2.0, due to its very high content of glutamic and aspartic acids. It also contains glucosamine bound through asparagine, and galactosamine bound through the hydroxyl group of serine or threonine. Glucose makes up about 4% of the total weight of the molecule, which is quite unusual since unmodified glucose, i.e., as glucose rather than, say, glucosamine, is not usually found in proteins at all. The protein is found in all brain areas, but is not evenly distributed. It is present in particularly high concentration in areas rich in ganglia, and in low concentration in white matter. About 3% of the total soluble protein in the brain is GP-350; it has been reported to be present in the synaptic membranes [137].

A few proteins with rather specific localization have been found recently. A protein now known as vesiculin has been observed as one of the four protein constituents of synaptic vesicles from the cholinergic synaptosomes of the *Torpedo* electric organ [138,139]. This protein has a molecular weight of 10,000 and is acidic, with an isoelectric point of about 3.5. The function of the protein is unknown. Suggestions about its role in vesicular function are complicated by data which show that the amount of this protein in the synaptic fraction does not change upon repeated stimulation, although the number and size of the vesicles themselves is markedly lower under such conditions.

The olfactory bulb protein discovered by Margolis [140,141] is found only in that specific portion of the nervous system. The protein has a minimum molecular weight of 20,000 and an isoelectric point of 4.7. It contains no carbohydrate. It comprises between 0.5 and 1.0% of the soluble protein found in the olfactory bulb, but its exact function has not yet been ascertained. Experiments attempting to show differences in the content of this protein in various strains or between male and female animals have been negative. Results of incorporation experiments suggest that the protein is synthesized by bipolar chemoreceptor neurons and then transported by axoplasmic flow to the olfactory bulb.

The visual system has also been the source of at least one unique protein [131,142]. The optic nerve of the pigeon contains a protein which disappears within 3 weeks after retinal ablation, perhaps due to the interruption of axoplasmic flow from the retina. The protein is of unknown function, but it has been shown to be specific to the optic nerve.

A calcium-binding protein has been purified from pig brain by Wolff and Siegel [143,144] and its binding properties thoroughly studied. It has a molecular weight of about 13,000 and is specific to brain. Three copper-containing fractions have been isolated from brain. One of these, cerebrocuprein I, has been highly purified. Whether these components are unique to brain is unknown.

The β-trace protein [145-149] , found in large amounts in cerebrospinal fluid, is relatively specific to the central nervous system although it is also found in plasma, urine, and in both male and female genital tissue. The protein has a molecular weight of 31,000, a sedimentation of 2.3S, and is one of the major constituents of the cerebrospinal fluid. Normal concentrations of this protein are 2.6 mg/100 ml of cerebrospinal fluid, and the protein makes up about 7% of the protein content of the fluid. As such, it is in much higher concentration in the cerebrospinal fluid than it is in plasma. Its concentration in the central nervous system is higher than in other tissues, and it is present in higher concentration in white matter than in gray. Low amounts of β-trace protein are found in peripheral nerves. The protein may be of substantial diagnostic value since it is found in higher than normal amounts in the urine of patients with brain damage, and in elevated levels in the cerebrospinal fluid of individuals suffering from cerebrovascular disorders.

A basic protein which seems to be present only in neurons has been reported [150] . This protein does not appear to be species specific since it was found in the brains of all the species studied. The protein was described as a histone with a molecular weight of about 18,000.

A specific ribonucleoprotein has been isolated from goat brain [151] . The molecular weight of this material is about 25,000, and the protein portion itself is about 19,000. The protein is present only in brain and has been found in astrocytoma cells in culture. The linkage between the nucleic acid and the protein appears to be covalent since the two are not dissociated by detergent or by extremes of pH.

Tubulin

That there may be many more brain-specific proteins than have up to now been found is indicated by the large number of reports which have not as yet been expanded into detailed studies. In addition to the unique brain proteins, there are a few other proteins which, while not unique to brain, have been studied in brain because of the uniquely neural function which they appear to serve.

In the first class is tubulin. The brain is the richest source of tubulin in the body. This protein [152-154] is the subunit for the formation of the long, unbranched structures seen in the neurons known as microtubules, which are discussed earlier. Such structures are abundant in neurons, but also are present in glia, sperm, and a variety of other cells. Tubulin, then, is the subunit of the tubules, and the protein makes up 15 to 40% of the soluble protein of the brain. The protein is a difficult one to work with, despite its abundance, because the property by which it is recognized and measured, namely, its ability to bind colchicine, is labile. Tubulin is thought to be a protein of molecular weight 110,000, with a sedimentation of 6S. It is made up of two nonidentical subunits, has an isoelectric point of 5.2 to 5.4, and may contain some small amount of carbohydrate but no lipid. The 6S tubulin binds two molecules of GTP. Also, the tubulin binds one molecule each of colchicine and vinblastine at sites different from the GTP sites, and apparently different from each other. It is these last two binding properties which have proved useful in the purification and in the study of the functions of tubulin. As mentioned above, the binding of colchicine is a labile property of the molecule under standard purification conditions. This binding can be stabilized by keeping the protein in strong sucrose solutions, on the order of 1M [155]. Vinblastine, on the other hand, binds to tubulin at a different site. Upon binding vinblastine in the presence of Mg^{2+}, the protein precipitates [156], and this property has been used as a tool in the purification. The binding of vinblastine causes an ordered aggregation of tubulin in vitro. Purified tubulin can be repolymerized under appropriate conditions [157]. The reassembly takes place above room temperature; GTP is required, as is Mg^{2+}, and also EGTA, a specific chelator of Ca^{2+} which seems to prevent the reassembly.

In vivo, colchicine and vinblastine cause a depolymerization of cytoplasmic microtubules, presumably as a consequence of the binding of these agents to tubulin [158]. The binding of colchicine or of vinblastine to tubulin has, in fact, become such a definitive property of the tubulin, and of the microtubular structures in which it participates, that any process which is disrupted by the administration of these agents is automatically classed as one which involves tubulin. Such is the case with axoplasmic transport which is disrupted by these agents [159,160], and the outgrowth of neurites from neuroblastoma in culture [161]. Some caution must be exercised, however, since colchicine and vinblastine are known to bind to proteins other than tubulin [162].

Another neural protein of interest is neurostenin. This protein was first observed by Berl and his coworkers [163-166], during a search for contractile proteins in the nervous system. Neurostenin is an actomyosin-like material which has contractile properties and a Ca^{2+},Mg^{2+}-stimulated ATPase activity, as does muscle actomyosin. The protein has been found also in synaptosomes [167], and it has been suggested that the contractile properties of the protein are involved in the release of transmitter substance [166].

The Proteins of Myelin

For purposes of completeness, it should be pointed out that the proteins of the myelin sheath are, of course, abundant and significant, if not unique, brain constituents. There is, as mentioned earlier, still some controversy about how many proteins are present in central nervous system myelin. Clearly, the two most accepted protein constituents of myelin are the A_1 basic protein and the proteolipid fraction. The A_1 [168], also known as P_1, has a molecular weight of 18,400 and makes up about 30% of the myelin protein. It is unique to the nervous system. Peripheral nerve myelin also contains A_1 and other related proteins as well. The A_1 protein is the basic protein of the myelin which is responsible, when injected peripherally, for the condition known as experimental allergic encephalomyelitis, which was discussed earlier. The structure of the A_1 has been investigated thoroughly, and it is known that it contains about 25% basic amino acids, a high content of proline and glycine, but only one tryptophan and no cysteine [169]. There is no lipid or carbohydrate associated with the protein, and the molecule has no discernable secondary structure. It can be heated at 95°C for 60 min and it is still antigenic. Another argument for the lack of secondary structure is its ready digestion by proteolytic enzymes, which suggests an unfolded conformation [170]. The A_1 has a number of interesting properties which have come to light fairly recently. It is the only known acceptor for a galactosaminyltransferase enzyme isolated from bovine submaxillary glands [171]. The acceptor site for the galactosamine is a threonine residue on the molecule, but the significance of this accepting function is not known. The protein has methylated arginine residue in its linear sequence [172,173], which is somewhat unusual. The A_1 is present on the outside surface of the membrane since it is easily extracted and it reacts so readily with the antibody elicited by A_1 injection. The disease itself is thought to be an autoimmune state mediated by T lymphocytes rather than by circulating antibodies [168].

Since the total sequence of the molecule is known, and since the conformation of the molecule seems unfolded, it makes sense to try to find some fragment of the linear sequence which will cause the symptoms produced by the total protein. Attempts in this direction have been partially successful but have also uncovered some unexpected complications [168]. A nine-amino acid fragment from the interior of the molecule has been shown to be effective in producing the characteristic lesion in the guinea pig. This fragment has been synthesized, as have a number of closely related peptides. To be effective it is known that the peptide must contain the single tryptophan, a glutamine, and must have a lysine or an arginine in the N-terminal position. The unexpected aspect is that the fragment which is effective in the guinea pig is not effective in the monkey. It seems surprising that a different portion of the molecule produces the same lesion in another species.

Very recently it has been shown that the ability to produce experimental allergic encephalomyelitis in the mouse is in some way genetically linked, in that some strains of mice can easily be induced to develop the condition while others seem quite resistant [174]. Perhaps genetic analysis will provide a powerful tool in the understanding of the mechanism of the disease.

Proteolipid is not a unique constituent of brain since proteolipid exists in all cellular membranes in the body. But since the proteolipid is a family of related proteins and not a homogeneous, structurally well-characterized entity, it is still something of a question whether the Folch-Lees proteolipid of myelin [175,176], and hence of the central nervous system, is identical to that of the peripheral nerves or of the other cell membranes. Current evidence involving the production of antibody to the myelin proteolipid and its use as an immunofluorescent reagent indicates that the myelin proteolipid is, indeed, unique to the myelin sheath and different from the proteolipids found in other cellular membranes [177].

The proteolipid of myelin is found, of course, mostly in the white matter of the brain. It is composed mainly of sulfur-containing amino acids and nonpolar amino acids such as glycine, alanine, leucine, and isoleucine. It is low in basic and acidic amino acids and high in tryptophan. It is soluble in chloroform-methanol mixtures whether or not it is combined with the lipids with which it is associated in the tissue. Unlike the basic protein of myelin, the proteolipid is quite resistant to proteolytic enzymes, except for the very active and nonspecific enzyme pronase.

The proteolipid complex, as isolated, can be dissociated by acidifying the chloroform-methanol extract and dialyzing out the lipid. The apoprotein is still soluble in the chloroform-methanol, but also becomes soluble in water. The protein has a high helical content, while the lipid is still associated with it but becomes nonhelical upon removal of the lipid. The proteolipid has no bound carbohydrate, nucleotide, or phosphate, but does have some covalently bound lipid in the form of fatty acids which are esterified to the protein chain and make up about 2 to 3% of the weight. The molecular weight of the apoprotein has been estimated most often at around 12,000, but other estimates exist.

References

1. S. H. Barondes and G. R. Dutton, in *Basic Neurochemistry* (R. W. Albers, G. J. Siegel, R. Katzman, and B. W. Agranoff, eds.), Little, Brown, Boston, 1972.
2. S. Roberts, in *Handbook of Neurochemistry,* Vol. Va (A. Lajtha, ed.), Plenum, New York, 1971.
3. B. Droz, in *Handbook of Neurochemistry,* Vol. II (A. Lajtha, ed.), Plenum, New York, 1969.
4. B. Droz and H. L. Koenig, in *Protein Metabolism of the Nervous System* (A. Lajtha, ed.), Plenum, New York. 1970.
5. A. Hamberger, C. Blomstrand, and T. Yanagihara, *J. Neurochem. 18:*1469. (1971).
6. B. Tiplady and S. P. R. Rose, *J. Neurochem. 18:*549 (1971).
7. Y. Takahashi, C. S. Hsu, and S. Honura, *Brain Res. 23:*284 (1970).
8. A. Lajtha and N. Marks, in *Handbook of Neurochemistry,* Vol. Vb (A. Lajtha, ed.), Plenum, New York, 1971.

9. A. Lajtha and E. Toth, *Biochem. Biophys. Res. Commun. 23*:294 (1966).

10. J. G. Wood and N. King, *Nature (London) 119*:56 (1971).

11. M. E. Smith, *J. Neurochem. 18*:739 (1971).

12. P. J. Garlick and I. Marshall, *J. Neurochem. 19*:577 (1972).

13. C. E. Zomzely, S. Roberts, D. M. Brown, and C. Provost, *J. Mol. Biol. 19*: 455 (1966).

14. A. T. Campagnoni and H. R. Mahler, *Biochemistry 6*:956 (1967).

15. C. Zomzely-Neurath, C. York, and B. W. Moore, *Arch. Biochem. Biophys. 155*:58 (1973).

16. C. Zomzely-Neurath, C. York, and B. W. Moore, *Proc. Natl. Acad. Sci. USA 69*:2326 (1972).

17. C. E. Zomzely, S. Roberts, C. P. Gruber, and D. M. Brown, *J. Biol. Chem. 243*:5396 (1968).

18. C. Vesco and A. Giuditta, *J. Neurochem. 15*:81 (1968).

19. S. H. Appel, W. Davis, and S. Scott, *Science 157*:836 (1967).

20. C. E. Zomzely, S. Roberts, and D. Rapaport, *J. Neurochem. 11*:567 (1964).

21. S. Roberts and B. S. Morelos, *J. Neurochem. 12*:373 (1965).

22. N. A. Peterson and C. M. McKean, *J. Neurochem. 16*:1211 (1969).

23. A. L. Rubin and K. H. Stenzel, *Proc. Natl. Acad. Sci. USA 53*:963 (1965).

24. H. Herschman, *J. Biol. Chem. 246*:7569 (1971).

25. G. Wiche, C. Zomzely-Neurath, and A. J. Blume, *Proc. Natl. Acad. Sci. USA 71*:1446 (1974).

26. I. D. Herriman and G. D. Hunter, *J. Neurochem. 12*:937 (1965).

27. C. B. Klee and L. Sokoloff, *Proc. Natl. Acad. Sci. USA 53*:1014 (1965).

28. L. C. Mokrasch, *J. Neurochem. 13*:49 (1966).

29. H. S. Barra, J. A. Rodriguez, C. A. Arce, and R. Caputto, *J. Neurochem 20*:97 (1973).

30. H. S. Barra, C. A. Arce, J. A. Rodriguez, and R. Caputto, *J. Neurochem. 21*:1241 (1973).

31. J. A. Burdman and L. Journey, *J. Neurochem. 16*:493 (1969).

32. J. A. Burdman, K. G. Haglid, and A. R. Dravid, *J. Neurochem. 17*:669 (1970).

33. A. R. Dravid and E. Wong, *J. Neurochem. 19*:2709 (1972).

34. E. Koenig and G. B. Koelle, *J. Neurochem. 8*:169 (1961).

35. D. H. Clouet and H. Waelsch, *J. Neurochem. 8*:201 (1961).

36. E. Koenig, *J. Neurochem. 12*:343 (1965).

37. M. Singer and M. M. Salpeter, *J. Morphol. 120*:281 (1966).

38. G. S. L. Appeltauer and E. E. A. Saa, *Exp. Neurol. 14*:484 (1966).

39. A. Lajtha, S. Furst, and H. Waelsch, *Experientia 23*:168 (1957).

40. D. F. Matheson, *J. Neurochem. 15*:179 (1968).

41. D. F. Matheson, *J. Neurochem. 15*:187 (1968).

42. L. Austin and I. G. Morgan, *J. Neurochem. 14*:377 (1967).

43. L. A. Autilio, S. H. Appel, P. Pettis, and P. L. Gambetti, *Biochemistry 7*: 2615 (1968).

44. M. W. Gordon and G. G. Deanin, *J. Biol. Chem. 243:*5222 (1968).
45. J. M. Gilbert, *J. Biol. Chem. 247:*6541 (1972).
46. J. D. Fernstrom and R. J. Wurtman, *Science 173:*149 (1971).
47. J. D. Fernstrom and R. J. Wurtman, *Science 174:*1023 (1971).
48. K. Aoki and F. L. Siegel, *Science 168:*129 (1970).
49. G. T. Pryor, L. S. Otis, M. K. Scott, and J. J. Colwell, *J. Comp. Physiol. Psychol. 63:*236 (1967).
50. M. Luxoro, *Nature (London) 188:*1119 (1960).
51. F. Orrego and F. Lipmann, *J. Biol. Chem. 242:*665 (1967).
52. D. A. Jones and H. McIlwain, *J. Neurochem. 18:*41 (1971).
53. S. P. R. Rose, *Nature (London) 215:*253 (1967).
54. J. Altman, *Nature (London) 199:*777 (1963).
55. F. T. Merei and F. Gallyas, *J. Neurochem. 11:*265 (1964).
56. P. Bobillier, F. Sakai, S. Seguin, and M. Jouvet, *J. Neurochem. 22:*23 (1974).
57. C. F. Baxter, S. Tewari, and S. Raeburn, *Adv. Biochem. Pharmacol. 4:*195 (1972).
58. M. K. Campbell, H. R. Mahler, W. J. Moore, and S. Tewari, *Biochemistry 5:*1174 (1966).
59. S. Tewari and C. F. Baxter, *J. Neurochem. 16:*171 (1969).
60. C. F. Baxter and S. Tewari, in *Protein Metabolism in the Nervous System* (A. Lajtha, ed.), Plenum, New York, 1970.
61. S. Gelber, P. L. Campbell, G. E. Diebler, and L. Sokoloff, *J. Neurochem. 11:*221 (1964).
62. C. B. Klee and L. Sokoloff, *J. Neurochem. 11:*709 (1964).
63. L. Sokoloff, in *Handbook of Neurochemistry,* Vol. Vb (A. Lajtha, ed.), Plenum, New York, 1971.
64. D. H. Clouet and M. Ratner, *Brain Res. 4:*33 (1967).
65. D. H. Clouet and M. Ratner, *J. Neurochem. 15:*17 (1969).
66. E. J. Simon and D. Van Praag, *Proc. Natl. Acad. Sci. USA 51:*877 (1964).
67. W. D. Noteboom and G. C. Mueller, *Mol. Pharmacol. 2:*534 (1966).
68. W. D. Noteboom and G. C. Mueller, *Mol. Pharmacol. 5:*38 (1969).
69. D. A. Rappoport, R. R. Fritz, and S. Yamagami, in *Handbook of Neurochemistry,* Vol. Vb (A. Lajtha, ed.), Plenum, New York, 1971.
70. S. S. Oja, *Ann. Acad. Sci. Fenn.,* Ser. A5, *131:*1 (1967).
71. R. J. Schain, M. J. Carver, J. H. Copenhaver, and N. R. Underdahl, *Science 156:*984 (1967).
72. T. C. Johnson and M. W. Luttges, *J. Neurochem. 13:*545 (1966).
73. T. C. Johnson, *J. Neurochem. 15:*1189 (1968).
74. T. C. Johnson and G. Belytschko, *Proc. Natl. Acad. Sci. USA 62:*844 (1969).
75. S. Yamagami and K. Mori, *J. Neurochem. 17:*721 (1970).
76. M. R. V. Murthy, in *Protein Metabolism of the Nervous System* (A. Lajtha, ed.), Plenum, New York, 1970.
77. M. R. V. Murthy, *Biochim. Biophys. Acta 119:*586 (1966).

78.   M. R. V. Murthy, *Biochim. Biophys. Acta 119:*599 (1966).
79.   M. R. V. Murthy, *Biochim. Biophys. Acta 166:*115 (1968).
80.   S. Bondy and S. Roberts, *Biochem. J. 109:*533 (1968).
81.   T. Itoh and J. H. Quastel, *Science 164:*79 (1969).
82.   A. Fellous, J. Francon, and J. Nunez, *J. Neurochem. 21:*211 (1973).
83.   B. E. Gilbert, B. K. Grove, and T. C. Johnson, *J. Neurochem. 19:*2835 (1972).
84.   S. S. Oja, *J. Neurochem. 19:*2057 (1972).
85.   N. Marks and A. Lajtha, in *Handbook of Neurochemistry,* Vol. Va (A. Lajtha, ed.), Plenum, New York, 1971.
86.   M. W. Kies and S. Schwimmer, *J. Biol. Chem. 145:*685 (1942).
87.   G. B. Ansell and D. Richter, *Biochim. Biophys. Acta 13:*87 (1954).
88.   A. V. Palladin, N. M. Polyakova, and V. K. Lishko, *J. Neurochem. 10:* 187 (1963).
89.   A. Lajtha, in *Regional Neurochemistry* (S. S. Kety and J. Elkes, eds.), Pergamon, London, 1961.
90.   N. Marks and A. Lajtha, *Biochem. J. 97:*74 (1965).
91.   J. M. W. Bouma and M. Graber, *Biochim. Biophys. Acta 89:*545 (1964).
92.   J. K. McDonald, S. Ellis, and T. J. Reilly, *J. Biol. Chem. 241:*1494 (1966).
93.   J. K. McDonald, B. B. Zeitman, T. J. Reilly, and S. Ellis, *J. Biol. Chem. 244:*2693 (1969).
94.   N. Marks and A. Lajtha, *Biochem. J. 97:*74 (1965).
95.   N. Marks and A. Lajtha, *Biochem. J. 89:*438 (1963).
96.   G. B. Ansell and D. Richter, *Biochim. Biophys. Acta 13:*92 (1954).
97.   Y. V. Belik, S. L. Smerchinska, Y. T. Terletska, and O. G. Grineko, *Ukr. Biokhim. Zh. 40:*543 (1968).
98.   G. Guroff, *J. Biol. Chem. 239:*149 (1964).
99.   P. J. Riekkinen and J. Clausen, *Brain Res. 15:*413 (1969).
100.   P. J. Riekkinen, J. Clausen, and U. Artsila, *Brain Res. 19:*213 (1970).
101.   G. I. Drummond and L. Duncan, *J. Biol. Chem. 243:*5532 (1968).
102.   M. Inoue, A. Kishimoto, Y. Takai, and Y. Nishizuka, *J. Biol. Chem. 252:* 7610 (1977).
103.   E. Adams and E. L. Smith, *J. Biol. Chem. 191:*651 (1951).
104.   S. S. Oja, *Ann. Acad. Sci. Fenn.,* Ser. A5, *131:*7 (1967).
105.   S. S. Oja and H. Oja, *J. Neurochem. 17:*901 (1970).
106.   E. R. Einstein, K. B. Dalal, and J. Csejtey, *Brain Res. 18:*35 (1970).
107.   E. R. Einstein, J. Csejtey, and N. Marks, *FEBS Lett. 1:*191 (1968).
108.   B. W. Moore, *Biochem. Biophys. Res. Commun. 19:*739 (1965).
109.   B. W. Moore and D. J. McGregor, *J. Biol. Chem. 240:*1647 (1965).
110.   B. W. Moore, in *Proteins of the Nervous System* (D. J. Schneider, ed.), Raven, New York, 1973.
111.   B. W. Moore, V. J. Perez, and M. Gehring, *J. Neurochem. 15:*265 (1968).
112.   P. Dannies and L. Levine, *Biochem. Biophys. Res. Commun. 37:*587 (1969).

113.   V. J. Perez, J. Olney, T. J. Cicero, B. W. Moore, and B. A. Balin, *J. Neuro-chem. 17:*511 (1970).
114.   T. J. Cicero, W. M. Cowan, B. W. Moore, and V. Sungzeff, *Brain Res. 18:* 25 (1970).
115.   P. Benda, J. Lightbody, G. Sato, L. Levine, and W. Sweet, *Science 161:* 370 (1968).
116.   N. Miani, G. DeRenzis, F. Michetti, S. Correr, C. Olivieri-Sanglacomo, and A. Caniglia, *J. Neurochem. 19:*1387 (1972).
117.   H. Hydén and B. McEwen, *Proc. Natl. Acad. Sci. USA 55:*354 (1966).
118.   P. Calissano, B. W. Moore, and A. Friesen, *Biochemistry 8:*4318 (1969).
119.   P. Calissano, in *Proteins of the Nercous System* (D. J. Schneider, ed.), Raven, New York, 1973.
120.   P. Calissano and A. D. Bangham, *Biochem. Biophys. Res. Commun. 43:* 504 (1971).
121.   H. Hydén and P. W. Lange, *Exp. Cell. Res. 62:*125 (1970).
122.   U. B. Singh and G. P. Talwar, *J. Neurochem. 16:*951 (1969).
123.   F. Michetti, N. Miani, G. DeRenzis, A. Caniglia, and S. Correr, *J. Neuro-chem. 22:*239 (1974).
124.   N. Miani, F. Michetti, G. DeRenzis, and A. Caniglia, *Experientia 29:*1453 (1973).
125.   B. W. Moore and V. J. Perez, in *Physiological and Biochemical Aspects of Nervous Integration* (F. D. Carlson, ed.), Prentice-Hall, Englewood Cliffs, N.J., 1968.
126.   H. P. Herschman and M. P. Lerner, *Nature (London) New Biol. 241:*242 (1973).
127.   E. Bock and J. Dissing, *Scand. J. Immunol. 4:*31 (1975).
128.   P. J. Marangos, C. Zomzely-Neurath, and C. York, *Biochem. Biophys. Res. Commun. 68:*1309 (1976).
129.   K. Warecka and H. Bauer, *Dtsche. Z. Nervenheilk. 189:*53 (1966).
130.   K. Warecka, H. J. Moller, H. M. Vogel, and J. Tripatzis, *J. Neurochem. 19:*719 (1972).
131.   D. J. Schneider, in *Proteins of the Nervous System* (D. J. Schneider, ed.), Raven, New York, 1973.
132.   L. F. Eng, J. J. Vanderhaeghen, A. Bignami, and B. Gerstl, *Brain Res. 28:* 381 (1971).
133.   C. T. Uyeda, L. F. Eng, and A. Bignami, *Brain Res. 37:*81 (1972).
134.   A. Bignami, L. F. Eng, D. Dahl, and C. T. Uyeda, *Brain Res. 43:*429 (1972).
135.   E. G. Brunngraber, in *Handbook of Neurochemistry,* Vol. I (A. Lajtha, ed.), Plenum, New York, 1969.
136.   A. van Nieuw Amerongen, D. H. van dan Eijnden, J. Heijlman, and P. A. Roukema, *J. Neurochem. 19:*2195 (1972).
137.   A. van Nieuw Amerongen and P. A. Roukema, *J. Neurochem. 21:*125 (1973).
138.   V. P. Whittaker, M. J. Dowdall, and A. F. Boyne, *Biochem. Soc. Symp. 36:*49 (1972).

139. G. Ulmar and V. P. Whittaker, *J. Neurochem. 22:*451 (1974).
140. F. L. Margolis, *Proc. Natl. Acad. Sci. USA 69:*1221 (1972).
141. F. L. Margolis and J. F. Tarnoff, *J. Biol. Chem. 248:*451 (1973).
142. M. Cuenod, P. Marko, and E. Niederer, *Brain Res. 49:*422 (1973).
143. D. J. Wolff and F. L. Siegel, *J. Biol. Chem. 247:*4180 (1972).
144. D. J. Wolff and F. L. Siegel, *Arch. Biochem. Biophys. 150:*578 (1972).
145. J. Clausen, *Proc. Soc. Exp. Biol. Med. 107:*170 (1961).
146. H. Link, *Acta Neurol. Scand.,* Suppl. 28, *43:*1 (1967).
147. J. E. Olsson and H. Link, *J. Neurochem. 20:*837 (1973).
148. J. E. Olsson and L. Nord, *J. Neurochem. 21:*625 (1973).
149. J. E. Olsson, H. Link, and B. Nosslin, *J. Neurochem. 21:*1153 (1973).
150. L. G. Thomasi and S. E. Kornguth, *J. Cell Biol. 35:*133A (1967).
151. N. C. Sharma and G. P. Talwar, *J. Neurochem. 20:*1625 (1973).
152. M. L. Shelanski and E. W. Taylor, *J. Cell Biol. 34:*549 (1967).
153. R. C. Weisenberg, G. G. Borisy, and E. W. Taylor, *Biochemistry 7:*4466 (1968).
154. M. L. Shelanski, in *Proteins of the Nervous System* (D. L. Schneider, ed.), Raven, New York, 1973.
155. R. P. Frigon and J. C. Lee, *Arch. Biochem. Biophys. 153:*587 (1972).
156. R. Marantz, M. Ventilla, and M. L. Shelanski, *Science 165:*498 (1969).
157. R. C. Weisenberg, *Science 177:*1104 (1972).
158. H. Wisniewski, M. L. Shelanski, and R. D. Terry, *J. Cell Biol. 38:*224 (1968).
159. J. O. Karlsson and J. Sjostrand, *Brain Res. 13:*617 (1969).
160. G. W. Kerutzberg, *Proc. Natl. Acad. Sci. USA 62:*722 (1969).
161. N. W. Seeds, A. G. Gilman, T. Amano, and M. W. Nirenberg, *Proc. Natl. Acad. Sci. USA 66:*160 (1970).
162. R. E. Fine and D. Bray, *Nature (London) New Biol. 234:*115 (1971).
163. S. Puszkin, S. Berl, E. Puszkin, and D. D. Clarke, *Science 161:*170 (1968).
164. S. Berl and S. Puszkin, *Biochemistry 9:*2058 (1970).
165. W. J. Nicklas, S. Puszkin, and S. Berl, *J. Neurochem. 20:*109 (1973).
166. S. Berl, S. Puszkin, and W. J. Nicklas, *Science 179:*441 (1973).
167. S. Puszkin, W. J. Nicklas, and S. Berl, *J. Neurochem. 19:*1319 (1972).
168. E. H. Eylar, in *Proteins of the Nervous System* (D. J. Schneider, ed.), Raven, New York, 1973.
169. E. H. Eylar, S. Brostoff, G. Hashim, J. Caccam, and P. Burnett, *J. Biol. Chem. 246:*5770 (1971).
170. E. H. Eylar and M. Thompson, *Arch. Biochem. Biophys. 129:*468 (1969).
171. A. Hagopian, F. C. Wastall, J. Whitehead, and E. Eylar, *J. Biol. Chem. 246:*2519 (1971).
172. E. H. Eylar, *Proc. Natl. Acad. Sci. USA 67:*1425 (1970).
173. P. R. Carnegie, *Nature (London) 229:*25 (1971).
174. D. L. Gasser, C. M. Newlin, J. Palm, and N. K. Gonatas, *Science 181:*872 (1973).
175. J. Folch and M. Lees, *J. Biol. Chem. 191:*807 (1951).

176.   J. Folch-Pi, in *Proteins of the Nervous System* (D. J. Schneider, ed.), Raven, New York, 1973.

177.   H. C. Agrawal, B. K. Hartman, W. T. Shearer, S. Kalmbach, and F. L. Margolis, *J. Neurochem. 28:*495 (1977).

16
CARBOHYDRATES

Glucose

Glucose is the major source of energy for the brain. Further, although the brain will metabolize any number of other materials, glucose, by several different and convincing lines of evidence, is the most adequate energy source for neural tissue [1-4]. The respiratory quotient for the metabolism of brain, measured in vivo or in perfusion, is close to 1.0, indicating a metabolism of carbohydrate. Also, the ratio of O_2 to glucose consumed by brain in vivo is on the order of 5.5, the theoretical ratio for pure glucose metabolism being 6.0. Although the brain makes up only about 2 to 3% of the total mass of the body, 25% of the glucose consumption of the intact animal is accounted for by neural tissue. Since the brain has little reserve carbohydrate, the major source of glucose for brain metabolism is the blood. Glucose consumption by brain under control conditions amounts to about 20 μmol g^{-1} hr^{-1}. There are some data available to suggest that mannose will also support brain function, but these observations may reflect the conversion of mannose to glucose elsewhere in the body [3]. Fructose has also been reported to be an adequate substrate, but other evidence does not support this contention [3]. Clearly other monosaccharides, such as galactose or the pentoses, are inadequate for the support of brain function. Under special conditions the brain can use certain lipids, such as β-hydroxybutyric acid, as an energy source, but this does not occur under normal circumstances. Less extensive work has been done on peripheral nervous tissue, but the available information indicates that glucose is the preferred substrate for the metabolism of these tissues also.

Although the rate of glucose utilization by brain is one of the highest in the body, and 20-fold greater than the average, this rate can be increased massively under certain conditions, particularly those involving intense brain activity [4]. This means, without doubt, that the enzymatic capacity of the brain to metabolize glucose is much greater than the amount measured under normal circumstances. Further, this suggests a tight but responsive control over the rate of glucose utilization. During periods of anoxia, ischemia, or intense brain activity, the utilization of glucose can increase by a factor of 20 or even more. Since the brain is using 20% of the body oxygen even normally, and thus, the rate at which oxygen is supplied to the brain can increase at most threefold [2], the increased utilization of glucose results in a marked increase in the production of lactate through glycolysis, but only a relatively modest increase in the rate of oxidative metabolism and CO_2 production. Thus, the production of lactate in brain, which is about 6 μmol g^{-1} h^{-1}, or about 15% of the glucose consumption, can rise substantially during emotional excitement, and can increase 30- to 100-fold during convulsions. As clearly shown by Lowry and his collaborators some years ago [5], the capacities of the glycolytic enzymes in brain are much greater than their normal rate of catalysis (Table 16.1). The control of these enzymes can also lead to decreased glucose utilization. For example, the action of the barbiturates in lowering glucose consumption by the brain under a variety of conditions has been thoroughly documented [6].

Under normal conditions the free glucose concentration of the brain is very low. The best data available indicate that the levels in the whole brain are on the order of 1.5 μmol/g of tissue in most species [3]. There is great variation in the values reported in the literature, and many lower values are available. There are probably two causes for the variability in the published data [3]. First, over many years, a number of different methods have been used to measure brain glucose levels. Second, and more significant, is the extremely rapid postmortem decline in the glucose levels due to continuing glycolytic activity. Unless special precautions are taken, such as instant freezing of the brain upon the death of the animal, glucose begins to disappear within seconds.

The glucose which is found in brain may be all extracellular. If the assumption is made that the extracellular fluid of the brain contains the same glucose concentration as the cerebrospinal fluid, and the further assumption that the extracellular space of brain amounts to 25 to 30% of the total brain, then the intracellular glucose concentration of the brain is close to zero. Both these assumptions seem quite reasonable, and, in fact, independent evidence [7] bears out the contention that the intracellular glucose concentration of brain is negligible.

Glucose Transport

If, as it seems, the intracellular content of glucose is zero, then the transport of glucose into the cell from the blood or the extracellular fluid must be rate limiting. That is, the metabolism of glucose must be faster than its transfer into the cells, a situation known to exist in muscle but, incidentally, not in the sympathetic ganglia where the tissue level is the same as the blood level [1]. The transport system for glucose, then, is worthy of study since it may play a role in the

Table 16.1 Concentration of Glycolytic Intermediates and Rates of Individual Enzymatic Steps in Mouse Brain

Substance	Cerebral concentration (μmol/g)	Enzyme (K_M, mM)	Rate (μmol/g fresh wt/hr)
Glucose	1.25–1.55	Hexokinase, 0.05	660
Glucose-6-phosphate	0.065–0.08	Phosphoglucomutase, 0.03	3,300
Fructose-6-phosphate	0.016	Phosphofructokinase,	540
Fructose-1,6-diphosphate	0.12	Aldolase, 0.01–0.04	310
Glyceraldehyde-3-phosphate and dihydroxyacetone phosphate	0.3	Triose phosphate isomerase	—
1,3-Diphosphoglycerate	—	Glyceraldehyde-3-phosphate dehydrogenase, 0.044	3,360
3-Phosphoglycerate	0.1–0.5	Phosphoglycerate kinase, 0.01	10,000
2-Phosphoglycerate	0.1–0.5	Phosphoglyceromutase, 0.24	2,340
Phosphoenolpyruvate	0.8	Enolase, 0.033	1,800
Pyruvate	0.1–0.2	Pyruvate kinase, 0.058	7,080
Lactate	2	Lactic dehydrogenase, 0.03	3,540

From O. H. Lowry and J. V. Passonneau, *J. Biol. Chem. 239:* 31 (1964).

overall regulation of glucose metabolism in the brain [8]. The information now available indicates that the transport of glucose into the brain in the living animal is mediated by a "facilitated diffusion" system [9]. That is, the uptake is by means of a carrier which is stereospecific and saturable, but does not transport glucose against a concentration gradient and, so, does not require energy. Such a system is different from the "active transport" systems for amino acids, for example, in that these latter will transport materials against their concentration gradients and do require a source of energy for the transport process. There is some evidence that certain sugars can be taken up actively in vitro by brain slices. The glucose carrier in the brain is fairly specific for glucose, but is inhibited, presumably competitively, by D-mannose. Some other monosaccharides, such as D-galactose, L-arabinose, and D-xylose, enter the brain readily, but D-fructose and D-ribose do not [10]. The transport of glucose is inhibited by the glucose analogs 3-O-methylglucose and 2-deoxyglucose. In fact, the intravenous administration of the competitor, 2-deoxyglucose, to monkeys decreases the level of consciousness and impairs the electrical activity of the brain within 1 to 3 min, presumably due, at least in part, to its ability to keep glucose out of the brain by competing for the transport carrier [11]. The K_M of the glucose carrier for its substrate in many species is on the order of 5 mM, which is near the blood concentration of glucose, and suggests that alterations in the blood level can alter the rate of transport. Stimulation of the brain is thought to increase the entrance of glucose into the brain; this seems necessary, a priori, if glucose entry is truly rate limiting and glucose utilization increases markedly during stimulation. The question of whether insulin increases glucose uptake into the brain remains a controversial one. There is evidence that glucose uptake into brain is not affected by systemically administered insulin, but is increased by insulin given intracerebrally. The possible effects, if any, of endogenous insulin release on brain glucose uptake remain unknown.

There has been some interest in the possibility that a specialized glucose transport system is present in the membrane of the synaptic ending [12,13]. The rationale for these experiments involves measurements which suggest that neurons have a higher oxidative metabolism than glia. The extra metabolic conversion of glucose could be due to the contribution of the specialized portion of the neuron, namely, the synapse. Indeed, it has been found that the synaptosomes will transport glucose into the interior, and that this transport system is different than that found in preparations from whole brain such as brain slices. The synaptosomal system is a high-affinity, carrier-mediated transport, with a K_M for glucose on the order of 0.24 mM. A second, lower-affinity system is also found in the synaptosomes and may represent the normal transport system found in all portions of the cell membrane. The observation that a special transport system for glucose is present in the synaptosome, coupled with the finding that the synaptosomes can oxidize glucose completely and produce ATP, indicates that this transport system is a real entity and may be present in order to serve the special energy needs of the synaptic portion of the cell.

Glycolysis

The metabolism of glucose in the brain proceeds almost exclusively through glycolysis. As will be mentioned below, the hexose monophosphate shunt pathway is present in brain and, in certain circumstances, seems to be active. But in the normal adult brain, very little metabolism of glucose is carried out by pathways other than glycolysis. The enzymes of the glycolytic pathway in brain are qualitatively the same as the enzymes which catalyze the comparable reactions in liver or in muscle. In fact, much of the basic work on the glycolytic enzymes was done using brain as a source of the enzymes, since they are all present and quite active.

The activity of hexokinase (Fig. 16.1) in the brain, when measured under optimal conditions, is 17 times greater than that of liver. In fact, hexokinase activity is higher in brain than in any organ in the body. Of the several forms of the enzyme known, the brain enzyme appears to be the type I, or electrophoretically "slow-moving."

A unique and much investigated property of brain hexokinase is its intracellular location. Unlike hexokinases from other tissues, which seem to be almost completely in the soluble portion of the cell, 90% of the hexokinase of brain is found in the mitochondria [14]. This binding is almost certainly not an artifact of the preparation. The brain enzyme is distributed anatomically like a mitochondrial and not like a glycolytic enzyme [15]. When extracted from the mitochondria the bound hexokinase can be shown to be similar, but not identical, to the soluble form. The similarities between the forms include their comparable kinetics, and similarities in their behavior upon chromatography and electrophoresis [16,17], as well as their apparent immunologic identity [18]. Some interesting differences can be shown, however, if the binding properties are examined. It has been shown that the bound form can be released from the mitochondria by treatment with glucose-6-phosphate and, under the appropriate con-

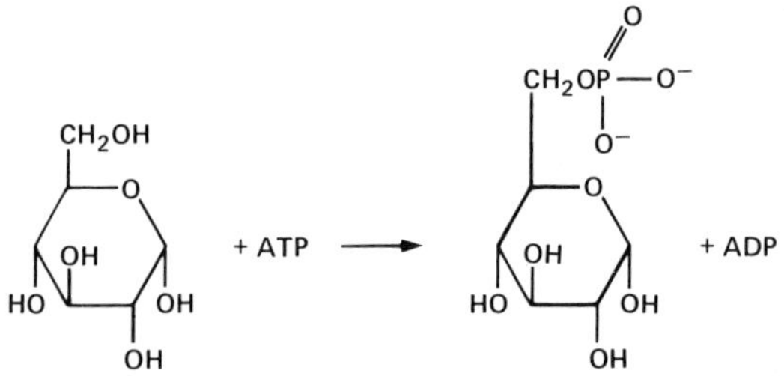

Figure 16.1   Hexokinase.

ditions, will readily rebind [19,20]. The soluble form under the same conditions will not bind to the mitochondria [21]. More recent data also shows that the developmental course of the two forms is substantially different [18]. In summary, then, there is good evidence to support the concept that the bound form of hexokinase is different than the free form.

It is known that hexokinase in brain is normally quite inactive compared to its total possible activity when assayed in the test tube. The best data to be had shows that only 3 to 5% of the available hexokinase activity is being utilized. There are a number of explanations available for this inhibited state, and one of them is that the enzyme bound to the mitochondria, probably by some ionic linkages, is masked or latent in this state. However, there are a number of other regulators of hexokinase known which might account for the low activity found. Among these inhibitors are the products, ADP and glucose-6-phosphate, and certain other compounds which appear to act allosterically. These latter include ATP, 3-phosphoglyceric acid, and cyclic AMP (cAMP). But the total inhibition exerted by these metabolites when present at concentrations comparable to those found in the brain does not account for the 90 to 95% inhibition of the activity known to be the usual state in vivo. Thus, the binding to the mitochondria also may have some meaningful influence on the activity in the intact brain. Whatever the mechanism, it is known that the rate of hexokinase activity can increase dramatically in response to physiological stimulation. But even this large reserve of activity can become limiting under extreme conditions. The total activity of hexokinase in the brain is not much in excess of the total level of glucose utilization under maximal electrical stimulation. Normally, however, the control of hexokinase activity by the various metabolites may provide one of the major control points for glycolytic activity in the brain [1,4].

The specificity of the enzyme in brain is comparable to that in other tissues. The K_M for glucose is on the order of 0.05 mM. Brain hexokinase will phosphorylate fructose and mannose as well. Also, the analog 2-deoxyglucose is a competitive inhibitor for glucose and is itself phosphorylated, which may also contribute to the marked neurological effect of this analog mentioned earlier. Hexokinase is the major monosaccharide kinase present in the brain, since glucokinase, the specific glucose-phosphorylating enzyme found in liver, is absent from the brain.

Phosphohexosisomerase of brain is similar to that of other tissues. The reaction catalyzed (Fig. 16.2) is a reversible one, with the equilibrium comewhat in the direction of glucose-6-phosphate. The reaction is inhibited by the analog 2-deoxyglucose, probably after the analog is phosphorylated to 2-deoxyglucose-6-phosphate by hexokinase [22]. The activity of this enzyme is high in all parts of the brain. The concentrations of its substrate and its product, glucose-6-phosphate and fructose-6-phosphate, are quite low in brain, being on the order of 0.07 and 0.01 μmol/g of tissue, respectively.

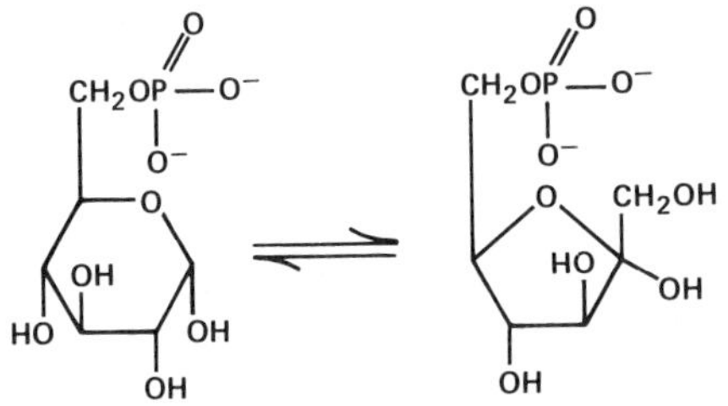

Figure 16.2   Phosphohexosisomerase.

A great deal of work has been done on phosphofructokinase (Fig. 16.3) from brain and from other sources since it is a complicated enzyme and one of the major regulating factors in the glycolytic scheme. The step catalyzed is irreversible, as are all the kinase reactions. The reaction is inhibited by ATP and Mg^{2+}, and by citrate, and is stimulated by NH_4^+, K^+, inorganic phosphate (P_i), AMP, cAMP, ADP, and fructose-1,6-diphosphate, the product. The relationship between the various effectors is a complex one. For example, the inhibition by ATP is decreased by the presence of the substrate, fructose-6-phosphate, or by other of the effectors such as cAMP, Mg^{2+}-ATP is less inhibitory than ATP itself. Citrate and ATP are not independent but seem to have a synergistic effect on the enzyme. In fact, it has been suggested by Lowry, who has worked extensively on this enzyme [23,24], that it has between 7 and 12 separate regulatory sites. The regulatory picture is augmented by the recent observation that the enzyme occurs in at least three isozyme forms in the various tissues, and that each of these forms has regulatory properties which differ quantitatively from those of the others. In brain, specifically, the molecular form C predominates, and this form responds in a regulatory sense chiefly to the relative amounts of the adenine nucleotides and inorganic phosphate present [25,26]. As will be mentioned later, the various alterations in the functional state of neural tissue cause changes in the levels of the various effectors of this enzyme which change the rate of this vital control step in most delicate ways.

The remainder of the glycolytic pathway contains no unusual enzymes. It can be seen from Table 16.1, however, that the activities of most of the remaining

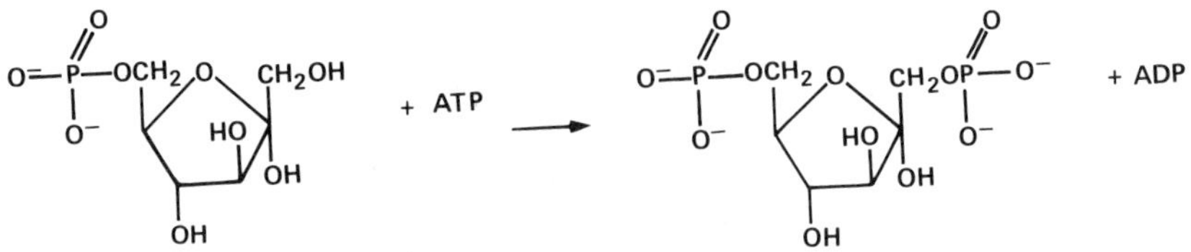

Figure 16.3   Phosphofructokinase.

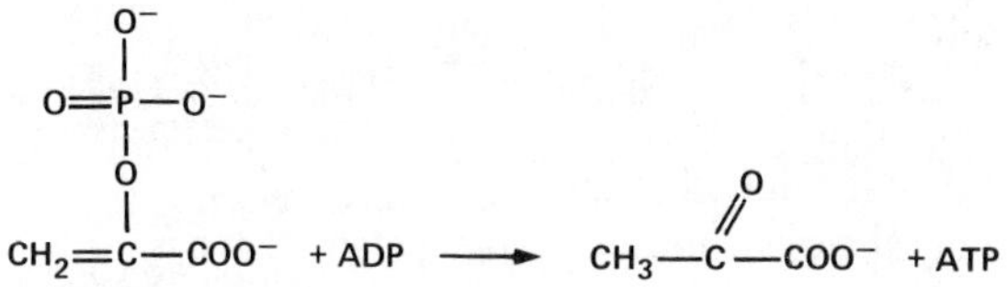

$$CH_2\!=\!\underset{\underset{\underset{O^-}{|}}{\overset{\overset{O}{\|}}{\underset{|}{O\!=\!P\!-\!O^-}}}}{C}\!-\!COO^- + ADP \longrightarrow CH_3\!-\!\overset{\overset{O}{\|}}{C}\!-\!COO^- + ATP$$

Figure 16.4   Pyruvate kinase.

enzymes are in marked excess to those in the early stages, notably hexokinase and phosphofructokinase, the regulatory steps. On the other hand, it must be remembered that the level of activity measured in vitro, as in the case of hexokinase, may have little to do with the true in vivo rate of the reaction.

Probably the pyruvate kinase reaction (Fig. 16.4) in brain, as in other tissues, is also a point of control for the glycolytic pathway [27,28]. The enzyme is inhibited by ATP, and again the Mg^{2+}:ATP ratio is important. Further, the enzyme is influenced by monovalent cations, K^+ stimulating and Na^+ inhibiting.

Lactic dehydrogenase (Fig. 16.5) is found in brain, and all five isozymes are present. The isozyme which predominates, the most rapid band electrophoretically (type I), is the isozyme most often identified with aerobic metabolism. In fact, the isozyme pattern found in the brain is similar to that observed in heart [29]. This isozyme distribution, however, is different at different stages of development. The type I form makes up only 10% in immature brain, but rises to 40% in the brains of adults. These observations support the general proposition that the synthesis of the various isozymes of lactic dehydrogenase depends upon, and is controlled by, the level of oxygen in the tissue. Also, the changes in the isozyme pattern of cerebral lactic dehydrogenase apparently correlate with the resistance of the animal to anoxia [2].

In general, then, the glycolytic enzymes are very active in the brain, usually several times more active in brain than in liver. Both neurons and glia have extremely high glycolytic rates. Early data from Lowry's group indicated that there were differences in the glycolytic complement of enzymes between the two cell types, but that the neurons from different portions of the brain were much alike in this regard. More recent work from the same laboratory in which many different types of neurons were studied [30] shows that, on the contrary,

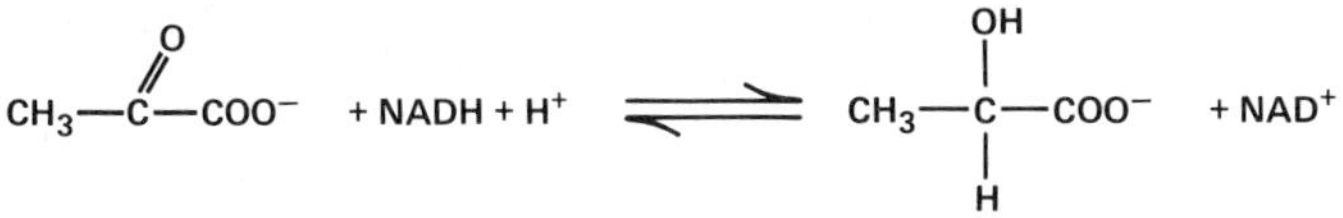

$$CH_3\!-\!\overset{\overset{O}{\|}}{C}\!-\!COO^- + NADH + H^+ \rightleftharpoons CH_3\!-\!\underset{\underset{H}{|}}{\overset{\overset{OH}{|}}{C}}\!-\!COO^- + NAD^+$$

Figure 16.5   Lactic dehydrogenase.

different neuronal cell types differ substantially in the levels of specific glycolytic
enzymes which they contain. The authors conclude that such measurements cast
some doubt on values obtained by bulk separations of neurons and glia, as well as
on the generality of values derived from large molluscan neurons. Peripheral
nerves have relatively low glycolytic rates, on the order of 10%, compared to neurons from the central nervous system [1]. As could be expected, perhaps, the
concentrations of the glycolytic intermediates in brain are lower than those in
most other tissues.

Gluconeogenesis

In contrast to the high levels of glycolytic enzymes in the brain, the enzymes
which enable the reverse process, gluconeogenesis, to proceed are absent from
brain, or else occur in very low levels. This pathway, which allows the conversion of pyruvate to glucose, is catalyzed by most of the same enzymes which
function in glycolysis, merely acting in the reverse direction. In a couple of key
steps, however, namely those which are irreversible in the glycolytic direction,
the glycolytic enzymes obviously cannot function. At these points entirely different enzymes are involved in the reverse reactions in other tissues. In brain, one
of these enzymes, fructose-1,6-diphosphate phosphatase, which catalyzes the reaction which circumvents the phosphofructokinase reaction, is absent. As for
the other unique enzymes in the pathway, pyruvate carboxylase activity is low
and glucose-6-phosphate phosphatase activity is negligible.

Hexose Monophosphate Shunt

The hexose monophosphate shunt pathway is also low in whole brain compared
to other tissues. It has been estimated that less than 1% of all the glucose metabolized by the brain is handled by this pathway [31,32]. This estimate is based,
in part, on the observation that the ratio of CO_2 produced from glucose labeled
in the 6 position to that produced from glucose labeled in the 1 position is about
1.0. Further, the activities of the enzymes of the shunt, when measured directly,
are very low in brain. Also consistent with this estimate is the observation that
relatively low levels of NADPH are present in brain. Nevertheless, this route of
metabolism may be of substantial importance in certain parts of brain or during
certain developmental periods. For example, the activity of the shunt is higher
in developing brain than it is in adult brain, and particularly high during myelination. The activity of the shunt enzymes can best be observed in heavily myelinated tracts [33]. Also, there are reports that the hexose monophosphate shunt
is relatively higher in medulla and in cerebral dispersions than it is in whole brain
[34]. Finally, there are a number of experiments which show that the metabolism of glucose by this route is stimulated markedly, as are the activities of the

enzymes which constitute the pathway, both in vivo and in vitro, by various kinds of electrical stimuli [35,36]. So, the hexose monophosphate shunt route of metabolism is definitely present in brain and its activity responds to a number of physiological and developmental stimuli.

It has been noted before, in another context, that the conversion of labeled glucose to labeled amino acids is very rapid in brain as compared to other tissues. The reasons for this circumstance should now be clear. The high glycolytic rate, the virtual absence of alternate pathways of glucose metabolism, the low rate of gluconeogenesis, and the low concentration of glycolytic intermediates, all contribute to a rapid and relatively complete funneling of glucose carbon into the Krebs cycle. In addition, the production of acetyl coenzyme A (CoA) from fatty acids is not normally a major function of brain. In fact, the reverse is true. Glucose is the precursor of almost all brain lipid, so there is little acetyl CoA in brain that does not come from glucose. The relatively large pool of glutamic acid in the brain, which is in rapid equilibrium with the α-ketoglutaric acid of the Krebs cycle, provides a trap for the glucose carbon passing through the cycle. All these factors account for the very active conversion by nervous tissue of labeled glucose into amino acids such as glutamic acid, aspartic acid, acetylaspartic acid, and γ-aminobutyric acid. In fact, as noted before, about 70% of the isotope present in the acid-soluble fraction of rat brain 20 min after the injection of carbon-labeled glucose is in the amino acid fraction. This is a much larger fraction than appears in the comparable compartment in other tissues under the same conditions. Eventually all the label appears as $^{14}CO_2$, as expected for the complete oxidation of glucose, so that the labeling of the amino acids is merely a temporary diversion in a rapidly turning over metabolic system.

Regulation of Glycolysis

There are a number of circumstances which are known to, or thought to, change the rate of glycolytic metabolism either relatively or absolutely in the brain. First, during the development of the animal, the reliance on anaerobic pathways seems to change. The immature brain has a low energy requirement and is relatively insensitive to anoxia. The energy requirement rises as the brain matures and the cells differentiate and form connections. Then the oxygen consumption of brain declines in senescence, but it is not clear whether this is a biochemical change or merely one related to reduced blood flow to the brain. The relative resistance to anoxia in young animals is well documented [37]. Generally it is assumed that this signifies a greater reliance on glycolysis in the young. Obviously, this would mean a relatively greater contribution of glycolysis in the young brain with a relative increase in aerobic pathways in the adult, but this does not indicate that the absolute rate of glycolysis changes in any way with maturation. Other explanations of the resistance to anoxia in the young are possible, such as a greater stability of the ATP pool in the young due to lowered levels of ATPases [38].

The Pasteur effect can be demonstrated in brain as well as in other tissues and the possible explanations for this effect in brain are the same as for other tissues. For example, the decreased glycolysis seen upon admission of oxygen to brain preparations could be due to a competition between the pathways for available cofactors such as NAD^+, or due to the inhibition of phosphofructo-kinase by the ATP produced in aerobic metabolism. One of the most tenable suggestions is that the two pathways, the aerobic and the anaerobic, compete for ADP and inorganic phosphate. If this latter is the case, the situation in brain is that inorganic phosphate is present in adequate levels, but that ADP is relatively low, so that competition for ADP is the most likely site at which the two pathways impinge.

Many other conditions influence the rate of glycolysis in brain. Administration of insulin, as mentioned before, increases the uptake of glucose by the brain [39-41] and, thus, the rate of glycolysis. Of course, excess insulin can cause hypoglycemia, which will be discussed below, and hypoglycemia due to any cause has severe consequences for the brain due to the complete reliance of the brain on glucose as a metabolic substrate. Diabetes, either experimentally induced in animals through the use of alloxan, or the human condition, has severe and poorly understood neurological consequences due, presumably, to the constant, high circulating levels of glucose, or the appearance in the brain of unnatural products of glucose metabolism.

Hypoglycemia

Hypoglycemia, whether produced by insulin injection or by any other cause, is followed by a number of neurological problems, including coma, convulsions, permanent brain damage, and death. The effects of acute hypoglycemia can be observed within 20 to 25 min after the lowering of the blood glucose. This appears to be about the time necessary for the depletion of the endogenous stores of glycolytic intermediates and glycogen, and for the complete utilization of any alternate substrates, such as β-hydroxybutyric acid, available to the brain from other sources. The effect of insulin, then, on brain function is secondary to its effect on glucose levels. In fact, there is little if any hard evidence for the direct effect of any hormone, including insulin, on glycolysis in brain.

Convulsions caused by insulin hypoglycemia, or by electrical stimulation, are accompanied by an increased glycolytic flux. The total high-energy phosphate content of the brain drops, as do the levels of glucose [42]. Concentrations of glycolytic intermediates in the brain are altered, and much more glucose is consumed [43].

Hypoxia

Hypoxia, like hypoglycemia, profoundly affects brain carbohydrate metabolism.
There is a marked increase in glycolysis and a concomitant increase in lactic acid.
Consequently, the pH of the brain drops due to the increased acid. There are
recent data which indicate that of the three possible control points in the glycolytic pathway, hypoxia causes an activation of the pyruvate kinase reaction and
of hexokinase, but not of the phosphofructokinase reaction [44].

Ischemia

Ischemia, the interruption of blood flow, and with it the interruption of the supply of glucose and of oxygen, cannot be tolerated by the brain for more than a
minute or two. Initially the extraction of glucose and O_2 from the blood increases and glucose levels in the tissue are rapidly depleted. Glycolysis and lactic
acid production increase also. Shortly thereafter the level of high-energy phosphate in the tissue drops, the lactic acid concentration and the CO_2 level rises,
and the pH drops. Except for the increase in CO_2 the changes are much like
those following hypoxia. Finally, there is local vasodilation, deterioration of the
normal ionic gradients in the tissue, and necrosis and death of the affected cells.
Enzymatically the activation of glycolysis in ischemia is thought to occur through
an activation of phosphofructokinase. The mechanism of this activation is most
likely the changes in the levels of the various effectors of the enzyme, especially
the drop in the ATP levels and the rise in ADP which accompany the cessation
of oxidative activity [6].

Finally, in starvation, the neurological involvement is much more gradual.
After about 24 hr the body and brain sources of carbohydrate are used up. The
brain, then, begins to metabolize other sources, such as the ketone bodies formed
from fatty acids in the liver. These seem to support brain function adequately,
but do not seem to be the preferred source of energy.

Glycogen

Normally there are low levels of glycogen in the brain. The best estimates available indicate that the glycogen content of normal brain ranges between 80 and
100 mg/100 g of fresh tissue [1,45], but values as low as 40 and as high as 160
mg have been reported in various species. The levels are lower than those generally found in liver, but higher than those found in muscle. The variation in values
reported, as in the case of glucose levels, is due to the rapid postmortem changes
which occur in glycogen stores in the brain. It has been shown that unless precautions are taken to prevent degradation, the levels of glycogen in the brain are

stable for only about 6 sec after the death of the animal [6]. Generally, it is felt that the glycogen levels remain constant until after the available glucose is used and then begin to drop. Although the glycogen content of the brain is quite low, glycogen is the largest single energy reserve available to nervous tissue, and it is generally thought to be an important component of brain metabolism. Thus, the changes in brain glycogen levels under various conditions have been thoroughly studied.

The capacity for glycogen synthesis in the brain is quite low [2]. Nevertheless, the enzymes of glycogen synthesis and glycogen degradation are present in brain and seem to be the same as those in other tissues. The accessory factors necessary for the cascade control of glycogen synthesis and breakdown, namely cAMP and the enzyme which synthesizes it, adenylate cyclase, are also present. Indeed, the various forms of glycogen phosphorylase and of glycogen synthetase have been observed in brain, the interconversions between them seem to be the same as in other tissues, and the regulating factors, cAMP and norepinephrine, interact with the various forms in much the same way as in other tissues [46]. The levels of the enzymes involved in the biosynthesis and the degradation have been measured by Lowry and Passonneau [47].

Brain glycogen has not been fully characterized as regards its molecular size range or its branching, but it seems to be similar in these characteristics to the glycogen of liver. It is found in all anatomical areas of the brain and in the brain of many species. Glycogen is also present in the spinal cord, in the sympathetic and sensory ganglia, and in peripheral nerve. Glycogen is found in the cytoplasm of both neurons and glia. It is present in certain brain tumors such as astrocytoma and spongioblastoma, but not in others such as oligodendrogliomata or medulla blastoma.

Radioactive glucose is readily incorporated into brain glycogen in vivo, even under conditions where no net change occurs in glycogen levels. This indicates that brain glycogen is in rapid equilibrium with cellular glucose and continues to turn over, as do most body polymers. In the mouse the turnover time for cerebral glycogen is on the order of 4.4 hr [48]. This turnover time can vary appreciably under various conditions and can be altered in some disease states, as will be discussed below.

The levels of glycogen in the brain can be influenced by the administration of a number of hormones. Whether these effects are primary or merely reflections of the effect of these hormones on the overall energy status of the body is not clear in most cases, but the latter is most likely. Nevertheless, it is important to note that the administration of epinephrine, especially in the presence of alcohol, to weaken the blood-brain barrier to the hormone, causes a decrease in the glycogen content of the brain. The level is stable, however, to the effects of adrenalectomy or chemical sympathectomy [49]. Insulin has a biphasic effect on the glycogen content. Administration of low levels causes a rise in the glyco-

gen content [40], but larger amounts which induce hypoglycemia are followed
by a drop in the glycogen content [45]. Infusions of glucose, with or without
insulin, raise brain glycogen levels [40]. Hydrocortisone treatment is reported
to raise brain glycogen and to lower the turnover time to 2.2 hr [48] in the
mouse. Adrenocorticotropic hormone administration lowers the content of glycogen in the brain [50].

Glycogen levels also respond to a number of exogenous stimuli. Glycogen
levels drop in response to hypoxia, ischemia [51], hypothermia, and hyperthermia [52]. Some reports of lowering due to starvation are available, although
most reports suggest that the levels are relatively stable to severe starvation, for
example, even 3 days starvation in the dog [53]. Generally, convulsions, however induced, cause a drop in glycogen levels. The exceptions seem to be convulsions caused by electrical stimuli, about which reports are mixed, convulsions
caused by audiogenic stimuli, which do not alter glycogen levels, and convulsions
caused by methionine sulfoximine, which cause an increase in brain glycogen
levels by decreasing the active form of glycogen phosphorylase [45,54]. The
administration of amphetamine causes a profound drop in glycogen levels [46].

Increases in glycogen levels occur upon various kinds of injury to the brain
[55-57]. When axons degenerate for any of a number of reasons, glycogen deposits can be seen in the glial cells which surround them. Also, glycogen deposits
can be found in the axoplasm of transected axon processes. Irradiation of the
brain causes deposits of glycogen to appear in astrocytes, and the same has been
reported in certain cases of severe liver disease. Cobalt intoxication has been
found to result in the deposition of glycogen in dendritic processes [50].

Glycogen levels also rise in response to anesthesia [40,58,59]. The turnover time of glycogen in the brain increases to double the normal value, and this
is found to be due to a decreased rate of glycogen breakdown rather than an increase in glycogenesis [48]. Some data are available which suggest that one effect of anesthetics is to increase glucose transport and to lower the levels of inorganic phosphate [40]. There are also numerous reports of increases in glycogen following the administration of barbiturates [40]. Various tranquilizers
also cause accumulations, as does administration of lithium [60], propranolol
[61], or caffeine [62]. Consistent with the observations on tranquilizers and
anesthesia, perhaps, are the findings that the brains of animals contain greater
levels of glycogen during hibernation than they do at other times.

Glycogen Storage Diseases

There are a number of disease states which involve the accumulation of glycogen.
Collectively these are known as the glycogen storage diseases [50,63]. Some few
of these involve alterations in the functioning of the central nervous system.
Since glycogen is in a state of dynamic equilibrium, all the glycogen storage dis-

Table 16.2 The Glycogen Storage Diseases

Glycogenosis type	Biochemical defect	Diagnostic features	Metabolic disturbance	Neurological lesion
I (von Gierke's)	Glucose-6-phosphatase	Hepatomegaly, hypo-glycemia	Hypoglycemia, lactic acidosis, ketoacidosis, hyperuricemia, lipidemia	Symptoms of hypo-glycemia, hypotonia, retardation
II (Pompe's)	Lysosomal acid maltase	Cardiomegaly, gen-eralized organomegaly	Usually none	Vomiting, amyotonia, retardation
III (Cori's)	Debranching enzyme system	Hepatomegaly	Milder than type I	Milder than type I
IV (Andersen's)	Branching enzyme	Cirrhosis	Liver disease	Retardation
V (McArdle-Schmid-Pearson)	Muscle phosphorylase	Muscle cramps	Glycogen deposition in skeletal muscle	Myopathy
VI (Her's)	Liver phosphorylase	Hepatomegaly	Milder than type I	Milder than type I
VII	Muscle phosphofructo-kinase	Muscle cramps, myoglobinuria	Glycogen deposition in skeletal muscle	Myopathy
VIII	Muscle phosphohexos-isomerase	Muscle fatigue, myoglobinuria	As for type VII	Myopathy
IX	Liver phosphorylase kinase	As for type VI	Milder than type I	Milder than type I

Table taken from Y. E. Hsia, in *Basic Neurochemistry* (R. W. Albers, G. J. Siegel, R. Katzman, and B. W. Agranoff, eds.), Little, Brown, Boston, 1972.

eases must involve either an increase in the rate of synthesis or a decrease in the rate of degradation. The major glycogen storage diseases have been fairly well studied, and the enzymatic basis for most of them is known.

There are at least nine glycogen storage diseases, classified types I through IX (Table 16.2). The one in which neurological involvement is most evident and retardation is most often seen is type II (Pompe's disease). Pompe's disease affects the nervous system in a rather specific manner. The disease is transmitted as an autosomal recessive. The condition is seen in infants, and they usually do not survive past 1 year of age. Dustlike glycogen deposits appear inside neurons. The sites most involved are the motor nuclei of the cranial nerves and the anterior horn cells. Deposits also occur in glia of the central nervous system and in the Schwann cells of the dorsal root ganglia. The affected individuals suffer from enlarged tongues and hearts and the usual cause of death is cardiac or respiratory failure. The neurological effects can be manifest in a number of different ways, but they are infrequent enough that the disease is sometimes classed as neurologically innocuous. The exact enzyme deficit has been shown to be the absence of an acid maltase, a lysosomal enzyme which hydrolyzes the α-1,4 linkage between glucose molecules. The enzyme is missing from all organs examined, and clearly its absence could lead to a lowered rate of glycogen breakdown and a substantial accumulation.

Pompe's Disease

The patient, a 4-month-old boy, was admitted for generalized failure to thrive. He was small, but not stunted, and had a normal developmental history up to a month before admission. The child appeared weak and hypotonic. He exhibited ptosis and facial palsy, and deep tendon reflexes were absent. The child's tongue was enlarged and protruded from his mouth. He could neither cry nor suck. The parents were normal, with no history of disorders, and the single sibling, a 3-year-old girl, was normal.

Upon clinical work-up the child showed an abnormal electromyogram, and an electrocardiogram exhibited a shortened P-R interval, a wide amplitude QRS, and left axis deviation. Chest X-ray revealed an enlarged, globular heart. Leukocytes from the child were assayed for acid maltase, and the enzyme was found to be absent.

After 2 months of hospitalization the child died of pneumonia. By this time the child had become nearly paralyzed. Upon autopsy glycogen deposits were found in the skeletal and heart muscle, and in the motor neurons of the brainstem and the spinal cord. The walls of the ventricles were thickened and massively infiltrated. The heart itself was enormously enlarged, and round. The liver, also, was moderately enlarged. Skeletal muscles were atrophied, and the myofibrils were largely destroyed by granular glycogen deposits [64].

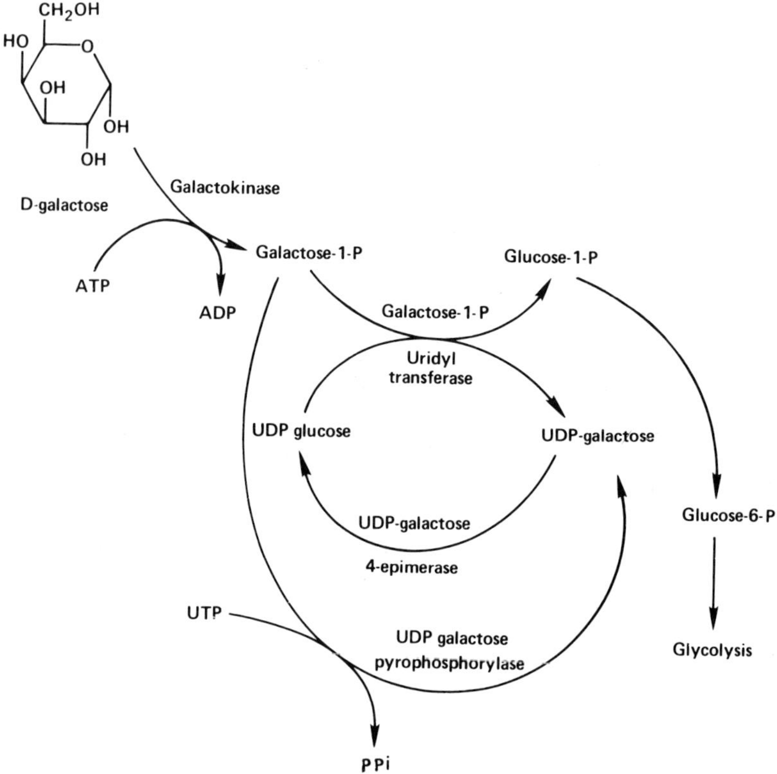

Figure 16.6 Enzymatic conversion of galactose to glucose.

Galactose and Galactosemia

The enzymes for the conversion of galactose to glucose are present in the brain. The normal pathway for this conversion in humans is shown in Fig. 16.6. The enzymes involved in this pathway were identified by Leloir [65] and characterized by Kalckar and his colleagues. The pathway is of special importance during infancy when large amounts of galactose are ingested as lactose from breast milk.

One of the known genetic deletions in this pathway has very severe consequences for the health of the infant. This condition is known as galactosemia and is caused by the lack of the enzyme galactose-1-phosphate uridyltransferase [66]. The absence of this enzyme, as seen in Fig. 16.6, effectively prevents the conversion of galactose to glucose by one of the possible metabolic routes. Ironically, the alternate pathway to UDP-glucose, via uridine diphosphate galactose pyrophosphorylase, is not available in infants since the enzyme does not appear

until relatively late in childhood. Unfortunately, by the time the enzyme does
appear it is no longer necessary since the damage to the developing nervous system
has already been done and is quite irreversible.

Galactosemia is transmitted as an autosomal recessive and occurs in about
1 in 70,000 children. As such it is one of the most prevalent of the genetically
based mental retardations. The disease, if untreated, results in severe permanent
retardation, cataract, and death. Diagnosis is now possible either prenatally or
postnatally using sensitive enzyme assay techniques [67] applied to material obtained from aminocentesis or on red or white blood cells or skin fibroblasts, all
of which have the enzyme in normal individuals. The treatment of the disease,
once diagnosed, is merely to leave galactose out of the diet of the child. This
treatment has proved to be quite effective in children diagnosed at birth.

The biochemical consequences of the condition include the accumulation
of galactose-1-phosphate and of galactitol, the sugar alcohol analog of galactose
(Fig. 16.7). It is not known what the exact agent of damage is in this condition,
but it is probably not galactitol since in another genetic lesion, namely, the much
rarer galactokinase deficiency, there is a build-up of galactitol in the system, but
no accumulation of galactose-1-phosphate. In this condition there is no retardation [63]. In any case, there is no information on the cause of the neurological
involvement. Perhaps the damage is caused by the accumulation of galactose-1-
phosphate, by the accompanying cerebral edema, by the abnormal energy metabo-

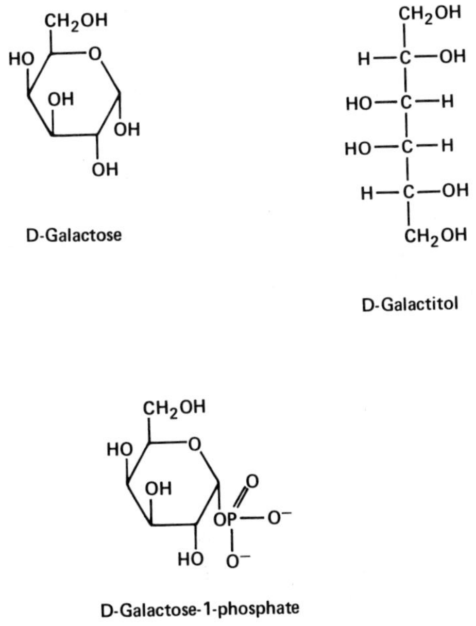

D-Galactose

D-Galactitol

D-Galactose-1-phosphate

Figure 16.7 Galactose and some
of its metabolites.

Table 16.3 Inherited Disorders of Monosaccharide Metabolism Associated with Neurological Dysfunction

Condition	Biochemical defect	Diagnostic features	Metabolic disturbances	Neurological lesions
Galactosemia	Galactose-1-phosphate uridyltransferase	Cataract, hepato-megaly, jaundice,	Dehydration, hypo-glycemia	Vomiting, lethargy, seizures, coma, retardation
Fructose intolerance	Fructose-1-phosphate aldolase	Gastrointestinal reaction to fructose ingestion, hypoglycemia, fructosuria	Hypoglycemia, hypo-phosphatemia, hyper-uricemia	Vomiting, drowsiness, coma, retardation
Fructose-1,6-diphosphatase deficiency	Fructose-1,6-diphos-phatase	Severe hypoglycemia	Hypoglycemia, keto acidosis, lactic acidosis	Vomiting, hypotonia seizures, retardation
Lysosomal acid phosphatase deficiency	Lysosomal acid phos-phatase	—	Hypoglycemia, acidosis	Vomiting, lethargy, hypotonic convulsions, retardation
Total acid phosphatase deficiency	All cellular acid phosphatases	—	Acidosis	Vomiting, lethargy, opisthotonos
β-Xylosidase deficiency	Lysosomal β-xylosidase	—	—	Seizures, choreo-athetosis, retardation

Table taken from Y. E. Hsia, in *Basic Neurochemistry* (R. W. Albers, G. J. Siegel, R. Katzman, and B. W. Agranoff, eds.), Little, Brown, Boston, 1972.

lism, or by some interference with the normal transport of glucose into the brain. Studies with galactose-fed experimental animals have served as indicators of a sort and have led to some interesting, but somewhat controversial conclusions about the mechanism of the damage in this disease. A number of other conditions associated with lesions in monosaccharide metabolism are listed in Table 16.3.

Galactosemia

The patient, an 8-month-old boy, was admitted suffering from jaundice and lethargy. Upon physical examination the child was found to have bilateral lenticular cataracts, and appeared to have enlargement of the liver and the spleen. The child was unresponsive and hypotonic, vomited frequently, and had been losing weight for several months. The parents had no family history of such complaint, and the child was their first.

Laboratory studies revealed galactosuria, proteinuria, generalized amino aciduria, anemia, and hypoprothrominemia, and assay of a hemolyzate of the red blood cells indicated a complete absence of the enzyme galactose-1-phosphate uridyltransferase. The child was diagnosed as a galactosemic and was immediately put on a milk-free diet.

Within 3 days the jaundice disappeared, and the galactosuria also was no longer in evidence. The child then began to gain weight and the enlarged liver and spleen showed signs of retraction. Even after several months the cataract persisted.

The child developed normally from a physical standpoint through the fifth year, when the family left the area. Observational and formal testing procedures indicated severe retardation and the beginnings of psychologic and social maladjustment. Two subsequent pregnancies were monitored by amniocentesis, and the fetuses were found to be normal. These measurements were verified by the birth of two normal children [64].

The accumulation of galactitol in the tissues of galactosemic patients indicates the presence in various tissues and in brain of the enzyme which reduces the sugar to the sugar alcohol. The same enzyme presumably accounts for the appearance of sorbitol in the tissues of diabetic individuals. These two alcohols originate from galactose and fructose, respectively, and are formed from them by reduction.

Mucopolysaccharides

The class of acid polymers known as mucopolysaccharides have been investigated heavily in recent years [68-71]. These polymers are present to the extent of about 1.7 mg/g of fresh brain and are fairly evenly distributed in various brain

Table 16.4 Composition of the Common Mucopolysaccharides in Brain

Polymer	Composition
	Repeating disaccharide
Hyaluronic acid	$(1 \rightarrow 4)$D-glucuronic acid $(1 \rightarrow 3)$ N-acetyl-D-glucosamine
Chondroitin-4-sulfate	$(1 \rightarrow 4)$D-glucuronic acid $(1 \rightarrow 3)$ N-acetyl-D-galactosamine 4-sulfate
Chondroitin-6-sulfate	$(1 \rightarrow 4)$D-glucuronic acid $(1 \rightarrow 3)$ N-acetyl-D-galactosamine 6-sulfate
Dermatan sulfate	N-acetyl glucosamine, glucuronic acid, L-iduronic acid
Heparan sulfate	Largely a repeating unit of N-acetyl glucosamine and glucuronic acid

areas, but are found in higher concentration in the gray matter than in the white since they are not found in myelin. These polymers are also present in peripheral nerve. The mucopolysaccharides are made up of repeating disaccharide units each of which consists of a hexuronic acid and a hexosamine, the latter usually carrying an N-acetyl group. In some of the polymers the 4-hydroxyl or the 6-hydroxyl of the acetylated hexosamine is esterified with sulfate. The composition of the major brain mucopolysaccharides is shown in Table 16.4. Fundamentally there are three polymers which are well-established brain constituents. These are hyaluronic acid, chondroitin-4-sulfate, and chondroitin-6-sulfate. There are reports of trace amounts of a number of other materials, including dermatan sulfate, heparan sulfate, and a galacturonic acid polymer. These latter are normally found in very small amounts in brain, but, as we shall see, can be in much greater quantities in certain disease states. These polymers are usually isolated covalently bound to proteins. The complex is known as mucoprotein. Recent evidence suggests that the mucopolysaccharide is joined to the protein through a glycosidic bond to a hydroxyl group of one of the serines of the protein. It is clear that the polymers are extracellular in nature.

The Mucopolysaccharidoses

There are a number of disease states that are now known to involve alterations in the metabolism of the mucopolysaccharides and are collected under the term mucopolysaccharidoses. A listing of the currently recognized conditions in this category of aberrations with the enzymatic lesion thought to be responsible is found in Table 16.5.

Table 16.5 The Mucopolysaccharidoses

Syndrome	Enzymatic lesion
Hurler's	α-L-Iduronidase
Hunter's	Sulfoiduronate sulfatase
Scheie's	α-L-Iduronidase
Sanfilippo A	Heparan sulfate N-sulfatase
Sanfilippo B	N-Acetyl-α-glucosamidase
Monoteaux-Lamy's	A sulfatase
β-Glucuronidase deficiency	β-Glucuronidase

Table taken from E. F. Neufeld and M. J. Cantz, *Ann. N.Y. Acad. Sci. 179:*580 (1971); G. Bach, F. Eisenberg, Jr., M. Cantz, and E. F. Neufeld,*Proc. Natl. Acad. Sci. USA 70:*2134 (1973).

The most prevalent and the most thoroughly studied of these is the Hurler's syndrome (MPSI), alternatively known as gargoylism. The physical deformities and the severe mental retardation which are characteristic of this disease make it one of the most easily recognized and distressing of the diseases afflicting children. The disease is transmitted as an autosomal recessive and occurs in about 1 in 100,000 children. The incidence, then, is low and does not constitute a major health problem. In fact, the total number of children born with mucopolysaccharidoses of all kinds is about 100 per year. Nevertheless, as is so often the case, a great deal of information about normal brain chemistry has been obtained from the study of these relatively rare conditions.

Hurler's syndrome was first described in 1919 [72]. Affected children are short in stature and suffer severe retardation, bony abnormalities, corneal opacities, grotesque ("gargoyl-like") facial features, and progressive mental and physical deterioration. Biochemically, the disease is grossly characterized by large increases in the urinary excretion of mucopolysaccharides, and infiltration of brain and other tissues by mucopolysaccharide. The mucopolysaccharide found is not the normally occurring complement, but mostly a mixture of the low molecular weight polymers dermatan sulfate and heparan sulfate. These polymers accumulate in cells and spill over into the urine. They do not, however, occur in cerebrospinal fluid, perhaps because they are so highly charged. Normally, these materials are present in very low amounts in the brain, but in the Hurler's syndrome they can constitute up to 2% of the lipid-free weight of that tissue. In addition to the deposition and excretion of mucopolysaccharides, there are increases of up to 10-fold in ganglioside deposition in the white matter of the brain. This deposition is fairly common in various neurological conditions, probably is quite nonspecific, and certainly is not the primary lesion in Hurler's. The composition of the myelin is generally normal, but may be isolated in low yield.

The work of Neufeld and her group [71,73-76] has provided an enzymatic basis for the understanding of the chemical changes in the mucopolysaccharidoses. It is now clear that the lesions are in the normal degradative mechanism for the mucopolysaccharides, hence the accumulation. These degradative enzymes normally reside in the lysosomes, and there is, in each of these diseases, a lack of one of these lysosomal degradative functions. In the specific case of the Hurler's syndrome the missing enzyme is an iduronidase, the lack of which prevents the degradation of mucopolysaccharides.

Hurler's Syndrome

The patient, a 3-year-old boy, was admitted for hydrocephalus. He had grotesque facies with a prominent brow, flattened nose, widely spaced teeth, wide nostrils, and large and protruding lips and tongue. Clouding of both corneas was apparent. His hands were broad and stubby, and he exhibited flexion contractures of the elbows and the fingers. Upon detailed examination he was found to have enlarged spleen and liver and bilateral inguinal hernia. He was below the third percentile in height. He could not walk without support and could not speak in full sentences. His mental age was estimated to be approximately 1 year.

Laboratory studies revealed a 10-fold increase above normal of the urinary excretion of mucopolysaccharides. Both dermatan sulfate and heparan sulfate were elevated. Cultured skin fibroblasts from the patient showed an accumulation of dermatan sulfate, an abnormal metachromasia, and an absence of the enzyme iduronidase. The patient's blood cells contained an unusually high proportion of granules.

During hospitalization a ventriculoatrial shunt was performed to arrest the increasing hydrocephalus. Brainstem hemorrhage developed as a complication. The patient was discharged 3 months postoperatively in a decerebrate state, quadriplegic and akinetic. The patient was readmitted 1 year later for respiratory infection and died on the fourth day of cardiac failure.

The mitral valve was obtained at necropsy within 12 hr after death. Upon histochemical investigation and electron microscope examination the cells were found to be distended and crowded with swollen collagen fibers. Two kinds of deposits were seen histochemically in the cells. Acid mucopolysaccharide was identified in some cells, and other cells were seen to have deposits of ganglioside [64].

A second and related condition, the Hunter syndrome (MPSII) [77], was for a long time considered to be merely a less severe variant of Hurler's. Indeed, the two conditions are clinically very much alike. Recent work, however, has clearly shown that the basic enzymatic lesion in the two diseases is different. Also, it is now clear that the conditions are genetically distinct, the Hurler's being an autosomal recessive, the Hunter's an X-linked recessive.

Both diseases are defects in the lysosomal mechanism for the degradation
of the mucopolysaccharides, and in both there is an abnormal build-up of the
dermatan and heparan sulfates. But the missing enzyme in the Hunter's syndrome
is a sulfatase acting on iduronic acid sulfate, an enzyme quite distinct from the
iduronidase missing in the Hurler's syndrome.

The definitive finding in the differentiation of these two conditions, and
in the classification of the mucopolysaccharidoses in general, was the observation
that cultured skin fibroblasts from individuals with the various diseases can correct in vitro. That is, when cells from either Hurler's or Hunter's patients are
grown in culture they reflect in vitro the condition of the patient in that they
accumulate mucopolysaccharides. If radioactive precursors are included in the
growth medium, the mucopolysaccharides become labeled. The rate at which
these labeled polymers are degraded in the cells is much lower than in normal
cells. If concentrates of normal urine or extracts from normal cultured fibroblasts are included in the medium, the turnover is much faster. Comparable extracts from Hunter's cells, placed in Hunter's culture do not alter the turnover
time. However, extracts from Hunter's cells placed in Hurler's culture do alter
the turnover time. In fact, when the two cell types are cocultured on the same
plate, both lines grow normally and do not accumulate mucopolysaccharide.
That is, one cell type complements or corrects the defect in the other cell type.
The simplest explanation of these data are that corrective factors are present in,
and are secreted by, one cell type and taken up by the other. When the corrective factors were purified and studied it turned out that these corrective factors
were, in fact, the enzymes missing in the two states [73,74]. The corrective factor secreted by the Hunter cells, which allowed the Hurler cells to grow normally
in culture, was the iduronidase genetically deleted in the Hurler patient. The use
of this in vitro curative technique has allowed the differentiation and classification of the mucopolysaccharidoses.

The Hunter's syndrome is less severe than Hurler's. Phenotypically, the
children appear quite similar. They have enlarged heads and facial abnormalities
not unlike those in Hurler's. There is, however, no evidence for corneal clouding
in these children. Hernia is common, as is hydrocephalus. The overall growth
rate is lower than normal, but not so much lower as in the Hurler's children.
Neurologically, the period of normal development is longer than in the Hurler's
children. Most of the patients learn to speak in simple sentences, and toilet
training-level learning is possible. Seizures are sometimes seen. Pathological
changes in these individuals are indistinguishable from those in the Hurler's.
Death often occurs at between 8 and 15 years of age from pulmonary infection
or as a complication of the seizures, but some patients have survived to age 50
[64].

The other of the mucopolysaccharidoses which is associated with retardation is Sanfilippo's syndrome (MPSIII). There are two recognized variants of this condition, Sanfilippo A and Sanfilippo B. These are clinically very difficult to distinguish. The enzymatic basis, however, has been shown to be different by the tissue culture assay technique described above. The lesion in Sanfilippo A seems to be the absence of an N-sulfatase, and in Sanfilippo B, the deletion of an N-acetylglucosaminidase. The conditions are characterized by a urinary excretion of mucopolysaccharides, mostly heparan sulfate. The concentration of mucopolysaccharides in the brain is normal, but the composition is very different, being mostly heparan sulfate. There is also an increase in the concentration of brain gangliosides. The children with Sanfilippo's syndrome have what appears to be a moderate Hurler's. Their heads are slightly enlarged, their noses slightly flattened. There is no evidence of corneal clouding, and the liver and spleen are only slightly enlarged. Mental development is normal up to between 1 and 4 years of age. After that time the child becomes difficult to control, speech deteriorates, and physical control is lost. Ultimately the children are bedridden and demented. The life span appears relatively normal. Genetic studies show that Sanfilippo's syndrome behaves like an autosomal recessive.

There are a number of other conditions in which mucopolysaccharides accumulate. Among these are the lipidosis, metachromatic leukodystrophy, which is characterized clinically by a progressive paralysis and dementia accompanied by demyelination, and Alzheimer's disease, a progressive dementia involving neuronal loss. In these conditions the accumulation does not appear to be the primary lesion. Discussion of these conditions is reserved to a later chapter.

I-cell disease (mucolipidosis II) is quite a different proposition, although also involving the accumulation of mucopolysaccharides, and presenting a clinical picture much like that of the mucopolysaccharidoses. This condition is a lethal autosomal recessive which is characterized by psychomotor retardation and physical abnormalities. It derives its name from the presence of cellular "inclusion" bodies which are, in fact, lysosomes filled with debris and membrane fragments. Apparently this condition involves a defect in the structure of the lysosomes which causes a release of a large number of the normally lysosome-bound hydrolytic enzymes and, of course, prevents their normal functioning. This leads to the accumulation of the polymeric materials they would normally destroy [78].

References

1. H. S. Marker and G. M. Lehrer, in *Basic Neurochemistry* (R. W. Albers, G. J. Siegel, R. Katzman, and B. W. Agranoff, eds.), Little, Brown, Boston, 1972.
2. R. Balazs, in *Handbook of Neurochemistry,* Vol. III (A. Lajtha, ed.), Plenum, New York, 1970.
3. H. S. Bachelard, in *Handbook of Neurochemistry,* Vol. I (A. Lajtha, ed.), Plenum, New York, 1969.

4. H. S. Bachelard, in *Handbook of Neurochemistry,* Vol. IV (A. Lajtha, ed.), Plenum, New York, 1970.

5. O. H. Lowry and J. V. Passonneau, *J. Biol. Chem. 239:*31 (1964).

6. O. H. Lowry, J. V. Passonneau, F. X. Hasselberger, and D. W. Schultz, *J. Biol. Chem. 239:*18 (1964).

7. H. S. Bachelard, *Biochem. J. 104:*286 (1967).

8. K. D. Neame, in *Handbook of Neurochemistry,* Vol. IV (A. Lajtha, ed.), Plenum, New York, 1970.

9. C. Crone, *J. Physiol. (London) 181:*103 (1965).

10. C. R. Park, L. H. Johnson, J. H. Wright, and H. Batsel, *Am. J. Physiol. 191:*13 (1957).

11. B. S. Meldrum and R. W. Horton, *Electroencephalogr. Clin. Neurophysiol. 35:*59 (1973).

12. I. Diamond and R. A. Fishman, *J. Neurochem. 20:*1533 (1973).

13. G. M. Heaton and H. S. Bachelard, *J. Neurochem. 21:*1099 (1973).

14. R. K. Crane and A. Sols, *J. Biol. Chem. 203:*273 (1953).

15. M. K. Johnson, *Biochem. J. 77:*610 (1960).

16. J. E. Wilson, *Biochem. Biophys. Res. Commun. 28:*123 (1967).

17. M. F. Thompson and H. S. Bachelard, *Biochem. J. 118:*25 (1970).

18. E. W. Kellogg, H. R. Knull, and J. E. Wilson, *J. Neurochem. 22:*461 (1974).

19. I. A. Rose and J. U. B. Warms, *J. Biol. Chem. 242:*1635 (1967).

20. J. E. Wilson, *J. Biol. Chem. 243:*3640 (1968).

21. H. R. Knull, W. F. Taylor, and W. W. Wells, *J. Biol. Chem. 248:*5414 (1973).

22. R. W. Horton, B. S. Meldrum, and H. S. Bachelard, *J. Neurochem. 21:*507 (1973).

23. O. H. Lowry and J. V. Passonneau, *J. Biol. Chem. 239:*31 (1964).

24. O. H. Lowry and J. V. Passonneau, *J. Biol. Chem. 241:*2268 (1966).

25. M. Tsai and R. G. Kemp, *J. Biol. Chem. 248:*785 (1973).

26. M. Y. Tsai and R. G. Kemp, *J. Biol. Chem. 249:*6590 (1974).

27. T. E. Duffy, S. R. Nelson, and O. H. Lowry, *J. Neurochem. 19:*959 (1972).

28. H. S. Bachelard, L. D. Lewis, U. Ponten, and B. K. Siesjö, *J. Neurochem. 22:*395 (1974).

29. J. S. Nisselbaum, D. E. Packer, and O. Bodansky, *J. Biol. Chem. 239:*2830 (1964).

30. T. Kato and O. H. Lowry, *J. Neurochem. 20:*151 (1973).

31. W. Sacks, *J. Appl. Physiol. 10:*37 (1957).

32. W. Sacks, *J. Appl. Physiol. 20:*119 (1965).

33. D. B. McDougal, D. W. Schultz, J. V. Passonneau, J. R. Clark, M. A. Reynolds, and O. H. Lowry, *J. Gen. Physiol. 44:*487 (1961).

34. S. S. Hotta, *J. Neurochem. 9:*43 (1962).

35. E. Giacobini and J. F. Jongkind, *Acta Physiol. Scand. 73:*255 (1968).

36. H. Kimura, K. Naito, K. Nakayawa, and K. Kuriyama, *J. Neurochem. 23:*79 (1974).

37. M. M. Cohen, in *Monographs in Neural Sciences,* Vol. 1 (M. M. Cohen, ed.), Karger, Basel, 1973.

38. G. Takagaki, *J. Neurochem. 23:*479 (1974).

39. O. J. Rafaelsen, *J. Neurochem. 1:*45 (1961).

40. S. R. Nelson, D. W. Schultz, J. V. Passoneau, and O. H. Lowry, *J. Neurochem. 15:*1271 (1968).

41. S. Eisenberg and H. S. Seltzer, *Metabolism 11:*1162 (1962).

42. L. J. King, O. H. Lowry, J. V. Passonneau, and V. Venson, *J. Neurochem. 14:*599 (1967).

43. J. L. Carl and L. J. King, *J. Neurochem. 17:*293 (1970).

44. H. S. Bachelard, L. D. Lewis, U. Ponten, and B. K. Siesjö, *J. Neurochem. 22:*395 (1974).

45. R. V. Coxon, in *Handbook of Neurochemistry,* Vol. III (A. Lajtha, ed.), Plenum, New York, 1970.

46. S. R. Nahorski and K. J. Rogers, *J. Neurochem. 23:*579 (1974).

47. O. H. Lowry and J. V. Passonneau, *J. Biol. Chem. 244:*910 (1969).

48. H. Watanabe and J. V. Passonneau, *J. Neurochem. 20:*1453 (1973).

49. J. V. Passonneau, E. A. Brunner, C. Molstad, and R. Passonneau, *J. Neurochem. 18:*2317 (1971).

50. J. H. Austin, in *Handbook of Neurochemistry,* Vol. VII (A. Lajtha, ed.), Plenum, New York, 1972.

51. P. Garfield, O. Lowry, D. Schultz, and J. Passonneau, *Neurochem. 13:*185 (1966).

52. T. Fujita, *Kobe J. Med. Sci. 13:*39 (1967).

53. S. E. Kerr and M. Ghantus, *J. Biol. Chem. 116:*9 (1936).

54. J. Folbergrova, *J. Neurochem. 20:*547 (1973).

55. R. L. Friede, *Exp. Neurol. 5:*89 (1962).

56. L. Guth and P. K. Watson, *Exp. Neurol. 22:*590 (1968).

57. L. Guth and P. J. Dempsey, *Exp. Neurol. 29:*152 (1970).

58. E. A. Brunner, J. V. Passonneau, and C. Molstead, *J. Neurochem. 18:*2301 (1971).

59. J. Folbergrova, O. H. Lowry, and J. V. Passonneau, *J. Neurochem. 17:*1155 (1970).

60. O. J. Rafaelsen, P. Plenge, and E. T. Mellerup, *Proc. Second Int. Mtg. Int. Soc. Neurochem.,* Milan, 1969.

61. C. J. Estler and H. P. T. Ammon, *J. Neurochem. 14:*799 (1967).

62. D. A. Hutchins and K. J. Rogers, *Br. J. Pharmacol. Chemother. 39:*9 (1970).

63. Y. E. Hsia, in *Basic Neurochemistry* (R. W. Albers, G. J. Siegel, R. Katzman, and B. W. Agranoff, eds.), Little, Brown, Boston, 1972.

64. L. B. Holmes, H. W. Moser, S. Halldorsson, C. Mack, S. S. Pant, and B. Matzilevich, *Mental Retardation,* MacMillan, New York, 1972.

65. L. F. Leloir, *Arch. Biochem. Biophys. 33:*186 (1951).

66. H. M. Kalckar, E. P. Anderson, and K. J. Isselbacher, *Biochim. Biophys. Acta 20:*262 (1956).

67. A. N. Weinberg, *Metabolism 10:*728 (1961).

68. R. U. Margolis, in *Handbook of Neurochemistry,* Vol. I (A. Lajtha, ed.), Plenum, New York, 1969.

69. K. Suzuki, in *Handbook of Neurochemistry*, Vol. VII (A. Lajtha, ed.), Plenum, New York, 1972.

70. V. A. McKusick, D. Kaplan, D. Wise, W. B. Hanley, S. B. Suddarth, M. E. Sevick, and A. E. Maumanee, *Medicine 44:*455 (1965).

71. E. F. Neufeld and M. J. Cantz, *Ann. N.Y. Acad. Sci. 179:*580 (1971).

72. G. Hurler, *Z. Kinderheilkd. 24:*220 (1919).

73. G. Bach, F. Eisenberg, Jr., M. Cantz, and E. F. Neufeld, *Proc. Natl. Acad. Sci. USA 70:*2134 (1973).

74. G. Bach, R. Friedmann, B. Weissman, and E. F. Neufeld, *Proc. Natl. Acad. Sci. USA 69:*2048 (1972).

75. H. Kresse and E. F. Neufeld, *J. Biol. Chem. 247:*2164 (1972).

76. M. Cantz, A. Chrambach, G. Bach, and E. F. Neufeld, *J. Biol. Chem. 247:* 5456 (1972).

77. C. Hunter, *Proc. R. Soc. Med. 10:*104 (1917).

78. E. F. Neufeld, *Proc. 9th Int. Cong. Biochem.*, Stockholm, 1973.

17
LIPIDS

Content and Composition

The lipid content of the brain is higher than that of any other tissue in the body, except for adipose tissue itself. In fact, about one-half of the dry weight of the brain is lipid. The high lipid content is due in large part to the presence, in mature brain, of the lipid-rich myelin sheath, which contains 30% of the brain lipid. This fact is illustrated by the finding that the lipid composition and content of the gray and the white matter of the brain are quite different. The lipid content of the white is, of course, much higher than that of the gray, on the order of 2 to 3 times, depending on the basis upon which it is calculated. In the immature brain, i.e., before myelination, the lipid content and composition of the gray and the intended "white" areas are the same. Thus, the regional differences, overall high lipid content, and unique lipid composition of the brain are due to the presence of myelin, and they appear during myelination.

The lipid composition of the brain is unique among the tissues, and the high concentration of lipid is undoubtedly crucial to the adequate functioning of the brain. Experimentally, the high lipid content of the brain influences many of the procedures used in the study of the tissue, e.g., the isolation of proteins from the brain, and special techniques must be employed in most cases. Indeed, many of the nonlipid materials isolated from the brain contain lipid in chemical or physical linkage.

The major lipid constituents of the brain are listed in Table 17.1. Basically there are three categories of lipid in the brain. These are the glycerophospholipids, the sphingolipids, and cholesterol. Since the lipid composition of brain changes gradually with age, and markedly with the onset of myelination, it is difficult to give typical values for the amounts of the various lipids. Nevertheless, the table gives the average values, derived from the formulations presented by Rouser and Yamamoto [1], for each of the major lipids in adult, male human brain. The overall lipid content, as a percentage of the fresh weight, is less than

Table 17.1   The Major Lipid Constituents of Human Whole Brain (Male, Age 20)

Constituent	Content (mmol/100 g fresh weight)
Total phospholipid	6.5
Phosphatidylethanolamine	2.5
Phosphatidylcholine	2.2
Cerebroside	2.0
Plasmalogen	1.5
Phosphatidylserine	1.3
Sphingomyelin	0.3
Sulfatide	0.2
Phosphatidylinositol	0.15
Ganglioside	0.1
Phosphatidic acid	0.05
Cholesterol	6.0

Table adapted from G. Rouser and A. Yamamoto, in *Handbook of Neurochemistry,* Vol. 1 (A. Lajtha, ed.), Plenum, New York, 1969.

4% at birth, rises to 11 to 12% at age 30 (or to about 50% of the dry weight), and drops to 8 to 9% by age 60. In addition to the components listed in the table, the brain contains a number of minor lipids, some of which are as yet uncharacterized. The brain contains little or no triglyceride, little or no free fatty acid, and, in the adult, very little if any cholesterol esters.

As mentioned above, the lipid composition of the white and the gray matter is quite different. This is due to the lipid content and composition of the myelin. The comparison between white and gray is shown in Table 17.2 [2].

Most of the lipid in the brain, and in other tissues as well, is contained in the cellular membranes. Each subcellular organelle of the brain has a unique lipid composition, but the various organelles such as the mitochondria or the nuclei from brain have lipid compositions similar to mitochondria or nuclei from other organs and from the brains of other species. In most instances, only the fatty acid composition of the various lipid classes vary among the different species. However, the plasma membrane and, consequently, the myelin sheath of brain cells, has a different lipid composition from membranes of cells from other organs. In addition, the myelins from different species differ in their lipid compositions.

Fatty Acids

Before discussing the complex lipids, a few words about the fatty acids in these lipids are appropriate. As mentioned before, the content of free fatty acids in

Table 17.2 A Comparison of the Lipid Composition of
White and Gray Matter of Normal Adult Human Brain

	Gray	White
Constituent	(% fresh weight)	
Total lipid	5.9	15.6
Cholesterol	1.3	4.3
Total phospholipid	4.1	7.2
Phosphatidylethanolamine	1.3	2.3
Phosphatidylcholine	1.6	2.0
Sphingomyelin	0.4	1.2
Phosphatidylinositol	0.16	0.14
Phosphatidylserine	0.5	1.2
Plasmalogen	0.7	1.8
Cerebroside	0.3	3.1
Sulfatide	0.1	0.9
Ganglioside	0.3	0.05

Table adapted from K. Suzuki, in *Basic Neurochemistry*, (R. W.
Albers, G. J. Siegel, R. Katzman, and B. W. Agranoff, eds.),
Little, Brown, Boston, 1972.

nervous tissue is very low, less than 1% of the dry weight of the tissue. Nevertheless, the brain has a substantial capacity to synthesize fatty acids from acetate.
Such synthesis can be readily demonstrated in brain preparations in vitro [3].
In the complex lipids the fatty acids are usually 16 to 24 carbons long, often unsaturated, and sometimes hydroxylated. As might be expected, the overall fatty
acid composition of the lipids in the brain changes with myelination, due to increases in the amount of long-chain saturated fatty acids [4]. Brain has a limited
ability to degrade fatty acids as well. It is clear, however, that fatty acids normally do not serve as a source of energy for the brain. In fact, the injection of
short-chain fatty acids into brain has a narcotic effect. The synthesis of fatty
acids increases during the period of myelination [5], and this increase can be
correlated with increases in brain acetyl coenzyme A (CoA) carboxylase, the enzyme which appears to be rate limiting for the synthesis of fatty acids in brain
and other tissues [6].

Some fatty acids can be incorporated into the brain from the diet. The
brain is relatively more resistant to such incorporation than are other tissues.
There is evidence however, that the fatty acid composition of brain can be altered by altering the diet, although there are no known behavioral or pathological consequences of such an alteration. The composition of the fatty acids of
the brain is also fairly resistant to starvation. There are reports that the compo-

sition of the fatty acids of the brain is constant during such stress, while the composition and content of fatty acids in the liver and the muscle does change [7]. Even specific dietary deficiencies such as a lack of dietary essential fatty acids does not alter the content or composition of the brain fatty acids [8,9]. However, other studies indicate that acute undernutrition, whether caused by limitation of food intake [10] or by severe vitamin deficiency [11,12], can cause changes in the fatty acid pattern of the brain and, indeed, inhibition of de novo fatty acid synthesis.

There are few conditions known which in humans have neurological involvement and appear to have their metabolic basis in faulty fatty acid metabolism. One is the syndrome of Sidbury and coworkers [13], which is hereditary and is characterized by lethargy and seizures. Clinically the disease involves acidosis and dehydration. An associated and distinctive body odor is caused by an accumulation of butyric and hexanoic acids in body fluids. Death usually occurs within the first few weeks after birth. The exact metabolic lesion is not known.

Refsum's disease is somewhat more completely characterized [14]. This condition is also hereditary, and is accompanied by involvement of the nervous system. Generally associated with this condition are signs of peripheral neuritis, retinitis pigmentosa, deafness, and cerebellar ataxia. The accumulation of phytanic acid (3,7,11,15-tetramethylhexadecanoic acid) (Fig. 17.1) in tissue is the distinctive feature here and appears to be caused by a defect in the oxidative metabolism of branched-chain fatty acids. A complicating observation in this study, and one which makes the connection between the phytanic acid accumulation and the neurological lesions seem quite tenuous, is the recorded occurrence of Refsum's syndrome [15], a condition similar to Refsum's disease, but lacking in the phytanic acid accumulation.

Phospholipids

The glycerophospholipids of nervous tissue include phosphatidic acid, the ethanolamine and serine phosphatides (cephalins), the choline phosphatides (lecithin), the plasmalogens, the glyceryl ether phosphatides of serine and ethanolamine, and the inositol phosphatides. Although the biosynthesis and degradation of these compounds will be outlined briefly below, it is appropriate to point out

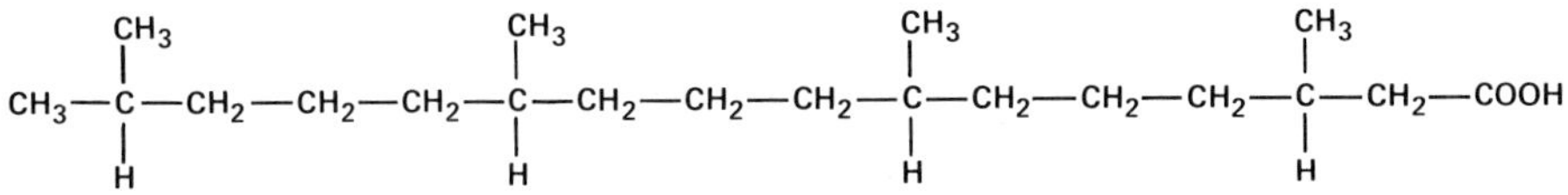

$$CH_3-\underset{\underset{H}{|}}{\overset{\overset{CH_3}{|}}{C}}-CH_2-CH_2-CH_2-\underset{\underset{H}{|}}{\overset{\overset{CH_3}{|}}{C}}-CH_2-CH_2-CH_2-\underset{\underset{H}{|}}{\overset{\overset{CH_3}{|}}{C}}-CH_2-CH_2-CH_2-\underset{\underset{H}{|}}{\overset{\overset{CH_3}{|}}{C}}-CH_2-COOH$$

Figure 17.1 Phytanic acid.

here that the pathways by which the major phospholipids are made and broken
down in brain are similar, if not identical, to those found in other tissues. The
biochemistry of the minor constituents, such as the plasmalogens and the glyceryl ether phosphatides, is as yet somewhat obscure.

The biosynthesis of phosphatidic acid is outlined in Fig. 17.2 [16]. There
are, in fact, two routes to the synthesis of this pivotal intermediate, the route
from the glycolytic intermediate, glyceraldehyde-3-phosphate, the so-called Korn-
berg and Pricer pathway, has been known for a number of years. The alternate
route from diglyceride and ATP was shown somewhat more recently by Hokin
and Hokin, and will be discussed below since the implications for brain functioning are quite substantial. Both pathways are found in brain, but the ability of
the latter pathway to respond to neuroactive materials is of particular interest to
the discussion here.

The biosynthesis of the choline, ethanoline, and serine phospholipids is
shown in Fig. 17.3 [16]. Again, as with the routes to phosphatidic acid, these
pathways are generally the same as those found in liver and other tissues. There
may be differences in the quantitative importance of each in brain as compared
to liver. For example, overall there is very little interconversion of the phospholipids in brain compared to those of liver. Primarily, choline is incorporated directly into phosphatidylcholine (lecithin) and ethanolamine directly into phosphatidylethanolamine. One qualitative difference between brain and liver seems
to be that there are no methylation reactions in brain for the conversion of phosphatidylethanolamine to phosphatidylcholine.

Phosphatidylcholine makes up more than 5% of the dry weight of the adult
human brain. The fatty acid composition of the lecithin depends upon the site
in the brain from which it is isolated. The major fatty acids are palmitic and
oleic acids, but others are present in lesser amounts. As a general rule, the saturated fatty acids are in the β position of the glyceride. It is known that the lecithin of myelin is different than that in other fractions of the brain, being richer
in oleic and stearic acids and poorer in palmitic, but no functional information
has become available from these structural observations.

Phosphatidylethanolamine and phosphatidylserine (the cephalins) have
fatty acid compositions which are quite distinct from the fatty acids found in
the lecithin present at the same site. Generally, in the cephalins, stearic and
oleic acids predominate. Ethanolamine cephalin is present in larger amounts
than serine cephalin. Lecithin and the cephalins are present in all membranes,
not only in myelin. Accordingly, the rise in the concentration of these lipids
during myelination is quite moderate.

Three inositol-containing phospholipids are found in the brain, the monophosphoinositides (MPI), diphosphoinositides (DPI), and triphosphoinositides
(TPI) [17]. These all contain L-myoinositol, as shown in Fig. 17.4. The figure
also presents the biosynthetic route to these compounds in brain and other

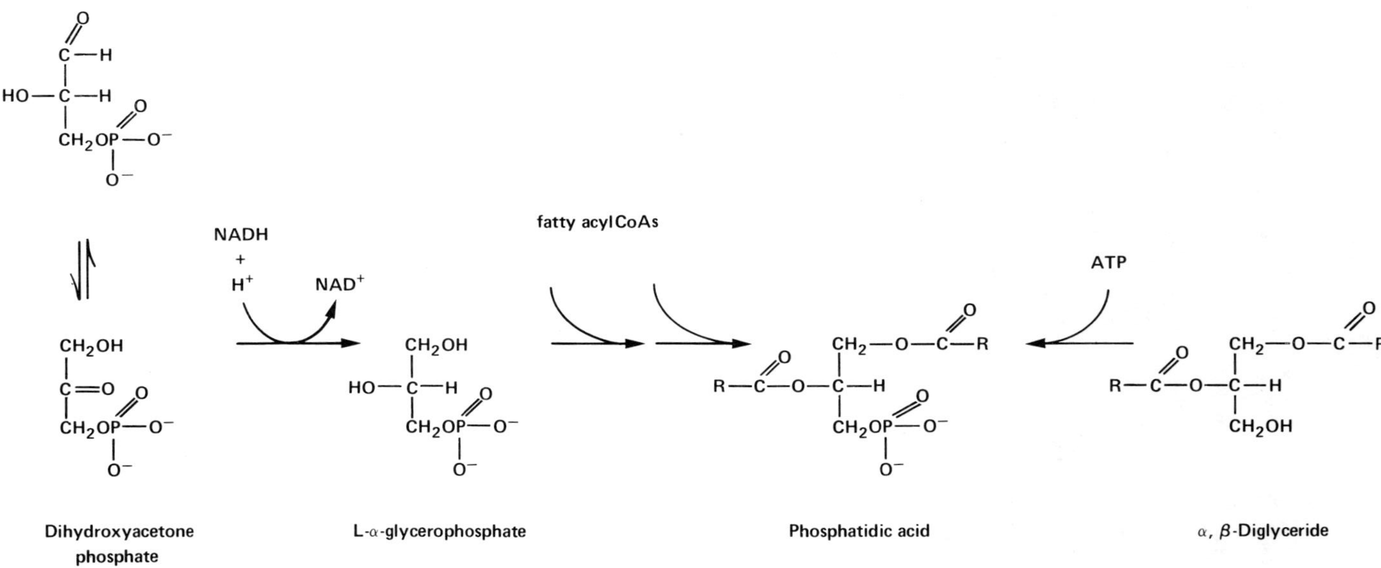

Figure 17.2   Biosynthesis of phosphatidic acid.

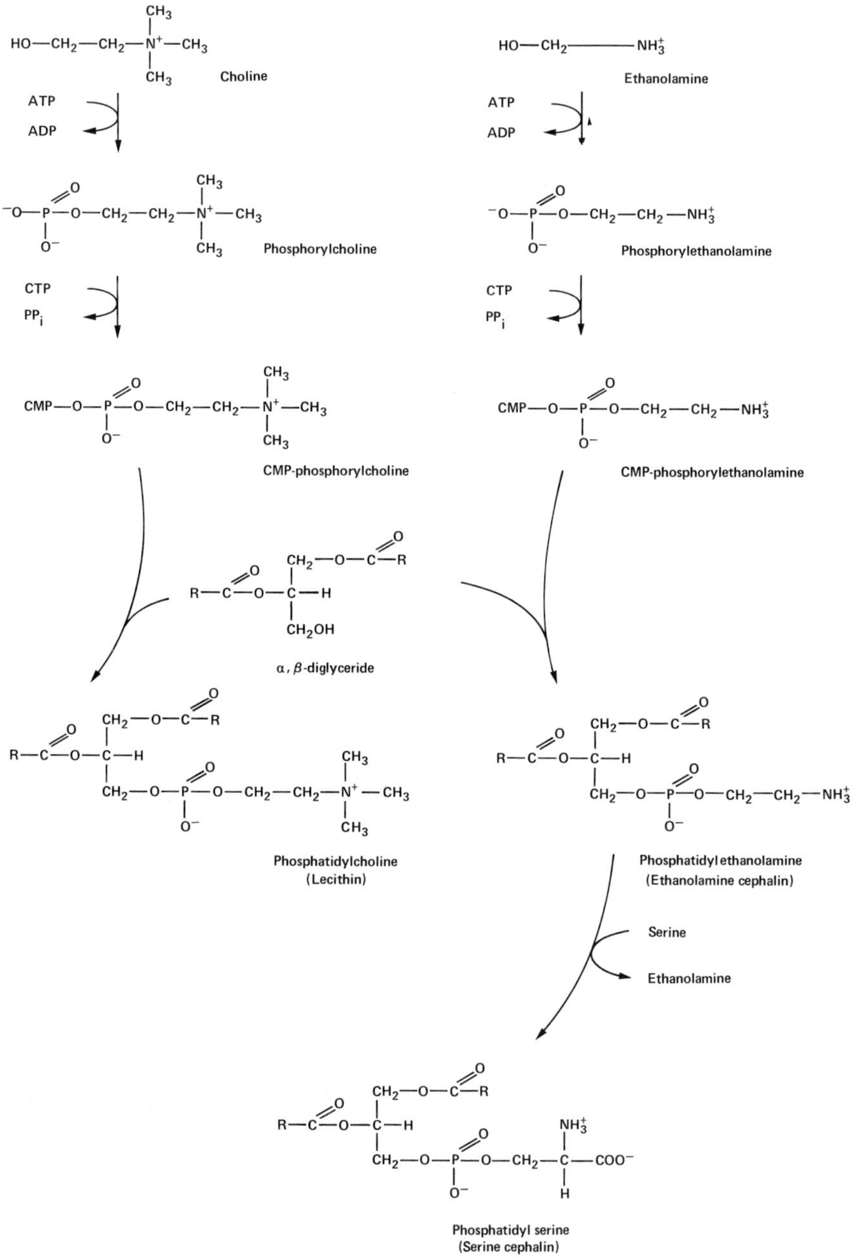

Figure 17.3   Biosynthesis of lecithin and the cephalins.

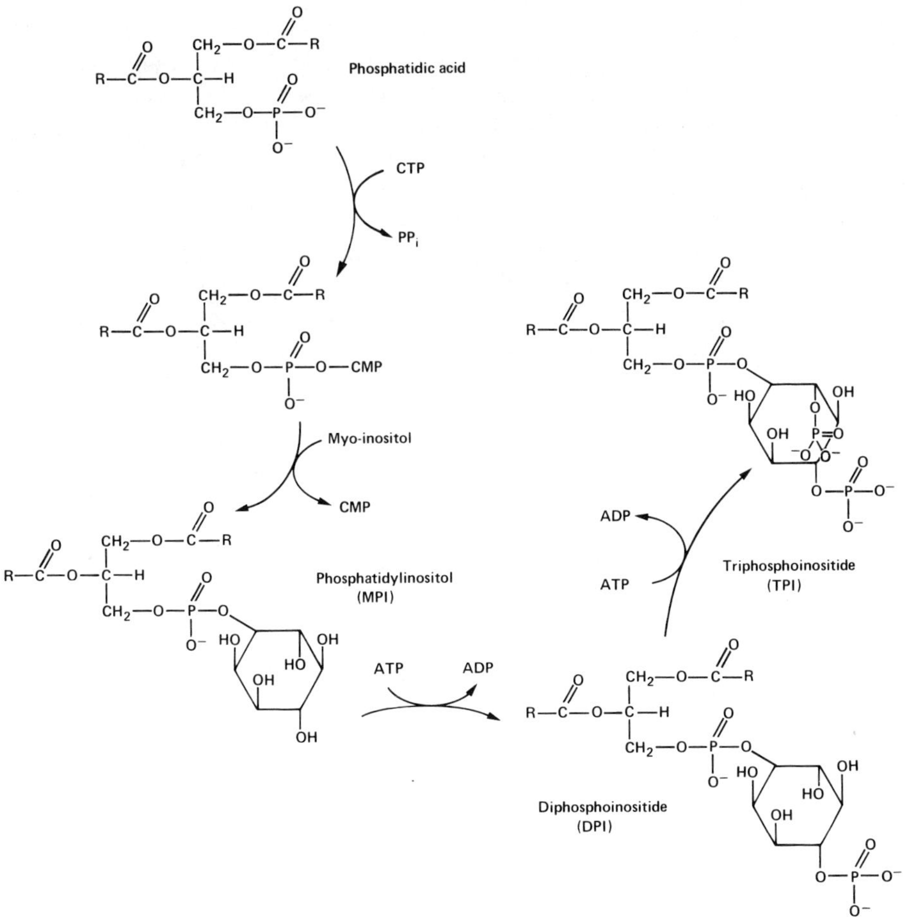

Figure 17.4   Biosynthesis of the phosphoinositides.

tissues. A tetraphosphate has been reported, but its existence is not generally
accepted. Some inositol is made from glucose in the brain, but most apparently
comes from dietary sources. The content of free inositol in the brain is on the
order of 7 μmol/g of fresh tissue, and the content in cerbrospinal fluid is quite
high. The total inositol in the brain, both free and bound into phospholipids,
is about 9 μmol/g so most of the inositol is found free. In many species the three
phosphoinositides occur within the range of 0 to 100, 0 to 10, and 25 to 60 μg/g
of tissue, respectively. The turnover of the phosphoinositides is the most rapid

of all the phospholipids in brain, a point which will be covered in some detail below. It is felt that the diphosphoinositides and triphosphoinositides are myelin constituents. The triphosphoinositide binds calcium ion more avidly than does ethylenediaminetetraacetate (EDTA). The unique metabolic and structural properties of the phosphoinositides have led to several theories about their possible function in the brain. One hypothesis is that they may have a role in the generation of the action potential of the neuron, a theory which is in some sense supported by the rapid turnover of these compounds in vivo [17].

The structures of the glyceryl ether phospholipids and the plasmalogens, the remaining well-characterized phospholipid components of the brain, are presented in Fig. 17.5. The brain is the richest source of these lipids. The glyceryl ether phospholipids make up only about 3% of the lipid phosphorus in the brain, whereas the plasmalogens are much more abundant, constituting between 18 and 30% of the cerebral phospholipids in various species. Relatively little is known about the biochemistry of the glyceryl ether phospholipids. More is known about the plasmalogens, including the observations that they are present primarily in myelin and generally contain ethanolamine as the nitrogenous component. The fatty acid portion is largely saturated and usually 18 carbons long, but also contains substantial amounts of 20 and 22 carbon acids. There is some information about the synthesis of plasmalogens in brain, and some data are available about "plasmalogenase," the initial degradative enzyme for these compounds [18]. The overall metabolic turnover of the ethanolamine plasmalogens is about the slowest of all the phospholipids discussed so far [19], as might be expected from its known localization in the myelin.

Glyceryl ether

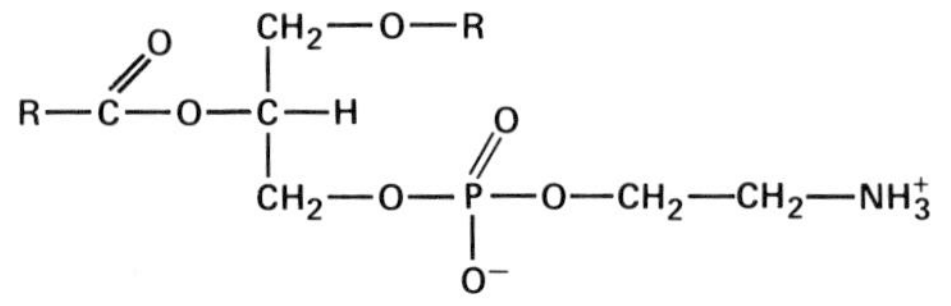

$$R-C(=O)-O-C(H)(CH_2-O-R)-CH_2-O-P(=O)(O^-)-O-CH_2-CH_2-NH_3^+$$

Plasmalogen

$$R-C(=O)-O-C(H)(CH_2-O-CH=CH-R)-CH_2-O-P(=O)(O^-)-O-CH_2-CH_2-NH_3^+$$

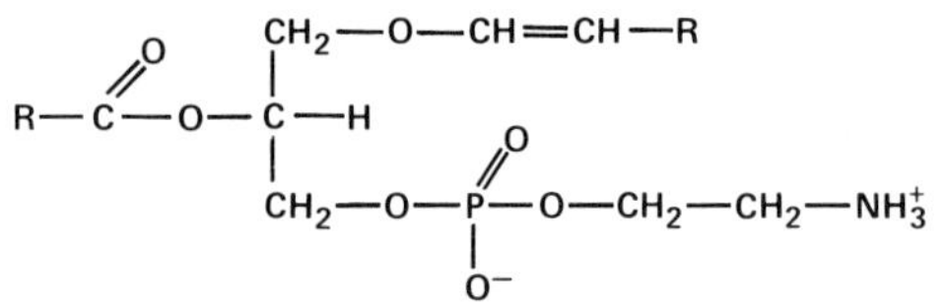

Figure 17.5   Structure of the glyceryl ethers and the plasmalogens.

The phospholipid-degrading enzymes, the phospholipases, act in the same manner on most, if not all of the phospholipids mentioned above. They seem to be nonspecific with regard to various substituents, i.e., choline, ethanolamine, serine, or inositol. Also, the nature of the fatty acid residues which esterify the remaining alcohol functions of glycerol seem to play little part in determining the rate of attack by these enzymes. There are at least four phospholipases known, and they are designated phospholipases A through D. More recently, the phospholipase A activity of brain has been fractionated into A_1 and A_2. Phospholipase A_1 is present in microsomes, and A_2 is present in mitochondria and is in much higher amounts in neurons than in glia [20,21]. All the phospholipases are calcium requiring, and all are fairly nonspecific. They will not act on the glyceryl ether bond, or on the vinyl ether bond found in the plasmalogens, but they will degrade the major phospholipids found in the brain. The nature of the action of these enzymes is indicated in Fig. 17.6.

It should be mentioned that phospholipase activities A and B are widely distributed in most tissues of animals. Phospholipase C is a bacterial enzyme but occurs in mammalian tissues also. Phospholipase D is primarily found in plants. Specifically, in brain both phospholipase A_1 and A_2 have been found. Phospholipase B also is known to be present in brain and to be in higher levels in gray matter than in white. Weak phospholipase C activity has also been observed in brain. It is of interest to note that phospholipases A_1 and A_2 activities increase markedly during degenerative conditions which affect the nerve, such as Wallerian degeneration, or the acute phase of degeneration associated with experimental allergic encephalomyelitis [20,21]. Such changes in response to changing conditions indicate the active nature of the role played by these catabolic enzymes.

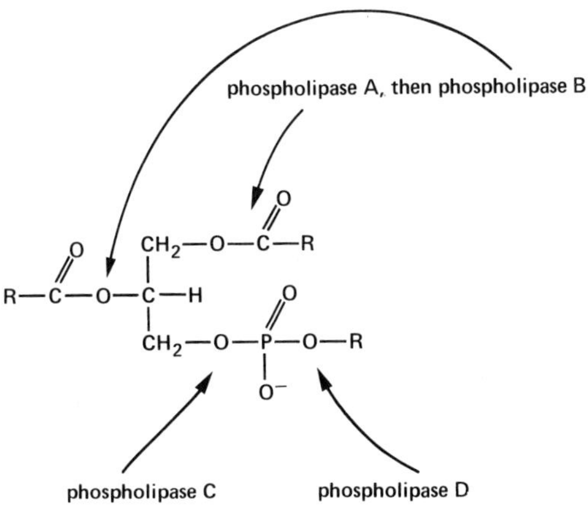

Figure 17.6   Degradation of the phospholipids.

The glyceride phospholipids are membrane constituents, and the metabolic turnover of some of them, notably phosphatidic acid and phosphomonoinositide, is quite rapid. Some few years ago a series of investigations by Hokin and Hokin culminated in the observation that when cation transport through cell membranes is stimulated, phosphate incorporation into phosphatidic acid and phosphomonoinositide is markedly increased [22,23]. Thus, when brain slices were incubated with acetylcholine and eserine, ^{32}P incorporation into these phospholipids is enhanced. This increase is due to a stimulation of the diglyceride kinase route of phosphatidic acid synthesis (Fig. 17.2) [24]. A large body of experimental work has been done on this phenomenon, which is now known as the "phospholipid effect." That is, chemical and physical stimuli, e.g., neurohormones, neurotransmitters, electrical impulses, increase ^{32}P incorporation into phosphatidic acid and phosphomonoinositide in secretory and nervous tissue. Although the exact role of this turnover in overall cellular function is not yet known, it is obviously related to membrane events. It is clear that, although cholinergic agents stimulate ^{32}P incorporation into synaptosomes in vivo [25] and in vitro [26], the effects of acetylcholine in stimulating the phosphate incorporation into the phospholipids of brain slices, as well as the effects of the other neuroactive substances, norepinephrine, dopamine, histamine, and serotonin, are general effects, in that they occur in all types of cells, glia as well as neurons [27]. Thus, it is considered unlikely that they are related solely to synaptic transmission, but are thought to be involved in more general aspects of membrane function. It was suggested by Hokin and by others that the phospholipids were involved in, or perhaps were the actual carrier molecules for, the transport of sodium ions out of the cell by the fundamental ion transport mechanism of all cells, the sodium pump. This is no longer considered likely, but it does seem clear that the rapidly turning over phosphatides, phosphatidic acid and phosphomonoinositide, are intimately involved in the membrane events initiated by acetylcholine and other neuroactive substances.

Sphingolipids

The sphingolipids include sphingomyelin and the glycolipids. The glycolipids can be further subdivided into the cerebrosides, the sulfatides, and the gangliosides. All contain sphingosine, the structure of which was determined by Carter et al. in 1947 [28] and the biosynthesis of which was worked out by Brady and his coworkers [29] and is illustrated in Fig. 17.7. Sphingosine forms the basic structure for all the sphingolipids, which are built by addition to the functional group of the molecule.

Sphingomyelin, isolated by Thudicum in 1884 and so called by him because of its sphinxlike or mysterious qualities, makes up about 10% of the phospholipid of the mature brain [30]. It is much more abundant in white matter than in gray because it is a constituent of myelin, and as such its metabolic turnover

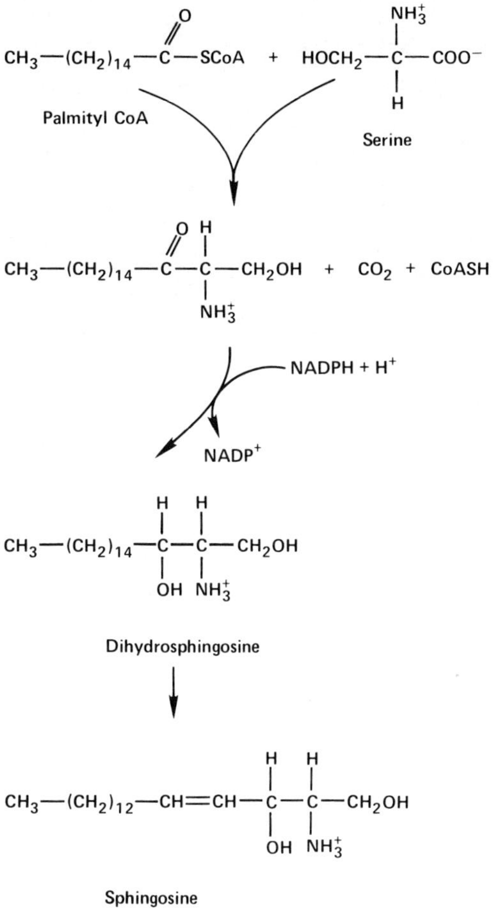

$$CH_3-(CH_2)_{14}-\overset{\displaystyle O}{\overset{\|}{C}}-SCoA \;+\; HOCH_2-\overset{\displaystyle NH_3^+}{\underset{\displaystyle H}{\overset{|}{\underset{|}{C}}}}-COO^-$$

Palmityl CoA   Serine

$$CH_3-(CH_2)_{14}-\overset{\displaystyle O}{\overset{\|}{C}}-\overset{\displaystyle H}{\underset{\displaystyle NH_3^+}{\overset{|}{\underset{|}{C}}}}-CH_2OH \;+\; CO_2 \;+\; CoASH$$

$$-\,NADPH + H^+$$

$$NADP^+$$

$$CH_3-(CH_2)_{14}-\overset{\displaystyle H}{\underset{\displaystyle OH}{\overset{|}{\underset{|}{C}}}}-\overset{\displaystyle H}{\underset{\displaystyle NH_3^+}{\overset{|}{\underset{|}{C}}}}-CH_2OH$$

Dihydrosphingosine

$$CH_3-(CH_2)_{12}-CH=CH-\overset{\displaystyle H}{\underset{\displaystyle OH}{\overset{|}{\underset{|}{C}}}}-\overset{\displaystyle H}{\underset{\displaystyle NH_3^+}{\overset{|}{\underset{|}{C}}}}-CH_2OH$$

Sphingosine

Figure 17.7   Biosynthesis of sphingosine.

rate is extremely slow. The biosynthesis of sphingomyelin was elucidated by
Kennedy and by Brady and their coworkers and is presented in Fig. 17.8. Although there are two possible routes for this biosynthesis, i.e., depending upon
which of the substituents, choline or fatty acids, is added first, it seems likely
that the pathway illustrated is the operant one in vivo. The alternate route,
through ceramide, which is acylated sphingosine, appears, in the test tube at
least, to start from the wrong isomer, one which does not occur naturally. The
fatty acid composition of the sphingomyelin in the brain, unlike that in other
tissues, changes as the animal matures. In fetal animals, on the order of 80% of

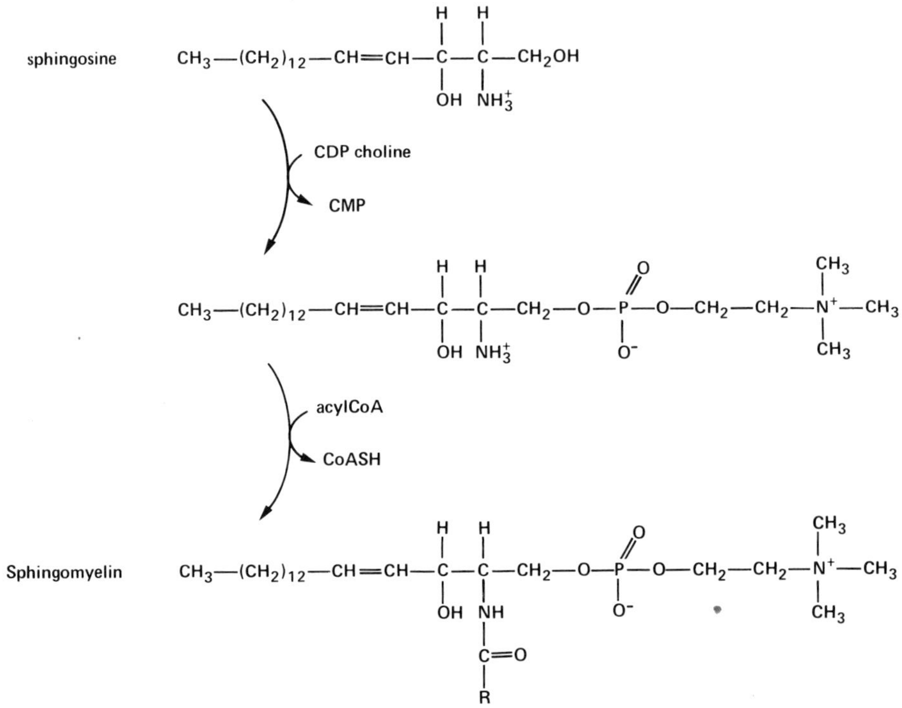
sphingosine
CH3—(CH2)12—CH=CH—C—C—CH2OH
H H
OH NH3+
CDP choline
CMP
CH3—(CH2)12—CH=CH—C—C—CH2—O—P—O—CH2—CH2—N+—CH3
acylCoA
CoASH
Sphingomyelin
C=O
R

Figure 17.8 Biosynthesis of sphingomyelin.

the fatty acids are 16 and 18 carbons long and saturated. After myelination the major fatty acids are on the order of 24 carbons long, and a substantial amount of unsaturation exists. The data suggest that the spingomyelin of the myelin has longer and less saturated fatty acids than the sphinomyelin in other sites in the brain. Consistent with that interpretation is the finding that the spingomyelin with long-chain fatty acids has a different metabolic turnover than sphingomyelin with shorter fatty acids [31].

The degradation of sphingomyelin also occurs in brain. Sphingomyelinase cleaves phosphorylcholine from the molecule, leaving N-acylsphingosines (ceramides). The same reaction is catalyzed by phospholipase C, but it seems clear that these two enzymes are separate entities. Brain possesses a separate and specific sphingomyelinase [32], as well as the less specific phospholipase C which will attack lecithin and the cephalins as well. The further catabolism involves ceramidase, an enzyme which deacylates ceramide to produce sphingosine plus a fatty acid.

In Niemann-Pick disease [33,34] there is a hereditary lack of sphingo-

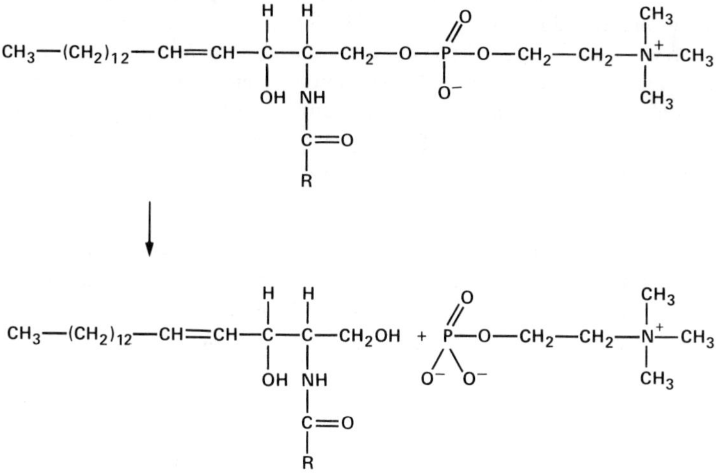

$$CH_3-(CH_2)_{12}-CH\!=\!CH-\underset{\underset{\displaystyle C=O}{\overset{\displaystyle |}{\underset{\displaystyle |}{NH}}}}{\overset{\displaystyle H}{\overset{\displaystyle |}{C}}}\;\;\underset{R}{}$$

Figure 17.9 Sphingomyelinase.

myelinase (Fig. 17.9) [35], and sphingomyelin accumulates in the liver and the spleen, and to a much lesser extent in the brain and cerebrospinal fluid. Such accumulation can also be seen in cells from Niemann-Pick patients grown in culture [36]. Accumulation of cholesterol esters in nervous tissue is a common accessory feature. The condition is transmitted as an autosomal recessive and occurs generally, but not exclusively, in Jews. The large amount of sphingomyelin which accumulates in various body sites is thought to originate in the stroma of erythrocytes, and is released during the normal cycle of red blood cell breakdown. The failure in the degradative pathway, i.e., the absence of sphingomyelinase, leads to the accumulation. The disease is accompanied by central nervous sytem damage, and severe retardation of both mental and physical development. Usually the affected children die before the age of 2.

Niemann-Pick Disease

The patient, an 8-week-old boy, was admitted suffering from severe diarrhea. He was lethargic, febrile, poorly developed, had yellowish olive skin, and did not eat well. Upon examination the child's spleen and liver were found to be markedly enlarged. Lymph node enlargement was also noted. Hemoglobin was normal, as were cultures of blood and stool. Roentgenograms showed vertebral abnormalities, and bone marrow studies revealed the presence of large vacuolated foam cells. A diagnosis of Niemann-Pick was made on the basis of the histologic examination of marrow, and on the results of enzyme assay of leukocyte extracts.

Upon reevaluation at 3 ½ months, the patient was below the third percentile in weight, head circumference, and body length. His lethargy had become more evident, and cherry-red spots were noted in his maculi. The child had a markedly protuberent abdomen. He seemed unresponsive and was most probably retarded. Shortly after evaluation he again had loose stools and vomiting, and passed blood by mouth and rectum. He died a few days past 4 months of age. Upon autopsy, foamy cells were noted in spleen, liver, marrow, and lungs.

There are four types of Niemann-Pick disease [37]. Type A is the most rapidly progressing and the most common. Prenatal diagnosis of the disease has been possible using cultured amniotic cells and specific enzyme assays with labeled sphingomyelin as substrate [38]. No treatment is yet available.

Cerebrosides

The cerebrosides are glycolipids which consist of sphingosine, a fatty acid attached to the amino group in amide linkage, and a hexose, either glucose or galatose, attached through its hemiacetal hydroxyl to the primary alcohol (Fig. 17.10). They are not phospholipids. They are polar compounds, but unlike the gangliosides which have many sugar residues, the cerebrosides, which have only one, are lipid soluble. Galactose is the preponderant monosaccharide in the cerebrosides.

The cerebrosides were isolated by Thudicum. They are present to the extent of 2 to 3% of the fresh weight of the brain of adults. They are myelin constituents and, as such, are in higher concentration in the white matter than in the gray. Also, as such, the concentration of cerebrosides in brain rises dramatically during the period of myelination, and the turnover of cerebrosides in adult brain is very low. The fatty acids of brain cerebrosides are primarily 24 carbons long and largely hydroxylated and unsaturated. Cerebrosides are also found in tissues

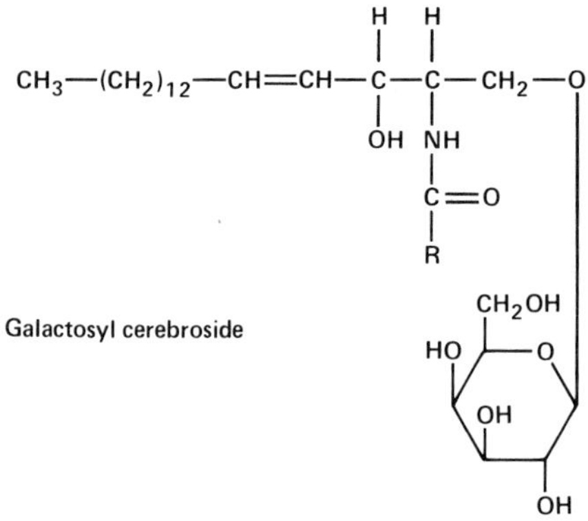

Galactosyl cerebroside

Figure 17.10 Structure of the cerebrosides.

other than the brain, but while the cerebrosides of mature, normal brain contain galactose almost exclusively, the immature or diseased brain and organs other than the brain contain a substantial amount of glucocerebroside. It seems likely that galactose is the exclusive sugar of the cerebrosides and that the glucocerebrosides are present only as precursors, or degradation products, of the gangliosides. This seems all the more likely in view of the observation that the enzyme transferring glucose to ceramide is present primarily in the neurons, the site of ganglioside synthesis, and the enzyme transferring galactose is primarily in the glia, the site of the synthesis of myelin [39].

There are alternate possible routes to the biosynthesis of the cerebrosides. Most of the evidence supports the pathway through psychosine [40], but a substantial number of studies indicate that ceramide can also be an intermediate in cerebroside synthesis [41]. It seems possible that both pathways are utilized in brain. In any case, the biosynthesis begins with sphingosine and takes place in the microsomes. The monosaccharide is presented as UDP-galactose, (or UDP-glucose in the formation of glucocerebroside). The fatty acid component is donated by the appropriate acyl CoA. The enzymes for the degradation of the cerebrosides are present in the lysosomes of normal tissues, and their absence leads to the genetic diseases discussed below. In addition to these now well-characterized enzymatic deficiencies, the cerebrosides accumulate in other conditions for which alterations in cerebroside metabolism may not be the primary lesion. For example, they are known to accumulate in the cerebrospinal fluid, but not the serum, of patients with multiple sclerosis.

Certain of the mouse mutants discussed in Chap. 5 have notably low levels of brain cerebroside. Specifically, the quaking and the Jimpy strains are deficient in this, as well as in other myelin components. As mentioned before these mice are quite abnormal and exhibit characteristic neurologic signs. Although any deficiency in the synthesis of myelin would probably lead to a lowering of all the constituents, there is some evidence that the Jimpy condition is due to a specific lesion in the synthesis of cerebroside.

The work of Brady has [42-44] shown that the genetic deletion of the lysosomal enzyme which cleaves the glucose from glycosyl cerebroside is the molecular cause of the condition known as Gaucher's disease. Also known as glucocerebroside lipidosis, this condition was originally described by the French physician Gaucher in 1882. A deficiency of glucocerebroside β-glucosidase (Fig. 17.11) leads to the deposition of cerebroside in various body sites, and to nervous system involvement and death. There are three forms of the disease known clinically. The adult form is characterized by splenomegaly, hepatomegaly, and erosion and fracture of the long bones. Large and debris-laden cells ("granular cells") appear in the bone marrow, the spleen, the liver, and the lymph nodes. These are large cells, 20 to 100 μm, the cytoplasm of which has a reticular pattern that looks like wrinkled tissue paper or crumpled silk. The infantile form

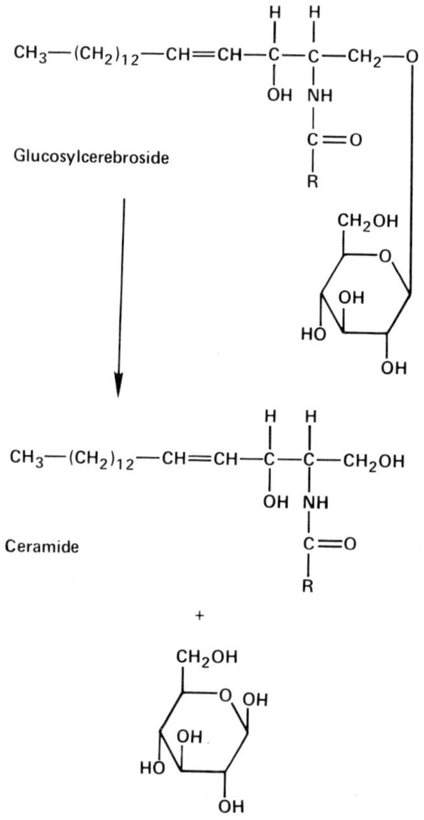

Glucosylcerebroside

Ceramide

Figure 17.11   Glucocerebroside β-glucosidase.

exhibits all the above symptoms, but also includes central nervous system changes, primarily neuronal degeneration. The juvenile form runs a rapid course, but usually does not involve central nervous system alteration. In all forms there is accumulation of glucocerebroside which normally constitutes 0.2% or less of the dry weight of the spleen but which in Gaucher's patients can reach 4% or more. The disease is generally transmitted as an autosomal recessive and can now be diagnosed in utero by enzymatic assay of cultured amniotic cells [45]. The source of the cerebroside which accumulates peripherally may be the glycolipid of erythrocytes or leukocytes which are destroyed in the normal course of events. The source of the cerebroside which accumulates in the brain is a little less obvious, but probably comes from the faulty breakdown of the gangliosides [42].

Gaucher's Disease

The patient, an 8-month-old white male, was admitted because of motor development regression. Abdominal protuberance was noted. A history revealed that development had apparently been normal up to about 6 months of age.

Upon physical examination liver and spleen were found to be massively enlarged. Stiffness and hypertonicity of the extremities were noted, and the deep tendon reflexes were brisk. Gaucher cells were found in the bone marrow, and serum phosphatase was markedly elevated. Gaucher's disease was diagnosed using specific enzyme assay on a skin biopsy.

By 12 months of age the patient could no longer sit unaided. Shortly thereafter the boy developed retroflexion of the head and spasticity, and he died at 14 months of febrile illness associated with thrombocytopenia.

Upon autopsy the liver and spleen were found to be overwhelmingly infiltrated with Gaucher cells, and scattered foci were found in the lymph nodes and the thyroid. The liver contained 77 mg of cerebrosides per gram dry weight, normal being no more than 10. The brain appeared grossly normal.

Initial attempts at a therapy of Gaucher's disease by enzyme replacement have proved encouraging [44]. Such experiments have involved the administration of purified β-glucosidase obtained from normal human placenta. Enzyme has been administered to certain children with Gaucher's disease who have enough enzyme activity to protect them from the central nervous system involvement, but who are likely to die from the accumulation of cerebroside in the liver and the spleen. It might be expected that the patients suffering from accumulations in the central nervous system would be unaffected by this type of treatment, since replacement enzyme would be unlikely to cross the blood-brain barrier. It is thought possible, however, that even these children might benefit from such treatment since circulating enzyme might lower the blood concentration of the relevant cerebrosides and thereby shift the equilibrium allowing cerebroside to leave the brain and be removed from the body.

A reduction in the activity of the corresponding degradative enzyme for the galactocerebrosides, i.e., galactocerebroside β-galactosidase (Fig. 17.12), leads to Krabbe's disease, otherwise known as globoid cell leukodystrophy. Children suffering from Krabbe's disease exhibit a progressive mental deterioration beginning at 2 to 5 months of age and rarely live beyond 1 year. There is a virtual absence of myelin in the brain, which points up the differing location and function of the glucocerebrosides and the galactocerebrosides. In Gaucher's disease, which is a derangement in glucocerebroside metabolism, there is relatively little myelin involvement since the glucocerebroside is not thought to be primarily a myelin

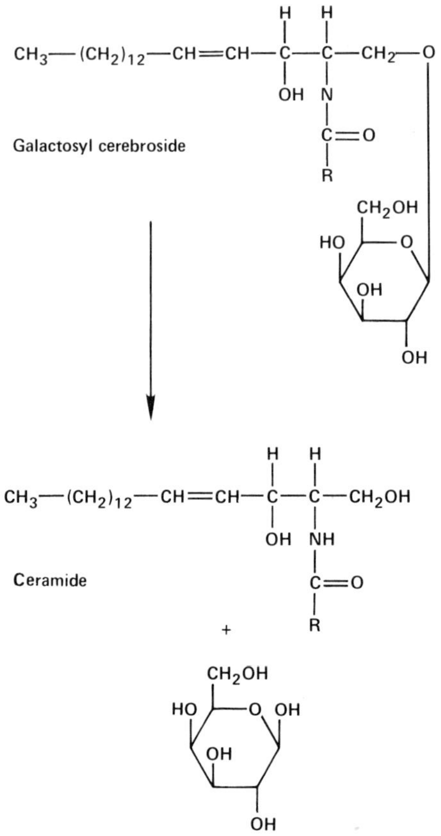

Galactosyl cerebroside

Ceramide

Figure 17.12   Galactocerebroside β-galactosidase.

constituent. In Krabbe's disease there is a demyelination, and total brain cerebrosides are lower. Multinucleated globoid cells appear in the white matter of
the central nervous system, and these cells contain galactocerebroside. The activity of galactocerebroside β-galactosidase is generally found to be reduced, but
not absent.

The disease is transmitted genetically as an autosomal recessive and usually
goes through three clinical stages [37]. In the first stage the children are irritable
and may have light-related or noise-related tonic spasms. Frequently there are
bouts of fever without infection. In stage two there is rapid and severe motor
deterioration. There is also abnormal posture, and the arms may be flexed while
the legs remain extended. Frequent episodes of high fever and intense sweating
occur. About half the patients have seizures. Pneumonia is also a frequent occurrence and a frequent cause of death. In the third stage the patients appear

decerebrate. They have no voluntary movements and are completely unresponsive. Upon autopsy these individuals have markedly atrophied brains due to a
virtual absence of white matter. Myelin may be totally absent.

Sulfatides

The sulfatides are galactocerebrosides with a sulfate esterifying the 3-hydroxyl
of galactose (Fig. 17.13). The sulfatides, like the cerebrosides, are lipid-soluble
compounds and are thought to be exclusively myelin constituents [40]. The
fatty acid composition of the sulfatides resembles that of the cerebrosides and
is quite unlike that of the other sphingolipids. The sulfate of phosphoadenosine
phosphosulfate (PAPS) is the precursor for the formation of the sulfatides from
galactocerebroside, and such sulfation can be demonstrated with brain preparations [46]. Since the sulfatide is a myelin constituent its concentration rises
dramatically, as does that of the galactocerebrosides, during the developmental
period encompassing myelination. It follows that the enzyme required for the
formation of the sulfatides, cerebroside sulfotransferase, is found in much higher
amounts in the oligodendroglia, the cells which form myelin, than in the neurons
[47].

The absence of the sulfatase which degrades the sulfatides causes metachromatic leukodystrophy [48], a hereditary condition associated with the
deposition and excretion of undegraded sulfatides. The disease was described
first in adults in 1921, and then in children in 1925. The enzyme cleaves the
sulfate from the galactose (Fig. 17.14), and this presumably initiates the degradation of the molecule. There are two components to the enzyme, one which is
heat stable and one which is heat labile. It has been shown that the heat-labile
portion of the sulfatase is identical with the long-known enzyme, arylsulfatase
A. In some individuals with this condition the enzyme protein is made but is

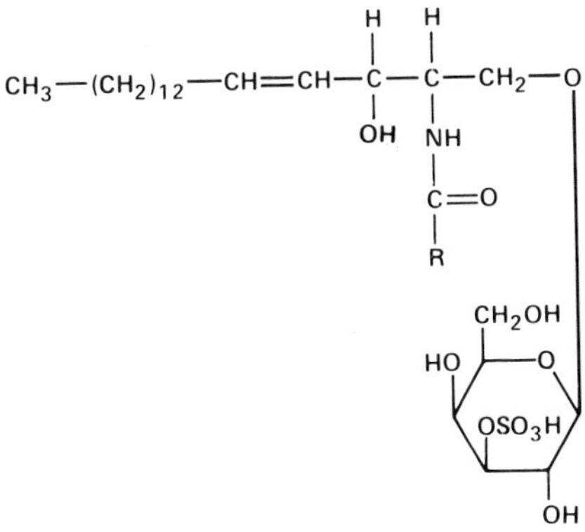

Figure 17.13   Structure of the sulfatides.

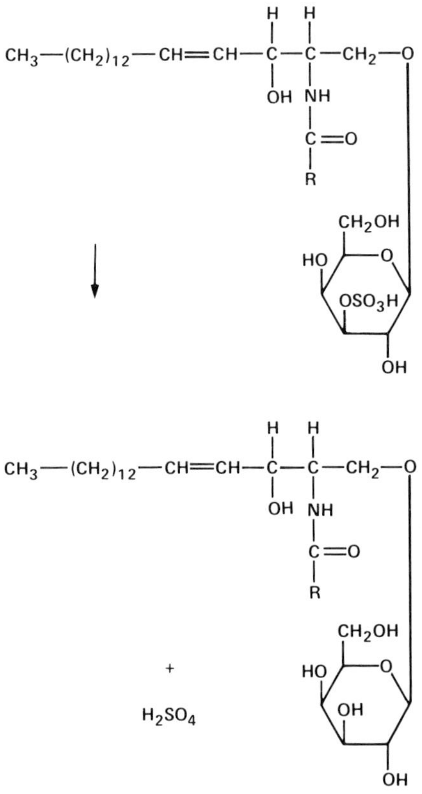

Figure 17.14   Sulfatide sulfatase.

not active [49]. Sulfatides accumulate in the brain, in peripheral nerve, and in kidney. There are two clinically recognized forms of the disease, the infantile and the adult. In the infantile form there are marked locomotion and speech problems, weakness, ataxia, hypotonus, and, finally, paralysis. The adult form seems to involve fewer of the physical and more of the mental problems, and the individuals suffer from psychological disturbances and dementia. The disease is transmitted as an autosomal recessive, and no sexual or racial preponderance has been associated with its occurrence. The disease can be definitively diagnosed by the presence of metachromatic granules in nerve biopsy, or by the absence of arylsulfatase A in leukocytes.

Metachromatic Leukodystrophy (Sulfatide Lipidosis)

The patient, a 2-year-old white male, was admitted suffering from weakness of the legs. Physical examination revealed no abnormal features other than diffi-

culty in walking and standing. The legs were hypotonic, and their movements
were ataxic. Deep tendon reflexes were slightly diminished. No obvious
peripheral neuropathy was detected, and there was no evidence of cerebellar
tumor. Some signs of reversal were detected within 10 days, and the boy was
placed on outpatient status with periodic visits.

The patient was readmitted 5 months later, now suffering incoordination of
the arms and difficulty in swallowing. Some signs of retardation were noted.
Extensive screening tests were carried out, and sulfatides were detected in the
patient's urine. A diagnosis of metachromatic leukodystrophy was confirmed
by direct assay of arylsulfatase A in leukocytes, and the parents were confirmed as carriers by similar assays in cultures of skin fibroblasts.

The patient died at 4 years and 3 months of age of acute kidney failure. Upon
autopsy the brain was shown to contain little myelin and an accumulation of
metachromatic lipids in the macrophages and glial cells. Metachromatic lipids
were also present in the kidney and the gallbladder.

A rather surprising and puzzling variant of this disease has been described.
This is known as MLD-MSD, metachromatic leukodystrophy with multiple sulfatase deficiencies [50,51]. There is a proven deficiency in this condition of
arylsulfatases A, B, and C, steroid sulfatase, and also of cerebroside sulfatase and
psychosine sulfatase, which may be manifestations of one of the arylsulfatases.
The multiple deficiency of these enzymes suggests that the actual lesion may be
in some common subunit or some common binding site. In any case, the afflicted
individuals suffer from enlarged liver, skeletal changes, and an accumulation of
the mucopolysaccharides and cholesterol sulfate, as well as of the sulfatides.
Clinically the condition resembles the mucopolysaccharidoses.

Gangliosides

Gangliosides are glycosphingolipids which contain sialic acid (N-acetyl neuraminic
acid) [52]. They are the most hydrophilic of the fatty acid derivatives of brain,
and are water soluble. The fatty acid of the gangliosides is primarily stearic acid.
All the major gangliosides contain the same neutral carbohydrate moiety [galac-
tosyl(1 → 3)-N-acetylgalactosaminyl-β-(1 → 4)glycosyl-β-(1 → 4)glycosyl], attached
to the primary hydroxyl of the sphingosine backbone. Sialic acid is attached to
the 3-hydroxyl of the second galactose, and one or more additional sialic acid
molecules may be attached to the terminal galactose. They are classified on the
basis of the number of sialic acid residues (monosialogangliosides, disialoganglio-
sides, and so on), and the monosialogangliosides are the most abundant.

The gangliosides were first described by Klenk [53] and named by him on
the basis of the observation that such compounds accumulated in the ganglia of
certain patients. These individuals are now known to have been suffering from

the ganglioside-accumulating condition, Tay-Sachs disease. It has been shown
that gangliosides are found in the brain and in many other tissues including spleen,
liver, and muscle. They occur in brain primarily in the gray matter and so are not
constituents of the myelin. They are in higher concentration in gray matter than
they are in other tissues. They make up approximately 2% of the dry weight of
the cerebral cortex in the human brain, but they are found in all brain areas. They
are also found in all species, in greater or lesser amounts, but are low or absent in
cerebrospinal fluid and blood plasma. The concentration of the gangliosides in
the human brain increases moderately with maturation and seems to be associated with increased arborization and synapse formation of the neurons in the
nervous system. The gangliosides are membrane components, and recently it has
been shown that they are concentrated in synaptosomal membranes. It is further
known that variations in functional activity of the brain are accompanied by alterations in the turnover of the gangliosides, suggesting that they have some role
in transmission [54].

The gangliosides are made biosynthetically by a pathway which has been
elucidated by the work of several groups. The precursors of the sialic acid residues include fructose, glutamine, and phosphoenol pyruvate (Fig. 17.15) [55,56].
The precursors of the carbohydrate portion are the individual sugar nucleotides,
and the monosaccharides are added to the sphingosine backbone individually,
rather than being condensed as an oligosaccharide first [57,58]. The biosynthesis
is thought to proceed in the Golgi apparatus. The whole pathway has been demonstrated in cell-free preparations from normal brain, as have all the enzymes necessary to catalyze the complete degradation of the gangliosides.

The studies of Brady [44] and of O'Brien [59] have elucidated the enzymatic lesions in a number of conditions known as gangliosidoses. As in the sphing-
olipidoses described above, these are lesions in the lysosomal degradative pathways
for these lipids. The absence of any of these enzymes leads to the pathological
accumulation of an undegraded ganglioside and to the state in humans characteristic of that accumulation. Tay-Sachs disease was the first of the lipid storage diseases to be recognized. Such patients were reported in the 1880s. During the
following years a number of separate but similar conditions were identified, so
that now there are six gangliosidoses, all of which involve the absence of a specific
degradative enzyme.

The lesion in Tay-Sachs disease has been shown to be the absence of hexosaminidase A, the enzyme which normally splits the bond between galactose and
N-acetylgalactosamine in GM_2 ganglioside (Fig. 17.16) [60]. The disease is characterized by severe retardation, muscular weakness, blindness, and death by age
3. As is usually the case in accumulation diseases, the children develop normally
for a period and then, presumably when the accumulation has reached a certain
point, the individual rapidly develops the characteristic symptoms. The disease

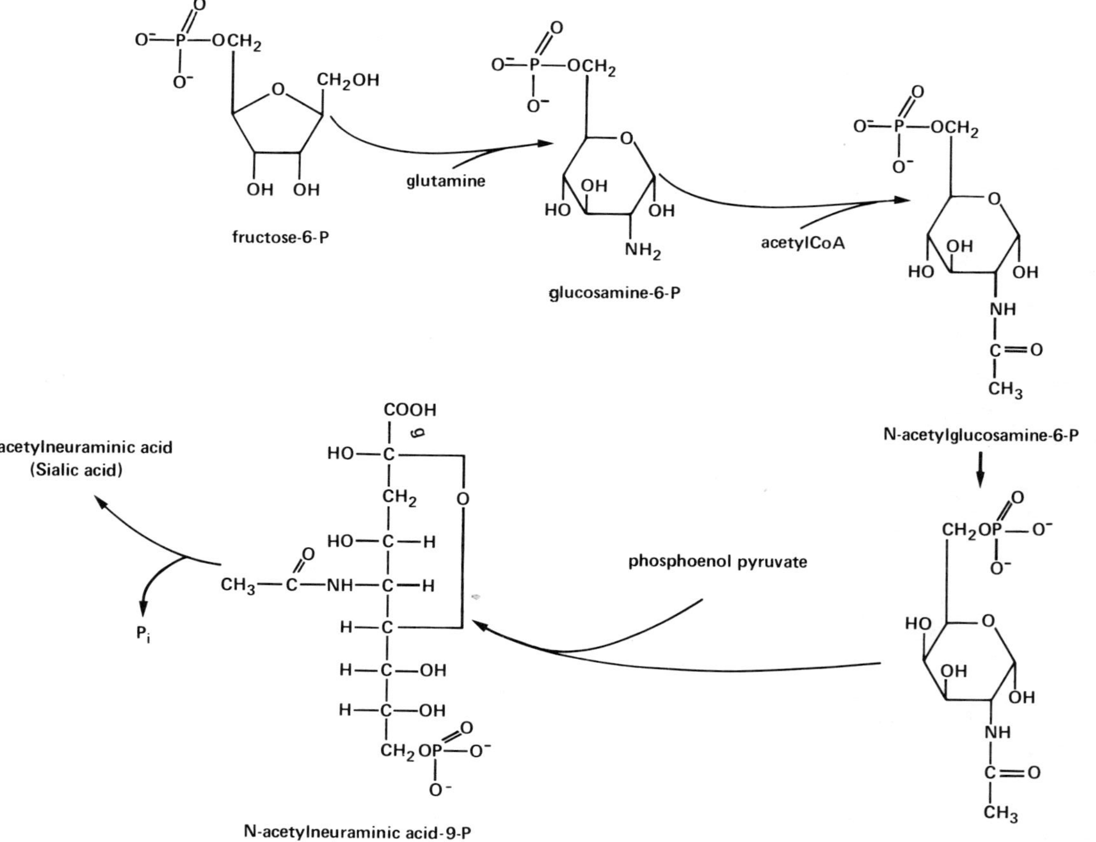

Figure 17.15   Biosynthesis of sialic acid.

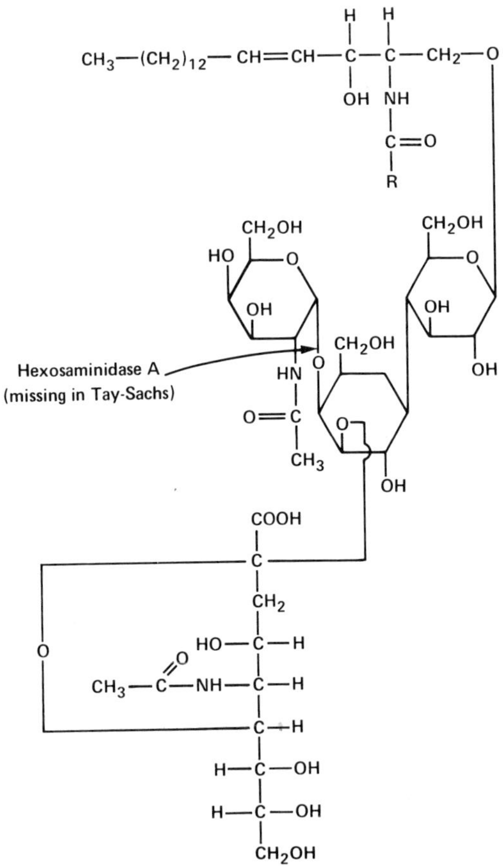

Figure 17.16   GM$_2$ ganglioside N-acetylgalactosaminidase.

is generally a recessive. It occurs in approximately 1 in 6000 children of Ashkenazic Jewish ancestry, and two-thirds of all affected children are Jews. The frequency in non-Jews is much less, although it has been reported in most races. Among Ashkenazic Jewish adults there is a Tay-Sachs gene frequency of 1.3%. Gangliosides accumulate in the brain and other organs. The neurons in the brain of these patients become swollen and contain round membranous bodies within the cytoplasm. These inclusions are more than 90% lipid, and they contain the GM$_2$ monosialoganglioside. A variant of Tay-Sachs is known in which both hexosaminidases A and B are missing. Patients so afflicted present a similar but much more rapidly progressing disease.

The exact metabolic lesion in Tay-Sachs is known, but a puzzling aspect of this condition is that the ganglioside which accumulates contains a bond which can readily be split by neuraminidase. Neuraminidase is fully active in the brains of Tay-Sachs patients, but the susceptible bond is not split. Thus, the exact substrate of the missing enzyme is not accumulated, but a molecule which could be degraded one step further before the block. The reason for this seeming anomaly is not known, and it is all the more puzzling since it is a unique situation among the gangliosidoses.

Tay-Sachs Disease (GM$_2$ Gangliosidosis, Familial Amaurotic Idiocy)

The patient, a 21-month-old girl, was admitted for observation following a series of generalized convulsions. Upon physical examination it was found that she drooled excessively, her head was markedly enlarged, and she had cherry-red macular spots. The child followed a moving light with her eyes, but her vision was estimated to be no more than 50%. Neurological work-up revealed exaggerated motor responses to sudden sound, which consisted of initial upward extension of the arms with flexion, extension of the legs, a startled facial expression, and occasionally a shrill cry. Her musculature was hypotonic, but her deep tendon reflexes were normal. Tay-Sachs was diagnosed by the absence of hexosaminidase A in her leukocytes.

During hospitalization the child deteriorated rapidly. She lost motor function, and developed hyperactive reflexes and a progressive rigidity. She developed aspiration pneumonia while confined and died at the age of 23 months. Upon postmortem examination the child's brain was found to be edematous and loaded with excess white matter. The neurons were distended and filled with granules. There was an accumulation of GM$_2$ ganglioside in the brain, kidney, and liver, which were the only tissues examined.

Accurate diagnosis of Tay-Sachs disease is now possible through the use of specifically labeled ganglioside substrate [61]. Prenatal diagnosis allows the early termination of pregnancy if necessary. Treatment, on the other hand, is so far not possible. Recently, purified human hexosaminidase A has been injected into a child suffering from Tay-Sachs [44]. Human enzyme was used to prevent sensitization of the patient to foreign protein. The enzyme was very rapidly cleared from the patient's bloodstream and taken up by the liver. The levels of blood globoside, which is also degraded by this enzyme, fell by about 40% within hours after the injection. However, the enzyme did not appear to cross the blood-brain barrier, so it could not gain access to the lipid stored in the neurons. The finding is described by the authors as disappointing, but not unexpected. Such treatment has proved more rewarding in gangliosidoses which are not accompanied by a neurological involvement.

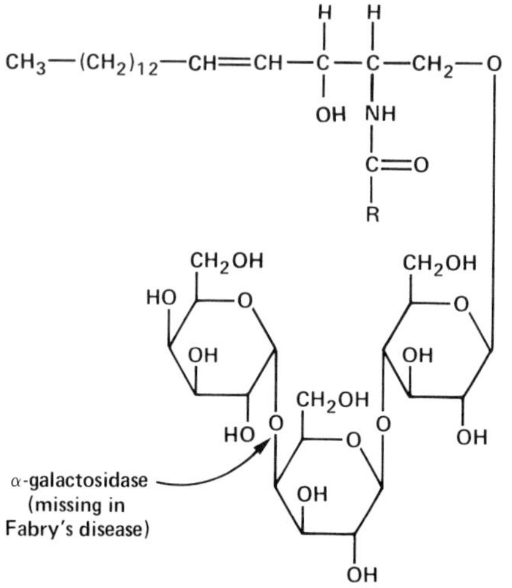

Figure 17.17 Ceramide trihexoside galactosidase.

A case in point is Fabry's disease, a gangliosidosis discovered in 1898. In this condition, which is transmitted as an X-linked recessive, there is an accumulation of cerebroside trihexoside (Fig. 17.17) in several body sites, but not in the brain. The kidneys are a major site of deposition, perhaps because the ganglioside is so water soluble, but deposition also occurs in intestine, lymph nodes, and other tissues. The disease is characterized by rash, skin lesions, and corneal opacity. The missing enzyme is an α-galactosidase. There is little or no neurological involvement, and death eventually results from kidney failure, most certainly as a consequence of the ganglioside accumulation. The missing enzyme has been purified from human placental tissue, and attempts have been made to replace the enzyme missing in Fabry's patients with this exogenous source [44]. It has been infused in patients, and they apparently tolerated the infusion quite well. The accumulating lipid in the bloodstream of the patients was significantly reduced, reaching, at least in one patient, normal levels.

Generalized gangliosidosis, a disease observed rather more recently is somewhat similar in its clinical presentation to Tay-Sachs disease, except that there is also substantial visceral involvement. The disease is transmitted as an autosomal recessive and is caused by the absence of a β-galactosidase (Fig. 17.18). Both the naturally occurring GM_1 ganglioside and ceramide tetrahexoside accumulate in neurons, and in spleen, liver, and kidney. There is also an excretion of certain abnormal mucopolysaccharide sulfates in the urine. Accompanying the mental retardation is severe liver enlargement, skeletal deformities, and usually red spots in the retina. There are two types of this gangliosidosis recognized, type I and type II. In type I the symptoms have some resemblance to those seen in Hurler's

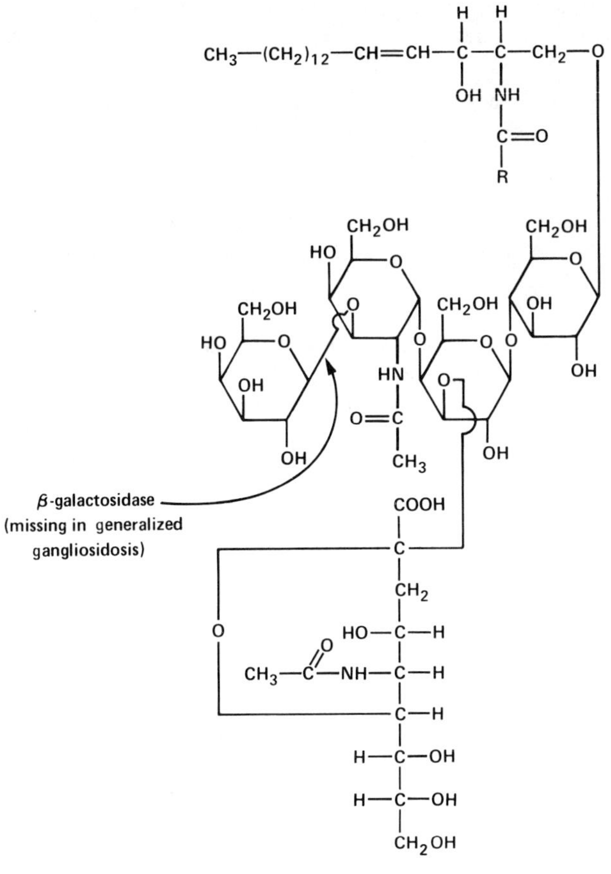

Figure 17.18 GM_1 galactoside galactosidase.

syndrome, and the course of the disease is quite rapid, with onset by 5 months and death usually by 2 years. In type II the symptoms appear much later, and the children can live to 5 years of age.

Generalized Gangliosidosis (GM₁ Gangliosidosis, Late Infantile Systemic Lipidosis)

The patient, an 8-month-old boy, was hospitalized for observation due to general poor development, centering on poor muscle tone and apparent mental retardation. Upon physical examination the child was seen to have coarse facial features with prominent brow and depressed bridge of the nose. The abdomen was enlarged, liver and spleen were palpable, there was a moderate skin rash, and a mild enlargement of the lymph nodes was found. Dorsolumbar kyphoscoliosis was observed, the fingers were short and stubby, and there was nontender enlargement of the wrists and ankles.

A urinary analysis revealed marked increase in urinary keratan sulfate. Diagnosis of generalized gangliosidosis was confirmed by direct assay for β-galactosidase in leukocytes.

The patient was observed for several months, and his condition worsened. He began to have mild tonic-clonic convulsions. Gradually he became unable to eat and had to be fed by tube. He finally became unresponsive, appeared both blind and deaf, and developed decerebrate rigidity. He died at the age of 15 months of bronchopneumonia.

Upon autopsy the following observations were recorded. There was marked visceral histiocytosis, lipid deposits in the neuronal cells, and vacuoles in the renal glomerular epithelial cells. Mucopolysaccharides were found in the cornea, and gangliosides were found in the retinal ganglion cells. The brain contained about 10 times the normal concentration of the GM_1 ganglioside.

Thus, there are a number of heriditary lipid storage diseases. They are presented in Fig. 17.19, taken from the work of Brady and his group. The lesions in the major gangliosidoses have been identified by constructing specifically labeled substrates and devising assays to detect the appropriate enzymes. The relationships between the various members of this family of genetic conditions, both the similarities as lesions in degradative lysosomal pathways, and the differences in the specific steps in the breakdown they involve, have been delineated. This group has also been concerned with the purification of the relevant enzymes and the attempts at amelioration or cure by enzyme replacement. Such work has clarified and systemitized these diseases so that now there is a firm understanding of the underlying enzymatic deletions. There now seems a possibility that a new chapter has begun with the recognition of the first of what may be a family of diseases involving the biosynthetic enzymes [58,62]. In any case, these investigations are among the most comprehensive and elegant in contemporary neurochemistry.

Cholesterol

Cholesterol, the third major "category" of lipid in the nervous system, is, of course not a family of lipids, but a single entity (Fig. 17.20). It is the only major sterol in the adult nervous system. It constitutes 4 to 5% of the fresh weight of the white matter of the spinal cord, 10% of the dry weight of the whole brain, and reaches a concentration of about 20 mg/g of fresh weight in humans. It is the single most concentrated lipid constituent of the brain [63].

The very high concentration of cholesterol in the brain and especially in the white matter is to be expected, since cholesterol is characteristically a constituent of membranes and the myelin is, first of all, a membrane. Consistent with its location is the observation that the major deposition of cholesterol occurs at the time of myelination. Also consistent is the finding that, in the healthy mature nervous system, there is very little if any turnover of cholesterol [64].

Esters of cholesterol are present only in trace amounts in healthy, adult brain tissue. They are present in substantial quantities in developing brain, and may constitute 20% of the total cholesterol in such tissue. They are also found in deteriorating nervous tissue, for example, during Wallerian degeneration, where the appearance of esters is a very early event [65], or in certain pathological states such as multiple sclerosis or Niemann-Pick disease, where the primary lesion is not in cholesterol metabolism. Upon regeneration the cholesterol esters are not converted back to cholesterol, but are discarded, and the cholesterol is replaced by de novo synthesis. Cholesterol esters also appear in high concentration in certain brain tumors. The presence of cholesterol esters, per se, does not seem to have deleterious consequences for the individual since it is known that there are conditions in which cholesterol esters are stored, and they are virtually without harmful effect.

Several other sterols occur in adult brain in trace amounts. Many of these are thought to be biosynthetic precursors of cholesterol. Again, in immature brain these are in much higher amount. Specifically, desmosterol, which is not detectable in mature, healthy brain, can be found in developing brain. Interestingly, desmosterol can also be observed in various brain tumors of the adult nervous system. Such observations suggest that tumors, like developing brain, synthesize cholesterol actively. This can be verified in direct studies of the incorporation of precursors into tumor tissue in vitro. It is of interest that, upon treatment with triparanol, an inhibitor of cholesterol synthesis, the content of desmosterol in the cerebrospinal fluid of patients with brain tumors is significantly higher than in normals. This finding may be of some practical importance in the diagnosis of brain tumors.

The biosynthetic pathway of cholesterol in brain seems to be similar, if not identical, to that in liver, and proceeds through mevalonate and squalene, although all the steps have not been formally shown to be present. The synthesis occurs in the microsomes and also requires soluble factors [66]. It is clear that the major source of brain cholesterol is through local synthesis, although small amounts can come from the bloodstream. The synthesis is active in young individuals and is demonstrable, but much lower, after myelination is complete. The synthesis seems to be primarily in the glia [67]. Glucose is the best precursor in brain, but this is true of all biosynthetic processes in brain since glucose is the preferred substrate for brain in general. Degradative enzymes for

DISEASE	SIGNS AND SYMPTOMS	MAJOR LIPID ACCUMULATION	ENZYME DEFECT
GAUCHER'S DISEASE	SPLEEN AND LIVER ENLARGEMENT, EROSION OF LONG BONES AND PELVIS, MENTAL RETARDATION ONLY IN INFANTILE FORM	GLUCOCEREBROSIDE — CERAMIDE / GLUCOSE	GLUCOCEREBROSIDE β-GLUCOSIDASE
NIEMANN-PICK DISEASE	LIVER AND SPLEEN ENLARGEMENT, MENTAL RETARDATION, ABOUT 30 PERCENT WITH RED SPOT IN RETINA	SPHINGOMYELIN — PHOSPHORYLCHOLINE	SPHINGOMYELINASE
KRABBE'S DISEASE (GLOBOID LEUKODYSTROPHY)	MENTAL RETARDATION, ALMOST TOTAL ABSENCE OF MYELIN, GLOBOID BODIES IN WHITE MATTER OF BRAIN	GALACTOCEREBROSIDE — GALACTOSE	GALACTOCEREBROSIDE β-GALACTOSIDASE
METACHROMATIC LEUKODYSTROPHY	MENTAL RETARDATION, PSYCHOLOGICAL DISTURBANCES IN ADULT FORM, NERVES STAIN YELLOW-BROWN WITH CRESYL VIOLET DYE	SULFATIDE — GALACTOSE-3-SULFATE	SULFATIDASE
CERAMIDE LACTOSIDE LIPIDOSIS	SLOWLY PROGRESSING BRAIN DAMAGE, LIVER AND SPLEEN ENLARGEMENT	CERAMIDE LACTOSIDE — GLUCOSE / GALACTOSE	CERAMIDELACTOSIDE β-GALACTOSIDASE

Disease	Deficient enzyme	Accumulated lipid	Symptoms
FABRY'S DISEASE	CERAMIDETRIHEXOSIDE α-GALACTOSIDASE	CERAMIDE TRIHEXOSIDE	REDDISH PURPLE SKIN RASH, KIDNEY FAILURE, PAIN IN LOWER EXTREMITIES
TAY-SACHS DISEASE	HEXOSAMINIDASE A	GANGLIOSIDE GM$_2$	MENTAL RETARDATION, RED SPOT IN RETINA, BLINDNESS, MUSCULAR WEAKNESS
TAY-SACHS VARIANT	HEXOSAMINIDASE A AND B	GLOBOSIDE (AND GANGLIOSIDE GM$_2$)	SAME AS TAY-SACHS DISEASE BUT PROGRESSING MORE RAPIDLY
GENERALIZED GANGLIOSIDOSIS	β-GALACTOSIDASE	GANGLIOSIDE GM$_1$	MENTAL RETARDATION, LIVER ENLARGEMENT, SKELETAL DEFORMITIES, ABOUT 50 PERCENT WITH RED SPOT IN RETINA
FUCOSIDOSIS	α-FUCOSIDASE	H-ISOANTIGEN	CEREBRAL DEGENERATION, MUSCLE SPASTICITY, THICK SKIN

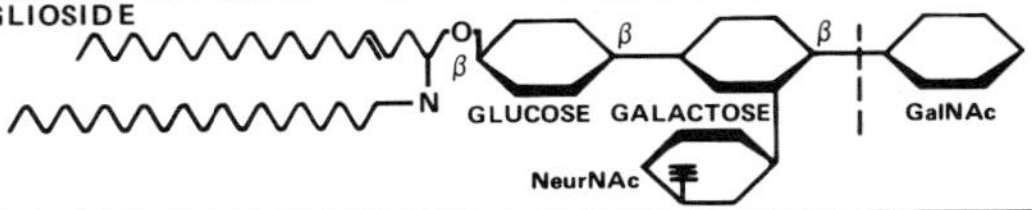

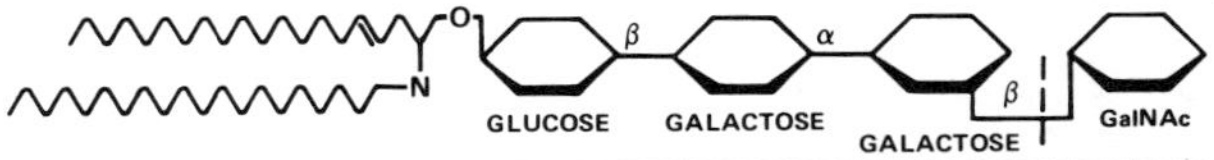

Structure labels (from the molecular diagrams):

- CERAMIDE TRIHEXOSIDE: GLUCOSE, GALACTOSE, GALACTOSE
- GANGLIOSIDE GM$_2$: GLUCOSE, GALACTOSE, GalNAc, NeurNAc
- GLOBOSIDE (AND GANGLIOSIDE GM$_2$): GLUCOSE, GALACTOSE, GALACTOSE, GalNAc
- GANGLIOSIDE GM$_1$: GLUCOSE, GALACTOSE, GalNAc, GALACTOSE, NeurNAc
- H-ISOANTIGEN: N-ACETYL GLUCOSAMINE, GALACTOSE, GALACTOSE, FUCOSE

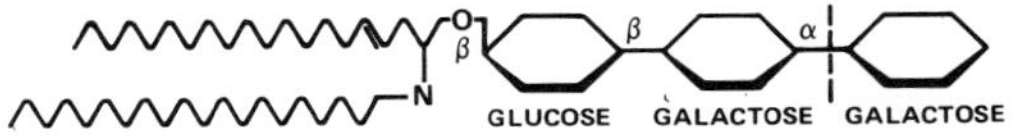

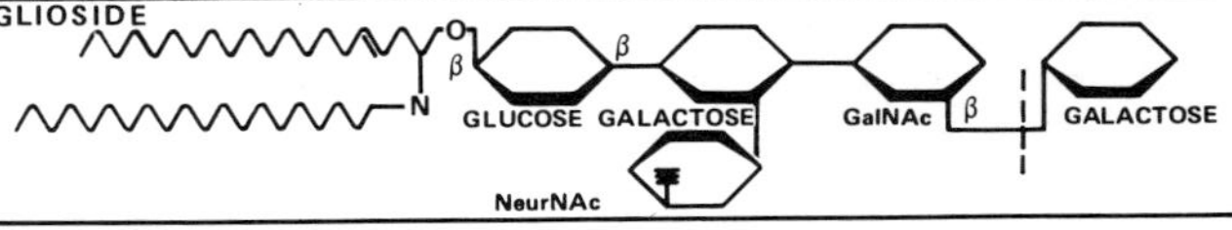

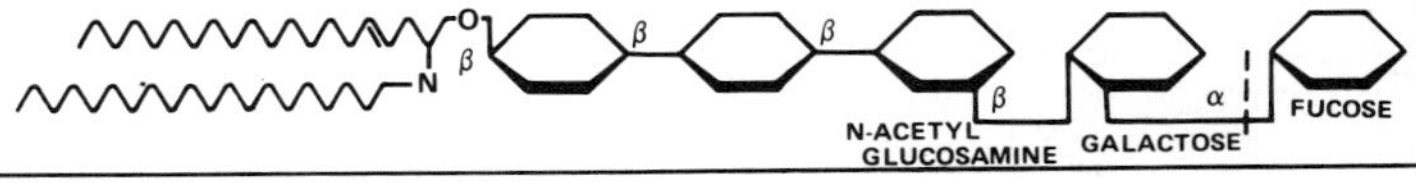

Figure 17.19   The chemistry of the hereditary lipid storage diseases.

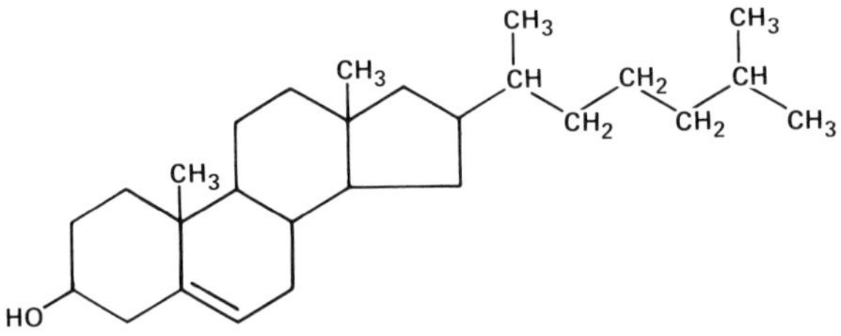

Figure 17.20   Cholesterol.

cholesterol do not seem to occur in the brain. So, what little cholesterol is discarded through normal turnover or during degenerative processes must find its way out of the brain and into the bloodstream.

References

1.   G. Rouser and A. Yamamato, in *Handbook of Neurochemistry* Vol. I (A. Lajtha, ed.), Plenum, New York, 1969.

2.   K. Suzuki, in *Basic Neurochemistry* (R. W. Albers, G. J. Siegel, R. Katzman, and B. W. Agranoff, eds.), Little, Brown, Boston, 1972.

3.   A. F. D'Adamo, Jr., in *Handbook of Neurochemistry,* Vol. III (A. Lajtha, ed.), Plenum, New York, 1970.

4.   J. A. Benjamins and G. M. McKhann, in *Basic Neurochemistry* (R. W. Albers, G. J. Siegel, R. Katzman, and B. W. Agranoff, eds.), Little, Brown, Boston, 1972.

5.   E. M. Carey, *J. Neurochem. 24:*237 (1975).

6.   M. S. Patel and B. L. Tinkonow, *J. Neurochem. 23:*309 (1974).

7.   C. D. Joel, C. A. Ellis, J. K. Lace, P. B. Joel, M. R. Swanson, and J. R. Stroemer, *J. Neurochem. 23:*23 (1974).

8.   I. Karlsson, *J. Neurochem. 25:*101 (1975).

9.   C. Alling, Å. Bruce, and I. Karlsson, *J. Neurochem. 23:*1263 (1974).

10.   I. Gross and J. B. Warshaw, *J. Neurochem. 25:*191 (1975).

11.   R. Rajalakshmi and H. L. Nakhasi, *J. Neurochem. 24:*979 (1975).

12.   S. E. Geel and P. M. Dreyfus, *J. Neurochem. 24:*353 (1975).

13.   J. B. Sidbury, Jr., E. K. Smith, and W. Harlan, *J. Pediatr. 70:*8 (1967).

14.   D. Steinberg, F. Q. Vroom, W. K. Engel, J. Cammermeyer, C. E. Mize, and J. Avigan, *Ann. Intern. Med. 66:*365 (1967).

15.   E. H. Kolodny, W. K. Hass, B. Lane, and W. D. Drucker, *Arch. Neurol. 12:*583 (1965).

16.   R. J. Rossiter and K. P. Strickland, in *Handbook of Neurochemistry,* Vol. III (A. Lajtha, ed.), Plenum, New York, 1970.

17. J. N. Hawthorne and M. Kai, in *Handbook of Neurochemistry*, Vol. III (A. Lajtha, ed.), Plenum, New York, 1970.

18. R. A. D'Amato, L. A. Horrocks, and K. E. Richardson, *J. Neurochem. 24:* 1251 (1975).

19. H. Woelk, K. Kanig, and K. Peiler-Ichikawa, *J. Neurochem. 23:*1057 (1974).

20. H. Woelk and K. Kanig, *J. Neurochem. 23:*739 (1974).

21. H. Woelk, K. Kanig, and K. Peiler-Ichikawa, *J. Neurochem. 23:*745 (1974).

22. L. E. Hokin, *Int. Rev. Cytol. 23:*187 (1968).

23. L. E. Hokin, in *Structure and Function of Nervous Tissue*, Vol. III (G. H. Bourne, ed.), Academic, New York, 1969.

24. L. E. Hokin and M. R. Hokin, *J. Biol. Chem. 234:*1387 (1959).

25. G. G. Lunt and M. R. Pickard, *J. Neurochem. 24:*1203 (1975).

26. J. Schacht, E. A. Neale, and B. W. Agranoff, *J. Neurochem. 23:*211 (1974).

27. A. A. Abdel-Latif, S. J. Yau, and J. P. Smith, *J. Neurochem. 22:*383 (1974).

28. H. E. Carter, W. P. Norris, F. J. Glick, G. E. Phillips, and R. Harris, *J. Biol. Chem. 170:*269 (1947).

29. R. O. Brady, J. V. Formica, and G. J. Koval, *J. Biol. Chem. 233:*1072 (1958).

30. A. Rosenberg, in *Handbook of Neurochemistry*, Vol. III (A. Lajtha, ed.), Plenum, New York, 1970.

31. A. Lastennet, L. Freysz, and P. Mandel, *J. Neurochem. 25:*185 (1975).

32. A. Roitman and S. Gatt, *Isr. J. Chem. 1:*190 (1963).

33. A. Niemann, *Jahrb. Kinderheilkd. 79:*1 (1914).

34. L. Pick, *Med. Klin. 23:*1483 (1927).

35. R. O. Brady, J. N. Kanfer, M. B. Mock, and D. S. Fredrickson, *Proc. Natl. Acad. Sci. USA 55:*366 (1966).

36. H. R. Sloan, B. W. Uhlendorf, J. N. Kanfer, R. O. Brady, and D. S. Fredrickson, *Biochem. Biophys. Res. Commun. 34:*582 (1969).

37. L. B. Holmes, H. W. Moser, S. Halldorsson, L. Mack, S. S. Pant, and B. Matzilevich, *Mental Retardation*, MacMillan, New York, 1972.

38. C. J. Epstein, R. O. Brady, E. L. Schneider, R. M. Bradley, and E. Shapiro, *Am. J. Hum. Genet. 23:*533 (1971).

39. N. Latovitzki and D. H. Silberberg, *J. Neurochem. 24:*1017 (1975).

40. N. S. Radin, in *Handbook of Neurochemistry*, Vol. III (A. Lajtha, ed.), Plenum, New York, 1970.

41. T. P. Carter and J. Kanfer, *J. Neurochem. 23:*589 (1974).

42. R. O. Brady, in *Handbook of Neurochemistry*, Vol. VII (A. Lajtha, ed.), Plenum, New York, 1972.

43. R. O. Brady, in *Basic Neurochemistry* (R. W. Albers, G. J. Siegel, R. Katzman, and B. W. Agranoff, eds.), Little, Brown, Boston, 1972.

44. R. O. Brady, *Chem. Phys. Lipids 13:*271 (1974).

45. E. L. Schneider, W. G. Ellis, R. O. Brady, J. R. McCulloch, and C. J. Epstein, *J. Pediatr. 81:*1134 (1972).

46. D. F. Farrell, *J. Neurochem. 23:*219 (1974).

47. J. A. Benjamins, M. Guarnieri, K. Miller, M. Sonneborn, and G. M. McKhann, *J. Neurochem.* *23:*751 (1974).

48. H. Jatzkewitz and E. Mehl, *J. Neurochem.* *16:*19 (1969).

49. D. Stumpf, E. Neuwelt, J. Austin, and P. Kohler, *Arch. Neurol.* *25:*427 (1971).

50. J. Austin, D. Armstrong, and L. Shearer, *Arch. Neurol.* *13:*593 (1965).

51. Y. Eto, S. Rampini, U. Wiesmann, and N. N. Herschkovits, *J. Neurochem.* *23:*1161 (1974).

52. L. Svennerholm, in *Handbook of Neurochemistry,* Vol. III (A. Lajtha, ed.), Plenum, New York, 1970.

53. E. Klenk, *Z. Physiol. Chem.* *273:*76 (1942).

54. J. T. R. Clarke, *J. Neurochem.* *24:*533 (1975).

55. G. W. Jourdian and S. Roseman, *Ann. N.Y. Acad. Sci.* *106:*202 (1963).

56. L. Warren, R. S. Blacklow, and C. W. Spearing, *Ann. N.Y. Acad. Sci.* *106:*191 (1963).

57. S. Roseman, *Chem. Phys. Lipids* *5:*270 (1970).

58. P. H. Fishman, *Chem. Phys. Lipids* *13:*305 (1974).

59. J. S. O'Brien, in *The Metabolic Basis of Inherited Disease,* 3rd ed. (J. B. Stanbury, J. B. Wyngaarden, and D. S. Fredrickson, eds.), McGraw-Hill, New York, 1972.

60. S. Okada and J. S. O'Brien, *Science* *165:*698 (1969).

61. E. H. Kolodny, R. O. Brady, and R. W. Volk, *Biochem. Biophys. Res. Commun.* *37:*526 (1969).

62. S. R. Max, N. K. Maclaren, R. O. Brady, R. M. Bradley, M. B. Rennels, J. Tanaka, J. H. Garcia, and M. Cornblath, *N. Engl. J. Med.* *291:*929 (1974).

63. R. Paoletti, E. Grossi-Paoletti, and R. Fumagalli, in *Handbook of Neurochemistry,* Vol. I (A. Lajtha, ed.), Plenum, New York, 1969.

64. A. N. Davison, in *Handbook of Neurochemistry,* Vol. III (A. Lajtha, ed.), Plenum, New York, 1970.

65. J. G. Wood and R. M. C. Dawson, *J. Neurochem.* *22:*631 (1974).

66. J. P. Jones, A. Rios, H. J. Nicholas, and R. B. Ramsey, *J. Neurochem.* *24:*117 (1975).

67. J. P. Jones, H. J. Nicholas, and R. B. Ramsey, *J. Neurochem.* *24:*123 (1975).

18
ELECTROLYTES

The inorganic constituents of the brain make up on the order of 1.1% of the wet weight of the brain, or about 5% of the dry weight [1]. The ash residue of the brains of most species consists primarily of Na^+, K^+, Ca^{2+}, and Mg^{2+}, which are the major cations of the brain and other tissues. The total cation content of fresh gray matter of most mammalian brains amounts to about 170 $\mu Eq/g$. The diffusable anions of the brain are Cl^-, HCO_3^-, SO_4^{2-}, PO_4^{3-}, and organic acids, primarily N-acetylaspartic acid. The diffusable intracellular anions, including the organic anions, make up only about one-half the amount necessary to neutralize the intracellular cations, thus the remaining intracellular anion component must be macromolecular, i.e., the anion sites on the proteins and, uniquely in the brain, the lipids in the cells. Na^+ and Cl^- are primarily extracellular components. The concentrations of the various inorganic electrolytes in the brain are indicated in Table 18.1.

In addition to the major cations of the brain there are a number of metal ions found in much smaller amounts. As shown in Table 18.1, there are small amounts of copper, iron, manganese, and zinc in the brain. The copper is found mainly in association with brain proteins known as cerebrocupreins. The iron exists partly as ferritin and partly in combination with the cytochromes. The importance of these minor constituents can best be seen by the consequences of their absence or overabundance, some of which will be discussed below. Further testimony to the importance of the correct ions in the correct proportion is given by the severe neural consequences sustained when the brain is exposed to metal ions which are not normally brain constituents, some of which will also be covered in subsequent pages.

The outstanding feature of electrolyte metabolism in the brain is the continuing unequal distribution of Na^+ and K^+. That this should be so in the brain is no surprise since electrical activity depends on differential ion concentrations between the inside of the cell and the fluid which surrounds it. The sodium concentration of the intracellular compartment is about one-eighth that found in

Table 18.1   Major Inorganic Constituents of Brain

Constituent	Content (μmol/g fresh tissue)
Na^+	57[a]
K^+	95[a]
Ca^{2+}	2.2[a]
Mg^{2+}	5.1[a]
Cl^-	45[a]
HCO_3^-	12[b]
Cu^{2+}	0.050[c]
Mn^{2+}	0.023[c]
Zn^{2+}	0.25[c]
Fe^{3+}	0.45[c]

[a]Cerebral cortex of the cat.

[b]Average of cerebral cortex and corpus callosum of the cat.

[c]Cerebral cortex of the rabbit.

Data presented by D. B. Tower, in *Handbook of Neurochemistry,* Vol. I (A. Lajtha, ed.), Plenum, New York, 1969.

Table 18.2   Ionic Content of the Intracellular Space of Cat Cerebral Cortex Compared to the Ionic Content of Cat Cerebrospinal Fluid

Constituent	Intracellular content, calculated (μmol/g)	Cerebrospinal fluid (μmol/g)
Na^+	18.0	141
K^+	166.0	3
Ca^{2+}	3.35[a]	1.2
Mg^{2+}	8.5[a]	1.2
Cl^-	9.0	130

[a]Most of the Ca^{2+} and Mg^{2+} is bound, so that trace concentrations in the intracellular fluid are likely to be fractions of these theoretical maxima.

Data presented by D. B. Tower, in *Handbook of Neurochemistry,* Vol. I (A. Lajtha, ed.), Plenum, New York, 1969.

the cerebrospinal fluid, whereas the potassium concentration is more than 50 times that found in the cerebrospinal fluid (Table 18.2). The differential distribution across the cell membrane in the brain, as in other tissues, requires a continuing supply of ATP, which is the specific, immediate, and sufficient energy source. In fact, the preservation of this gradient may be the major energy-consuming reaction in the cell. The inevitable consequence of an interruption of the energy supply is an equilibration of the Na^+ and K^+ levels across the membrane.

The Sodium Pump and ATPase

The differential distribution of sodium and potassium across the cell membrane of neural cells requires, as in other tissues, the continuing operation of the sodium pump. This sodium pump appears to be closely related to, or identical with, the Mg^{2+}-dependent, Na^+,K^+-activated, ouabain-inhibitable ATPase of the cell membrane. This is an enzyme which hydrolyzes the terminal phosphate of ATP and which is specifically stimulated by the ions transported. The ATPase, while present in most, if not all, tissues, has been studied most extensively in excitable tissue such as brain, nerve, and the electric organ of the eel [2-4], which are extremely good sources. Secretory organs such as kidney and choroid plexus also contain high activity. The presence of a functioning sodium pump in nervous tissue was first proposed by Hodgkin and Keynes in 1955 [5]. The suggestion that the sodium pump of membranes would manifest itself in broken cell preparations as a sodium-dependent ATPase was based on experiments in which ATP restored sodium transport in poisoned cells. The suggestion was quickly followed by the observation of just such an enzyme in crab nerve by Skou in 1957 [6,7].

A great deal of work has now been done on this enzyme. As with other membrane enzymes, study of the ATPase has been difficult due to the continuing inability to prepare soluble, active, homogeneous preparations. Also, study of this ATPase has been complicated by the presence of other, unrelated ATPases in the cell. Nevertheless, a fair amount of information is available about its action and the nature of the enzyme protein. The enzyme is membrane bound, and in broken cell preparations it is found in the microsomal fraction. It hydrolyzes ATP quite specifically. ATP hydrolysis requires Na^+ and is stimulated by K^+. As with other ATPases, the presence of Mg^{2+} is obligatory. The enzyme is activated by low concentrations of detergents. It is normally associated with phospholipids in the membrane [8,9], is inactivated by phospholipase, and is stimulated under certain conditions by exogenous phospholipids. The molecular weight of the enzyme is a matter of some controversy and has been variously estimated at 250,000 and 1,000,000, and at several values in between, depending on the techniques used in the estimation. There seem to be two distinct subunits [10]. The requirement for Na^+ and for Mg^{2+} are absolute and specific. The stimulation by K^+ can be more or less duplicated with NH_4^+, Rb^+, Cs^+, and Li^+, in decreasing order of effectiveness. Ca^{2+}, Cu^{2+}, and F^- are inhibitors.

Many models have been proposed for the mechanism of the hydrolysis. Most of the available information is consistent with a model in which the hydrolysis occurs in three steps, as shown in Fig. 18.1 [2]. In the first step, which is thought to be the magnesium-requiring and sodium-requiring portion, the ATP

1. $E + ATP \xrightarrow[\text{Mg}^{2+}]{\text{Na}^+} E \sim P + ADP$

2. $E \sim P \longrightarrow E\text{-}P$

3. $E\text{-}P \xrightarrow{\text{K}^+} E + P$ **Figure 18.1**   Action of Na^+, K^+-dependent ATPase.

is split and the phosphate is linked covalently to either the enzyme itself or to
some other membrane phosphoprotein found associated with the ATPase. There
is reasonable chemical evidence to indicate that the linkage involves the γ-carboxyl
of a glutamate in the protein to which the phosphate becomes attached in a phos-
phate-carboxyl anhydride linkage. The second step involves a conformational
shift in which the energy of the phosphate anhydride is used to perform mechanical work. The third step is a hydrolysis of the phosphate which requires K^+ and
is inhibited by low concentrations of ouabain and other cardiac glycosides, an
inhibition which is discussed in detail below. The enzyme also catalyzes, under
certain conditions, an ADP-ATP exchange. The observation of such an exchange
was one of the first clues to the mechanism, namely to the covalent linkage of
the phosphate. The exchange requires Na^+ and Mg^{2+}, as would be predicted from
the mechanism indicated above, and ATPase inhibitors such as N-ethylmaleimide,
BAL (dimercaprol), arsenite, and oligomycin apparently activate this exchange,
presumably by inhibiting the K^+-dependent hydrolysis step.

The evidence that the Na^+, K^+-ATPase is responsible for the transport of
Na^+ and K^+ is admittedly indirect. It is, however, quite strong and widely accepted. Large amounts of data indicate that the detailed kinetic properties of
the ATPase are identical with the kinetics of the transport process itself. Indeed,
the half-maximal concentrations of Na^+ and K^+ required for both processes are
similar. The ions which will substitute for K^+ in the action of the ATPase will
also be transported across the membrane, and in the same order of effectiveness.
Finally, the cardiac glycosides inhibit both the transport process and the ATPase
in the same low concentrations, and in the same order of effectiveness.

The distribution of the enzyme is also in accord with its role in transport.
That is, although most, if not all, cells exhibit both Na^+, K^+-dependent ATPase
activity, the levels of the enzyme present in membrane preparations from a wide
variety of tissues correlate relatively well with the level of the Na^+ pump as measured in the whole cells.

In the brain this correlation between distribution and transport has some
interesting aspects. One of these involves the observation that glia generally have
a higher K^+ content [1], a higher level of the ATPase [11-13], and a more active
transport of Na^+ and K^+ than do neurons [14]. Such data support the concept

that one of the functions of the glia, and specifically of the astroglia, is to serve as a buffer of sorts and to absorb the potassium released into the intracellular spaces of the brain when the neuron with which the astroglia is associated fires. The purpose would be to prevent adjacent neurons from being subjected to wide fluctuations in their ionic environment. Such a concept is widely supported [15-17]. In fact, it has been stated in a recent paper that the "weight of circumstantial evidence favours the concept that astrocytes may occupy a central or key role in modulating the external ionic milieu of the neurons they surround" [18].

Another aspect of the distribution of the ATPase is its relatively high activity in the membranes of synaptosomes [19-21]. This enrichment has suggested to many that there must be some special role of the ATPase in synaptic transmission and nerve conduction. There is no hard evidence for this, and although the sodium pump is the agency by which the membrane is polarized, i.e., by the disproportion of monovalent cations across it, the pump is not thought to be involved in the depolarization or repolarization associated with nerve conduction.

As mentioned above, cardiac glycosides inhibit the sodium pump and the ATPase. They do so in a specific manner and at very low concentrations, generally 1×10^{-6} M or lower. The action of ouabain, the most studied of the glycosides (Fig. 18.2), is known to take place at the outer surface of the cell membrane. The glycoside seems to bind tightly, but noncovalently, to the ATPase, and K^+ appears to compete for the binding site [22,23].

Again, although the action of the cardiac glycosides on the ATPase is common to all cells that transport sodium, there are quite special consequences when the brain is exposed to such inhibitors. Systemic administration of these compounds has little or no neurological effect, since they do not readily cross the blood-brain barrier. Direct application of ouabain, or systemic administration to young animals with a poorly developed blood-brain barrier, does have severe consequences [24,25]. These consequences appear to be directly related to the

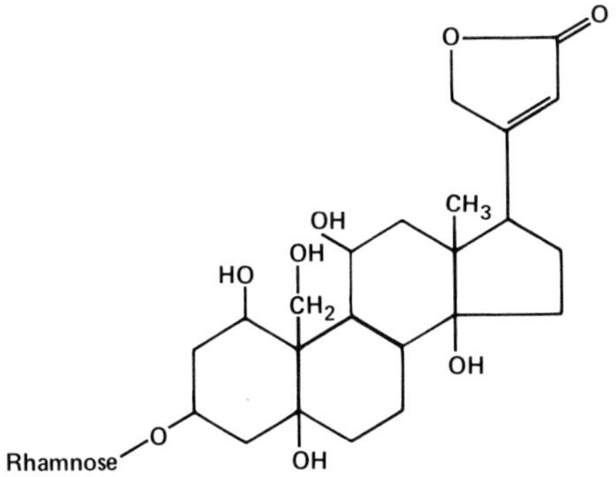

Figure 18.2   Ouabain.

inhibition by ouabain of the N^+,K^+-ATPase and, thereby, the sodium pump. Intracerebral administration of ouabain leads to leakage of K^+ from the tissue into the cerebrospinal fluid, followed by violent and frequently fatal convulsions. The physical cognate of the convulsions is a condition known as status spongiosus [26], which is due to a selective swelling of the astrocytes [27,28]. This astrocytic swelling would seem to follow from the information indicated above that the astrocytes are the relevant entities as regards the uptake of K^+ after neuronal firing and have high levels of the Na^+,K^+-ATPase.

A great number of materials in addition to ouabain cause convulsions in animals. Most of them have no effect on the Na^+,K^+-ATPase. Thus, obviously, the mechanism of all convulsions is not inhibition of this enzyme. On the other hand, there are a number of systemic convulsants, among them ethacrynic acid, mercuric acetate, copper ions, and aluminum ions, whose mechanisms of action do seem to involve the ATPase [29,30].

The administration of ouabain has been considered by some to be closely analogous to the situation in clinical epilepsy. Tower has suggested "that the final common expression of epileptogenicity is likely to be such an impairment which creates the characteristic instability of the excitable membrane" [1]. Experimentally it has been shown that samples of human epileptogenic cerebral cortex exhibit abnormally high Na^+ and abnormally low K^+ levels. Also, they cannot take up K^+ or extrude Na^+ in vitro as well as normal tissue [31,32]. It is attractive to implicate the sodium pump in the etiology of epilepsy. Such a model would include a faulty enzyme in certain glia. These glia would normally remove the extracellular K^+ released during neuronal firing. Since they do not fully execute their function, the threshold voltage needed to initiate action potential in the relevant neurons is quite low and the neurons are easily or spontaneously excitable.

In any case, the ouabain model has been a useful one. Convulsions have also been produced by the injection of slivers of cobalt into the brains of rats. Such injection is followed by a condition of slow onset but prolonged duration which involves the appearance of an epileptic electroencephalogram pattern, a lowered seizure threshold to chemical convulsants, and spontaneous convulsions. This model has been explored to some extent, but there is some doubt as to whether it has relevance to epilepsy since substantial nonspecific tissue destruction is associated with such treatment [33,34].

It has also been tempting to implicate faulty ion transport in the etiology of cerebral edema [35]. Edema usually involves fluid accumulation in the intercellular space of the subcortical white matter. The fluid which accumulates is high in sodium and chloride and contains albumin, which characteristically indicates that the fluid is an exudate of plasma. Edema can result from many causes, including, but not limited to, wounds, tumors, and various chemicals such as derivatives of tin. A form of edema occurs following episodes of circulatory arrest.

Therapy for edema involves either surgery or the administration of hyperosmolar agents such as urea. Steroids such as dexamethasone or cortisone have also been used successfully. The edematous swelling correlates with increases in the extra-cellular K^+ concentration, and a selective swelling of the astroglia [36].

In any case, reception, conduction, and transmission of the nerve impulse, indeed, the very core of neuronal functioning, depends directly upon normal op-eration of the membrane processes which maintain a differential distribution of Na^+ and K^+ across the excitable neural membrane. A failure of energy metabo-lism due to O_2 deprivation (ischemia) produces a catastrophic loss of this function.

This is not to say that the monovalent cations, Na^+ and K^+, have no other function in the economy of the brain. Clearly they are involved in a number of aspects of brain metabolism. Although the exact mechanism is obscure, there is substantial evidence that sodium ions are implicated in the transports of amino acids and carbohydrates. Probably the Na^+,K^+-ATPase is involved also. It has been shown that Na^+ is required for amino acid transport into both brain tissue [37] and peripheral nerve in vitro [38]. It is also known that ouabain inhibits amino acid transport into brain. Almost identical statements can be made about the transport of glucose, although there have not been such detailed studies re-garding sugar transport into the brain. Nevertheless, although there are reports of other ATPases, e.g., the Ca^{2+}-stimulated ATPase or the HCO_3^--stimulated ATPase [39], both from the mitochondria, and both presumably involved in the transport of the ions by which they are stimulated, there are no reports of ATP-ase stimulated by amino acids or carbohydrates. Similarly, it is well known that the monovalent cations are involved in the uptake of γ-aminobutyric acid [40, 41] and of norepinephrine [42,43] into the synaptosomes. There is no informa-tion as yet as to the mechanism of this involvement.

It is not appropriate to reveiw all the involvements of the monovalent ca-tions here, since they are discussed in other, more relevant sections. But it must be mentioned that the ionic requirements for in vitro protein synthesis are well known, and that they are quantitatively stricter in brain than in other tissues. High levels of K^+ are inhibitory to brain protein synthesis both in vivo and in vitro [44,45]. Also, it is well known that K^+ has marked and immediate effects on respiration and glycolysis in neural preparations. These latter effects are most certainly complex and multitargeted since it has been shown that a number of isolated enzymes in the energy economy of neural and other tissues are influ-enced in various ways by the presence of sodium and/or potassium ions.

Lithium

The effects of lithium ion deserve special mention here [46], since its influence may be dependent in some way on its interaction with sodium transport and the sodium pump. Lithium will substitute weakly for potassium in stimulating the

ATPase [6,47], and has been reported to compete with sodium for equilibration across the cell membrane [48]. On the other hand, lithium itself equilibrates very slowly into the brain and distributes about equally between blood and brain at equilibrium, quite differently than either sodium ions or potassium ions, and recent studies suggest that lithium accumulates in the particulate fraction [49].

Early work indicated that lithium is extremely toxic, especially when the Na^+ content of the body is below normal, as, for example, during treatment with diuretics or under various forms of salt restriction. Upon administration of lithium, tremors appear, followed by vomiting, twitching, vertigo, ataxia, asthenia, and drowsiness. In moderate doses, however, and with appropriate attention to the sodium levels in the body, lithium administration has proved to be a remarkably successful therapy in the treatment of the manic phase of manic-depressive illness. Thus, careful treatment with levels of the order of 1 Eq/liter of blood causes a reduction of the irritability, and a lowering of the restlessness, excitement, and dysphoria characteristic of the mania. It is quite a specific treatment since it seems to have no effects, in these dosages, on normal individuals, on other types of emotional illness, or even on the depressive aspect of the bipolar manic-depressive. Lithium treatment has been used in a prophylactic sense against relapses into bipolar disease.

Although the effectiveness of lithium in treating these specific mental disorders is well established [50], numerous studies have failed to elucidate its mode of action. In addition to the possible effects on the sodium pump and cation distribution, lithium has been reported to have effects on glycolysis and on mitochondrial respiration [51], and to cause alterations in the pattern of high-energy phosphates [52]. But the area which has attracted the most interest is the possible role of lithium in the metabolism and release of the amines. This is the case because of the prevailing impression that the amines are in some integral way involved in emotional disorders. There have been, in fact, many reports on the ability of lithium to inhibit amine release [53], to deplete amine stores [54], to increase amine turnover and synthesis [55-57], and so on. There are also any number of negative reports on the same subjects. Thus, although it is widely suspected that lithium influences the emotional state through its action on amine metabolism or utilization, no coherent experimental picture has emerged which would explain all the actions of lithium on the brain.

A number of divalent cations are now known to be of special importance in neural functioning. A number of others have been shown to alter neuronal function when administered. The major divalent cations in the brain, as mentioned before, are Mg^{2+} and Ca^{2+}. The enzymatic roles of Mg^{2+} are well known and center about the function of the phosphate transferase enzymes. Not only is Mg^{2+} required for the action of the cation transport ATPase described above, but for all the phosphate-transferring enzymes in the glycolytic and oxidative pathways, as well as in the pathways of nucleic acid metabolism. The contractile protein of the brain, neurostenin, has a Mg^{2+}-dependent ATPase action [58], as do the other contractile proteins in other cells.

The role of calcium ions is perhaps not so well characterized, but may be as widespread and seems, enzymatically, more diverse. Calcium has a number of actions as enzyme cofactor, including activity in the action of proteinases in the brain [59,60], and as cofactors for enzymes which cause base exchange into phospholipids [61]. Also, calcium ion appears to be a pivotal part of a number of membrane processes and may be involved in the fundamental structural aspects of neural and other membranes [62-64]. Recent studies have implicated calcium ion in release of neurotransmitters [65], axonal conduction and excitability [66], and the binding of opiates to their receptors [67]. Calcium ions are accumulated by mitochondria and by nuclei, and are bound specifically to a number of soluble proteins in the nervous system, including the S-100 [68-70]. These latter binding processes, however, are of as yet unknown significance.

Chronic administration of large amounts of manganese causes psychiatric and neurological disturbances. The symptoms are similar to those seen in Parkinson's disease [71]. The similarity extends to the observation that manganese poisoning, like Parkinson's disease, causes a deficit of dopamine [72], and the administration of dopa ameliorates both conditions. The detailed mechanism of the depletion of dopamine has not yet been reported.

The neurological impact of administration of various heavy metals has been explored, but incompletely. It is known that certain derivatives of tin cause edema, and poisoning with lead seems to involve inhibition of certain of the energy-yielding reations in the brain. But these materials cause extensive damage, and it is probably the case that the neurological and other problems are due to interference with a number of biological processes.

Copper

The case of copper deficiency and copper excess is somewhat better understood. It is suspected that a genetically-based inability to maintain the body level of copper results in what is variously called Menkes' steely-hair disease or Menkes' kinky-hair disease [73]. This is a sex-linked recessive condition which is characterized by growth retardation and cerebral and cerebellar degeneration. Localized neuronal destruction has been observed. Mitochondrial malformations are found in brain and liver. There is some feeling that the condition results from a lesion in the absorption of copper from the intestine [74], but there are conflicting reports on this point. A model of this condition has been produced in laboratory animals [75], and these animals have enlarged, abnormal mitochondria in the brain and the liver. The oxidation by these particles is lower than normal by 30%, and they have a much lower content of the various cytochromes. A naturally occurring condition in grazing animals may parallel the condition in humans, since it has been found that copper deficiency in cows, as well as in lambs and guinea pigs, causes neurological abnormalities [76].

The other side of this coin is what has been called hepatolenticular degeneration, or Wilson's disease. This condition, first described in 1912 [77], is an hereditary aberration of copper metabolism transmitted as an autosomal recessive. In such individuals, serum copper levels are low and urinary excretion of copper is high. Depositions of copper appear in many parts of the brain, and copper-containing proteins are found which are different than the cerebrocupreins normally present in brain. Although early mental development is normal there is a progressive dementia, with paranoid delusions and episodes of bizarre behavior. The biochemical defect is as yet unknown, but one theory is that ceruloplasmin, the normal copper-carrying protein of the plasma, is altered or deficient and that copper is more loosely bound to albumin instead. This loose binding facilitates excess deposition in tissues. Administration of dimercaprol or of D-penicillamine, both of which are copper chelators, has provided a successful therapy in many cases.

Wilson's Disease

The patient, a 36-year-old white male, was admitted with slow slurring dysarthria punctuated by long stridulus inspiration. He had bilateral Kayser-Fleischer rings, with arcus senilis on top, but no nystagmus. A history revealed the appearance of shaking of the left hand 3 years before admission, followed 1 or 2 months later by similar shaking of the right hand. The tremor had increased, gradually affecting his arms and legs and was accompanied by side-to-side rotation of the head. A year before his admission his speech had become slurred and he developed generalized headaches. He was found to have mild jaundice, and palpation indicated a moderate enlargement of the liver.

The diagnosis of Wilson's disease was confirmed by decreased plasma copper (60 μg/100 ml), decreased ceruloplasmin (9 mg/100 ml), and, upon liver biopsy, vastly increased copper content (80 μg/g wet weight of tissue). Administration of [67]Cu showed a markedly prolonged retention of isotope in the body.

The patient was given 4 ml of a 5% solution of dimercaprol subcutaneously 4 times a day for 1 day, twice a day for 2 days, and once a day for 3 days. This course was repeated at intervals 4 times. On each occasion nausea and vomiting occurred on the first day, and drowsiness followed each course of treatments. On each occasion there was marked inprovement. He was able to walk unsupported, and the tremor of the upper limbs was less marked. By the time of discharge 8 months after admission he was able to leave the hospital and take long walks in the neighborhood. His improvement was maintained, however, for only 2 or 3 months after discharge.

References

1. D. B. Tower, in *Handbook of Neurochemistry*, Vol. I (A. Lajtha, ed.), Plenum, New York, 1969.
2. G. J. Siegel and R. W. Albers, in *Handbook of Neurochemistry*, Vol. IV (A. Lajtha, ed.), Plenum, New York, 1970.
3. P. D. Swanson and P. D. Stahl, in *Basic Neurochemistry* (R. W. Albers, G. J. Siegel, R. Katzman, and B. W. Agranoff, eds.), Little, Brown, Boston, 1972.
4. R. W. Albers, *Ann. Rev. Biochem. 36:*727 (1967).
5. A. L. Hodgkin and R. D. Keynes, *J. Physiol. (London) 128:*28 (1955).
6. J. C. Skou, *Biochim. Biophys. Acta 23:*394 (1957).
7. J. C. Skou, *Physiol. Rev. 45:*596 (1965).
8. S. S. Goldman and R. W. Albers, *J. Biol. Chem. 248:*867 (1973).
9. W. L. Stahl, *Arch. Biochem. Biophys. 56:*154 (1973).
10. J. Kyte, *J. Biol. Chem. 246:*4157 (1971).
11. S. P. R. Rose and A. K. Sinha, *J. Neurochem. 16:*1319 (1967).
12. H. Haljamäe and A. Hamberger, *J. Neurochem. 18:*1903 (1971).
13. F. A. Henn, H. Heljamäe, and A. Hamberger, *Brain Res. 43:*437 (1972).
14. H. K. Kimelberg, *J. Neurochem. 22:*971 (1974).
15. H. M. Gershenfeld, F. Wald, J. A. Zadunaisky, and E. D. P. deRobertis, *Neurology 9:*410 (1959).
16. R. K. Orkand, J. G. Nicholls, and S. W. Kuffler, *J. Neurophysiol. 29:*788 (1966).
17. R. S. Bourke and K. M. Nelson, *J. Neurochem. 19:*663 (1972).
18. T. H. Gill, O. M. Young, and D. B. Tower, *J. Neurochem. 23:*1011 (1974).
19. R. W. Albers, G. Rodriguez de Lores Arnaiz, and E. deRobertis, *Proc. Natl. Acad. Sci. USA 53:*557 (1965).
20. H. F. Bradford, E. K. Brownlos, and D. B. Gammach, *J. Neurochem. 13:*1283 (1966).
21. A. Y. Sun, G. Y. Sun, and T. Samirajski, *J. Neurochem. 18:*1711 (1971).
22. H. Yoshida, T. Nukada, and H. Fujisawa, *Biochim. Biophys. Acta 48:*614 (1961).
23. P. D. Swanson and H. McIlwain, *J. Neurochem. 12:*877 (1965).
24. G. C. Santesson and L. Strindberg, *Scand. Arch. Physiol. 35:*51 (1918).
25. A. Rizzolo, *Arch. Farmacol. Sper. Sci. Affini 50:*16 (1929).
26. A. Bignami and G. Palladini, *Nature (London) 209:*413 (1966).
27. A. Stefanelli, G. Palladini, and L. Ieradi, *Experientia 21:*717 (1965).
28. K. Renkawek, G. Palladini, and L. A. Ieradi, *Brain Res. 18:*363 (1970).
29. D. J. Brown and W. E. Stone, *J. Neurochem. 20:*1461 (1973).
30. G. Venturini, *J. Neurochem. 21:*1147 (1973).
31. D. B. Tower, *Neurochemistry of Epilepsy* Thomas, Springfield, Ill., 1960.
32. D. B. Tower, *Epilepsia 6:*183 (1965).
33. W. A. Hunt and C. R. Craig, *J. Neurochem. 20:*559 (1973).
34. R. J. Cenedella and C. R. Craig, *J. Neurochem. 21:*743 (1973).

35.   D. B. Tower, in *Basic Neurochemistry* (R. W. Albers, G. J. Siegel, R. Katz-
      man, and B. W. Agranoff, eds.), Little, Brown, Boston, 1972.
36.   A. Ames, III, R. L. Wright, M. Kowanda, J. M. Thurston, and G. Majno,
      *Am. J. Pathol. 52:*437 (1968).
37.   A. Lajtha and H. Sershen, *J. Neurochem. 24:*667 (1975).
38.   D. D. Wheeler, *J. Neurochem. 24:*1197 (1975).
39.   H. K. Kimelberg and R. S. Bourke, *J. Neurochem. 20:*347 (1973).
40.   D. L. Martin, *J. Neurochem. 21:*345 (1973).
41.   J. R. Simm, D. L. Martin, and M. Krell, *J. Neurochem. 23:*981 (1974).
42.   D. F. Bogdanski, A. Tissari, and B. B. Brodie, *Life Sci. 7:*419 (1968).
43.   R. W. Colburn, F. K. Goodwin, D. L. Murphy, W. E. Bunney, and J. M.
      Davis, *Biochem. Pharmacol. 17:*957 (1968).
44.   J. Krivanek, *J. Neurochem. 17:*531 (1970).
45.   J. Krivanek, *J. Neurochem. 23:*1255 (1974).
46.   M. Schou, in *Handbook of Neurochemistry,* Vol. VI (A. Lajtha, ed.),
      Plenum, New York, 1972.
47.   W. N. Aldridge, *Biochem. J. 83:*527 (1962).
48.   M. Schou, *Acta Pharmacol. Toxicol. 15:*85 (1958).
49.   S. Christensen, *J. Neurochem. 23:*1299 (1974).
50.   F. K. Goodwin and R. L. Sack, in *Frontiers in Catecholamine Research*
      (E. Usdin and S. Snyder, eds.), Pergamon, Oxford, 1973.
51.   A. R. Krall, *Life Sci. 6:*1339 (1967).
52.   L. J. King, J. L. Carl, E. G. Archer, and M. Castellant, *J. Pharm. Exp.
      Ther. 168:*163 (1969).
53.   R. L. Katz, T. N. Chase, and I. J. Kopin, *Science 162:*466 (1968).
54.   H. Corrodi, K. Fuxe, T. Hokfelt, and M. Schou, *Psychopharmacologia 11:*
      345 (1967).
55.   D. N. Stern, R. R. Fieve, N. H. Neff, and E. Costa, *Psychopharmacologia
      14:*315 (1969).
56.   J. J. Schildkraut, M. A. Logue, and G. A. Dodge, *Psychopharmacologia
      14:*135 (1969).
57.   K. Greenspan, M. S. Aronoff, and D. F. Bogdanski, *Pharmacology 3:*129
      (1970).
58.   C. Mahendran, W. J. Nicklas, and S. Berl, *J. Neurochem. 23:*497 (1974).
59.   G. Guroff, *J. Biol. Chem. 239:*149 (1964).
60.   M. Hamon, S. Bourgoin, F. Hery, and J. Glowinski, *J. Neurochem. 23:*849
      (1974).
61.   A. Gaiti, G. E. DeMedio, M. Brunetti, L. Amaducci, and G. Porcellati, *J.
      Neurochem. 23:*1153 (1974).
62.   F. Brink, *Pharmacol. Rev. 6:*243 (1954).
63.   A. M. Shanes, *Pharmacol. Rev. 10:*59 (1958).
64.   A. M. Shanes, *Pharmacol. Rev. 10:*165 (1958).
65.   B. Katz and R. Miledi, *Nature (London) 215:*651 (1967).
66.   I. Tasaki, A. Watanabe, and T. Takenaka, *Proc. Natl. Acad. Sci. USA 48:*
      1177 (1962).

67. C. B. Pert and S. Snyder, *Proc. Natl. Acad. Sci. USA 70:*2243 (1973).
68. P. Calissano, B. W. Moore, and A. Friesen, *Biochemistry 8:*4318 (1969).
69. P. Calissano, in *Proteins of the Nervous System* (D. J. Schneider, ed.), Raven, New York, 1973.
70. P. Calissano and A. D. Bangham, *Biochem. Biophys. Res. Commun. 43:* 504 (1971).
71. G. C. Cotzias, P. S. Papavasiliou, J. Ginos, A. Steck, and S. Duby, *Ann. Rev. Med. 22:*305 (1971).
72. E. Bonilla and M. Diez-Ewald, *J. Neurochem. 22:*297 (1974).
73. J. H. Menkes, M. Alter, G. K. Steigleder, D. R. Weakley, and J. H. Sung, *Pediatrics 29:*764 (1962).
74. D. M. Danks, P. E. Campbell, J. Walker-Smith, B. J. Stevens, J. M. Gillespie, and J. Bloomfield, *Lancet 1:*1100 (1972).
75. J. R. Prohaska and W. W. Wells, *J. Neurochem. 25:*221 (1975).
76. E. J. Underwood, *Trace Elements in Human and Animal Nutrition,* 3rd ed., Academic, New York, 1971.
77. S. A. K. Wilson, *Brain 34:*295 (1912).

19
VITAMINS

Since the metabolism of the brain is, in most vital respects, no different qualitatively than that of any other tissue, the adequate functioning of brain requires optimal levels of all the vitamin coenzymes. Deficiencies of most essential nutrients, then, will eventually produce severe neurological complications. In fact, as discussed in Chap. 20, generalized malnutrition during certain stages of development is widely thought to produce irreversible cellular changes and probably just as irreversible behavioral changes. On the other hand, the special quantitative aspects of brain metabolism, such as its extreme reliance on glucose as an energy source, or its highly developed amino acid pathways, suggest that the brain may be especially sensitive to deficiencies in certain of the vitamins and their coenzyme forms. Such is the case, and, in fact, a neurological heirarchy of the vitamins can be drawn. That is, it is clear that neurological consequences ensue much more rapidly and severely upon depletion of certain coenzymes than of others. Roughly speaking, such a hierarchy of vitamins would be, in decreasing order of importance: thiamine (B_1), niacin, pyridoxal (B_6), B_{12}, folic acid, and pantothenic acid. In deficiencies of riboflavin (B_2), biotin, vitamin C, and the fat solubles, neurological problems are not predominant.

Thiamine

Among the major functions of thiamine pyrophosphate is to serve as coenzyme for the oxidative decarboxylation(s) of pyruvate and α-ketoglutarate (Fig. 19.1) and for the carbon rearrangements catalyzed by the enzyme transketolase (Fig. 19.2) in the hexose monophosphate shunt. These two pathways are of singular importance because of the exclusive use, by brain, of glucose as an energy source, and because the hexose monophosphate shunt provides reduced pyridine nucleotides for fatty acid biosynthesis. Thus, the sensitivity of the nervous system to any deficiency of vitamin B_1 is predictable, and, indeed, neurological malfunction is among the earliest signs of thiamine deficiency.

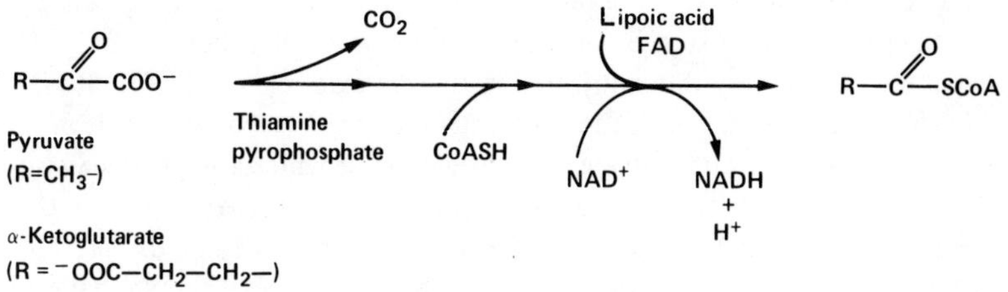

Figure 19.1   Pyruvate and α-ketoglutarate dehydrogenases.

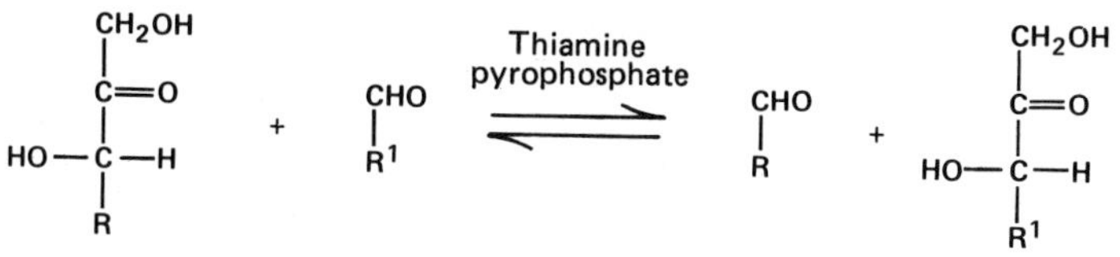

Figure 19.2   Transketolase.

Clinically, these neurological manifestations can be divided into those attending a mild deficiency ("neurasthenic syndrome") and those observed in a more pronounced one (Wernicke's syndrome). In mild deficiency there is lassitude, irritability, and other moderate behavioral abnormalities which can be completely reversed by the administration of thiamine. A more severe and prolonged deficiency, complicated usually by external aggravating factors such as alcoholism, produces peripheral neuritis, ataxia, loss of equilibrium, oculomotor palsy, nystagmus, paralysis, and a group of changes in mental functioning known collectively as Korsakoff's syndrome. These include confusion, disorientation with respect to time and place, and a specific loss of recent memory out of proportion to alterations in other cognitive functions. In these individuals the administration of thiamine is not fully curative. Permanent and irreversible damage results, and pathological changes occur in the gray matter. These include hemorrhage, gliosis, neuronal degeneration, and a proliferation of the blood vessels.

Biochemically, the brain is reasonably resistant to these changes. A loss of 80% of the thiamine content of the brain can occur before neurological signs appear [1], and a repletion to 30% of normal is usually enough to restore normal function. In thiamine deficiency glutathione levels in the body drop, as do levels of thiamine pyrophosphate in blood cells, and the content of pyruvate and lactate of the brain rises. Of the enzymes which use the coenzyme form of thiamine the most sensitive to depletion is transketolase [1-3]. In fact, the measure-

ment of transketolase activity in the red blood cells is probably the most reliable
and most sensitive indicator of a B_1 deficiency. Pyruvate and α-ketoglutarate
dehydrogenases are not very much lowered in thiamine deficiency [1,4]. It is
reasonable to expect alterations in the lipid patterns of the brain as a result of
thiamine deficiency, because pyruvate dehydrogenase provides acetyl coenzyme
A (CoA) for lipid synthesis, and also because transketolase is a participant in the
pathway which provides NADPH for the biosynthesis of the fatty acids. Nevertheless, no alterations in the lipid patterns of the brain have been shown, beyond
those which can be ascribed to the generalized malnutrition which accompanies
B_1 deficiency. The deficient animals have brain lipid patterns similar to those of
pair-fed littermates [5] due to retardation of the normal developmental process.
Similarly, reports that B_1 deficiency is accompanied by decreases in brain levels
of acetyl CoA have as yet not been universally confirmed [6].

The enzymes which require thiamine are unevenly distributed in the nervous system. Pyruvate dehydrogenase is present in highest concentration in the
gray matter, and is low in white matter. Transketolase is present in highest concentration in spinal cord and in brainstem, and is lowest in cerebral cortex. Brain
is moderately rich in thiamine derivatives. Of the total thiamine, at least 90% is
present as the coenzyme form, thiamine pyrophosphate. Only very low levels of
free thiamine are present. The remainder, or about 10% of the total thiamine, is
in the form of thiamine triphosphate, a derivative which is discussed below.

Experimentally, the role of thiamine, and the consequences of a thiamine
deficiency, have frequently been studied using thiamine antagonists. The analog
which has been most useful in this regard has been pyrithiamine (Fig. 19.3), the
mechanism of its action being an inhibition of the conversion of thiamine to its

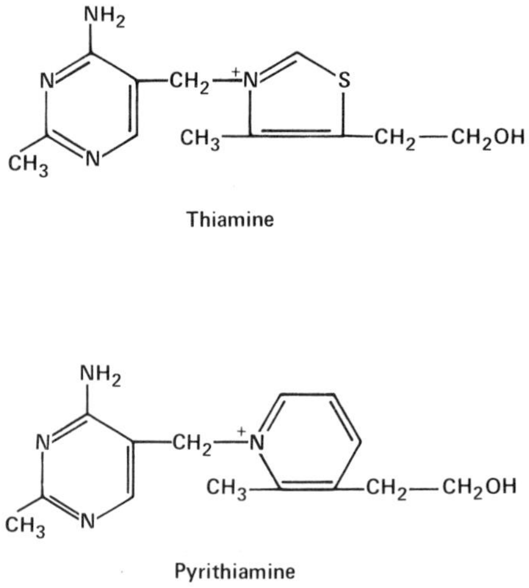

Thiamine

Pyrithiamine

Figure 19.3   Thiamine and
pyrithiamine.

coenzyme form. Generally speaking, the consequences of the administration of pyrithiamine, along with a dietary restriction of thiamine, are merely the expected aggravation of the thiamine deficiency. In other words, the administration of pyrithiamine merely increases the severity of the B_1 deficiency in the expected ways. There are some few studies, however, which suggest that pyrithiamine has some actions which are perhaps not merely extensions of its B_1 antagonism. Indeed, pyrithiamine produces the neurological symptoms of the thiamine deficiency and has, as well, a profound and irreversible effect on the electrical activity of nerve preparations.

This observation forms part of the rationale used to understand the involvement of the thiamine derivative, thiamine triphosphate, in the condition known as Leigh's disease, or more formally, subacute necrotizing encephalomyelopathy [7]. This condition, which was first recognized in 1951 [8], resembles the Wernicke's encephalomyopathy clinically in several respects. It is a fatal hereditary condition, with neurological manifestations, which has been observed in 40 or 50 individuals. During the first year of life there is a marked loss of weight, vomiting, weakness, and psychomotor retardation. Seizures are also seen. There are necrotic lesions found in the brainstem and the spinal cord.

On the other hand, there are some differences between Leigh's disease and Wernicke's encephalomyopathy. In Leigh's disease the thiamine levels of blood are normal, transketolase is not lowered, and normal doses of thiamine are not effective as therapy. Just the opposite is true of Wernicke's. Further, the lesions in Leigh's disease are in the basal ganglia and substantia nigra, and the mamillary bodies are never involved. Again, the reverse is true in Wernicke's. Clearly the two are independent lesions.

Some clue as to the biochemistry involved here is available from the observation that, in five such children, thiamine triphosphate (TTP) is absent from the brain [7]. Normally, as mentioned above, this derivative comprises some 10% of the total thiamine in brain. The enzyme which catalyzes the synthesis of TTP from ATP and thiamine pyrophosphate (TPP) has been found in normal brain tissue [9]. It has been reported that an inhibitor of this enzyme is present in the blood, urine, and cerebrospinal fluid of the children suffering from Leigh's disease [10]. This inhibitor, which has been shown to be a glycoprotein with a molecular weight on the order of 40,000 or more, is labile upon standing, and disappears from the urine when large doses of thiamine are administered [7]. The presence of this protein in the urine has been used as a diagnostic tool for the condition. The administration of large amounts of thiamine is said to be of positive but temporary value in this condition. It seems fair to say that the picture of Leigh's disease drawn in these studies is not universally accepted.

Be that as it may, there is some evidence to support the postulate [11,12] that thiamine or one of its phosphorylated forms has a function in nervous conduction. The major points of this argument are the following [7]:

1. Pyrithiamine produces irreversible effects on nerve preparations, and these effects seem to be due to displacement of thiamine itself.
2. Electrical stimulation releases thiamine from intact nerve preparations.
3. Thiamine phosphatases and phosphotransferases which regulate the levels of thiamine phosphoesters appear to be located in the membranes [13].
4. Thiamine restores the action potential in an ultraviolet irradiated nerve.

These and other arguments have led to the postulate that TPP or TTP occupies a site on the membrane and that cyclic dephosphorylation and rephosphorylation promotes the passage of ions, probably sodium, across the membrane. There is little direct evidence to support this postulate.

Niacin

The neural problems which arise in niacin deficiency are reasonably well known due to the widespread information on the condition known as pellagra, which is substantially due to a deficiency of exogenous and endogenous niacin. Neurological symptoms are among the earliest seen in this condition. Initially, there is depression and anxiety, dizziness, confusion, and a difficulty in sleeping. Later symptoms include apprehension and hallucinations along with delirium. This later phase is accompanied by structural changes in the brain, most prominently a chromatolytic degeneration of the large pyramidal cells of the motor cortex, the Betz cells, and a degeneration of peripheral nerves.

The functions of the pyrimidine nucleotides in nervous tissue are no different than those they serve in other tissues. The concentration of nicotinic acid derivatives in brain is quite high, and most of the nicotinic acid is present as the pyridine nucleotides. The route of synthesis of the nucleotides is no different in the brain than in the liver and other tissues, as has been demonstrated by intracisternal administration of nucleotide precursors.

Pyridoxal

The dependence of brain on normal amino acid metabolism makes predictable the severe and immediate neurological consequences of vitamin B_6 deficiency. Vitamin B_6, after all, is involved in a host of amino acid metabolic reactions, many of which are common to all tissues, but some of which, like the γ-aminobutyric acid (GABA) pathway, are restricted to brain. Among the neurological consequences of B_6 deficiency are lowered seizure threshold and convulsions, especially in preweanling animals, extreme irritability, and ataxia. Learning and conditioning deficits have also been reported. Isonicotinic acid hydrazide (Fig. 19.4), an antitubercular drug which was widely used some years ago, also lowers

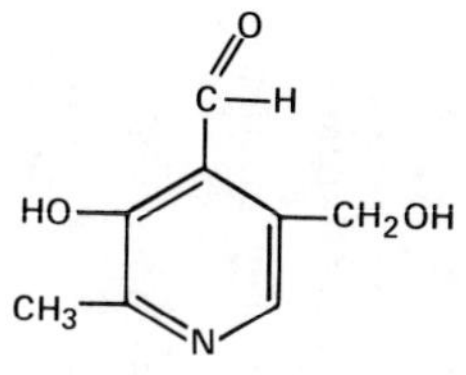

Pyridoxal

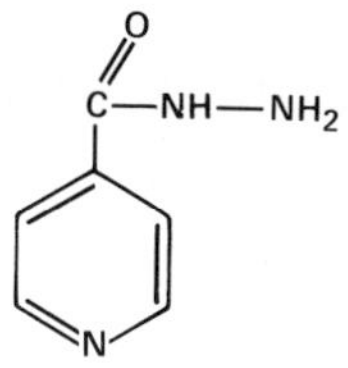

C—NH—NH$_2$

Isonicotinic acid hydrazide   **Figure 19.4** Pyridoxal and isonicotinic acid hydrazide.

tissue levels of the coenzyme form of vitamin B$_6$, pyridoxal phosphate, by inhibiting its formation from B$_6$, a conversion which has been shown to take place in the
brain as well as in other organs. Administration of isonicotinic acid hydrazide has
been reported to lead to psychotic episodes in certain patients, and to produce
convulsions in animals.

The various enzymes for which pyridoxal phosphate serves as cofactor are
not all equally susceptible to loss of this cofactor during B$_6$ deficiency. Nor are
they all equally sensitive to the various B$_6$ antagonists. A good example of such
a situation is the enzymes of the GABA pathway. Both the enzyme which
forms GABA and the one which destroys it are pyridoxal phosphate-requiring
enzymes. The former, glutamic acid decarboxylase, is the more sensitive of the
two to inhibition by thiosemicarbazide, a B$_6$ inhibitor, which causes a drop in
GABA levels. The latter, GABA transaminase, is more sensitive to hydroxylamine,
also a B$_6$ inhibitor, and GABA levels rise. A general finding is that the decarboxylases, which have relatively low affinities for the cofactor, are usually the enzymes depleted first. Thus, in simple B$_6$ deficiency, GABA levels are reduced
due to a selective loss of pyridoxal phosphate from the decarboxylase for glutamic acid.

More recently, a new class of inherited disease states has been described in
which an inborn lesion in the utilization of vitamin B$_6$ exists. Such individuals
sometimes respond to a high and continuing regimen of the vitamin. These con-

Table 19.1 Vitamin-Responsive Inherited Disorders Affecting the Nervous System

Vitamin	Disorder	Therapeutic dose	Enzymatic defect	Neurologic manifestations
Thiamine (B_1)	Branched-chain keto-aciduria	5-20 mg	Branched-chain ketoacid decarboxylase	Lethargy, coma
	Lacticacidosis	5-20 mg	Pyruvate carboxylase	Mental retardation
	Pyruvicacidemia	5-20 mg	Pyruvate dehydrogenase	Cerebellar ataxia
Pyridoxine (B_6)	Homocystinuria	> 25 mg	Cystathionine synthase	Mental retardation, cerebrovascular accident, psychoses
	Infantile convulsions	10-50 mg	Glutamic acid decarbox-ylase	Seizures
	Xanthurenicaciduria	5-10 mg	Kynureninase	Mental retardation
Cobalamin (B_{12})	Methylmalonicaciduria	1.000 μg	5-Deoxyadenosyl-B_{12} synthesis	Lethargy, coma
	Methylmalonicaciduria, homocystinuria, and hypomethioninemia	> 500 μg	5-Deoxyadenosyl-B_{12} and methyl-B_{12} synthesis	Developmental arrest, cerebellar ataxia

Folic acid	Megaloblastic anemia	<0.05 mg	Intestinal malabsorption	Mental retardation
	Formiminotransferase deficiency	>5 mg	Formiminotransferase	
	Homocystinuria and hypomethioninemia	>10 mg	N^5,N^{10}-Methylenetetrahydrofolate reductase	Schizophrenia
Biotin	β-Methylcrotonyl-glycinuria	5–10 mg	β-Methylcrotonyl-CoA carboxylase	Mental retardation
	Propionicacidemia	5–10 mg	Propionyl-CoA carboxylase	Lethargy, coma
Nicotinamide	Hartnup's disease	>400 mg	Intestinal malabsorption of tryptophan	Cerebellar ataxia

Table reprinted from L. E. Rosenberg, in *Brain Dysfunction in Metabolic Disorder,* Vol. 53 (F. Plum, ed.), Res. Publ. Res. Assoc. Nerv. Ment. Disease, Raven, New York, 1974.

ditions have been termed the "vitamin dependency" states, since without a constant administration of pharmacological levels of the vitamin the individual returns to the original condition.

Such a vitamin-dependent condition was first reported in 1954 [14] in a child with seizures who was more or less accidentally found to respond to a 10-fold normal dose of B_6. The condition has been studied in detail by Scriver [15]. It was suggested that the condition is due to a defect in glutamic acid decarboxylase [16], and this was subsequently confirmed [17]. Presumably the enzyme is altered in some way that prevents the normal binding of its coenzyme. Administration of very high levels of vitamin then permits this altered enzyme to function in a normal manner. Accordingly, it has been shown that early diagnosis and treatment of these vitamin-dependent children allows them to develop normally.

There are now several of these vitamin-dependent conditions known [18, 19]. A few of them affect the nervous system (Table 19.1). Some, like Hartnup's disease, are discussed above in a different context (Chap. 16). Most are quite rare. The causes of the various conditions, on a molecular basis, are quite diverse. In some cases it is thought that there is a defect in the transport of the vitamin into the various cells. In others there seems to be a feeble enzyme in the biosynthesis of the cofactor. Also there are conditions, like the glutamic acid decarboxylase lesion mentioned above, in which the apoenzyme portion is defective in its ability to bind the cofactor. In all such cases it is fairly obvious why large excesses of the vitamin would correct the defect and allow adequate functioning.

Vitamin B_{12}

There are severe neurological problems associated with a deficiency of vitamin B_{12} whether due to pernicious anemia or other causes. Among the earliest are a numbness and a tingling of the hands and feet, ataxia, and poor muscular control. Agitation and depression follow, accompanied by confusion, moodiness, and a poor recall. Eventually, individuals suffer delusions, hallucinations, and psychotic episodes.

The primary lesion appears to be one of demyelination throughout the nervous system. Various studies have recorded degeneration of the spinal cord, the optic nerve, the cerebral white matter, or the peripheral nervous system. The deficiency causes a decrease in the conversion of methylmalonyl CoA to succinyl CoA, a reaction catalyzed by a B_{12}-requiring enzyme (Fig. 19.5), and also the terminal reaction in the catabolism of propionyl CoA, a product of the breakdown of fatty acids with an odd number of carbon atoms. It may be that the inability of the body to metabolize propionyl CoA causes unnatural lipids to be incorporated into myelin [20]. Thus, a lesion in the metabolism of propionic acid could explain the myelin problems.

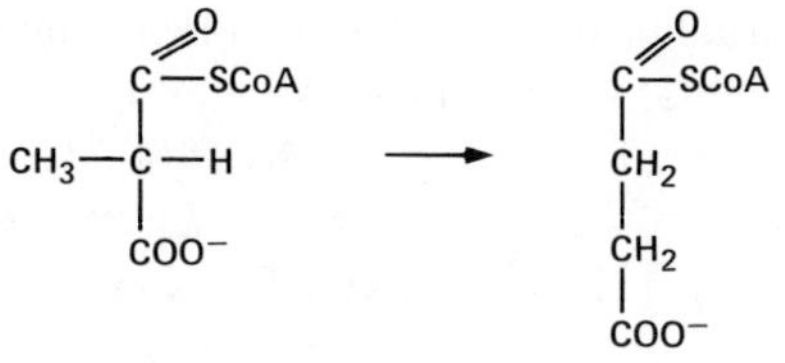

$$\text{Methylmalonyl CoA} \longrightarrow \text{Succinyl CoA}$$

Methylmalonyl CoA

Succinyl CoA

Figure 19.5   Methylmalonyl CoA mutase.

A continuing deficiency of B_{12} also lowers the conversion of homocysteine to methionine and, as a consequence, lowers methionine levels of the tissues, including the brain. A continuing deficiency of methionine, and consequently of methyl groups, could play a role in the mental problems encountered in the deficiency and in a deficiency of folate to be discussed below, since both B_{12} and folate are involved in the methionine-forming reaction.

The excretion of methylmalonic acid, then, is a clinical indicator of the B_{12} status of the individual. In recent years a number of children with hereditary defects in different aspects of B_{12} utilization has been observed, some of whom have also been shown to respond to large doses of B_{12} and been termed "B_{12}-dependent." The first, reported in the later 1960s, were children who excreted huge amounts of methylmalonic acid, but did not have the usual hematologic signs of B_{12} deficiency. It is now clear that the excretion of methylmalonic acid in the urine is due to the malfunction of the enzyme which converts methylmalonyl CoA to succinyl CoA. The methylmalonic acid can arise from a number of metabolic precursors in the body, including thymine, valine, or propionyl CoA, which, in turn, can come from the catabolism of certain amino acids, the oxidation of odd-numbered fatty acids, or the breakdown of the side chain of cholesterol. Equally clearly, the failure of the B_{12}-requiring enzyme to function can be caused by a number of different problems in the handling of B_{12} by the individual. These may include, in addition to a dietary deficiency of B_{12}, a deficiency or malfunction of intrinsic factor, a glycoprotein which serves to transport B_{12} through the intestine; the ileal receptor, which is the site of B_{12} attachment to the cells of the ileum; or of either of the transcobalamines, proteins which are involved in the transport of B_{12} through the blood stream. In addition to these there can also be lesions in the enzymatic synthesis of the B_{12} cofactor(s), or defects in the mutase itself. If the alterations in the transports of B_{12} or the synthesis of the B_{12} coenzyme are such that the affinity for substrate is lowered, then large, pharmacologic levels of B_{12} are corrective. When the relevant transport system or synthetic enzyme is missing or irreversibly altered then administration of B_{12} in any amount is useless.

To date there have been lesions reported in all the transport entities, intrinsic factor, ileal receptor, and both of the transcobalamines. In addition, there have been at least two different lesions reported in the biosynthesis of the B_{12} coenzymes. The most comprehensive and illuminating case study is that of a boy under the care of Rosenberg and his colleagues [21-23] whose defect appears to be in the synthesis of the deoxyadenosyl B_{12} cofactor serving the mutase. The child excreted methylmalonic acid, but had no problems associated with the sulfur amino acids. Administration of pharmacologic doses of B_{12} was effective, and the child developed normally. In another patient, studied by Mudd and his coworkers [23-25], there was methylmalonic aciduria, but in addition the patient had homocystinuria, cystathionemia, and hypomethionemia. Such a pattern suggested a defect not only in the deoxyadenosyl B_{12} coenzyme, but also in the methyl B_{12} coenzyme functioning in the biosynthesis of methionine. The patient had poor muscle tone, microcephaly, and suffered convulsions. Vitamin B_{12} in high doses was not effective, and the patient died at 7 ½ weeks of age. There is some thought that this second patient had a lesion in the uptake of B_{12} into the cell.

Folic Acid

Many of the neurological aspects of a folate deficiency are similar to those of B_{12} deficiency [26]. This is due to the fact that the methylation reaction for the formation of methionine requires both B_{12} and folate and can be shut down by the absence of either. On the other hand, there are some differences due to the fact that only B_{12} deficiency affects methylmalonyl CoA mutase. Prenatal deprivation of folic acid can lead to retardation in the child. Such children are found on occasion to have pathological changes in the brain, but never to suffer the characteristic demyelination found in B_{12} deficiency. In children, congenital defects in the absorption of folate cause mental retardation, megaloblastic anemia, and convulsions in the second decade of life. Two specific defects of folate utilization have been seen clinically. Several patients with a deficiency of formiminotransferase and one patient with a deficiency of methyltetrahydrofolate transmethylase have been reported. These patients have high serum folate levels because of an inability to utilize the vitamin. They suffer retardation, electroencephalogram abnormalities, cortical atrophy, and dilation of the ventricles. Interestingly, it has been reported that a deletion of dihydrofolate reductase, the enzyme which reduces the folate after its utilization in the formation of thymidine triphosphate, does not result in neurological abnormalities. In adults, the neurological consequences of folate deficiency are not as drastic as they are in children, causing only minor irritability and forgetfulness. Not surprisingly, prolonged treatment with folic acid antagonists will frequently produce neurological damage.

Some more recent observations suggest a role for folic acid in the etiology of the affective disorders. The crucial role of one-carbon transfer in the production of the various amine derivatives, and the suspected role of these amines in the regulation and alteration of mental states in humans, has led to some interesting avenues of inquiry. Briefly, it has been shown that a family of enzymes requiring folate derivatives as cofactors is involved in the methylation of various biogenic amines [27-29]. It has been proposed, and some evidence has been presented in support, that these folate-catalyzed methylations are significant in various disease states, including depression and schizophrenia [30].

Finally, there are some neurological consequences, early on, of a deficiency of pantothenic acid, but these are not prominent features of the deficiency. Perhaps this is related to the high concentration of pantothenic acid and of the enzymes necessary for the synthesis of CoA in the brain, and to the resistance of the brain to a depletion of pantothenic acid. Pantothenic acid deficiency has, however, been reported to result in lesions in the peripheral nervous system, and to a numbness and a tingling of the hands and the feet known as the burning foot syndrome. A severe deficiency causes paralysis, convulsions, and coma. Peripheral demyelination is the major pathological observation. The exact biochemical basis of the problem in pantothenic acid deficiency is not known. Studies involving the deficiency of pantothenic acid and the lipid composition of the brain did not reveal any specific differences. The only changes seen were those caused by the generalized malnutrition which is a consequence of the deficiency [31].

There is no substantial literature on any neurological consequences resulting from deficiencies of biotin, vitamin C, or the fat-soluble factors. It seems likely, upon casual consideration, that severe deficiencies of any of these vitamins would eventually cause malfunction of the nervous system. The absence of specific reports suggest only that there are, in fact, no prominent or early neurological symptoms.

References

1. P. M. Dreyfus and S. E. Geel, in *Basic Neurochemistry* (R. W. Albers, G. J. Siegel, R. Katzman, and B. W. Agranoff, eds.), Little, Brown, Boston, 1972.
2. H. G. Sie, J. N. Nigam, and W. H. Fishman, *Biochim. Biophys. Acta 50:* 277 (1961).
3. D. W. McCandless and S. Shenker, *J. Clin. Invest. 47:*2268 (1968).
4. R. E. Koeppe, R. M. O'Neal, and C. H. Hahn, *J. Neurochem. 11:*695 (1964).
5. S. E. Geel and P. M. Dreyfus, *J. Neurochem. 24:*353 (1975).
6. S. F. Reynolds and J. P. Blass, *J. Neurochem. 24:*185 (1975).
7. J. R. Cooper and J. H. Pincus, in *Inborn Errors of Metabolism* (F. A. Hommes and C. J. van den Berg, eds.), Academic, London, 1973.
8. D. Leigh, *J. Neurol. Neurosurg. Psychiatr. 14:*216 (1951).

9.   Y. Itakawa and J. R. Cooper, *Biochim. Biophys. Acta 158:*180 (1968).

10. J. R. Cooper, J. H. Pincus, Y. Itakawa, and K. Piros, *N. Engl. J. Med. 283:* 793 (1970).

11. A. von Muralt, *Vitam. Horm. (N.Y.) 5:*93 (1947).

12. A. von Muralt, *Ann. N.Y. Acad. Sci. 98:*499 (1962).

13. Y. Itakawa and J. R. Cooper, *Biochim. Biophys. Acta 196:*274 (1970).

14. A. D. Hunt, Jr., J. Stokes, Jr., W. W. McCrory, and H. H. Stroud, *Pediatrics 13:*140 (1954).

15. C. R. Scriver, *Pediatrics 26:*62 (1960).

16. C. R. Scriver and D. T. Whelan, *Ann. N.Y. Acad. Sci. 166:*83 (1969).

17. T. Yoshida, K. Tada, and T. Arakawa, *Tohoku J. Exp. Med. 104:*195 (1971).

18. C. R. Scriver and L. E. Rosenberg, *Amino Acid Metabolism and Its Disorders,* Saunders, Philadelphia, 1973.

19. L. E. Rosenberg, in *Brain Dysfunction in Metabolic Disorders,* Vol. 53 (F. Plum, ed.), Res. Publ. Assoc. Res. Nerv. Ment. Dis., Raven, New York, 1974.

20. E. P. Frenkel, *Proc. The American Society for Clinical Investigation, 63rd Ann. Mtg.,* Oxford, New York, 1971.

21. L. E. Rosenberg, A. C. Lilljequist, and Y. E. Hsia, *N. Engl. J. Med. 278:* 1319 (1968).

22. Y. E. Hsia, A. C. Lilljequist, and L. E. Rosenberg, *Pediatrics 46:*497 (1970).

23. L. E. Rosenberg and M. J. Mahoney, in *Inborn Errors of Metabolism* (F. A. Hommes and C. J. van den Berg, eds.), Academic, New York, 1973.

24. S. H. Mudd, H. L. Levy, and R. H. Abeles, *Biochem. Biophys. Res. Commun. 35:*121 (1969).

25. S. H. Mudd, H. L. Levy, and G. Morrow, III, *Biochem. Med. 4:*193 (1970).

26. V. Herbert and G. Tisman, in *Biology of Brain Dysfunction,* Vol. 1 (G. E. Gaull, ed.), Plenum, New York, 1973.

27. P. Laduron, *Nature (London) New Biol. 238:*212 (1972).

28. S. P. Banerjee and S. H. Snyder, *Science 182:*74 (1973).

29. S. H. Snyder, S. P. Banerjee, H. I. Yamamura, and D. Greenberg, *Science 184:*1243 (1974).

30. E. H. Reynolds, J. M. Preece, J. Bailey, and A. Coppen, *Br. J. Psychiatr. 117:*287 (1970).

31. R. Rajalakshimi and H. L. Nakhasi, *J. Neurochem. 24:*979 (1975).

20
MALNUTRITION

Within the last decade or so there has been a substantial sociological and humanitarian interest in the worldwide problem of infant malnutrition. A major question in this area has been whether the lower protein and caloric intake known as malnutrition or undernutrition can cause inadequate or improper brain functioning, and further whether such lowered cognitive function is permanent or can be reversed by subsequent nutritional therapy. More simply put, the question is whether childhood malnutrition will result in permanently lowered intelligence. A great deal of money and effort has been expended in a search for the answer, and some of the experimental designs have been not only exotic, but on the border of the unethical. Nevertheless, although the answer seems apparent, absolute proof that malnutrition lowers intelligence has been hard to obtain. This is due in large part to the difficulty in interpreting the major sociological and familial factors which invariably attend such experiments [1-5].

Work with experimental animals is less complicated and has provided a fairly satisfactory picture. There is no question at the moment that undernutrition in experimental animals during their development produces irreversible changes in the physical structure and the chemical composition of the brain. There is also data, and fairly persuasive data, that the behavior of such animals is permanently different than that of their adequately nourished littermates. Certainly, these studies have brought to light a great deal of fundamental information on the course of brain development.

Vulnerable Periods

The work of Dobbing and his collaborators has led to the concept of the vulnerable periods in the development of the brain [6,7]. In the specific case of the insult of malnutrition it is now known that a nutritional deprivation during the early life of the animal leads to irreversible alterations in the brain, while a comparable stress during adult life has no such permanent consequences. Carrying

the concept further, Dobbing has suggested that this vulnerable period involves
that period of most rapid brain growth, or, in other words, the "growth spurt."
Overall, during this period there is the most rapid increase in the fresh weight of
the brain. A closer inspection shows that the growth spurt is made up of a number of different periods of rapid growth. That is, there is first a period of rapid
neuronal multiplication, followed by a period of neuronal growth, then a period
of glial multiplication, and, finally, the period of rapid myelination. These periods follow in the same order in all species, but they may not be occurring simul-
taneously in all the various parts of the brain. Indeed, the cerebellum seems the
most vulnerable part of the brain of the infant rat because it is the part which is
growing the fastest in the immediate postnatal period. There is also some caution
necessary when considering the various species since the sequence of development of the brain occurs at different periods relative to the birth of the individual (Fig. 20.1). For example, the rat is born before the period of neuronal multiplication, and so the vulnerable period occurs after birth. Humans and pigs are
born somewhere during the period of glial development and multiplication. The
guinea pig is born fully myelinated, and so the vulnerable period occurs in utero.

Accepting all these cautions, the concept of the vulnerable period in brain
development is a most valuable concept and extends to the understanding of all
the many insults known to affect the developing brain, not just to undernutrition. Indeed, the necessity for treating the aminoacidurias as early as possible,
and the futility of instituting treatment after the period of myelination, testify
to the validity of the Dobbing concept.

The consequences of undernutrition for the developing rat brain depend to
some extent on the timing, duration, and severity of the deprivation. But rats
exposed to the normal preweaning undernutrition regimen, i.e., the placing of
12 to 18 pups with one mother [8], eventually can be shown to have microcephaly, fewer and smaller neurons, a delay in the proliferation and migration
of the glia, a reduction in myelin formation [9], a lowered protein content, and
a deficiency of certain specific chemical and enzymatic components of the brain
[10,11].

The concept of the vulnerable period carries the corallary that the brain
can not "catch up." There is apparently a series of carefully timed changes
which, if not executed in proper sequence and at the proper time with respect
to the rest of the life cycle of the animal, can never be effectively completed.
This is reflected in the finding that the sequence of developmental milestones,
such as eye opening, is delayed in malnourished animals. Thus, if rats are malnourished for the first 3 weeks of their lives, no amount of adequate or overnu-
trition later in life will correct the physical and mental deficits incurred.

The structural defects described above have been shown, primarily through
the work of Winick and his associates, to have chemical cognates. Winick's picture of the developmental stages of the brain is much like that of Dobbing. In

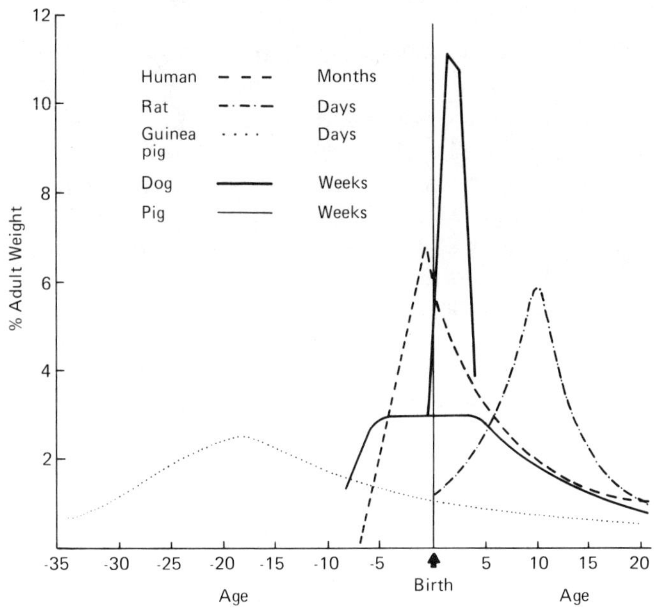

Figure 20.1   Rate curves of brain growth in relation to birth in different species. Values are calculated at different time intervals for each species. [Taken from J. Dobbing, in *Chemistry and Brain Development* (R. Paoletti and A. N. Davison, eds.), Plenum, New York, 1971.]

Winick's formulation [12,13], the development is divided into four stages. These are cellular multiplication, cellular multiplication and cellular growth, cellular growth, and finally, cessation of growth. Obviously the stages merge into each other, and different brain regions undergo the stages at different times in the early life of the animal. Nevertheless, such a framework is quite useful for describing some of the alterations produced by malnutrition.

Since most of the cells in the brain are diploid, each cell contains about the same amount of DNA. Consequently, the DNA content of a given weight of brain is proportional to the number of cells present. An increase in the concentration of DNA in any part of the brain indicates, then, cell multiplication. An increase in the protein/DNA ratio, on the other hand, indicates a growth of cells but not multiplication. Using measures such as these it can be said, generally, that the cell division in the normal rat brain stops at about day 21, whereas substantial cell growth continues until about day 99

Thus, early malnutrition interferes with cell multiplication [14,15], while malnutrition later in the life of the animal interferes with cell growth. Accord-

ing to Winick, interference with cell multiplication cannot be repaired in later
life no matter what nutritional regimen is instituted. Malnutrition in the adult,
which interferes with the growth of the cells, can be repaired if adequate nutrition is reinstated.

In sum, then, animal experiments indicate that early malnutrition results
in smaller brains, fewer cells, decreased lipid synthesis, neuronal and glial degeneration, decreased arborization of neurons [16], and a delay of the appearance
of certain enzymes. There seems to be agreement on the fact that all cell types
are affected by a restriction of the diet. Essentially, all the biosynthetic processes of the cells are involved, including the synthesis of lipids, proteins, and
nucleic acids [17-19]. Many enzyme levels are changed, some few reversibly,
and the overall pattern of amino acid metabolism and compartmentation is altered [20]. What little data there are in humans are consistent with this picture,
e.g., the DNA content of brain from malnourished children is substantially lower
than normal (Fig. 20.2) [21].

That there are permanent behavioral changes in animals malnourished in
early life is fairly well established. Gross behavioral abnormalities have been
shown, as have alterations in problem-solving ability. Some of the reported effects include apathy, reduced exploratory behavior and motor performance,
lessened problem-solving ability, heightened irritability, fearfulness, and reduced
motivation toward food [22-25]. Indeed, there are studies which show that
some alterations in behavior persist in the progeny of animals which had been
undernourished at birth [26]. Extrapolation to humans probably still requires
a great deal of caution.

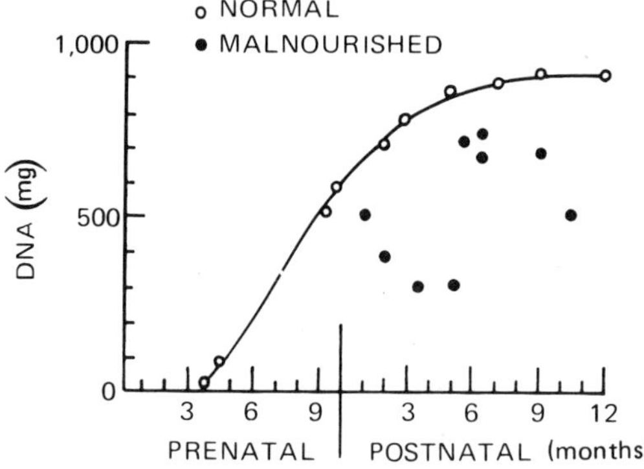

Figure 20.2 Number of cells in normal and malnourished human brain. [Taken
from M. Winick and P. Rosso, in *Biology of Brain Dysfunction,* Vol. I (G. E.
Gaull, ed.), Plenum, New York, 1973.]

References

1. H. G. Birch, *Am. J. Public Health 62:*773 (1972).
2. R. E. Klein, C. Yarbrough, R. E. Lasky, and J. P. Habicht, in *Early Malnutrition and Mental Development* (J. Cravioto, L. Hambraeus, and B. Vahlquist, eds.), Swedish Nutrition Foundation Symposia XII, Almquist and Wiksell, Uppsala, Sweden, 1974.
3. M. S. Read, *J. Am. Diet. Assoc. 63:*379 (1973).
4. M. S. Read, *J. Am. Diet. Assoc. 63:*386 (1973).
5. M. S. Read, in *The Growth and Development of the Brain,* IBRO Satellite Symposium, New Delhi, India, October 14-17, 1974.
6. J. Dobbing, in *Chemistry and Brain Development* (R. Paoletti and A. N. Davison, eds.), Plenum, New York, 1971.
7. J. Dobbing, in *Handbook of Neurochemistry,* Vol. VI (A. Lajtha, ed.), Plenum, New York, 1971.
8. E. M. Widdowson and R. A. McCance, *Proc. R. Soc. London,* Ser. B. *152:*88 (1960).
9. J. Dobbing and E. M. Widdowson, *Brain Res. 88:*357 (1965).
10. B. P. F. Adlard and J. Dobbing, *Br. J. Nutr. 28:*138 (1972).
11. B. P. F. Adlard, J. Dobbing, and A. Lynch, *Biochem. J. 130:*12P (1972).
12. M. Winick and P. Rosso, in *Biology of Brain Dysfunction,* Vol. 1 (G. E. Gaull, ed.), Plenum, New York, 1973.
13. M. Winick, in *The Nervous System,* Vol. 2 (D. B. Tower, ed.), Raven, New York, 1975.
14. M. Winick and A. Nobel, *J. Nutr. 89:*300 (1966).
15. M. Winick, *J. Pediatr. 74:*667 (1969).
16. B. G. Cragg, *Brain 95:*143 (1972).
17. J. W. T. Dickerson, J. Dobbing, and R. A. McCance, *Proc. R. Soc. London,* Ser. B, *166:*396 (1967).
18. D. A. Cheek, J. E. Graistone, and R. D. Rowe, *Am. J. Physiol. 217:*642 (1969).
19. A. E. R. de Gugliemone, A. M. Soto, and B. H. Duvilanski, *J. Neurochem. 22:*529 (1974).
20. M. K. Roach, J. Corbin, and W. Pennington, *J. Neurochem. 22:*521 (1974).
21. M. Winick and P. Rosso, *Pediatr. Res. 3:*181 (1969).
22. S. Frankova and R. H. Barnes, *J. Nutr. 96:*477 (1968).
23. S. Frankova and R. H. Barnes, *J. Nutr. 96:*485 (1968).
24. J. Altman, K. Sudarshan, G. D. Das, N. McCormick and D. Barnes, *Dev. Psychobiol. 4:*97 (1970).
25. J. L. Smart, *Psychiatr. Neurol. Neurochir. 74:*443 (1971).
26. R. H. Barnes, S. R. Cunnold, R. R. Zimmerman, H. Simmons, R. B. MacLeod, and L. Krook, *J. Nutr. 89:*399 (1966).

21
RESPIRATION AND BIOENERGETICS

It is abundantly clear that only glucose can fully support normal brain function [1,2]. If the glucose concentration of the blood is lowered, as in insulin hypoglycemia, drowsiness, stupor, and finally coma result, accompanied by convulsions and electroencephalogram changes. Only glucose can restore consciousness. Many other materials can be oxidized by the brain, more or less well in a quantitative sense, but none serve the complete needs of the brain. A number of materials which are readily oxidized by brain preparations in vitro are not adequate substrates in vivo, perhaps because of inadequate transport into the brain. Some reports indicate that mannose can serve as a substrate in the absence of glucose, but mannose is not normally found in the bloodstream. A number of other metabolites which have been thought to serve the function of glucose in the intact animal, e.g., restoring consciousness during hypoglycemic coma, are now known to do so by being rapidly converted to glucose in other organs before being presented to the brain.

Oxidation of the following substances has been demonstrated using brain slices: lactate, pyruvate, succinate, α-hydroxybutyrate, acetoacetate, ethanol, and certain amines. Slight oxidation of fumarate, malate, galactose, and cephalin has also been observed. Comparable brain preparations will not oxidize mannitol, arabinose, xylose, sucrose, α-aminobutyrate, fatty acids, gluconate, or glycerol. Even those substrates which are oxidized well by brain when present alone will generally not stimulate the uptake of oxygen when adequate glucose is also present.

Ketone Bodies

The oxidation of the ketone bodies, β-hydroxybutyrate and acetoacetate, requires separate discussion here. The metabolism of β-hydroxybutyrate proceeds via the pyridine nucleotide-linked enzyme, β-hydroxybutyrate dehydrogenase. The acetoacetate formed is then converted to its coenzyme A (CoA) derivative

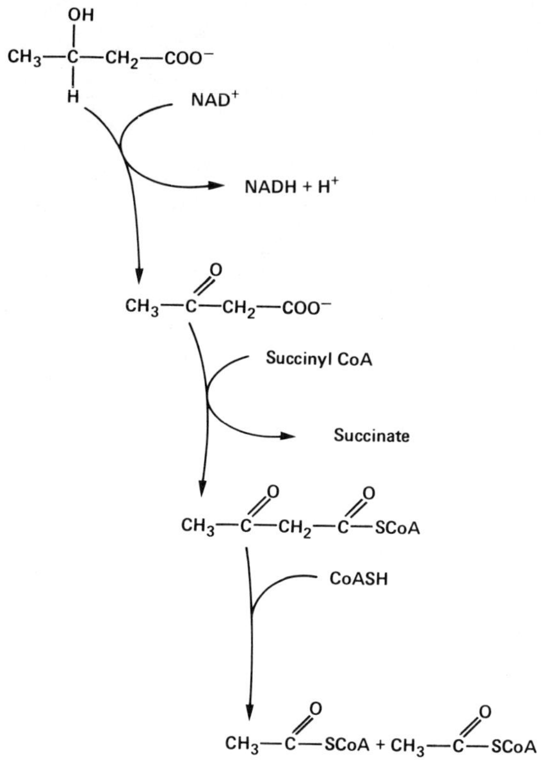

Figure 21.1   The metabolism of β-hydroxybutyrate by brain.

by reaction with succinyl CoA, and then is cleaved by a coenzyme A-dependent thiolase to give two molecules of acetyl CoA (Fig. 21.1). β-Hydroxybutyrate can serve as a primary energy source for brain under certain conditions, and it is oxidized at a rate comparable to that seen in other tissues. The enzyme β-hydroxybutyrate dehydrogenase has been studied in brain by Sokoloff [3,4] and by others. The enzyme is present in fetal brain, and its activity rises about fivefold in the first few days after birth. It seems likely that the rise is due to the fact that maternal milk is ketogenic, and thus ketone bodies are an important source of nourishment during early life. The level of the enzyme in brain falls after weaning and slowly declines to the adult level (Fig. 21.2). Under abnormal conditions, such as starvation, blood glucose levels drop, the ketone bodies in the bloodstream of adults rise, and the brain can again oxidize them. Under such conditions there may also be slight increases in the level of β-hydroxybutyrate dehydrogenase in

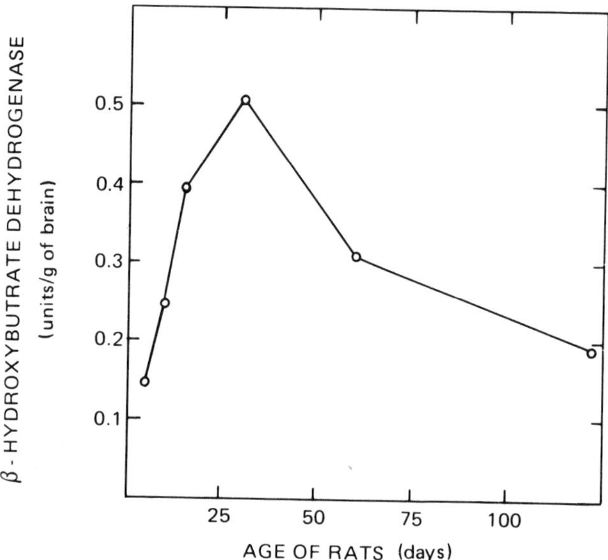

Figure 21.2   The developmental course of β-hydroxybutyrate dehydrogenase. [Figure from C. B. Klee and L. Sokoloff, *J. Biol. Chem. 242:*3880 (1967).]

the brain [5]. The enzymes are mitochondrial, localized in the inner mitochondrial membrane. The utilization of the ketone bodies is initiated by their appearance in the blood, and the rate of their metabolism appears to be in proportion to the concentration of each which is present [6]. Thus, β-hydroxybutyrate is the most important for brain metabolism because it is in the highest concentration in the bloodstream.

Energy Consumption

The brain has one of the most active rates of metabolism of all the organs of the body. It takes up oxygen at the highest rate, and in normal man this rate is about 3.5 ml of O_2 per 100 g of brain per minute or about 50 ml/min in a 70-kg man. The glucose uptake amounts to about 5.5 mg/100 g of brain per minute, or about 77 mg/min. The cerebral blood flow is about 57 ml/100 g of brain per minute, or about 798 ml/min in a 70-kg adult male [7]. The oxygen uptake for the whole body of such an individual is about 250 ml/min, so the brain, which makes up about 2% of the total mass, consumes under normal resting conditions about 20% of the total oxygen consumed by the body. Further, it accounts for about 20% of the cardiac output of the body and about 65% of the total glucose consumed.

This enormous energy consumption is almost exclusively devoted to sustaining and restoring neuronal membrane potentials through the maintenance of ion transport which allows proper neuronal functioning in conduction and transmission. The energy consumption of brain does not change appreciably from the waking to the sleeping state. Some increases in the rates of O_2 consumption have been observed during periods of REM sleep [8]. Experiments in which the energy consumption of the brain has been measured during periods of mental activity, such as during the solving of mathematical problems, have failed to show any increase. Also, there have been attempts to measure the consumption of energy by the brains of individuals suffering from various psychiatric disorders, but all such studies have been negative in that no differences could be found between these individuals and normals. Reduced oxygen consumption has been observed, along with decreased cerebral blood flow, in senile individuals. Such changes are most likely a result of the increased capillary resistance found in the elderly.

Extreme changes in the level of consciousness do produce changes in the energy consumption. Persons in coma have a lowered consumption of energy by brain. Individuals under anesthesia also exhibit lowered energy consumption by brain. On the other side of the coin, most kinds of convulsions are accompanied by increases in the oxygen uptake of the brain. Some convulsions, however, are caused by a lowering of energy reserves in the brain and a decreased oxygen consumption, so increased oxygen uptake is apparently not a necessary concomitant of all convulsions.

Ischemia

An interruption of blood flow to the brain (ischemia), and hence a loss of oxygen and glucose, is followed within seconds in the adult by a loss of consciousness. Irreversible brain damage occurs after a few minutes of such deprivation, accompanied, of course, by a complete loss of function. Any number of investigators have studied the sequence of changes which follow circulatory arrest [9-11]. The nerve cell changes caused by ischemia can be recognized by a shrinkage of the neurons, a disappearance of the Nissl bodies, a collapse of the nucleus, a swelling of the boutons of the cells which synapse with the degenerating neuron, and a retraction of the cytoplasmic membranes, leaving a clear zone between the affected area and the surrounding tissue. This series of neuronal changes has been seen in the early stages of hypoxia and trauma as well as in the first period after ischemia.

The neurons which are most sensitive to the effects of the ischemia may be detected in 6 hr or less after the interruption of the nutrient supply. Altered staining is seen within 1 to 2 days, and the faint outlines of the necrotic neurons remain visible up to 7 to 10 days. Not too much is known about the mechanism by which the biochemical foundation of these cells is irreversibly destroyed, but an early change appears to be the destruction of the lysosomes and the liberation of hydrolytic enzymes within the cells.

Few, if any, of the consequences of ischemia are reversible. Eventually the tissue undergoes dissolution. Gross observation of large areas of affected tissue shows that between 8 and 48 hr after the interruption of the oxygen supply the damaged tissue, both gray and white, swells, and the cortex is altered in color. The white matter loses its smooth texture and becomes uneven or granular. After several days the necrotic tissue becomes liquified. In the course of the next several weeks, the tissue disappears, and is replaced by a fluid-filled cavity.

Developmental Changes

Thus, the consequences of oxygen deprivation in the adult are rapid and catastrophic. The metabolism of the adult brain is highly aerobic. This is not true in the infant. It is startling that whereas adult animals die in 1 or 2 min when placed in nitrogen, fetal animals can survive without distress for periods up to 40 or 50 min. It has been estimated [12] that the survival of a fetal animal in nitrogen is some 45 times longer than that of an adult under comparable circumstances. This observation is a measure of the reliance of the fetal and infant brain on glycolysis rather than on aerobic metabolism.

During development, then, the brain undergoes a transition from a metabolism based largely on anaerobic glycolysis to one based almost solely on aerobic catabolism. A series of enzymatic events shepherd this change. During the first month of life in the rat there is a marked rise in the cerebral oxygen consumption, an increase in the activities of the citric acid cycle enzymes, an increase in the levels of the respiratory enzymes, and an increase in the number of mitochondria present [13,14]. There are also increases in the activities of the glycolytic enzymes of brain and an absolute increase in the rate of glycolysis, but it is small in relation to the rise in the components of aerobic metabolism [15].

Methods of Study

The respiratory characteristics of brain have been studied at a number of different levels. In vivo studies have been most important, and indeed, there are some who feel that in vivo studies provide the only reliable index of true brain energy metabolism. Procedures have been developed which allow the monitoring of the concentrations of O_2 and various metabolites flowing into and out of the brain of primates. Such arterial-venous (A-V) differences, coupled with a knowledge of the rate of blood flow, can give unique information about the energy metabolism of the brain in the living animal under various conditions. Unfortunately, sampling of the venous outflow in lower animals is a very difficult problem. But in humans and monkeys such studies have provided fundamental data on the overall metabolism of glucose, the changes in the utilization of oxygen during sleep and in response to various physiological changes, and the alterations in metabolism during development and aging [16].

Brain slices have also been widely studied, but such preparations may reveal more about the metabolic potential of brain to carry on specific metabolic steps than they do about the actual rates of various pathways in the living animal. Thin slices of brain show a substantial and a linear uptake of oxygen for several hours in the presence of glucose. Such slices will also exhibit a marked Pasteur effect. However, the level of oxygen uptake in the most carefully prepared slices is only about one-half of that shown by a comparable amount of brain in vivo. Electrical stimulation of the slices increases their oxygen uptake about 100%, to levels approximating those observed in vivo. These observations have led some investigators to conclude that the in vitro metabolism reflects respiratory rates found in depressed or basal nervous tissue, e.g., as seen in deep coma. Electrical stimulation in vitro, then, appears to simulate in vivo conditions leading to respiratory rates comparable to those seen in fully functional tissue [17].

Mitochondria from the brain are discussed in detail in Chap. 3. Preparations of brain mitochondria were, in fact, used for some of the early and fundamental studies on the dependence of respiration on inorganic phosphate. They provide, however, no special advantage for such studies, since they resemble liver mitochondria in their overall behavior, and are more difficult than liver mitochondria to prepare pure and functional due to the presence of synaptosomes and myelin fragments in the crude mitochondrial fraction.

Tricarboxylic Acid Cycle

The specific enzymes of the tricarboxylic acid cycle are present in brain and are similar if not identical to those in liver, although no detailed mechanistic characterization of any of them has been done. Indeed, it has been possible to write recently that "... amazingly little is known of the biochemistry of the TCAC (tricarboxylic acid cycle) in the nervous tissue" [18]. The rate of each enzyme has been measured, and the concentration of each intermediate is known. In addition to the enzymes of the citric acid cycle per se, it is known that the citrate cleaving enzyme (ATP:citrate oxaloacetate lyase) is present and quite active. It has been suggested that this enzyme provides acetyl CoA for the synthesis of brain fatty acids [19]. The carboxylation reactions necessary for replenishing the TCA cycle, the conversion of pyruvate to oxalacetate catalyzed by the pyruvate carboxylase, is also active in brain. It has been estimated that about 10% of the pyruvate produced in brain is carboxylated. Phosphoenolpyruvate carboxykinase, an important enzyme in gluconeogenesis, has very low activity in brain, where gluconeogenesis is a minor pathway [20]. The shuttle mechanism which allows the mitochondria of the brain to oxidize extramitochondrial NADH, namely the interconversion of α-glycerol phosphate and dihydroxyacetone in both the mitochondria and the cytoplasm, is known to occur in nervous tissue. The mitochondrial processes of electron transport and oxidative phosphorylation in brain differ only in detail from those in mitochondria from other tissues. The brain is inordinately high in creatine phosphate and creatine kinase, and the latter is found bound to the mitochondria. Creatine phosphate, then, is an important storage form of energy in the brain as it is in the muscle.

As regards the overall pattern of energy metabolism in the brain, certain general statements can be made. The γ-aminobutyric acid shunt, described in detail in Chap. 11, seems to participate in about 10% of the oxidative metabolism of the brain. The Pasteur effect, i.e., the inhibition of glucose utilization by the admission of oxygen, can be demonstrated in brain, and it probably has the same mechanism as in other tissues, namely the inhibition of phosphofructokinase by ATP. The pentose cycle accounts for some, but not much, of the overall glucose metabolism of the brain, on the order of 1 to 10%.

The effects of raising the potassium concentration on brain respiration in vitro have been known for a number of years [20-22]. The addition of 0.1 M KCl to brain slices increases their rates of glycolysis and of respiration. Addition of comparable amounts of NaCl has no such effect. Deletion of Ca^{2+} from the medium has an effect similar to that of adding KCl [20]. The mechanism of the KCl effect is not known, but it is specific to brain, and it probably reflects the increased metabolic activity associated with repolarization of the neuronal membranes.

The catastrophic consequences of anoxia have been described before. Various degrees of hypoxia cause alterations, albeit milder, in the enzymatic balance of the reactions in the brain, e.g., hexose monophosphate shunt metabolism rises, citrate cleavage drops. But the overall affinity of brain mitochondria for oxygen appears to be greater than that of liver mitochondria [23], and much higher than the affinities exhibited by the various soluble hydroxylases, such as those for tyrosine and tryptophan. Thus, the brain fares well during periods of mild hypoxia. High pressures of oxygen, on the other hand, produce convulsions and are frequently fatal. The mechanism of the injury here is obscure, but it is known from in vitro studies that high concentrations of oxygen will inhibit sulfhydryl enzymes, depress oxidative metabolism, and lower the brain stores of ATP and creatine phosphate.

Finally, hypercapnia, or high concentration of CO_2 (10 to 30%), has a number of deleterious consequences [24,25]. Mild hypercapnia decreases cerebral excitability. More extensive exposure increases excitability and leads to seizures. Severe hypercapnia causes anesthesia. These changes are accompanied by a number of marked alterations in the carbohydrate and amino acid metabolism of the brain. Decreases occur in most glycolytic and Krebs cycle intermediates. Glutamic and aspartic acids also decrease. Brain glucose levels rise, and there is intracellular acidosis. Levels of creatine phosphate and ATP decrease, but since all the nucleotide levels drop there does not seem to be an overall change in the "energy charge" of the brain.

References

1. H. E. Himwich and L. H. Nalium, *Am. J. Physiol. 90:*384 (1929).
2. H. E. Himwich and J. F. Fazekas, *Endocrinology 21:*800 (1937).
3. C. B. Klee and L. Sokoloff, *J. Biol. Chem. 242:*3880 (1967).
4. L. Sokoloff, *Ann. Rev. Med. 24:*271 (1973).
5. I. Pull and H. McIlwain, *J. Neurochem. 18:*1163 (1971).
6. O. E. Owen, A. P. Morgan, H. G. Kemp, J. M. Sullivan, M. G. Herrera, and G. F. Cahill, Jr., *J. Clin. Invest. 46:*1589 (1967).
7. L. Sokoloff, in *Basic Neurochemistry* (R. W. Albers, G. J. Siegel, R. Katzman, and B. W. Agranoff, eds.), Little, Brown, Boston, 1972.
8. D. R. Brebbia and K. Z. Altshuler, *Science 150:*1621 (1965).
9. O. H. Lowry, J. V. Passonneau, F. X. Hasselberger, and D. W. Schultz, *J. Biol. Chem. 239:*18 (1964).
10. F. W. Schmal, E. Betz, H. Talke, and H. J. Hohorst, *Biochem. Z. 342:*518 (1965).
11. U. Muller, W. Isselhard, D. H. Hinzen, and E. Geppert, *Pfluegers Arch. Gesamte Physiol. Menschen Tiere 32:*168 (1970).
12. T. E. Duffy, S. J. Kohle, and R. C. Vannucci, *J. Neurochem. 24:*271 (1975).
13. H. E. Himwich, in *Developmental Neurobiology* (W. A. Himwich, ed.), Thomas, Springfield, Ill., 1970.
14. A. N. Davison and J. Dobbing, in *Applied Neurochemistry* (A. N. Davison and J. Dobbing, eds.), F. A. Davis, Philadelphia, 1968.
15. G. Takagaki, *J. Neurochem. 23:*479 (1974).
16. S. S. Kety, in *The Neurosciences,* Vol. I (D. B. Tower, ed.), Raven, New York, 1975.
17. H. McIlwain, in *Basic Mechanisms of the Epilepsies* (H. H. Jasper, A. A. Ward, and A. Pope, eds.), Little, Brown, New York, 1969.
18. S. Z. Cheng, in *Handbook of Neurochemistry,* Vol. Va (A. Lajtha, ed.), Plenum, New York, 1971.
19. A. F. D'Adamo and A. P. D'Adamo, *J. Neurochem. 15:*315 (1968).
20. C. A. Ashford and K. C. Dixon, *Biochem. J. 29:*157 (1935).
21. F. Dickens and G. D. Greville, *Biochem. J. 29:*1468 (1935).
22. K. A. C. Elliott and F. Bilodeau, *Biochem. J. 85:*93 (1962).
23. J. B. Clark, W. J. Nicklas, and H. Degn, *J. Neurochem. 26:*409 (1976).
24. L. Sokoloff, *Anesthesiology 21:*664 (1960).
25. D. M. Woodbury and R. Kahler, *Anesthesiology 21:*686 (1960).

CHEMICAL PHYSIOLOGY OF NERVE AND BRAIN

22
NEURONS AND GLIA
AND THE NEURON-GLIA UNIT

Although several types of each occur in the nervous system, the two broad classes
of cells, the neurons and the glia, can be discussed with regard to their general
properties. They can be compared with respect to their growth characteristics
and enzymatic content, and, again broadly, their functions can be specified.
Finally, the relationship between the neurons and their surrounding glia can be
discussed.

Generally speaking, in humans, the total complement of neurons is present
at birth and does not increase in number during the lifetime of the individual.
Large numbers of neurons in the central nervous system die daily after the age of
35 or so, and these neurons are not replaced. Although neurons do not divide,
some have a limited, but very limited, capacity to regenerate or repair their processes, for example after injury. The ability to regulate and extend this regenera-
tive function is most important in the search for approaches to the extensive
problem of spinal cord injury and the resulting irreversible paralysis.

Glia, on the other hand, multiply extensively during the lifetime of a human. Glia proliferate during normal development and especially so in response
to a number of pathological influences. Experimental injuries of all kinds, nerve
transsections, and certain of the demyelinating diseases, cause an increased metabolic activity in the existing glia and an increase in the number of glia. In the
central nervous system the astrocytes seem the ones specifically affected, and the
microglia also in certain conditions, e.g., Wallerian degeneration.

Chemically, the neurons have more RNA, more protein, more ATP, and
less lipid than the glia. Their total nucleic acid to protein ratio is higher [1].
According to Hydén, neuronal RNA is higher in adenine and lower in guanine
than is glial RNA [2]. The neurons contain more ribosomal RNA than do the
glia, and the neurons have a higher rate of RNA and protein synthesis [1]. The
respiratory activity of the two cell types is roughly equivalent, but the neurons
show a greater preference for glutamate oxidation than do the glia. Consistent
with this characteristic is the observation that the neurons maintain a larger

amino acid pool than the glia [3] , and specifically a higher content of glutamate. Also consistently, the neurons have a much higher content of glutamate dehydrogenase, on the order of fourfold, than do the glia [3] . There are other enzymatic differences, e.g., the neurons are higher in hexokinase, phosphoglucoisomerase, lactic acid dehydrogenase, malic acid dehydrogenase, and some of the transaminases [4] . These differences, though, are quantitative only, and none of these enzymes can be considered to be neuronal markers in the sense that they occur exclusively in the neurons. The enzymes for transmitter synthesis are the closest to neuronal markers, and these occur only in the particular neurons served by that transmitter. In that regard, the specific acetylcholinesterase occurs only in the neurons. In the glia a related but much less specific enzyme, pseudocholinesterase or butyrylcholinesterase, is present. The protein known as 14-3-2, now shown to be brain-specific enolase [5,6] , is found only in the neurons.

The glia, as a class, are obviously lower in protein, RNA, and ATP concentration, and higher in lipid than the neurons. The lipid content has been shown to resemble that of early myelin, in that it is low in sphingomyelin and high in cerebroside [7] . The cholesterol content of the glia is much greater than that of the neurons, and the phosphatides are similarly elevated. The glia synthesize protein and RNA, and there is also a substantial DNA synthesis, especially following injury. The glia exhibit a metabolic preference for the oxidation of succinate and pyruvate in contrast to glutamate. Enzymatically, the glia have a higher content of glucose-6-phosphate dehydrogenase, 6-phosphogluconate dehydrogenase, isocitrate dehydrogenase, succinate dehydrogenase, and cytochrome oxidase [4] . Again, none of these can be considered glial markers. Indeed, the significance of these differences is somewhat obscure since these enzymes are thought to be in substantial excess of their true activities and true needs for the metabolism of the cell. The most significant enzymatic difference between neurons and glia is the presence, as mentioned before, of 50 to 100 times higher carbonic anhydrase activity in the glia. The significance of this distribution of carbonic anhydrase is not known, although there are some theories, but the enzyme is most useful as a glial marker. In the same sense that the enzymes synthesizing neurotransmitter substances are present in the relevant neurons, the unique constituents of myelin should be found only in the relevant glia. The enzyme 2,3-cyclic nucleotide 3′-phosphohydrolase is apparently found only in myelin and, thus, should be a marker for the myelin-forming glia. The protein S-100, which is of unknown function and has no known enzymatic activity, is found mostly in glia.

Functionally, the neurons are, of course, the communicating elements of the nervous system. In general, they carry on a rapid metabolism, transport a number of cell constituents by axoplasmic flow for relatively long distances through the cell, and synthesize a number of specific molecules such as transmitter substances and membrane constituents. They function by the transmission of an all or none action potential down the axonal membrane to the synapse.

This results in the release of transmitter substance across the synapse onto a receptor. This, in turn, initiates an electrical response in the next cell. The mechanism of this transmission is discussed in detail in Chap. 23.

The Functions of Glia

The function of the glia are not so clearly understood. Although a number of experiments have been done, and a number of functions have been ascribed to the glia, as described below, the only completely accepted function for any glia population is that of myelination. Clearly, the oligodendroglia in the central nervous system, and the Schwann cells in the peripheral system, support and insulate the axons by forming and maintaining a myelin sheath around them. Listed below are some of the other functions which the glia may serve [1,8,9].

1. Until fairly recently it was felt by some that the brain has very little extracellular space, and that nutrients and other necessary materials, such as ions and water, must pass through the glia to reach the neurons. Now it is generally accepted that there is a substantial extracellular space, and that nutrients can pass through extracellular channels to reach the neurons.

2. Some investigators feel that the membranes of the astroglial, which line the capillaries of the brain, constitute one aspect of the blood-brain barrier.

3. Several observations suggest that the glia can absorb potassium, specifically the potassium which is released upon depolarization of the neuron during neuronal firing [10,11]. The role of the glia in this regard would be as a "spatial buffer" to remove the extracellular potassium and prevent the depolarization of surrounding neurons.

4. It is likely that the glia play a more general role in regulating the ionic environment of the neurons, and rational mechanisms can be written, based on the carbonic anhydrase content of the glia, for the movement of chloride in exchange for bicarbonate.

5. Glia often border and cluster about the synapse, and there is direct proof that glia will take up neurotransmitters, expecially γ-aminobutyric acid [12,13], and perhaps in this way terminate its action at the synaptic site. Certainly such a function could be seen to contribute to the regulation of the microenvironment of the synaptic region.

6. The microglia, in particular, are thought to be scavenger cells in the nervous system.

In addition to these rather specific functional suggestions, there is the general feeling that the glia fabricate nutrients for the neurons, perhaps store energy reservoirs like glycogen, and even synthesize and transfer growth factors, enzymes, and RNAs to the neurons. Indeed, it has been possible to write recently [14] "In development, glial cells play roles in neuronal migration, axonal guidance, and possibly synaptic selection." Nevertheless, it is still the case that myelination is the only thoroughly proven function of the glia.

The Neuron-Glia Unit

The close and even vital relationship between neurons and the surrounding glia has suggested to some that the neuron and its associated glia make up a discrete and perhaps symbiotic unit. The concept was first enunciated by Hydén [15], and, although much of the specific original postulate is still not widely supported experimentally, there is ample other evidence to suggest that the concept is a valid one.

Hydén and his colleagues described a series of reciprocal changes which occur between certain neurons and their supporting glia during various functional states. They suggested that the glia support the neurons directly by a transfer of nutrients and macromolecules. For example, by the analysis of single cells through the use of ingeneous micromethods, Hydén and his coworkers obtained data [16, 17] showing that upon vestibular stimulation of rabbits, the RNA and protein content of certain neurons rises while that of the associated glia drops. The activity of cytochrome oxidase was also found to rise in the neurons and drop in the associated glia. In other studies the composition of the RNA in various neurons was found to change upon stimulation. Since opposite compositional changes occurred in the surrounding glia it was suggested that there could have been a transfer of RNA between the cells. In any case, these and other data have been interpreted in support of the idea that the glia could serve in this manner to nourish, regulate, or even program the neuron.

By way of qualification it is important to mention a couple of reservations about the Hydén studies. The dissection techniques used by Hydén involve cutting out the neuronal cell body from a block of tissue and designating the remaining material glia. It is known, however, that the remaining material also contains the processes of the neuron. Thus, the data could reflect transfer within rather than between cells [1]. That is, the RNA or other materials could be moving from the perikaryon down the axon rather than into the glia. Also, the techniques used to measure these changes, especially the changes in RNA, are quite difficult and specialized and have not yet been duplicated in other laboratories. In other words, the measurements made by Hydén still await confirmation by other investigators.

Since the work of Hydén, much controversy has surrounded the subject of glia-to-neuron transfer of macromolecules. Droz [18] has concluded that such transfer, if it exists at all, is quantitatively a very small proposition. The experiments of Lasek and Gainer [19,20] suggest that, at least in squid giant axon, a transfer of proteins does occur across the axonal membrane. When these giant fibers were separated from their neuronal cell bodies and incubated with radioactive amino acids, radioactive protein appeared in the axon. Since the associated glia were capable of protein synthesis, but the axon itself was not, the conclusion that proteins were transferred from glia to axon appears inescapable.

In any case, there is substantial evidence that the glia do relate in detail to the adjacent neurons, and that a communication exists between them. The neurons may signal the glia by the release of potassium ion which accompanies neuronal firing. This signal may cause the release of various glia-made substances. The neurons may also signal the initiation of myelin formation by the glia. The degeneration of the axon may in some way trigger glia proliferation, with consequent gliosis. Support for some of these interactions come from studies in cell culture where it is clear that neurons do better in cultures which contain or have been conditioned by glia [21,22]. In these cases it seems that the glia secrete proteins which are taken up by, and promote the well-being of, the neurons. So whether or not there is actual transfer of macromolecules from the glia to the neuron, there probably is a meaningful neuron-glia unit operating at the biochemical, anatomical, and functional level [15,23,24].

References

1. S. P. R. Rose, in *Handbook of Neurochemistry,* Vol. II (A. Lajtha, ed.), Plenum, New York, 1969.
2. H. Hydén and A. Pigon, *J. Neurochem. 6:*57 (1960).
3. S. P. R. Rose, *J. Neurochem. 15:*1415 (1968).
4. J. Clausen, in *Handbook of Neurochemistry,* Vol. I (A. Lajtha, ed.), Plenum, New York, 1969.
5. E. Bock and J. Dissing, *Scand. J. Immunol. 4:*31 (1975).
6. P. J. Marangos, C. Zomzely-Neurath, and C. York, *Biochem. Biophys. Res. Commun. 68:*1309 (1976).
7. A. N. Davison, M. L. Cuzner, N. C. Banik, and J. Oxberry, *Nature (London) 212:*1373 (1966).
8. S. Varon, *Exp. Neurol. 48:*93 (1975).
9. S. Varon and M. Saier, *Exp. Neurol. 48:*135 (1975).
10. J. G. Nicholls and S. W. Kuffler, *J. Neurophysiol. 27:*645 (1964).
11. R. K. Orkand, J. G. Nicholls, and S. W. Kuffler, *J. Neurophysiol. 29:*788 (1966).
12. M. J. Neal and L. L. Iversen, *Nature (London) New Biol. 235:*217 (1972).
13. R. Schon and L. L. Iversen, *Brain Res. 42:*503 (1972).
14. S. Varon, in *The Nervous System,* Vol. I (D. Tower, ed.), Raven, New York, 1975.
15. H. Hydén, in *The Neurosciences* (G. C. Quarton, T. Melnechuk, and F. O. Schmitt, eds.), Rockefeller University Press, New York, 1967.
16. H. Hydén, in *Neurochemistry* (K. A. C. Elliott, I. H. Page, and J. H. Quastel, eds.), Thomas, Springfield, Ill., 1962.
17. H. Hydén and P. W. Lange, *J. Cell Biol. 13:*233 (1962).

18. B. Droz, in *The Nervous System,* Vol. I (D. Tower, ed.), Raven, New York, 1975.

19. R. J. Lasek, H. Gainer, and J. L. Barker, *J. Cell Biol.* 74:501 (1977).

20. H. Gainer, I. Tasaki, and R. J. Lasek, *J. Cell Biol.* 74:524 (1977).

21. P. A. Burnham and S. Varon, *Proc. Natl. Acad. Sci. USA 69:*3556 (1972).

22. D. Monard, F. Solomon, M. Kentsch, and R. Gysin, *Proc. Natl. Acad. Sci. USA 70:*1682 (1973).

23. D. Bodian, in *The Neurosciences* (G. C. Quarton, T. Melnechuk, and F. O. Schmitt, eds.), Rockefeller University Press, New York, 1967.

24. F. O. Schmitt, in *The Neurosciences* (G. C. Quarton, T. Melnechuk, and F. O. Schmitt, eds.), Rockefeller University Press, New York, 1967.

23
THE CHEMICAL CONTROL OF NEURONAL GROWTH: The Nerve Growth Factor

Certain neurons in the peripheral nervous system require the presence of a protein known as the nerve growth factor for their survival and development. Specifically, the nerve growth factor is required by sympathetic neurons apparently throughout the entire life span of the cell, and by at least some of the sensory neurons during an early period in their development. As far as is currently known, there are no other neuron-directed factors which impinge in the same way on other types of neurons in the peripheral system, nor any such factors for any central nervous system neurons. It is quite possible that other such factors exist, but they have yet to be detected.

The original observations on the nerve growth factor were made, perhaps fortuitously, during experiments in which a tumor, sarcoma 180, was implanted in chick embryos. Unexpectedly the tumors became innervated by fibers from certain spinal ganglia. When comparable tumors were implanted on the chorioallantoic membrane, thus preventing any direct contact between tumor and nerve, the same result was obtained, indicating that the tumor was elaborating a diffusable factor. These latter observations in the early 1950s by Levi-Montalcini and her coworkers, and subsequent work by her group and many others, has led to a large body of information on this diffusable material, now known as the nerve growth factor [1-9].

Attempts by Cohen [10,11] to fractionate the tumor and to isolate the factor, involved the use of snake venom as a lytic agent. The venom itself proved to be a more potent source of the factor than the tumor. Subsequent survey of various tissues revealed the submaxillary gland of the adult male mouse as by far the richest source. Nerve growth factor activity in this tissue is at least 100-fold greater than in the other tissues in the mouse. Only the adult male has such large amounts, and only the mouse, since the submaxillary glands of other animals are quite low in activity. Castration of the adult male lowers the content of nerve growth factor in the gland, and administration of testosterone to the female raises the content, but these changes are related to the development of the gland itself and not specifically to its content of the nerve growth factor. Thus, snake venom and the submaxillary gland of the adult male mouse are the current sources of this material, with the latter being the most useful. There is no information as to the significance of the high levels of nerve growth factor in this gland of this species but it is probably made there and not merely stored. Administration of radioactive nerve growth factor does not lead to accumulations in the gland [12], providing evidence against storage. Explants of submaxillary gland will make nerve growth factor [13], thus providing evidence for biosynthesis.

Chemical Properties

The nerve growth factor from the salivary gland can be isolated in two forms. A high molecular weight form, known as the 7S [14], contains three different types of subunits, the α, the β, and the γ. The β contains all the nerve growth factor activity. The function of the other subunits is not known. The γ has been shown to have proteolytic activity [15]. By a somewhat different procedure the β subunit can be isolated [16], albeit slightly modified. This preparation is known as the 2.5S and is the form which is currently most frequently used. The 2.5S has a molecular weight of 26,000 and an isoelectric point of 9.3. It is composed of two subunits held together by noncovalent bonds, and these chains are almost identical, except for some minor proteolytic modification, which probably occurs during the purification [17]. It is of great current interest that the linear sequence of nerve growth factor (Fig. 23.1) has some homology with the sequence of proinsulin. This correlation, observed by Bradshaw and his colleagues [18], who also did the sequence work on the nerve growth factor [19], has opened new avenues of thinking about the mechanism of the action of nerve growth factor on its target cells, and the possible similarity between it and the mode of action of insulin.

Methodology

The methodology available for the assay of the nerve growth factor has been a serious and continuing problem in the study of this protein. It has not been particularly difficult to isolate and purify, and, indeed, preparations of the 7S form are commercially available. The quantitative measurement of its activity has been much more difficult. The first, and still the most used assay involves the ability of the nerve growth factor to induce outgrowth of processes in explants of dorsal root ganglia obtained from embryonic chickens (Fig. 23.2) [20,21]. The results of the assay are scored from 1^+ to 4^+, with the 3^+ response representing 1 biological unit of nerve growth factor. A biological unit is about 15 ng, so the assay measures, and the nerve growth factor is active, in the range of 10^{-11} to 10^{-12} M. The assay, however, is rather tedious, subject to substantial variation due to differences in various batches of culture reagents or investigators, and above all, only

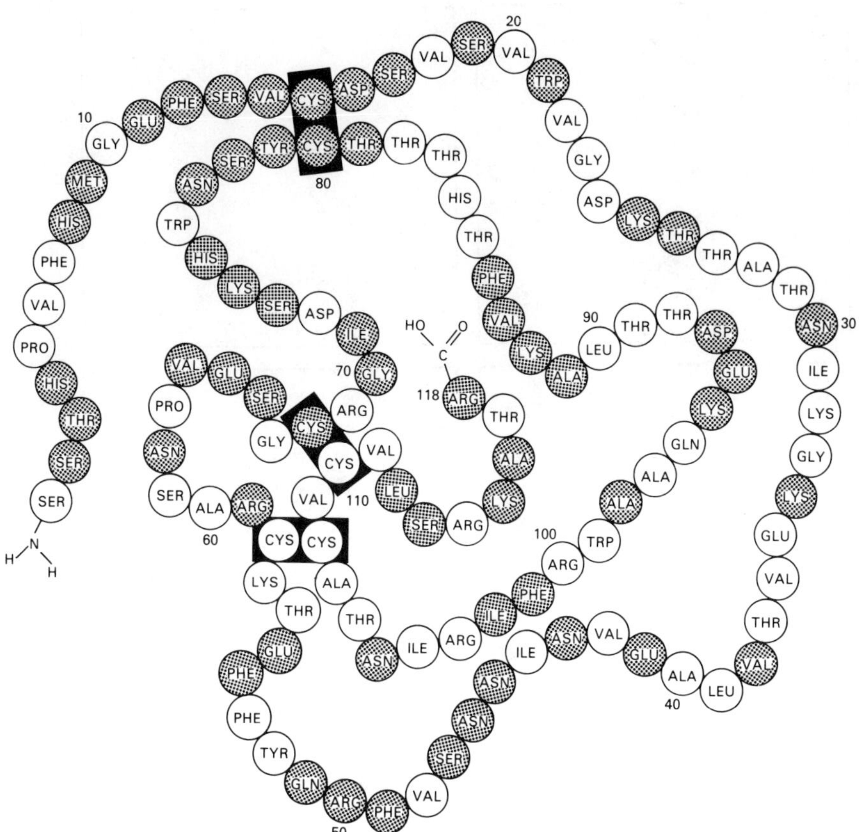

Figure 23.1   The linear sequence of mouse nerve growth factor and its homology with proinsulin. The cross-hatched residues are those which are either the same as corresponding residues in human proinsulin or are considered to be favored amino acid substitutions. [Adapted from the work of W. A. Frazier, R. H. Angeletti, and R. A. Bradshaw, *Science 176:*482 (1972).]

quantitative. There are other assays, but none are simple. At least three different immunoassays have been devised [22-24], and an assay based on the specific re-association of the various subunits [25]. Perhaps the most ingenious is a method based upon the chemical linkage of the nerve growth factor to the head of a bacteriophage [26]. The unknown amount of nerve growth factor is reacted with a known amount of a specific antibody to the nerve growth factor. A known amount of the chemically modified phage is added. The remaining antibody reacts with some of the phage, thus neutralizing it. The mixture is then plated onto

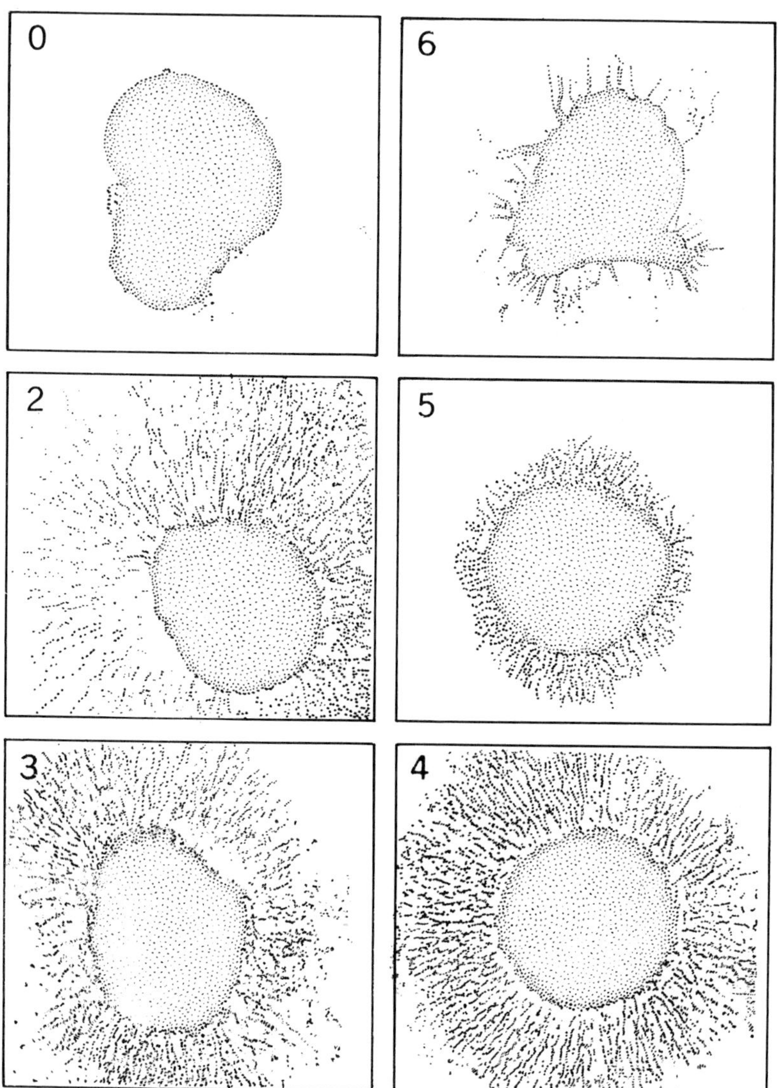

Figure 23.2   Fiber outgrowth from explanted dorsal root ganglia of chick embryo. Ganglia were cultured for 20 hr in the presence of no (0) nerve growth factor or increasing amounts (2 to 6) of nerve growth factor. Phase contrast, magnification ×100. [Adapted from S. Varon, in *Methods of Neurochemistry*, Vol. 3 (R. Fried, ed.), Dekker, New York, 1972.]

an appropriate bacterial plate, and the unneutralized phage causes a proportionate number of plaques to appear. Inventive though the assay is, it remains, as are all the methods, difficult and time consuming compared to usual enzyme assays. An assay for the nerve growth factor which would provide a fast and quantitative result would be most desirable.

Biosynthesis

Although the largest amount of nerve growth factor in the male mouse, at least, is clearly in the salivary gland and, as discussed above, the evidence indicates that it is made there and merely stored, it is not known if this is the only source in the body. There are experiments in the literature [27] in which the salivary glands of the mouse were removed and the levels of nerve growth factor in the serum were measured at various times after the surgery. The result of this experiment was that the levels in the blood dropped immediately after the surgery, but rose slowly in the succeeding weeks. The interpretation of the experiment is that the salivary glands are the major site of nerve growth factor biosynthesis, but that other tissues will make some. The data further indicate that if the major source is removed some kind of regulatory mechanism signals, or permits, the other tissues to pick up the burden of the synthesis. Direct studies on the biosynthesis have supported this picture. It has been shown that nerve growth factor can be made in culture by several different types of cells [6,13,26,28-30], indicating that many or all tissues can synthesize the molecule. Thus, the interpretation of the in vivo experiments seems to be a valid one.

Immunosympathectomy

Regardless of the site of its biosynthesis, it is possible to deprive an animal of the effects of the nerve growth factor by immunological means. It was shown some years ago that the injection of nerve growth factor into a horse or a sheep produced a potent antiserum. Subsequent administration of this antiserum to young rats causes what is known as a immunosympathectomy; that is, the sympathetic neurons deteriorate, with repeated injection of the antiserum, in a matter of days [31-33]. Presumably this is because of the constant need that these neurons have for the presence of the nerve growth factor, although there are several possible mechanisms by which the antibody might act [34,35]. Since these neurons have no capacity for regeneration, the sympathetic system is permanently missing from these animals. A total immunosympathectomy is not produced, primarily because of the need for the injection of very large volumes of the antiserum, but better than 90% of the system, as judged morphologically and biochemically, can be deleted by this means. The purification of the nerve growth factor antibody by means of affinity chromatography [36,37] should allow the production of a completely sympathectomized animal.

Receptors

Receptors for the nerve growth factor are present on the surface of the target organs, such as the sensory or the sympathetic ganglia [38-40]. These receptors bind the nerve growth factor specifically and with high affinity. As such they are much like the surface receptors found for insulin and other peptide hormones on their respective targets. There has been some initial disagreement in the literature about the nature of the binding and the tissue distribution of these binding sites. The explanation for the disagreement is not obvious, but there is some difference in the method by which the various groups prepare the radioactively labeled nerve growth factor used in these studies, and this may account for the differences. In any case, it seems well established that nerve growth factor is bound to the surface of its target organs, and it follows that nerve growth factor, like other peptide hormones, influences further cellular events as a consequence of its binding to this site. Indeed, it has been shown that nerve growth factor can produce some of its characteristic effects, namely outgrowth of neurites, even though it is bound to large beads or to virus, either of which should make it unable to enter the cell [26,41].

Retrograde Transport

Another line of experimentation, however, suggests that nerve growth factor can and does indeed enter the cell. Injection of nerve growth factor labeled with radioactive iodine into the anterior chamber of the eye of a rat or a mouse leads, after several hours, to the appearance and the accumulation of radioactive nerve growth factor in the superior cervical ganglion on the injected side of the animal [42-46]. Since the superior cervical ganglia has nerve terminals in the iris it is clear that the nerve growth factor reaches the ganglia by retrograde movement up the axon from the synaptic ending to the cell body. Inhibition of the appearance of radioactivity by colchicine, which blocks axonal flow, supports this conclusion. The retrograde transport is specific for nerve growth factor in that injection of proteins of size or charge similar to that of nerve growth factor does not lead to comparable accumulations in the ganglia. Even nerve growth factor is not transported if the tryptophan residues have been oxidized with N-bromosuccinimide, an oxidation which destroys its biological activity. Although in these experiments a substantial amount of the nerve growth factor leaks out of the eye into the bloodstream and reaches the ganglia by this route, there is some evidence that the nerve growth factor which arrives by retrograde transport is physiologically active in that it has its characteristic actions on the injected side ganglia, above and beyond that caused in both ganglia by the nerve growth factor which escapes from the eye and is carried through the blood. Thus, it is not clear whether the site of action of nerve growth factor is at the surface of the cell, in the interior, or both. Perhaps the last is the most likely because of the many different actions of nerve growth factor. It is of interest that insulin, whose structural relationship to nerve growth factor is mentioned above, has recently been shown to bind to specific receptors on liver nuclei [47], and nuclear receptors for nerve growth factor itself have recently been reported [48].

Mechanism of Action

Whatever the site or sites of nerve growth factor binding turn out to be, there is no doubt that the factor influences a large number of the processes in the cell either directly or indirectly (Fig. 23.2). It induces hypertrophy and hyperplasia

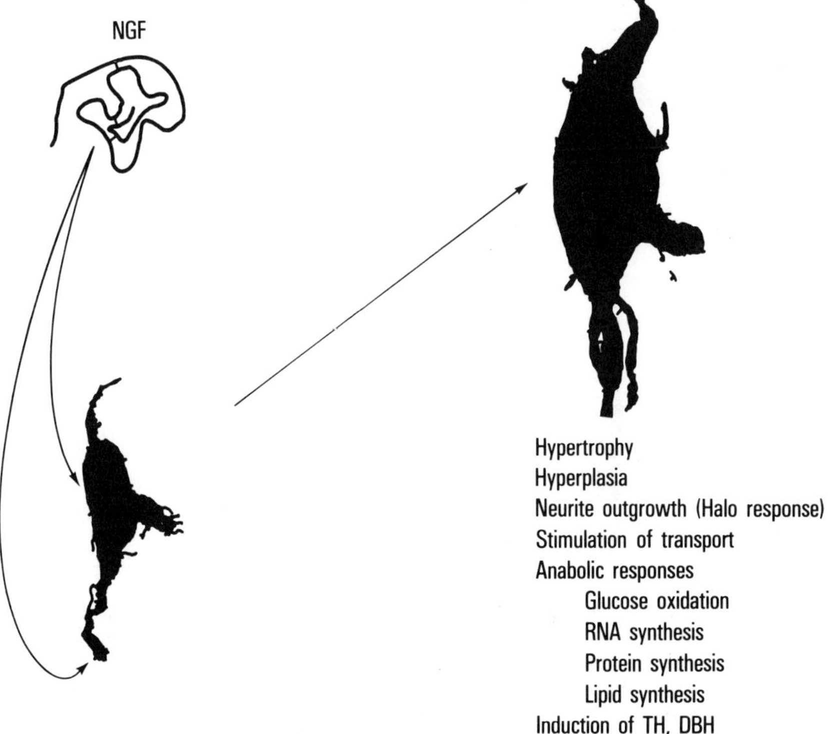

Figure 23.3   The actions of nerve growth factor in sympathetic ganglia.

in responsive ganglia. It produces an increase in many of the anabolic activities
of the cells, including a stimulation of the oxidation of glucose, the synthesis of
lipid, the synthesis of protein and of RNA, and the transport of amino acids and
nucleosides into the cell. It also promotes or permits the outgrowth of processes
[49-54] . These general effects have been known for a number of years. More
recently, however, it has been shown that the nerve growth factor has possibly
an even more specific action in sympathetic ganglia. When small doses of nerve
growth factor are given to young rats there is a rather rapid increase in the activities of two of the enzymes involved in the biosynthesis of norepinephrine
[55] , the transmitter substance of the sympathetic neurons. These two enzymes,
tyrosine hydroxylase and dopamine β-hydroxylase, increase both in total and in
specific activity (Fig. 23.4). The specificity of this effect is underlined by the
observation that the specific activities of a number of closely related enzymes,
such as dopa decarboxylase, monoamine oxidase, and dihydropteridine reductase

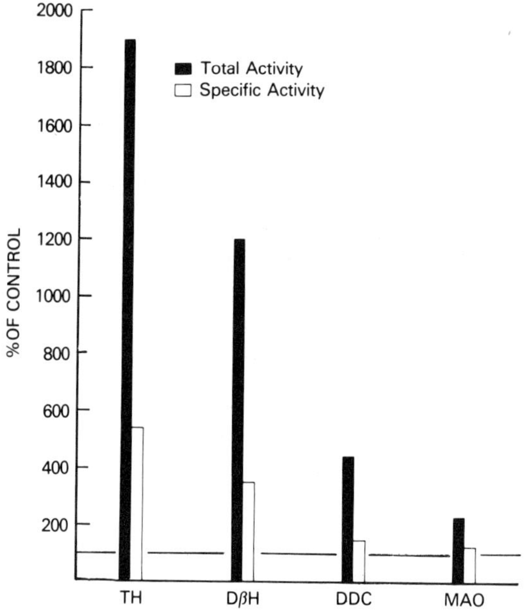

Figure 23.4   Effect of nerve growth factor on enzymes involved in synthesis and
metabolic degradation of norepinephrine. Newborn rats were treated with 10
μg/g of nerve growth factor for 10 days. The specific activity of all enzymes
studied in the superior cervical ganglia is expressed as percentage of controls.
[Adapted from H. Thoenen, P. U. Angeletti, R. Levi-Montalcini, and R.
Kettler, *Proc. Natl. Acad. Sci. USA 68:* 1598 (1971).]

[56] , do not rise. The effect on the activities of these enzymes is the most specific effect of nerve growth factor known at the present time, and shows that, by
whatever mechainsm, direct or indirect, the factor has an influence on the transcription of certain genes in sympathetic neurons.

The concept that nerve growth factor acts only on the sympathetic and
embryonic sensory ganglia is pretty well established. Nevertheless, there are a
number of other systems which respond in one way or another to nerve growth
factor. Binding of nerve growth factor to neuroblastoma [57] , to brain synaptosomes [58] , and to most peripheral tissues [59] has been observed, and this
binding is quite specific and presumably, therefore, meaningful in terms of function. Binding of the nerve growth factor antibodies to glia from target sensory
ganglia has been observed [60] . Vesicles from adrenal medulla appear to have
nerve growth factor receptors [61] , and the surface properties of tectal cells appear to change in response to large amounts of the nerve growth factor [62] .
Nerve growth factor inhibits the synthesis of a chondromucoprotein in explants
of embryonic chick cartilage rudiments [63] . Recently it has been shown that
a clonal line of pheochromocytoma, a tumor of the adrenal medulla, can be induced to form processes with nerve growth factor [64] . This may be the most
interesting system currently available since it is one of the only cells which responds in vitro in a specific way to the factor and, most important, does not die
in its absence. The effects of nerve growth factor on the central nervous system
are discussed below. In any case it does no longer seem true to state that the target cells for the nerve growth factor are only in the sympathetic and the sensory
ganglia.

The mechanism of action of the nerve growth factor is not known. It would
seem to be the most parsimonious postulate that the membrane binding of the
factor initiates a chain of events which eventually leads to the large number of
different actions which arc observed. It has been suggested by Varon that the
anabolic stimulations caused by nerve growth factor all stem from membrane alterations which cause, in turn, a stimulation of the various transport systems
serving the cell [5] . In this view the action of the factor in making more precursor available for the synthesis of protein, nucleic acid, lipid, and so on, would
then cause these processes to accelerate, and from these increases would flow all
the other expressions of increased cellular activity. To recommend this line of
reasoning are the experiments on transport which show that increases in such
membrane functions are quite fast and precede other responses in the cell, especially those involving increased synthesis of macromolecules. It seems more difficult to explain the outgrowth of processes on this basis, although some recent
studies show that even this response could depend not on some selective action
of nerve growth factor, but only on the ability of a cell to lay down an appropriate matrix in vitro [65] . Certainly, the enhanced formation of this matrix could
be a consequence of the increased anabolic activity of the cell.

Perhaps the most difficult of the many actions of nerve growth factor to attribute solely to a general membrane action are the apparently selective increases in the activities of tyrosine hydroxylase and dopamine β-hydroxylase. At least for tyrosine hydroxylase this selective increase in activity is now known to be due truly to a selective increase in the synthesis of the enzyme [66]. It would seem likely that at least this action is due to a gene-level action, rather than to a generalized membrane effect, and the recent demonstration that nerve growth factor produces an increase in cyclic AMP (cAMP) levels in rat superior cervical ganglia in vitro [67] may provide the link here. It should be noted, however, that other investigators [68,69] have failed to demonstrate a change in cAMP levels upon the addition of nerve growth factor to sensory ganglia in vitro, so the role of cAMP in the action of nerve growth factor, if any, is still not clear.

Thus, nerve growth factor reacts with membrane receptors, and probably has a membrane level effect on its target tissues, possibly one which results in a stimulation of the transport mechanisms of the cells. It may, in addition, have a gene-level effect mediated directly through nuclear receptors, to which the nerve growth factor gains access through retrograde transport or exocytosis, or indirectly through some second messenger or sequence of second messengers. Whatever its locus it probably stimulates the cell to run its normal program rather than inducing any change in its fundamental properties. Varon has written that the action of nerve growth factor may only be to allow the target tissues to survive, and that additional nerve growth factor would merely "lead the target tissues to anticipate or even exceed the performances they would otherwise display at their normal maturational stage" [5]. Thus, nerve growth factor maintains its target cells in a viable state and, perhaps in addition, intensifies aspects of their genetic program of development. The detailed molecular mechanism of either of these actions remains to be solved.

It has been suggested that nerve growth factor acts by binding to tubulin [70]. Indeed, nerve growth factor does show an intriguing affinity for this protein. However, it seems clear that in systems where nerve growth factor exhibits a rather early and specific action on neurite outgrowth or on tyrosine hydroxylase activity, its action on tubulin content and synthesis is seen at a later time and along with its general action on the content and synthesis of other cellular proteins [71-73]. At this point the balance of experimental evidence does not identify tubulin content or synthesis as the primary site of the action of nerve growth factor, although the possibility remains that nerve growth factor has an early and specific action on the polymerization of tubulin from already existing subunits.

Central Nervous System

The observations on norepinephrine-containing sympathetic neurons have prompted inquiry into the possible role of nerve growth factor in the main-

tainence, development, and even the regeneration of the catecholaminergic neurons in the central nervous system. Nerve growth factor probably does not enter
the central nervous system of adult animals, due to the blood-brain barrier. There
is no direct evidence, either, for the synthesis of nerve growth factor in the brain.
Thus, there is no information on the way in which nerve growth factor would
get to the brain to act. Finally, the systemic administration of the antiserum does
not have any apparent consequences for the function of the central nervous system. In view of these findings there should not seem to be any obvious reason
to look for an action of the factor on the brain. Nevertheless, there are clearly
specific binding sites for nerve growth factor in the brain [58]. Also, there are
recent publications which show that lesioned catecholaminergic neurons in the
central nervous system display increased axonal sprouting when nerve growth
factor is injected intraventricularly or directly into the brain substance [74,75].
The finding that nerve growth factor induces the pivotal enzyme ornithine decarboxylase in the brain [76], just as it does in sympathetic ganglia [77], provides a biochemical basis for these histochemical observations. Regeneration in
the nervous system being, obviously, one of the central problems in clinical
neurobiology, these findings are most provocative.

Clinical Implications

Another clinical aspect of the studies on nerve growth factor revolves around the
possible occurrence of high blood levels of nerve growth factor in individuals
with various brain tumors. It was reported some years ago that patients with
neuroblastoma had elevated levels of nerve growth factor in their sera [78]. This
finding could not be confirmed [79,80]. Subsequently there has been a report
of elevated values in individuals with von Recklinghausen's disease, a benign dis-
seminated tumor [81]. This finding, in turn, could not be confirmed [82]. In
this latter paper, elevated levels of nerve growth factor were reported in persons with central acoustic neuroma, a tumor which has been found to occur
with great frequency in certain families. In sum, the question of whether the
nerve growth factor is associated with the onset or the development of certain
brain tumors is still very much open.

The reason for the discrepancies is probably the problem of nerve growth
factor methodology referred to above. It may also relate to the presence of inhibitors of nerve growth factor function in vitro which have been observed in
tissues and sera [58,83,84]. The first problem, then, is whether these reports
are true at all. If nerve growth factor does, in fact, increase in the sera of patients with brain tumors, it is important to determine its value as a diagnostic
tool. It is also important to find out if the increased levels are a cause or an effect of the tumor. If they are, in fact, the cause, such increases may have pre-
dictive value. An intriguing aspect of the report on nerve growth factor levels
in the acoustic neuroma is the observation that children at risk who are below

the age of onset had increased levels of nerve growth factor even though no tumors had yet appeared. On the contrary, children beyond the age of onset in whom no tumors had occurred, and who were, by inference then, not to be afflicted, had normal nerve growth factor levels.

The study of nerve growth factor is a central one in neurobiology. Its mechanism of action remains unknown. Its exact function, be it viability, differentiation, or trophic action, is unknown. Its clinical significance is equally unknown. Perhaps most significant, it is not clear whether this factor is some unique aspect of the life cycle of certain neurons, or only the first known of a class of factors involved in the development of all kinds of neurons. The rapidly growing literature on the other similar growth factors, epidermal growth factor, mesenchymal growth factor, fibroblast growth factor, glial factor, and so on, suggest that the latter is the case.

References

1.   R. Levi-Montalcini, *Harvey Lect. 60:*217 (1966).
2.   R. Levi-Montalcini and P. U. Angeletti, *Physiol. Rev. 48:*534 (1968).
3.   I. Schenkein, in *Handbook of Neurochemistry,* Vol. Vb (A. Lajtha, ed.), Plenum, New York, 1971.
4.   H. Thoenen, *Pharm. Rev. 24:*255 (1972).
5.   S. Varon, *Exp. Neurol. 48:*75 (1975).
6.   R. A. Bradshaw and M. Young, *Biochem. Pharmacol. 25:*1445 (1976).
7.   S. Varon, in *The Nervous System,* Vol. I (D. B. Tower, ed.), Raven, New York, 1975.
8.   W. C. Mobley, A. C. Server, D. N. Ishii, R. J. Riopelle, and E. M. Shooter, *N. Engl. J. Med. 297:*1096, 1149, 1211 (1977).
9.   M. W. Yu, J. Lakshmanan, and G. Guroff, in *Essays in Neurochemistry and Neuropharmacology,* Vol. III (M. B. H. Youdim, W. Lovenberg, D. F. Sharman, and J. R. Lagnado, eds.), John Wiley, New York, 1978.
10. S. Cohen and R. Levi-Montalcini, *Proc. Natl. Acad. Sci. USA 42:*571 (1956).
11. S. Cohen, *J. Biol. Chem. 234:*1129 (1959).
12. R. H. Angeletti, P. U. Angeletti, and R. Levi-Montalcini, *Brain Res. 46:*421 (1972).
13. E. A. Berger and E. M. Shooter, *Fed. Proc. 35:*1684 (1976).
14. S. Varon, J. Nomura, and E. M. Shooter, *Biochemistry 6:*2202 (1967).
15. L. A. Greene, E. M. Shooter, and S. Varon, *Biochemistry 8:*3735 (1969).
16. V. Bocchini and P. U. Angeletti, *Proc. Natl. Acad. Sci. USA 64:*787 (1969).
17. J. B. Moore, W. C. Mobley, and E. M. Shooter, *Biochemistry 13:*833 (1974).
18. W. A. Frazier, R. H. Angeletti, and R. A. Bradshaw, *Science 176:*482 (1972).
19. R. H. Angeletti and R. A. Bradshaw, *Proc. Natl. Acad. Sci. USA 68:*2417 (1971).

20. R. Levi-Montalcini, H. Meyer, and V. Hamburger, *Cancer Res. 14:*49 (1954).
21. S. Varon, J. Nomura, and E. M. Shooter, in *Methods of Neurochemistry,* Vol. 3 (R. Fried, ed.), Dekker, New York, 1972.
22. D. G. Johnson, P. Gorden, and I. J. Kopin, *J. Neurochem. 18:*2355 (1971).
23. I. A. Hendry, *Biochem. J. 128:*1265 (1972).
24. R. Angeletti and E. Vigneti, *Brain Res. 33:*601 (1971).
25. H. D. Shine and J. R. Perez-Polo, *J. Neurochem. 26:*513 (1976).
26. J. Oger, B. G. W. Arnason, N. Pantazis, J. Lehrich, and M. Young, *Proc. Natl. Acad. Sci. USA 71:*1554 (1974).
27. I. A. Hendry and L. L. Iversen, *Nature (London) 243:*500 (1973).
28. M. Young, J. Oger, M. H. Blanchard, H. Asdourian, H. Amos, and B. G. W. Arnason, *Science, 187:*361 (1975).
29. R. A. Murphy, N. J. Pantazis, B. G. W. Arnason, and M. Young, *Proc. Natl. Acad. Sci. USA 72:*1895 (1975).
30. G. P. Harper, F. L. Pearce, and C. A. Vernon, *Nature (London) 261:*251 (1976).
31. S. Cohen, *Proc. Natl. Acad. Sci. USA 46:*302 (1960).
32. R. Levi-Montalcini and B. Booker, *Proc. Natl. Acad. Sci. USA 46:*384 (1960).
33. E. Zaimis, (ed.), *Nerve Growth Factor and its Antiserum,* Athlone, London, 1972.
34. M. T. Sabatini, A. Pellegrino, and E. deRobertis, *Exp. Neurol. 12:*4 (1965).
35. S. Varon, C. Raiborn, and S. Norr, *Exp. Cell Res. 88:*247 (1974).
36. J. T. Tomita and S. Varon, *Neurobiol. 1:*176 (1971).
37. K. Stoeckel, C. Gagnon, G. Guroff, and H. Thoenen, *J. Neurochem. 26:* 1207 (1976).
38. S. P. Banerjee, S. H. Snyder, P. Cuatrecasas, and L. Greene, *Proc. Natl. Acad. Sci. USA 70:*2519 (1973).
39. W. A. Frazier, L. F. Boyd, and R. A. Bradshaw, *J. Biol. Chem. 249:*5513 (1974).
40. K. Herrup and E. M. Shooter, *Proc. Natl. Acad. Sci. USA 70:*3884 (1973).
41. W. A. Frazier, L. F. Boyd, and R. A. Bradshaw, *Proc. Natl. Acad. Sci. USA 70:*2931 (1973).
42. I. A. Hendry, K. Stoeckel, H. Thoenen, and L. L. Iversen, *Brain Res. 68:* 103 (1974).
43. K. Stoeckel, U. Paravicini, and H. Thoenen, *Brain Res. 76:*413 (1974).
44. I. A. Hendry, R. Stach, and K. Herrup, *Brain Res. 82:*117 (1974).
45. U. Paravicini, K. Stoeckel, and H. Thoenen, *Brain Res. 84:*279 (1975).
46. K. Stoeckel, G. Guroff, M. Schwab, and H. Thoenen, *Brain Res. 109:*271 (1976).
47. I. D. Goldfine and G. J. Smith, *Proc. Natl. Acad. Sci. USA 73:*1427 (1976).
48. R. Y. Andres, I. Jeng, and R. A. Bradshaw, *Proc. Natl. Acad. Sci. USA 74:* 2785 (1977).
49. P. Angeletti, A. Liuzzi, R. Levi-Montalcini, and D. Gandini-Attardi, *Biochim. Biophys. Acta 90:*445 (1964).

50. P. Angeletti, D. Gandini-Attardi, G. Toschi, M. L. Salvi, and R. Levi-Montalcini, *Biochim. Biophys. Acta 95:*111 (1965).

51. G. Toschi, E. Dore, P. Angeletti, R. Levi-Montalcini, and C. deHaen, *J. Neurochem. 13:*539 (1966).

52. A. Liuzzi, P. Angeletti, and R. Levi-Montalcini, *J. Neurochem. 12:*705 (1965).

53. P. Angeletti, A. Liuzzi, and R. Levi-Montalcini, *Biochim. Biophys. Acta 84:*778 (1964).

54. R. Levi-Montalcini and P. Angeletti, in *Organogenesis* (R. L. de Haen and H. Ursprung, eds.), Holt, New York, 1965.

55. H. Thoenen, P. U. Angeletti, R. Levi-Montalcini, and R. Kettler, *Proc. Natl. Acad. Sci. USA 68:*1598 (1971).

56. B. Nikodijevic, M. W. Yu, and G. Guroff, *J. Neurochem. 28:*851 (1977).

57. R. Revoltella, L. Bertolini, and M. Pediconi, *Exp. Cell Res. 85:*89 (1974).

58. A. Szutowicz, W. A. Frazier, R. A. Bradshaw, *J. Biol. Chem. 251:*1516 (1976).

59. W. A. Frazier, L. F. Boyd, A. Szutowicz, M. W. Pulliam, and R. A. Bradshaw, *Biochem. Biophys. Res. Commun. 57:*1096 (1974).

60. S. Varon, C. Raiborn, and S. Norr, *Exp. Cell Res. 88:*247 (1974).

61. O. Nikodijevic, B. Nikodijevic, O. Zinder, M. W. Yu, G. Guroff, and H. B. Pollard, *Proc. Natl. Acad. Sci. USA 73:*771 (1976).

62. R. Merrell, M. W. Pulliam, L. Randono, L. F. Boyd, R. A. Bradshaw, and L. Glaser, *Proc. Natl. Acad. Sci. USA 72:*4270 (1975).

63. G. S. Eisenbarth, M. K. Drezner, and H. E. Lebovitz, *J. Pharm. Exp. Ther. 192:*630 (1975).

64. A. S. Tischler and L. A. Greene, *Nature (London) 258:*341 (1975).

65. S. B. Mizel and J. R. Bamburg, *Dev. Biol. 49:*20 (1976).

66. P. MacDonnell, N. Tolson, and G. Guroff, *J. Biol. Chem. 252:*5859 (1977).

67. B. Nikodijevic, O. Nikodijevic, M. W. Yu, H. Pollard, and G. Guroff, *Proc. Natl. Acad. Sci. USA 72:*4769 (1975).

68. D. C. Haas, D. B. Hier, B. G. W. Arnason, and M. Young, *Proc. Soc. Exp. Biol. Med. 140:*45 (1972).

69. W. A. Frazier, C. E. Ohlendorf, L. F. Boyd, L. Aloe, E. M. Johnson, J. A. Ferrendelli, and R. A. Bradshaw, *Proc. Natl. Acad. Sci. USA 70:*2448 (1973).

70. R. Levi-Montalcini, R. Revoltella, and P. Calissano, in *Recent Progress in Hormone Research,* Vol. 30 (R. O. Greep, ed.), Academic, New York, 1974.

71. K. M. Yamada and N. K. Wessells, *Exp. Cell Res. 66:*346 (1971).

72. K. Stöckel, F. Solomon, U. Paravicini, and H. Thoenen, *Nature (London) 250:*150 (1974).

73. S. B. Mizel and J. R. Bamburg, *Neurobiology 5:*283 (1975).

74. B. Bjerre, A. Björklund, and U. Stenevi, *Brain Res. 60:*161 (1973).

75. U. Stenevi, B. Bjerre, A. Björklund, and W. Mobley, *Brain Res. 69:*217 (1974).

76. M. E. Lewis, J. Lakshmanan, K. Nagaiah, P. C. MacDonnell, and G. Guroff, *Proc. Natl. Acad. Sci. USA 75:*1021 (1978).

77. P. C. MacDonnell, K. Nagaiah, J. Lakshmanan, and G. Guroff, *Proc. Natl. Acad. Sci. USA 74:*4681 (1977).

78. J. A. Burdman and M. N. Goldstein, *J. Natl. Cancer Inst. 33:*123 (1964).

79. A. H. Bill, E. S. Seibert, J. B. Beckwith, and J. R. Hartmann, *J. Natl. Cancer Inst. 43:*1221 (1969).

80. M. Waghe, S. Kumar, and J. K. Steward, *J. Pediatr. Surg. 5:*14 (1970).

81. I. Schenkein, E. D. Bucker, L. Henson, F. Axelrod, and J. Dancis, *N. Engl. J. Med. 290:*613 (1974).

82. D. C. Siggers, S. H. Boyer, and R. Eldridge, *N. Engl. J. Med. 292:*1134 (1975).

83. B. E. C. Banks, D. V. Banthrope, K. A. Charlwood, F. L. Pearce, and C. A. Vernon, *Nature (London) 246:*503 (1973).

84. S. Varon, C. Raiborn, and P. A. Burnham, *Neurobiology 4:*317 (1974).

24

THE MECHANISM OF CONDUCTION AND TRANSMISSION: The Biochemistry of Acetylcholine

Excitability, the property which allows a neuron to conduct an impulse down the axon and transmit it to another cell is, after all, the fundamental property which distinguishes a neuron from other cells. The exact molecular mechanism by which this is accomplished is still largely unknown. A consistent and detailed picture of certain of the events can be drawn, however, taking into account several lines of evidence. Most of the information available involves transmission by the cholinergic neuron, but as data on other systems accumulate, it appears that transmissions by various kinds of neurons and by different chemical transmitters are in principle quite similar.

Membrane Potential

In the resting state the neuronal membrane is polarized; that is, there is an electrical potential across it in the amount of about -70 to -90 mV. This potential exists because of the uneven distribution of sodium and potassium ions between the intracellular contents and the extracellular fluid. This distribution in the resting nerve is created by the permeability characteristics of the membrane itself, and by the constant pumping of sodium out of the cell by the Na^+,K^+-dependent ATPase in the cell membrane. The presence of ouabain, an inhibitor of the pump, causes irreversible membrane depolarization.

When a neuron is stimulated to transmit an impulse either by chemical or electrical means, the membrane is depolarized. That is, the resting potential of the nerve cell drops to a critical or threshold level. The first event which occurs upon discharge of this resting potential is the abrupt inward flow of Na^+ which is complete in some fraction of a millisecond. This influx of Na^+ corresponds to the rising phase of the action potential. Following the Na^+ influx, K^+ ions leave the cell, as do Cl^-. These effluxes are slower processes than the influx of Na^+, but still occur during the stimulus-induced period of increased permeability. The K^+ and Cl^- outflows correspond to the falling phase of the action potential. Dur-

410

ing this period the neuron does not respond to further stimulation, i.e., it is refractory. After the action potential has been triggered and propagated down the
axon, the initial permeability characteristics of the membrane and the resting
potential are restored by redistribution of the ions at the expense of high-energy
phosphate, but seemingly by a mechanism different that the Na^+,K^+-dependent
ATPase which maintains the ionic disequilibrium in the resting neuron [1].

   The mechanism by which the impulse is transmitted down the axon is
largely unknown. The propagated wave moves along the nerve because the axonal
membrane is capable of depolarizing and repolarizing in adjacent small segments.
The action potential is conducted along the axon as a propagated wave to the terminal portion.

Transmitter Release

Upon reaching the synaptic bouton, the action potential results in a depolarization of the nerve terminal. The depolarization initiates a series of chemical
changes which, in turn, result in the transmission of the impulse to another cell
[2]. The transmission is accomplished by the liberation of a transmitter substance into the synaptic cleft. The transmitter substance, in the present case
acetylcholine, is contained in the vesicles of the presynaptic bouton. When the
action potential reaches the presynaptic terminal, some of the vesicles fuse with
the membrane at specific release sites and liberate their contents by exocytosis
into the synaptic cleft. It is estimated that each of these vesicles release between
5000 and 50,000 molecules of acetylcholine. The best information available,
mostly from electron microscope studies of neurons at various stages of transmitter release [3], indicates that the vesicles reform at the inside surface of the
membrane at a point removed from the release site. The membrane is said to
"flow" to this point. The vesicles which have formed at the inside surface are
coated with a protein material. They appear to migrate through the cell and
coalesce to form cisturnae. New vesicles then emerge, uncoated and full of transmitter, and the cycle is repeated (Fig. 24.1).

   Calcium ions are taken up by the presynaptic ending during depolarization.
There is substantial evidence that these Ca^{2+} ions are required for the release of
the transmitter [2]. In some way they couple the axonal action potential to
synaptic transmitter release. The mechanism of this Ca^{2+}-dependent, stimulus-
evoked release of the transmitter is not known, but the observation that vesicles
isolated from stimulated tissues contain less acetylcholine and more Na^+ and Ca^{2+}
than normal supports the idea that the release involves an ion-exchange mechanism in which ions from the extracellular fluid displace acetylcholine bound to a
macromolecular, anionic matrix [4].

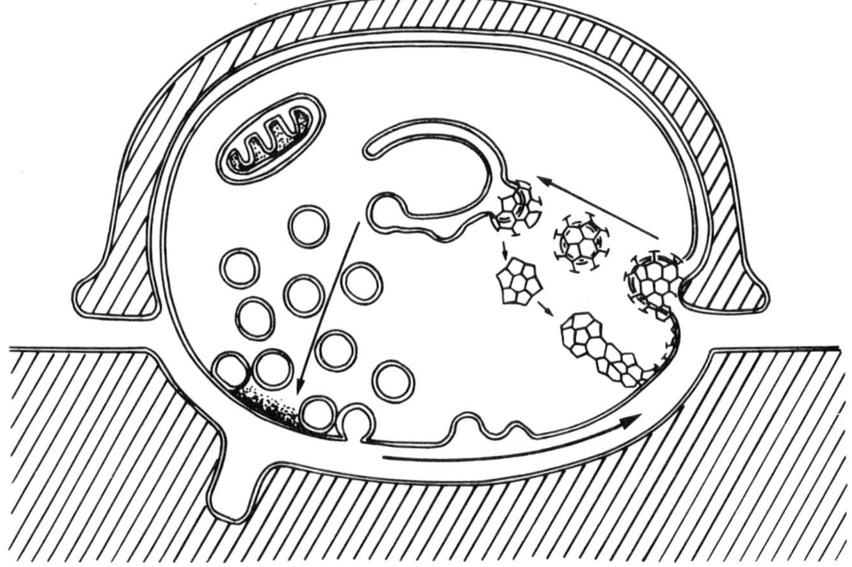

Figure 24.1 Diagrammatic summary of the path of synaptic vesicle membrane recycling: synaptic vesicles discharge their content of transmitter by coalescing with the plasma membrane at specific regions adjacent to the muscle. Then, equal amounts of membrane are retrieved by coated vesicles arising from regions of the plasma membrane adjacent to the Schwann sheath. Finally, the coated vesicles lose their coats and coalesce to form cisternae which accumulate in regions of vesicle depletion and slowly divide to form new synaptic vesicles. [Figure from J. E. Heuser and T. S. Reese, *J. Cell. Biol. 57:*315 (1973).]

Acetylcholine Receptor

The diffusion of acetylcholine across the 20-nm synaptic cleft is followed by its binding to specific receptors on the postsynaptic cells. There are two types of receptors for acetylcholine known, the nicotinic and the muscarinic. The nicotinic receptors are found at the neuromuscular junction of the skeletal muscle, in the autonomic ganglia, and in the electroplax of the electric eel or the torpedo fish. They are characterized pharmacologically by the property that they are stimulated by nicotine and blocked by curare. The muscarinic receptors are found in visceral autonomic cells, in smooth muscle, and in brain, and are stimulated by muscarine and blocked by atropine. They, of course, are both stimulated by acetylcholine.

In recent years substantial progress has been made in understanding the nicotinic acetylcholine receptor [5]. The use of specific agonists and antagonists which interact reversibly with the receptor in whole-cell preparations has provided a description of the electrical properties of the receptor and of the dose responses of the various agents [6]. Using the electroplax of the eel, which is a remarkably rich source of the nicotinic receptor, purely cholinergic in its innervation and about 4% synaptic elements by weight, it has been possible, in addition to prepare vesicles which retain the ion permeability of the original electroplax and can be used to study the permeability changes to Na^+ and K^+ in response to the addition of acetylcholine or its antagonists [7-10].

The electric organ of the ray, *Torpedo marmorata,* is a remarkably good system for the study of cholinergic mechanisms. The ray possesses two such organs, one on each side of its head. These organs consist of stacks of electrocytes, cells embryologically derived from muscle. They receive massive, purely cholinergic innervation from nerve trunks which originate from cell bodies in the electric lobes on the upper surface of the brainstem. The arrangement is like an overgrown mass of muscle motor endplates without the contractile mechanism. The tissues can generate shocks of about 25 to 30 V. From each fish 500 g or more of electric tissue can be obtained [4].

The isolation of the homogeneous receptor has been the target of a number of research groups. The work has been facilitated by the use of certain snake venoms which are irreversible inhibitors of the receptor, most frequently the α-bungarotoxin from the venom of the Formosan krait. This toxin is typical of the highly neurotoxic polypeptides of molecular weights between 7 and 8000 which bind irreversibly to the receptors. Upon iodination the toxin can be shown to bind to a component in the membranes of electroplax, and this component can be extracted using a nonionic detergent. The binding of the toxin can be prevented or reduced by the simultaneous presence of either agonists or other antagonists of the binding of acetylcholine. The binding of the radioiodinated neurotoxins has also been used to study the amount and the distribution of the receptors in other systems, such as the distribution of the receptors on normal and on denervated muscle cells [11], or the number of receptors in the central nervous system [12].

The purification of the nicotinic acetylcholine receptor has been accomplished by a number of groups. The sources of choice have usually been the electric organs of either the ray, *Torpedo,* or the eel, *Electrophorus,* but receptors have also been purified from vertebrate and invertebrate muscle, and from rat brain [2,5,13-16]. The methodology generally has involved extraction of the organ or the membrane with a salt solution, followed by solubilization of the membranes with the nonionic detergent, Triton X-100. The receptor is then further purified by affinity chromatography, usually on a dextran column containing bound cholinergic ligand or neurotoxin. The amount of the nicotinic receptor per milligram tissue, as determined by the binding of labeled neurotoxin, in the various sources is as follows:

Torpedo: 1000 fmol
Electrophorus: 45–450 fmol
Rat diaphragm: 8.5 fmol
Rat brain: 2–5 fmol

There is substantial agreement in the current literature concerning the properties of the receptor as isolated from various organs. Generally, the receptor appears to have a molecular weight of approximately 200,000 per binding site. There is evidence that the binding site is on a smaller subunit of approximately 40,000 molecular weight. The amino acid composition of the receptor is unremarkable. The binding of acetylcholine has a K_D of between 3×10^{-9} and 1×10^{-10} [16] and seems to be straightforward in that no cooperativity is observed. Exactly how the protein fits into the membrane matrix is not known, but there is evidence that it is a highly hydrophobic protein and may be in the class known as proteolipid. All the receptors isolated appear to be glycoproteins. Injection of the purified receptor into animals produces a flaccid paralysis similar to myasthenia gravis [17,18]. The isolation of the muscarinic receptor has been much more difficult [19] and has not progressed to nearly the same point as outlined above for the nicotinic receptor. Thus, there is no direct information as to the structural differences between the two types of receptors.

The combination of acetylcholine with its receptor initiates or permits a change in the permeability of the postsynaptic cell membrane to small ions, which depolarizes that cell. This depolarization leads to the unique response exhibited by the postsynaptic cell, be it contraction as with muscle, generation of electric current, as with the electroplax, or, indeed, to further nerve impulse transmission, as with the neurons in the sympathetic ganglia.

Action Potential

The action potential, which is initiated by the combination of acetylcholine with the nicotinic receptor, is a rapid one lasting only a few milliseconds. So far there is no information concerning the intracellular events which mediate the permeability changes supporting this action potential. The action potential accompanying the combination of acetylcholine with the muscarinic receptor is of much longer duration. Similar "slow" responses occur at catecholamine synapses as well. The intracellular events supporting these "slow" responses have now been studied. There is some evidence that they are mediated by changes in cyclic nucleotide levels in the postsynaptic cells, and perhaps by phosphorylation of membrane proteins by cyclic nucleotide-dependent protein kinases.

The Cyclic Nucleotide Hypothesis

The first studies to suggest cyclic nucleotide-mediated synaptic transmission
were prompted by the earlier observation that the action of catecholamines on
various metabolic systems, especially on glycogen metabolism in the liver, were
mediated by changes in the cyclic AMP (cAMP) "second messenger." Thus, it
made sense to look for similar changes in the action of catecholamines in neuronal
transmission. This approach was further encouraged by the observation that
adenylate cyclase activity was extremely high in the brain and, indeed, was highest in fractions containing synaptic membranes [20,21]. The early experiments
of Kakiuchi and Rall [22,23], then, solidified this line by showing that the addition of low concentrations of norepinephrine to slices of rabbit brain led to the
accumulation of cAMP. Indeed, exposure of the slices to either a high K^+ concentration or to electrical stimulation produced comparable rises in the nucleotide levels [24,25].

The Sutherland criteria, which are widely accepted, can be used to determine whether a given transmitter in a given system acts through a cAMP as
second messenger. These criteria are:

1. The transmitter should stimulate adenylate cyclase in broken cells.
2. The transmitter, in physiological concentrations, should increase cAMP
    levels in the cells.
3. cAMP itself should potentiate the action of the transmitter.
4. cAMP alone should mimic the action of the transmitter.

Based upon these criteria it now seems possible that cyclic nucleotides mediate
intracellular events at many different synapses [26,27]. cAMP is probably the
second messenger for the β-adrenergic synapses in the pineal, and for the Purkinje
cells in the cerebellum. A norepinephrine-stimulated adenylate cyclase can be
observed in cerebellum preparations. Dopamine transmission in the interneurons
of the superior cervical ganglia also seems to be mediated by cAMP, and a dopamine-stimulated adenylate cyclase has been demonstrated in this tissue, as well
as in the caudate nucleus. There is some evidence that transmission by histamine,
serotonin, and even glutamate is cAMP mediated. The muscarinic acetylcholine
receptor seems also to be linked to cyclic nucleotide metabolism, but this appears
to be due to increases in cGMP, rather than in cAMP.

Perhaps the superior cervical ganglion of the rabbit has been most studied
in this regard, primarily by Greengard and his coworkers [27-29]. Such tissue
has the advantages of simple organization and known transmitter chemistry. The
principle synapses in this tissue are the excitatory postganglionic nicotinic acetylcholine junctions formed between the presynaptic spinal cord fibers and the

large postganglionic neurons. Transmitter release at these junctions initiates a
fast excitatory postsynaptic potential. In addition, there are other cells, termed
the small intensely fluorescent cells (SIF cells), which also receive preganglionic
cholinergic input. These cells appear to have muscarinic receptors. These seem
to be interneurons modulating the preganglionic input by synapsing on the postganglionic neurons and producing a slow inhibitory postsynaptic potential. These
latter synapses are dopaminergic. Electrical stimulation of the ganglion in vitro
produces large increases in the cAMP content. A combination of inhibitor studies
and experiments involving the application of exogenous dopamine showed that
the cAMP increases were due to events at a dopaminergic synapse. The finding
of a dopamine-stimulated adenylate cyclase in the ganglia was consistent with
the hypothesis. It was further observed that the application of cAMP or of its
membrane-permeable derivatives produced the same type of alterations in the
membrane potential produced by the application of dopamine. Finally, cyto-
chemical localization of the increase in cAMP in the ganglia showed that the increases occurred in cells which could be identified as the postganglionic neurons.
Thus, the story appears complete in that the release of dopamine by the SIF
cells is followed by its binding to the dopamine receptor on the postganglionic
neuron. This receptor is, or is linked to, the dopamine-sensitive adenylate cyclase
which produces increased amounts of cAMP in the neurons. These increases in
cAMP may be responsible for the slow inhibitory postsynaptic potential which is
a characteristic of the action of the SIF cells.

A parallel series of experiments has been done in order to investigate the
chemical mediation involved in the generation of the slow excitatory postsynaptic
potential which is also seen upon stimulation of the ganglia. This potential is initiated by the presynaptic input on the SIF cells themselves via muscarinic acetylcholine receptors. By a strategy similar to that described above it has been
shown that the combination of acetylcholine with these receptors and with muscarinic receptors on the postganglionic neurons themselves produces increases in
cGMP in these cells and that this combination may be responsible for the generation of the characteristic electrical response in the cells. An overall model of the
ganglion in terms of its cellular responses has been presented by Greengard and
seems consistent with the accumulated information (Fig. 24.2). Studies in other
tissues, while not yet so extensive, are compatible with this model.

The characteristic action of cAMP in many systems is the stimulation of
protein kinases which phosphorylate cellular proteins and thus execute the actions
of the hormones for which cAMP is the second messenger. It has been suggested
that the action of the cyclic nucleotides in the neurons in which they act is to promote a comparable phosphorylation of membrane proteins. These membrane
proteins, in turn, may control the permeability of the membrane to Na^+ and K^+,
and phosphorylation of them alters that permeability. While this postulate is
still not supported by any substantial amount of experimental data, it is consis-

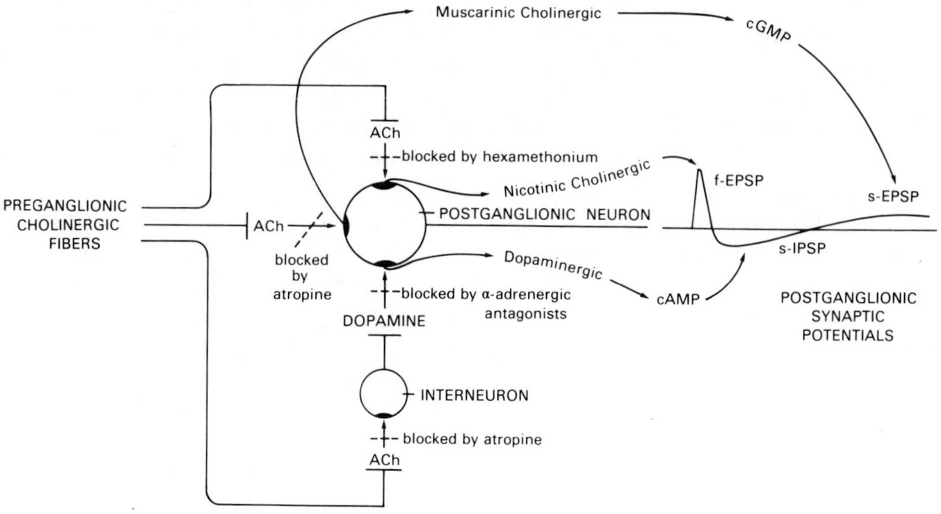

Figure 24.2   A schematic illustration of the principal synaptic junctions in the mammalian superior cervical ganglion and the postganglionic synaptic potentials thought to be related to each. The proposed role for cAMP in mediating the slow inhibitory postsynaptic potential produced by dopamine, and the proposed role for cGMP in mediating the slow excitatory postsynaptic potential associated with muscarinic cholinergic transmission, are indicated. [Reprinted from P. Greengard, in *Advances in Cyclic Nucleotide Research*, Vol. 5 (G. I. Drummond, P. Greengard, and G. A. Robison, eds.), Raven, New York, 1975.]

tent with what data there are, and is also consistent with the increasing knowledge of the action of cyclic nucleotides in other systems.

It should be pointed out that, while there is no question that the combination of transmitter and receptor produces large increases in cyclic nucleotide levels in many postsynaptic cells, the role of the nucleotides in the generation of the postsynaptic action potentials is not fully accepted. A number of investigators, using various systems, have not been able to duplicate the experiments in which the administration of cyclic nucleotide derivatives themselves generate action potentials. Further, some problems have arisen with the details of the pharmacology of the Greengard model. So, the direct link between increased nucleotide levels and the action potential in the postsynaptic cell remains hypothetical.

Depolarization

The combination of transmitter and receptor, whether mediated by changes in intracellular cyclic nucleotide levels and subsequent protein phosphorylation or not, initiates the permeability changes in the membrane which result in depolarization. There have been many theories about the mechanism of this membrane alteration. The description by Hodgkin and Huxley of the action potential in terms of changes in Na^+ and K^+ conductances in the membrane has been widely and immediately accepted since its first presentation in the early 1950s [30,31]. The theory requires the presence in the membrane of channels for the penetration of sodium and potassium ions, and although the physical evidence is not overwhelming, the description of these channels has been very carefully and finely considered. Thus, of the sodium ion pore it has been written that "The pore has a rectangular hole 3.1 X 5.1 Å formed by a ring of oxygen atoms. This hole is both a pathway for ion flow and the selectivity filter which determines which ions can flow. The narrow part of the pore is further supposed to be very short and to bear a single negative charge. The charge attracts cations and repels anions" [32]. The theory also requires the presence of a "gating current" which protects the opening from the surrounding cations. When the current is discharged there is a rush of Na^+ through the geometrically correct pore. There are thought to be separate and selective pores for the passage of potassium. The use of tetrodotoxin as a selective blocker of the sodium pore [33] has strengthened this picture of conduction.

An alternate and comprehensive theory, originated by Tasaki [34], the "two-stable state theory," proposes that the excitation of the membrane is due to the existence of macromolecules in the membrane which can exist in either of two stable conformational states, separated by an unstable, higher-energy intermediate conformation. In one of these conformations the membrane is in the resting state, in the other it is in the excited state. The resting conformation is held together by divalent cation (Ca^{2+}) bridges. As the ratio between the divalent and monovalent cation at the membrane changes, the bridges are broken and the macromolecule becomes unstable. At high monovalent cation concentration the macromolecule flips into the excited but stable state. The influx of sodium ions through the membrane is visualized as a consequence of, and not a prerequisite for, the excitation of the membrane.

In either case, the excitation of the membrane caused by the combination of transmitter and receptor causes a brief set of conformational or chemical changes in the membrane leading to the influx of sodium ions. This influx is due either to the opening of specific channels or to alterations in the general properties of the membrane. The exact molecular details of this rapid and massive change in the permeability of the nerve membrane to sodium ions are not known.

Acetylcholine Cycle

The biochemistry of the transmitter substance, acetylcholine, on the other hand, is well understood and has been described as the "acetylcholine cycle" (Fig. 24.3).

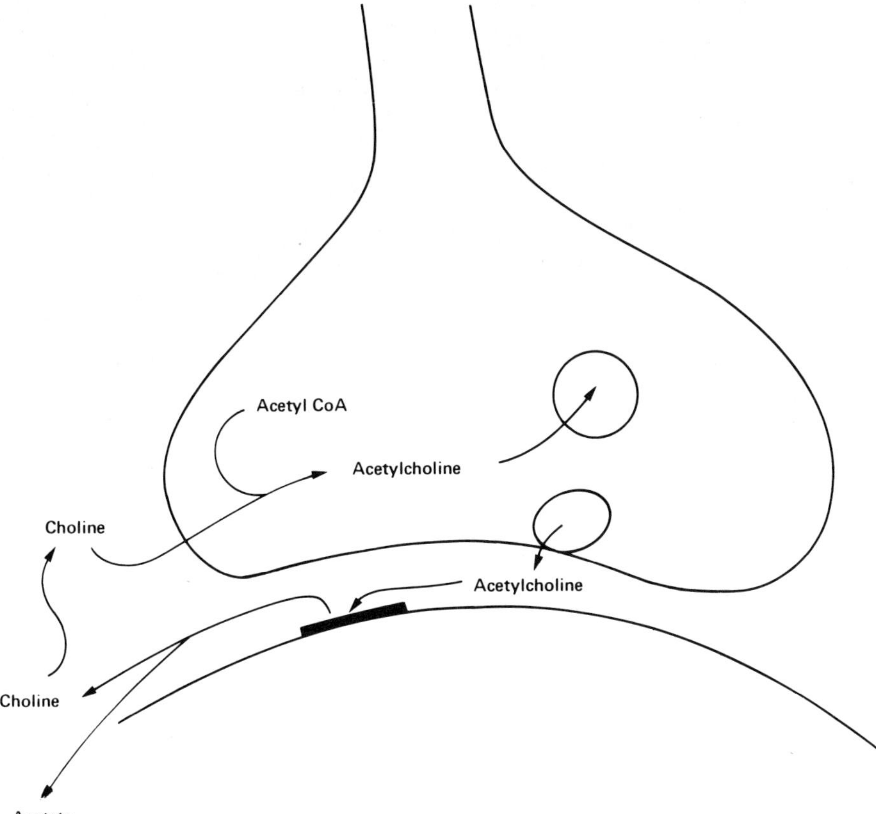

Figure 24.3   The acetylcholine cycle.

The various steps include the uptake of choline into the cell, the synthesis of acetylcholine, the transport of acetylcholine into the synaptic vesicle, the release of the contents of the vesicle, the binding of the acetylcholine to its receptor, the release of the acetylcholine from the receptor, and finally, the destruction of the transmitter by the hydrolysis of the acetylcholine to acetate and choline. The uptake of the choline into the presynaptic terminal initiates the process once again.

Choline Uptake

Choline can not be made in nervous tissue [35,36], so the uptake of choline from the bloodstream is a necessary part of the synthesis of acetylcholine. In recent years it has become clear, moreover, that the uptake of choline into the

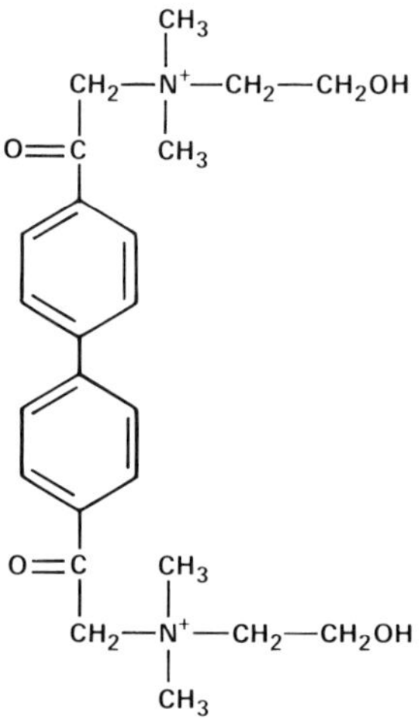

Figure 24.4   The structure of hemicholinium.

presynaptic terminal is a carefully regulated process and may in some way govern
acetylcholine synthesis. There appear to be two systems for the active transport
of choline across the neuronal membrane [37,38]. The high-affinity system is
found in synaptic membranes, is sodium dependent, and has a K_M for choline of
about 3.5 μM. It is energy requiring and is inhibited by hemicholinium (Fig. 24.4),
which appears to be a specific inhibitor of the uptake of choline by this route. A
high proportion of the choline transported by this system is converted to acetylcholine in the synaptic bouton. There is also a low-affinity system in neuronal
and nonneuronal membranes, with a "K_M" of about 40 to 100 μM. This system is
not dependent on the presence of sodium and does not seem to participate in the
synthesis of acetylcholine. It may provide choline for the synthesis of phospholipid. It seems possible that the high-affinity system for the uptake of choline can
be rate-limiting in the synthesis of acetylcholine [39,40].

Choline Acetyltransferase

The synthesis of acetylcholine is catalyzed by choline acetyltransferase (CAT)
(Fig. 24.5). This enzyme is localized in the synaptosomal portion of the cell. It

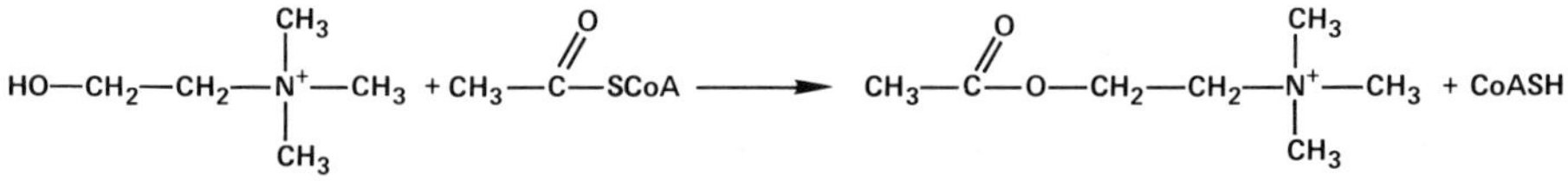

Figure 24.5   Choline acetyltransferase (CAT).

$$CH_3-\overset{\overset{\displaystyle O}{\|}}{C}-O-CH_2-CH_2-\overset{\overset{\displaystyle CH_3}{|}}{\underset{\underset{\displaystyle CH_3}{|}}{N^+}}-CH_3$$

Figure 24.6   Acetylcholine.

is a soluble enzyme found outside the vesicles. It was discovered in the early
1940s [41,42] and catalyzes the transfer of acetate from acetyl CoA to choline
[43]. It is found only in the nervous system. It has been partially purified from
several sources, including squid head, fly brain, guinea pig brain, rat brain, cow
striatum, and from placenta of several species. It is a fairly stable enzyme with
a pH optimum between 7.0 and 9.5. Molecular weights of between 50,000 and
89,000 have been reported. It is absolutely specific for choline, but not for
acetyl CoA. Purifications of several thousand fold have been reported, but the
preparations are still not homogeneous. The placenta is the richest source of the
enzyme in vertebrates, although it is not certain that the enzyme is the same as
that in the nervous system. The function of CAT in placenta and of its product,
acetylcholine, which is also present, is not clear, especially since the placenta is
not innervated. The inavailability of homogeneous preparations of CAT has
made the work on this important enzyme somewhat slow. It has been difficult
to prepare antibody, and attempts which have been made with the preparations
available have met with limited success [44]. Another important difficulty in the
study of this enzyme has been the continued lack of potent and specific inhibitors.

Acetylcholine itself [43] (Fig. 24.6) is present in a concentration of $13\,\mu M$
in the mouse brain and about $200\,\mu M$ in ganglia. It is a strong quaternary base
and is stable enough at pH 4 that it can be autoclaved. The transfer of acetylcholine from the synaptosomal cytoplasm where it is formed to the interior of
the vesicle from which it is released seems to be an active process, and the transfer occurs against a concentration gradient [45]. The concentration of acetylcholine in the vesicles of vertebrate brain is on the order of 0.15 M and in vesicles
from the electric organ of *Torpedo* is approximately 0.6 M

Acetylcholinesterase

After the acetylcholine has been released into the synaptic cleft, has bound to
the receptor, and has produced its action on the postsynaptic cell, it is released
from the receptor. Its action is terminated by the hydrolytic activity of acetylcholinesterase (AChE) (Fig. 24.7). This enzyme, present in or on the synaptic
membrane, has been extensively studied. It is a large enzyme (MW 230,000 to
250,000) with subunit structure. When purified and crystallized from the electric organ of the eel the enzyme exhibits four subunits of molecular weight
64,000 each, but subunit weights of between 57,000 and 80,000 have been reported. It has a K_M of about 3×10^{-4} M and a pH optimum of somewhere

$$CH_3-\overset{\overset{\displaystyle O}{\parallel}}{C}-O-CH_2-CH_2-\underset{\underset{\displaystyle CH_3}{|}}{\overset{\overset{\displaystyle CH_3}{|}}{N^+}}-CH_3 + H_2O \longrightarrow HO-CH_2-CH_2-\underset{\underset{\displaystyle CH_3}{|}}{\overset{\overset{\displaystyle CH_3}{|}}{N^+}}-CH_3 + CH_3-COOH$$

Figure 24.7   Acetylcholinesterase (AChE).

around 8.25. There is substantial evidence that there are isozymes of acetylcholinesterase, but the physical basis of these isozymes is unknown. There are also
several reports of high molecular weight aggregates of AChE which seem to form
during the purification.

The active site of the enzyme has been mapped, mostly through the work
of Wilson [46-48] , and the nature of the active groups in the catalysis are understood. The active site includes at least one anionic attachment for the binding
of the quaternary nitrogen of the acetylcholine, and an esteratic site which is the
hydrolytic entity. The anionic site has a pK_a of about 6.5 and is probably a histidine residue. The esteratic site probably contains a histidine residue also,
a serine hydroxyl, and a carboxyl group. The serine is acetylated during the hydrolysis, and this serine hydroxyl is also the site of attack of the irreversible inhibitor, diisopropylfluorophosphate, discussed in detail below. According to the
model proposed by Wilson, the anionic site binds the quaternary nitrogen, the
esteratic site hydrolyzes the ester bond with concomitant acetylation of the enzyme. This acetyl enzyme is then readily hydrolyzed by water (Fig. 24.8).

There are a number of inhibitors of acetylcholinesterase (Fig. 24.9). Some
are more or less reversible, and these include other quaternary nitrogen compounds which inhibit competitively. Physostigmine (eserine) and neostigmine
are longer-acting compounds, which act by leaving the slowly hydrolyzed carbamyl group on the serine at the esteratic site of the enzyme. These antagonists
initially potentiate the action of acetylcholine in that the transmitter cannot be
inactivated by the normal hydrolytic action. Eventually the results of administration of these acetylcholinesterase inhibitors is a total block of transmission.
The central nervous system effects of such administration include electroencephalogram changes, convulsions, headache, emotional lability, insomnia, and errors
in problem solving. The overall changes appear to be an effect on arousal mechanisms and mimic, in some respects, psychotic illness.

A totally different class of inhibitor is found among the organophosphorus
compounds and is typified by diisopropylfluorophosphate (DIFP) (Fig. 24.9).
This irreversible antagonist acts by stoichiometrically organophosphorylating the
serine at the esteratic site. The linkage formed cannot be split by water. The
consequence is an irreversible inhibition of the AChE which is rapidly lethal. A
number of similar poisons have been exploited for use as nerve gas weapons and

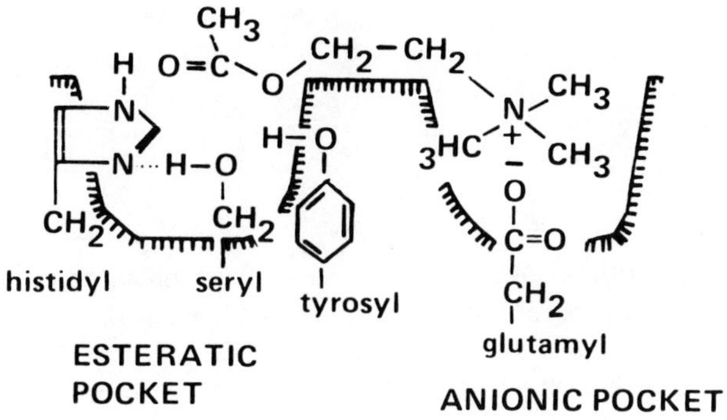

Figure 24.8 The two pocket model of the active site of acetylcholinesterase. A molecule of acetylcholine occupies the site with an orientation which allows the relatively positive carbonyl carbon (C=O) to interact with the nucleophilic oxygen on the activated hydroxyl group of serine. Activation is by hydrogen bonding to the imidazole nitrogen as shown. [Figure from A. R. Mann, in *Biology of Cholinergic Function* (A. M. Goldberg and I. Hamin, eds.), Raven, New York, 1976.]

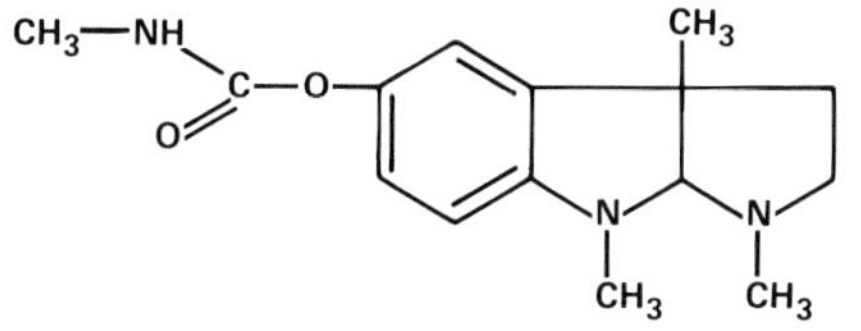

Physostigmine (eserine)

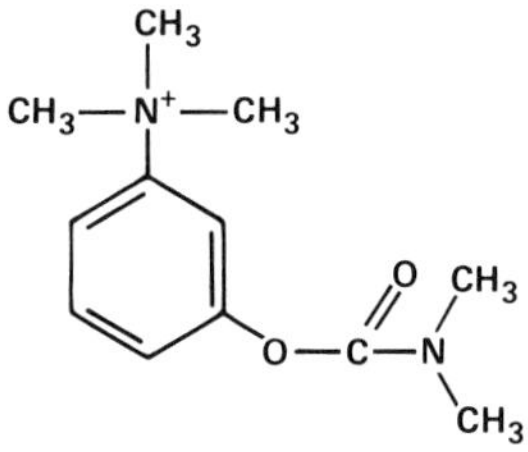

Neostigmine

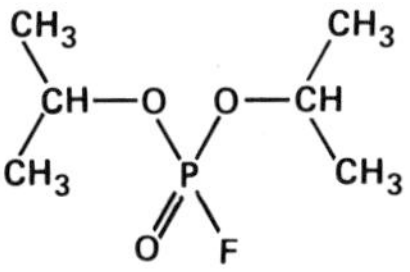

Diisopropylfluorophosphate

Figure 24.9 Inhibitors of acetylcholinesterase.

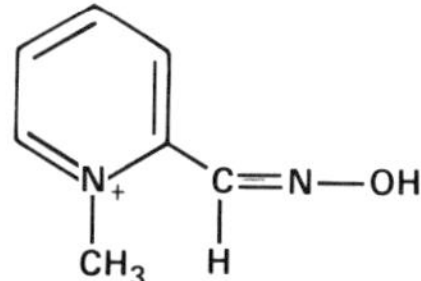

Figure 24.10   The structure of 2-pyridine aldoxime methiodide.

as insecticides. Having studied the chemistry of the active site of the enzyme, Wilson predicted that the organophosphorylated enzyme could be reactivated by nucleophiles more potent than water. By mapping the interatomic distances of the site, he was able to "construct" the first effective antidote to the organophosphorus poisons. This compound, 2-pyridine aldoxime methiodide (PAM) (Fig. 24.10) was attracted to the active site of AChE, fitted the active site exactly, and reversed the action of DIFP both in vivo and in vitro.

After the hydrolysis of acetylcholine by AChE, the choline is taken up into the presynaptic cell and the whole cycle begins again.

Acetylcholinesterase is present primarily in nervous tissue, but lower levels are present in other organs. Another enzyme, the pseudocholinesterase, butyrylcholinesterase or serum cholinesterase, is present in glia and in organs other than brain, and hydrolyzes acetylcholine. The two enzymes can be differentiated, however, by inspection of a number of characteristics of their action. They have different rates against several substrates and are inhibited by different inhibitors. For example, the acetylcholinesterase acts on acetylcholine, somewhat more slowly on acetyl β-methylcholine, very slowly on butyrylcholine, and not at all on benzoylcholine. The pseudocholinesterase acts on butyrylcholine, on benzoylcholine, but not on acetyl β-methylcholine. The acetylcholinesterase is inhibited by bis-quaternary ammonium compounds; the pseudocholinesterase is inhibited by low levels of organophosphates. Both are inhibited by physostigmine; both act on acetylcholine. The acetylcholinesterase is inhibited at high levels of substrate, whereas the pseudocholinesterase is not. The function of this pseudocholinesterase is not known.

Inhibitors of the Cycle

A number of other inhibitors of the "acetylcholine cycle" are known, and it is pretty clear at which point each of them inhibits transmission. Hemicholinium inhibits the uptake of choline, which seems to be rate-limiting in the synthesis of the transmitter. Inhibitors of the choline acetyltransferase are rare. The release of the transmitter seems to be inhibited by conditions in which calcium ion concentration is low or magnesium ion concentration is high. The presence of chelators will inhibit release. Cholinesterase inhibitors include, as mentioned above, physostigmine (eserine) and neostigmine, as well as the organophosphate

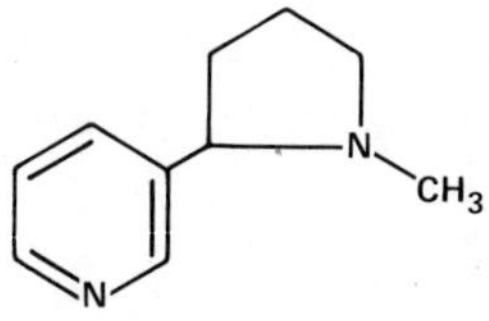

Nicotine

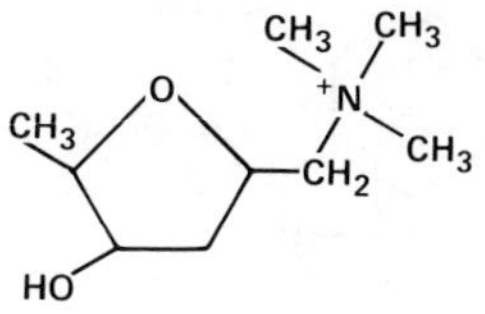

L-Muscarine

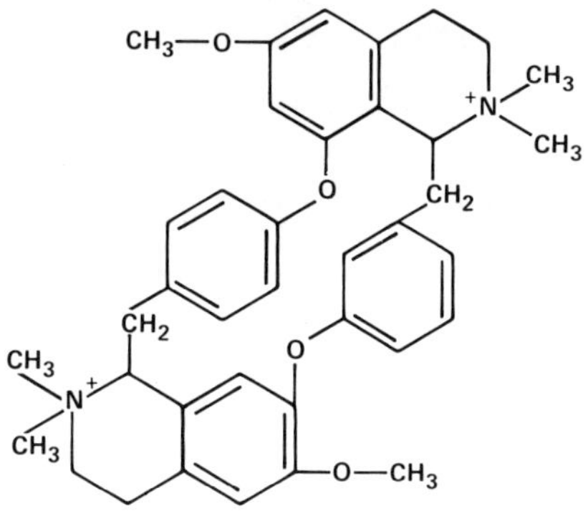

D-Tubocurarine

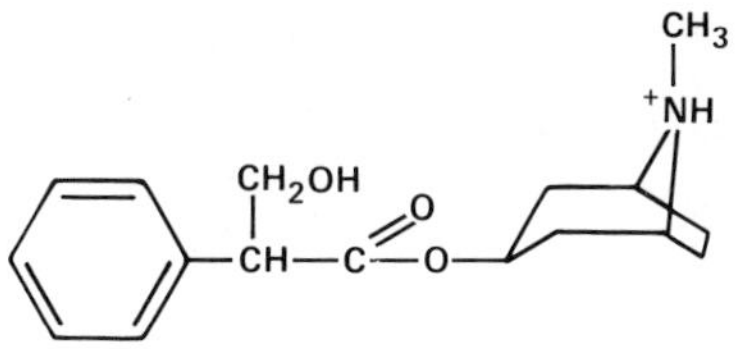

Atropine

Figure 24.11 Inhibitors of the acetylcholine receptors.

poisons. Bicuculline, an inhibitor of the γ-aminobutyric acid (GABA) receptor, also inhibits AChE [49]. The nicotinic and muscarinic acetylcholine receptors are blocked by d-tubocurarine and by atropine, respectively (Fig. 24.11). In addition, the nicotinic receptor is blocked by α-bungarotoxin, a peptide from cobra venom. The process of conduction is blocked by tetrodotoxin (Fig. 24.12), a poison from puffer fish, which is thought to bind specifically to the sodium pore and inhibit the passage of Na^+ through its specific channels [33]. Batrachotoxin [50] (Fig. 24.12), on the other hand, causes depolarization by enhancing the permeability of the membrane to sodium. Tityustoxin, from scorpion venom, causes depolarization and release of acetylcholine [51]. Botulinum toxin, an extremely toxic protein from the anaerobic organism, *Clostridium botulinum,* blocks transmitter release perhaps by blocking exocytosis [52]. Tetanus toxin appears to act at the same site as does botulinum toxin. Finally, ouabain inhibits the transmission mechanism irreversibly by blocking the Na^+, K^+-ATPase which is responsible for maintaining the membrane in a polarized state.

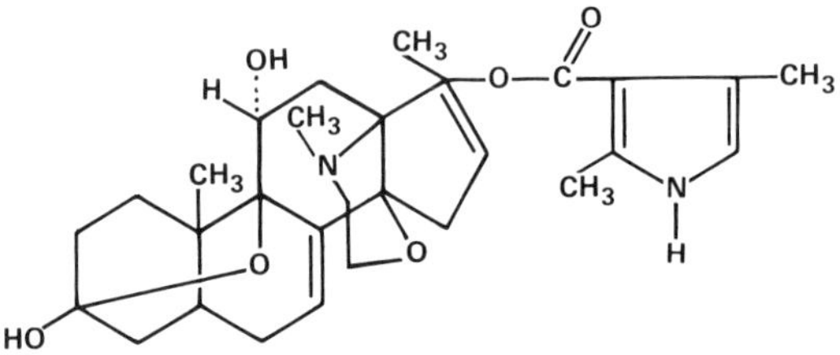

Batrachotoxin

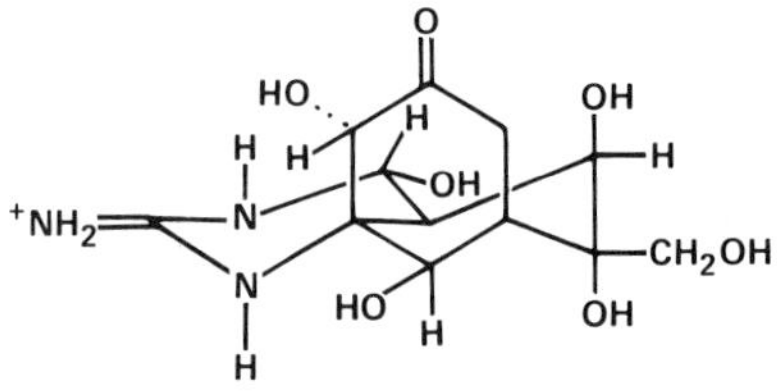

Tetrodotoxin (TTX)

Figure 24.12 Other inhibitors of the acetylcholine cycle.

Criteria of Transmitter

Acetylcholine is the only substance for which a transmitter role has been un-equivocally established and, even then, only in a very few systems. There are a number of other materials, however, for which a transmitter role is widely ac-cepted. These include, in roughly decreasing order of likelihood, GABA, nor-epinephrine, dopamine, glycine, serotonin, glutamic acid, aspartic acid, epineph-rine, histamine, substance P, purines, and peptides. In order to resolve the con-flicting reports and claims, there has been, by more or less general agreement, a set of criteria adopted for the assigning of a transmitter function to a given can-didate in a given system. These criteria are as follows: (1) The substance, and the enzymes which are involved in its biosynthesis, should be localized to or en-riched in, the synaptic portion of the neuron. (2) The substance itself should be further localized to the synaptic vesicles. (3) The substance should be selectively released, by a Ca^{2+}-dependent mechanism, upon stimulation of the neuron. (4) There should be a mechanism for the rapid inactivation of the substance. (5) The substance should mimic the effects of the natural material when applied ionto-phoretically to the appropriate postsynaptic cells.

γ-Aminobutyric Acid

The evidence that GABA is an inhibitory transmitter at the crustacean neuro-muscular junction is quite strong [53,54]. The evidence for a similar role for γ-aminobutyric acid in the vertebrate nervous system is also fairly strong, but the complete experimental proof is more difficult to obtain. In the crayfish stretch receptor it can be shown that γ-aminobutyric acid fulfills many of the criteria which justify assigning it a transmitter role [53]. It is released upon stimulation of the inhibitory nerves, and the release is Ca^{2+}-dependent. GABA mimics the effects of such stimulation when it is applied iontophoretically. Antagonists which prevent the binding of γ-aminobutyric acid, notably bicucul-line, inhibit the effects of nerve stimulation. Finally, there are high-affinity up-take systems which might serve to inactivate the material after it has been re-leased from its receptor. A hydrophobic protein which binds γ-aminobutyric acid with a dissociation constant of 8 μM and has a great affinity for bicucul-line has been isolated from crustacean muscle and may be the GABA receptor [55]. Evidence for a similar function of γ-aminobutyric acid in the vertebrate nervous system includes the following observations. γ-Aminobutyric acid is local-ized in the synaptic portions of the cell, as is glutamic acid decarboxylase I (GAD I). It is released from certain neurons in the brain, specifically the basket cells and the Purkinje cells of the cerebellum, the intrahippocampal neurons, the cells of the retinal ganglion layer, and the intracortical interneurons of the cortex, by stimulation, and from slices of the brain by high K^+ or by electrical stimulation, and this release is Ca^{2+}-dependent. GABA is taken up by a high-affinity Na^+-dependent transport system, thus providing a mechanism for its synaptic inac-tivation. Finally, a binding proteins for γ-aminobutyric acid, perhaps the GABA receptor, has been isolated from mammalian brain [56]. Additionally, inhibitors of the binding of γ-aminobutyric acid, such as bicuculline or picrotoxin, produce convulsions when injected into vertebrates [57]. The action of γ-aminobutyric acid as transmitter is to increase the permeability of the receptor cell to Cl^- ion. This increase in Cl^- permeability hyperpolarizes the membrane of the receptor cell, thus inhibiting the action of excitatory transmitters and resulting in an in-hibition of the response of the receptor cell.

Norepinephrine

Norepinephrine is the transmitter substance in the sympathetic nervous system and in several tracts in the older, or lower, parts of the mammalian central ner-vous system [58-60]. The biosynthetic pathway for norepinephrine is discussed in Chap. 11. The enzymes in the pathway are found in the synaptic portion of the neurons, as is the transmitter itself. It is released from the vesicles into the synaptic cleft by a Ca^{2+}-dependent mechanism upon stimulation of the cell, and reacts with a postsynaptic receptor. There are two types of receptors for nor-epinephrine; the α receptor causes vasoconstriction, muscle contraction, dilation of the pupils, and excitation of the endocrine glands. Pharmacologically, the α receptor can be distinguished by its ability to respond to phenylephrine and to be inhibited by phenoxybenzamine (Fig. 24.13). The β receptor is involved in vasodilation, relaxation of the smooth muscle, and increases in the rate and the force of the heartbeat. It is responsive to isoproterenol and inhibited by pro-pranolol (Fig. 24.13). The β-adrenergic receptor has been solubilized, using digi-toxin, from frog erythrocytes [61]. The release of norepinephrine from its re-ceptor is followed by inactivation of the transmitter. The mechanism of inac-

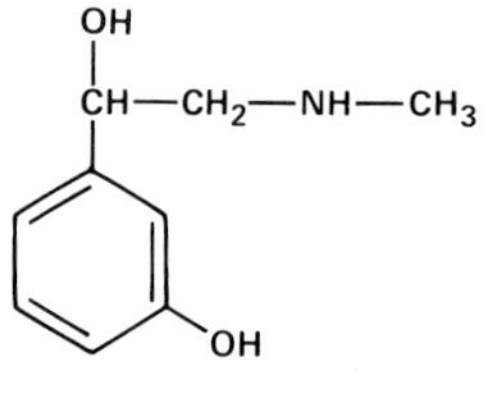

Phenylephrine

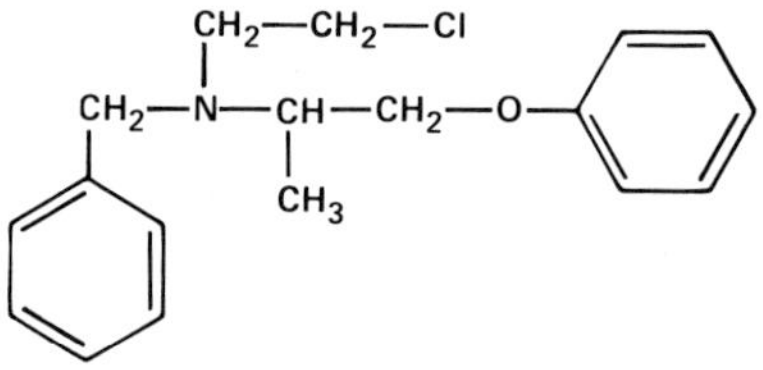

Phenoxybenzamine

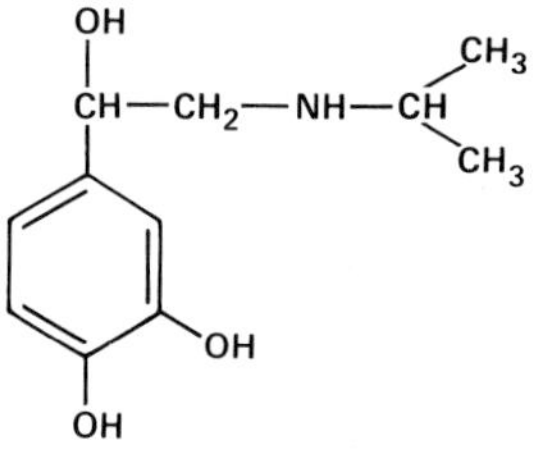

Isoproterenol

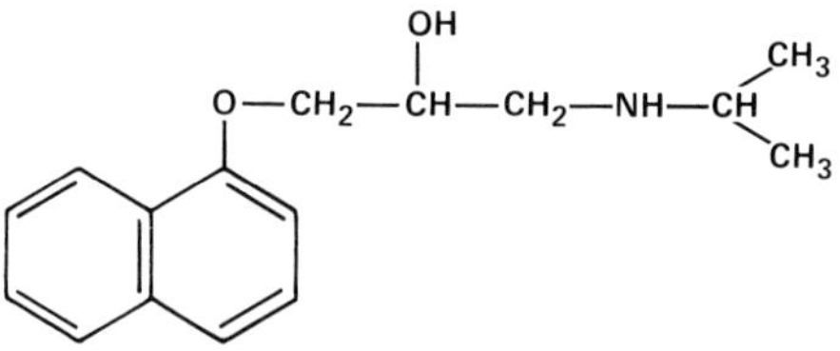

Propranolol

Figure 24.13   Agonists and antagonists of norepinephrine receptors.

tivation is active reuptake into the presynaptic cell. This is the clearest difference
between the general mode of action of acetylcholine as transmitter, where inactivation is by hydrolysis, and of the catecholamines, where the mode of inactivation is by reuptake. The work of Axelrod [62-64], and now of others, has shown
this process to be the important and limiting reaction in the transmitter action
of norepinephrine. The portion of the norepinephrine which is not recaptured
by the presynaptic cell is released into the circulation and is destroyed by the
enzymatic mechanism for the degradation of norepinephrine discussed earlier.
The action of norepinephrine in depolarizing the postsynaptic cell, at least in
some systems, is perhaps mediated, as discussed above, by cAMP.

Dopamine

A number of neurons in the peripheral and the central nervous system appear to
use dopamine as transmitter. Among these are the interneurons of the sympathetic ganglia [28], the neurons in the caudate nucleus [65], and some neurons
in the nucleus accumbens and the olfactory tubercle. Presumably, the cycle of
biosynthesis, vesicle release, receptor binding, and reuptake are the same as for
norepinephrine, but little has been done on these details in dopaminergic systems. It is known that there are dopamine receptors as well as a dopamine-
sensitive adenylate cyclase, in the caudate nucleus [66,67].

Other Transmitters

There are three amino acids for which transmitter roles have been reasonably well
established [53,54]. These are glycine, glutamic acid, and aspartic acid. The special problem with establishing a transmitter role for an amino acid is that they,
unlike the other candidate transmitters, have well-known and important metabolic roles in the cell. Their occurrence in the cell, and their metabolic response
to changes in the environment of the cell, make it difficult to cull out the transmitter role from the other roles. Nevertheless, it is generally accepted that glycine is the major inhibitory transmitter in the spinal cord and the brainstem [54];
that is, it serves as the transmitter substance for the inhibitory interneurons. As
such it hyperpolarizes the membrane of the postsynaptic cell, thus desensitizing
it to excitatory transmitters. Glycine satisfies most of the criteria needed to establish a transmitter role. About 25% of the synapses in the spinal cord are glycine synapses. There is a high-affinity, Na^+-dependent uptake of glycine into
spinal cord synaptosomes [68]. The transmitter function of glycine can be specifically inhibited by strychnine [69], and strychnine can be used to label the
glycine receptor [70]. Glycine is released upon stimulation of the appropriate
spinal neurons [71].
    Unlike glycine, which is inhibitory, the acidic amino acids, glutamic acid

and aspartic acid, are excitatory. But here, especially, the problems of proving
the transmitter function for so ubiquitous and metabolically active a substrate
as glutamate have proved to be difficult. Nevertheless, it is clear that glutamic
acid is a cerebral excitant [72-74]. In brain and spinal cord the application of
glutamate increases the rate of neuronal firing. Glutamate depolarizes the neurons, and the intracellular Na^+ concentration increases. Glutamate is released
upon electrical stimulation of the brain. The best evidence that glutamate is a
transmitter in the mammalian central nervous system is found with the granule
cells of the cerebellum, which are, quantitatively, the most prominent neurons
in the brain. Kainic acid, which is a "conformationally restricted" analog of
glutamate, is a specific agonist of the transmitter action of glutamate [75], and
binds to brain synaptosomal membranes [76]. Both glutamic and aspartic acids
raise the concentration of cAMP in brain slices to which they have been added
in high concentrations. There is a high-affinity uptake system for glutamate [77].
For aspartic acid, also, there is a high-affinity uptake system [78], a Ca^{2+}-depen-
dent release, and a hyperactivity produced in young animals when the compound
is administered systemically. High-affinity binding proteins for both glutamic
and aspartic acids have been found in the central nervous system of vertebrates,
and these proteins are strongly hydrophobic [79,80].

Proof for the transmitter role of a number of other candidate substances
is incomplete but suggestive. Serotonin appears to be a transmitter in certain
tracts in the older regions of the vertebrate central nervous system and produces
increases in cAMP levels in brain slices. Histamine, when added to brain slices,
also raises the cAMP concentration markedly. Taurine fulfills some of the criteria as an inhibitory transmitter in brain and retina [81]. Adenosine seems to
have receptors in brain slices which are linked in some way to adenylate cyclase.
Substance P has been implicated as a transmitter substance in certain spinal neurons [82]. Finally, good evidence is available to suggest that both purines [83]
and peptides [84] have transmitter roles.

Thus, about a dozen different substances have been implicated as transmitter in various systems. All fulfill some of the criteria generally accepted as
necessary for the assignment of transmitter function. Acetylcholine is the only
one which fulfills all the criteria. It is likely, however, that all these materials
have transmitter function, i.e., that they can serve as communicant between a
presynaptic and postsynaptic cell. Possibly there are many others that will eventually be recognized as having such function.

Seizures

The various forms of epileptic seizure appear to be caused by disruptions in the
orderly processes of nerve transmission. A wide variety of seizure types has been
observed and classified [85,86]. A large number of chemical and physiological

stimuli are known which cause convulsions in experimental animals and in humans. Indeed, the causes of seizures are so numerous that they encompass practically all disorders which affect brain function. The seizures themselves are sudden, recurrent, and transient disturbances in mental or physical functioning resulting from the excessive discharging of a population of neurons. Epileptic individuals exhibit normal brain and body metabolism between seizures. Extensive in vitro study, over many years, of focal tissue from epileptic brain revealed that the tissue has a normal oxygen consumption, a normal CO_2 production, and normal rates of aerobic and anaerobic metabolism. There are no discernible differences in the rates of synthesis of acetylcholine, or in the content of the acetylcholinesterase. There is some difference in the ability of the tissue to recapture K^+ lost during in vitro incubation. These observations and others, together with the general observation of astroglial gliosis at the epileptic focus, have led to the hypothesis that the fundamental lesion in the discharge is an inability of the glia of the epileptic individual to deal effectively with the K^+ released into the intracellular space during the firing of the neurons [87]. This hypothesis does not seem to be supported by all the data [85]. Some more recent experiments [88] have suggested that the lesion is more likely to be in the membrane of the epileptic neurons, but the exact difference between these membranes and the normal remains obscure. In view of the dramatic nature of such seizures, and the widespread distribution of the problem, the lack of solid metabolic or structural information with which to understand the epileptic condition is disappointing.

References

1. L. G. Abood, in *Basic Neurochemistry* (R. W. Albers, G. J. Siegel, R. Katzman, and B. W. Agranoff, eds.), Little, Brown, Boston, 1972.
2. G. B. Koelle, in *The Nervous System,* Vol. I (D. B. Tower, ed.), Raven, New York, 1975.
3. J. E. Heuser and T. S. Reese, *J. Cell Biol. 57:*315 (1973).
4. V. P. Whittaker, *Trends in Biochemical Sciences 1:*172 (1976).
5. A. Karlin, in *The Nervous System,* Vol. I (D. B. Tower, ed.), Raven, New York, 1975.
6. J. P. Changeaux and T. R. Podleski, *Proc. Natl. Acad. Sci. USA 59:*944 (1968).
7. M. Kasai and J. P. Changeaux, *J. Membr. Biol. 6:*1 (1971).
8. M. Kasai and J. P. Changeaux, *J. Membr. Biol. 6:*24 (1971).
9. M. Kasai and J. P. Changeaux, *J. Membr. Biol. 6:*58 (1971).
10. J. Cartand, E. L. Benedetti, M. Kasai, and J. P. Changeaux, *J. Membr. Biol. 6:*8 (1971).
11. C. Y. Lee, L. F. Tseng, and T. H. Chin, *Nature (London) 215:*1177 (1967).
12. P. M. Salvaterra and W. J. Moore, *Biochem. Biophys. Res. Commun. 55:* 1311 (1973).

13. M. E. Eldefrawi and A. T. Eldefrawi, *Ann. N.Y. Acad. Sci. 264:*183 (1975).

14. E. DeRobertis and J. Schacht (eds.), *Neurochemistry of Cholinergic Receptors,* Raven, New York, 1974.

15. J. P. Changeaux, M. Kasai, and C. Y. Lee, *Proc. Natl. Acad. Sci. USA 67:* 1241 (1970).

16. P. M. Salvaterra and H. R. Mahler, *J. Biol. Chem. 251:*6327 (1976).

17. J. Patrick and J. Lindstrom, *Science 180:*871 (1973).

18. J. Patrick, J. Lindstrom, B. Culp, and J. McMillen, *Proc. Natl. Acad. USA 70:*3334 (1973).

19. N. J. M. Birdsall and E. C. Hulme, *J. Neurochem. 27:*7 (1976).

20. E. W. Sutherland, T. W. Rall, and T. Menon, *J. Biol. Chem. 237:*1220 (1962).

21. E. deRobertis, G. Rodriguez DeLores Arnaiz, M. Alberici, R. W. Butcher, and E. W. Sutherland, *J. Biol. Chem. 242:*3487 (1967).

22. S. Kakiuchi and T. W. Rall, *Mol. Pharmacol. 4:*367 (1968).

23. S. Kakiuchi and T. W. Rall, *Mol. Pharmacol. 4:*379 (1968).

24. A. Sattin and T. W. Rall, *Fed. Proc. 26:*707 (1967).

25. S. Kakiuchi, T. W. Rall, and H. McIlwain, *J. Neurochem. 16:*485 (1969).

26. J. W. Daly, *Life Sci. 18:*1349 (1976).

27. K. G. Walton, in *Essays in Neurochemistry and Neuropharmacology,* Vol. 2 (M. B. N. Youdim, W. Lovenberg, D. F. Sharman, and J. R. Lagnado, eds.), John Wiley, New York, 1977.

28. P. Greengard, D. A. McAfee, and J. W. Kebabian, in *Advances in Cyclic Nucleotide Research,* Vol. 1 (P. Greengard and G. A. Robison, eds.), Raven, New York, 1975.

29. P. Greengard, in *Advances in Cyclic Nucleotide Research,* Vol. 5 (G. I. Drummond, P. Greengard, and G. A. Robison, eds.), Raven, New York, 1975.

30. A. L. Hodgkin and A. F. Huxley, *J. Physiol. (London) 117:*500 (1952).

31. R. D. Keynes, in *The Nervous System,* Vol. 1 (D. B. Tower, ed.), Raven, New York, 1975.

32. B. Hille, *J. Gen. Physiol. 59:*637 (1972).

33. C. Y. Kao, *Pharmacol. Rev. 18:*997 (1966).

34. I. Tasaki, in *The Nervous System,* Vol. 1 (D. B. Tower, ed.), Raven, New York, 1975.

35. J. Bremer and D. M. Greenberg, *Biochim. Biophys. Acta 37:*173 (1960).

36. G. B. Ansell and S. Spanner, *Biochem. J. 110:*201 (1968).

37. T. Haga and H. Noda, *Biochim. Biophys. Acta 291:*564 (1973).

38. H. I. Yamamura and S. H. Snyder, *J. Neurochem. 21:*1355 (1973).

39. P. Lefresne, P. Guyenet, J. C. Beaujouan, and J. Glowinski, *J. Neurochem. 25:*415 (1975).

40. J. R. Simon, S. Atweh, and M. J. Kuhar, *J. Neurochem. 26:*909 (1976).

41. D. Nachmansohn and A. L. Machado, *J. Neurophysiol. 6:*397 (1943).

42. D. Nachmansohn, H. M. John, and H. Waelsch, *J. Biol. Chem. 150:*485 (1943).

43. L. T. Potter, in *Handbook of Neurochemistry,* Vol. IV (A. Lajtha, ed.), Plenum, New York, 1970.

44. J. Rossier, *J. Neurochem. 26:*549 (1976).
45. M. E. Mattson and R. D. O'Brien, *J. Neurochem. 27:*867 (1976).
46. I. B. Wilson, F. Bergmann, and D. Nachmansohn, *J. Biol. Chem. 186:*781 (1950).
47. I. B. Wilson, *J. Biol. Chem. 190:*111 (1951).
48. I. B. Wilson, in *The Mechanism of Enzyme Action* (W. D. McElroy and B. Glass, eds.), Johns Hopkins Press, Baltimore, 1954.
49. E. Breuker and G. A. R. Johnston, *J. Neurochem. 25:*903 (1975).
50. E. X. Albequerique, M. Sasa, B. P. Avner, and J. Daly, *Nature (London) New Biol. 234:*93 (1971).
51. M. V. Gomez, M. E. M. Dai, and C. R. Diniz, *J. Neurochem. 20:*1051 (1973).
52. I. Kao, D. B. Drachman, and D. L. Price, *Science 193:*1256 (1976).
53. E. Roberts and R. Hammerschlag, in *Basic Neurochemistry* (R. W. Albers, G. J. Siegel, R. Katzman, B. W. Agranoff, eds.), Little, Brown, Boston, 1972.
54. S. H. Snyder, in *The Nervous System,* Vol. 1 (D. B. Tower, ed.), Raven, New York, 1975.
55. E. deRobertis and S. Fiszer de Plazas, *J. Neurochem. 23:*1121 (1974).
56. S. Fiszer dePlazas and E. deRobertis, *J. Neurochem. 25:*547 (1975).
57. D. R. Curtis, A. W. Duggan, D. Felix, G. A. R. Johnston, and H. McLennan, *Brain Res. 33:*57 (1971).
58. L. L. Iversen, in *Handbook of Neurochemistry,* Vol. IV (A. Lajtha, ed.), Plenum, New York, 1970.
59. S. Snyder, in *Basic Neurochemistry* (R. W. Albers, G. J. Siegel, R. Katzman, and B. W. Agranoff, eds.), Little, Brown, Boston, 1972.
60. F. E. Bloom, in *The Nervous System,* Vol. I (D. B. Tower, ed.), Raven, New York, 1975.
61. M. G. Caron and R. J. Lefkowitz, *J. Biol. Chem. 251:*2374 (1976).
62. J. Axelrod, H. Weil-Malherbe, and R. Tomchick, *J. Pharmacol. Exp. Ther. 127:*251 (1959).
63. L. G. Whitby, J. Axelrod, and H. Weil-Malherbe, *J. Pharmacol. 132:*193 (1961).
64. P. B. Molinoff and J. Axelrod, *Ann. Rev. Biochem. 40:*465 (1971).
65. O. Hornykiewicz, *Pharmacol. Rev. 18:*925 (1966).
66. H. Nyback and G. Sedvall, *J. Pharmacol. Exp. Ther. 162:*294 (1968).
67. J. W. Kebabian, G. L. Petzold, and P. Greengard, *Proc. Natl. Acad. Sci. USA 69:*2145 (1971).
68. Arregui, W. J. Logan, J. P. Bennett, and S. H. Snyder, *Proc. Natl. Acad. Sci. USA 69:*3845 (1972).
69. D. R. Curtis, A. W. Duggan, and G. A. R. Johnston, *Exp. Brain Res. 12:*547 (1971).
70. A. B. Young and S. H. Snyder, *Proc. Natl. Acad. Sci USA 70:*2832 (1973).
71. J. Hopkin and M. J. Neal, *Br. J. Pharmacol. 42:*215 (1971).
72. D. R. Curtis, J. W. Phillis, and J. C. Watkins, *J. Physiol. (London) 150:*656 (1960).
73. D. R. Curtis and J. C. Watkins, *J. Physiol. (London) 166:*1 (1963).

74. K. Krnjevic and J. W. Phillis, *J. Physiol. (London) 165:*274 (1963).
75. G. A. R. Johnston, D. R. Curtis, J. Davies, and R. M. McCulloch, *Nature (London) 248:*804 (1974).
76. J. R. Simon, J. F. Contrera, and M. J. Kuhar, *J. Neurochem. 26:*141 (1976).
77. A. M. Benjamin and J. H. Quastel, *J. Neurochem. 26:*431 (1976).
78. L. P. Davies and G. A. R. Johnston, *J. Neurochem. 26:*1007 (1976).
79. E. deRobertis and S. Fiszer dePlazas, *J. Neurochem. 26:*1237 (1976).
80. S. Fiszer dePlaza and E. deRobertis, *J. Neurochem. 27:*889 (1976).
81. R. Schmid, W. Sieghart, and M. Karobath, *J. Neurochem. 25:*5 (1975).
82. S. Konishi and M. Otsuka, *Brain Res. 65:*397 (1974).
83. G. Burnstock, *Pharmacol. Rev. 24:*509 (1972).
84. G. Zetler, *Biochem. Pharmacol. 25:*1817 (1976).
85. E. S. Goldensohn and A. A. Ward, Jr., in *The Nervous System,* Vol. 2 (D. B. Tower, ed.), Raven, New York, 1975.
86. E. S. Goldensohn, in *The Nervous System,* Vol. 2 (D. B. Tower, ed.), Raven, New York, 1975.
87. A. P. Fertziger and J. B. Ranck, Jr., *Exp. Neurol. 26:*571 (1970).
88. D. A. Prince and M. J. Gutnick, *Trans. Am. Neurol. Assoc. 96:*88 (1971).

25
THE EFFECT OF VARIOUS DRUGS ON BRAIN METABOLISM

There are several classes of drugs which enhance, inhibit, or distort neural processes. According to the classification used by Sourkes [1], these can be grouped as neuroleptics or major tranquilizers, anxiolytic sedatives or minor tranquilizers, antidepressants, and psychodysleptics or hallucinogens. In each of these categories there are drugs which differ widely in their chemistry. But they can be collected broadly this way since there are substantial similarities in their actions. In addition to these so-called psychotropic compounds, consideration must be given to the local and general anesthetics, the convulsants and the anticonvulsants, the salicylates, alcohol, and the narcotic analgesics. As will become obvious, relatively little is known about the mechanism of action of most of these materials. A great deal of biochemical, pharmacological, and clinical research has been done on the drugs which influence the brain, and a number of metabolic effects have been observed. But even in the cases where a number of alterations in metabolism are known to occur, it is far from certain that the effects observed are responsible for the physiological or behavioral changes in the individual. Nevertheless, the importance of these drugs for clinical work, and their potential for unraveling brain processes, make a discussion of the available information mandatory.

General Anesthetics

The general anesthetics, such as diethyl ether or chloroform, are chemically inert. Indeed, there are no specific chemical groupings which are common to the general anesthetics as a class. They do not bind to any specific chemical sites on the cells; they do not bind covalently at all. There would seem to be no need, then, to inspect the chemical properties of these materials in order to understand their action. More relevant to such an understanding is a consideration of their physical properties [2]. Early work suggested that anesthetic potency was directly and simply related to the lipid solubility of the anesthetic. The relationship, however, is now known to be more complex since correlations of the anesthetic properties

Table 25.1 Correlation Between Minimum Anesthetic Concentration (MAC) and Oil/Gas Partition Coefficient

Gas	MAC (atm)	Oil/gas partition coefficient (atm^{-1})	MAC $\times$ oil/gas
CF_4	26	0.073	1.90
SF_6	4.9	0.293	1.44
N_2O	1.9	1.40	2.63
Xe	1.2	1.90	2.26
C_3H_8	0.18	11.8	2.06
$CHCl_3$	0.0077	265	2.08

Table taken from S. L. Miller, E. I. Eger, II, and C. Lundgren, *Nature (London)* *221:*468 (1969).

of a compound with its lipid-water partition are far from consistent. Indeed, there are several physical properties of the general anesthetics which correlate more or less well with their anesthetic potency. Among these are polarizability, boiling point, hydrate dissociation pressure, oil/gas coefficient, and molecular weight or molecular volume. The oil/gas coefficient, i.e., the amount of anesthetic which dissolves in olive oil at 1 atm of pressure seems to provide the best correlation with potency (Table 25.1), but even this is not a completely reliable index.

The biochemical effects observed during general anesthesia include a reduced O_2 consumption, a decrease in cerebral lactic acid levels, an increase in brain glycogen, an increase in creatine phosphate, and a slight increase in the ATP concentration in the brain. In short, under general anesthetic conditions there appears to be an inhibition of the energy-requiring reactions of the cell and, thus, an increase in the levels of energy-rich metabolites. Upon prolonged exposure or exposure to high concentrations of anesthetic there is irreversible damage and a consequent decrease in energy-producing reactions and energy-rich metabolites. But under clinically acceptable anesthetic conditions, the biochemical picture is consistent with reduced brain function and, thus, reduced need for energy.

In vitro studies of the mechanism of anesthetic action have concentrated on the responses in brain slices. It has been shown that various anesthetics, in concentrations approximating those producing anesthesia, moderately inhibit the oxygen uptake of brain slices metabolizing glucose in a physiological medium. When the slices are stimulated electrically the oxygen consumption doubles, and the additional consumption is quite sensitive to inhibition by anesthetic. For example, 2 mM chloretone inhibits the oxygen uptake of unstimulated slices by 15%. When the slices are stimulated electrically the oxygen uptake increases, and this increase is blocked by 50% by the same 2 mM concentration of chloretone.

The fact that by other criteria, electrically stimulated slices seem to resemble functioning brain more closely than do unstimulated, or so-called silent slices, makes their increased sensitivity to anesthetic even more significant physiologically.

Anesthetics act on excitable membranes, and this action appears to be responsible for the effect of the drug. They are known to block axonal conduction, perhaps by dissolving in the membrane and clogging the aqueous channels through which the ions flow. At the synapse, anesthetics prevent the permeability change caused by the combination of transmitter and receptor. The exact molecular mechanism is not known, but it is clear that anesthesia causes inhibition of certain synaptic patheays in the central nervous system. The number of neurons firing per unit time is less than normal. This reduction in function results in a lowered utilization of energy sources and a reduced oxygen consumption. The brain is in a resting, nonenergy-utilizing state. But it should be clear that the reduced oxygen consumption and the increase in available energy sources are a consequence of anesthesia, not a cause of it.

Local Anesthetics

Most of the clinically useful local anesthetics are related in chemical structure to procaine [3] (Fig. 25.1). All are cationic and appear to react with the anions, such as the phospholipids and proteins, in the cellular membrane. The common denominator in the action of all the local anesthetics is their ability to block, in a reversible fashion, the electrical activity of nerve fibers without altering the membrane potential or causing a generalized toxicity. They appear to lower the permeability of the membrane to sodium ions. It seems a reasonable hypothesis that local anesthetics compete for the sites on the membrane which bind calcium ions and that these sites are directly involved in controlling the permeability of the membrane to Na^+ and K^+ [4-7].

Convulsants

Any number of insults cause convulsions. Included among these are hypoxia, hyperoxia, hypoglycemia, hyperammonemia, certain vitamin deficiencies, cerebral injury, electroshock, narcotic withdrawal, various epileptic conditions, and any number of chemicals. Indeed, the wide variety of chemical agents which induce seizures makes it unlikely that any general mechanism underlying convulsive activity will be found. There are, however, some general statements which can be made without the biochemical changes which occur during convulsion. Broadly speaking, these are reciprocal to the changes associated with anesthesia. High-energy phosphates in the brain, such as ATP and creatine phosphate, decrease precipitously, as do energy stores such as glycogen. The levels of lactic

Procaine
(Novocaine)

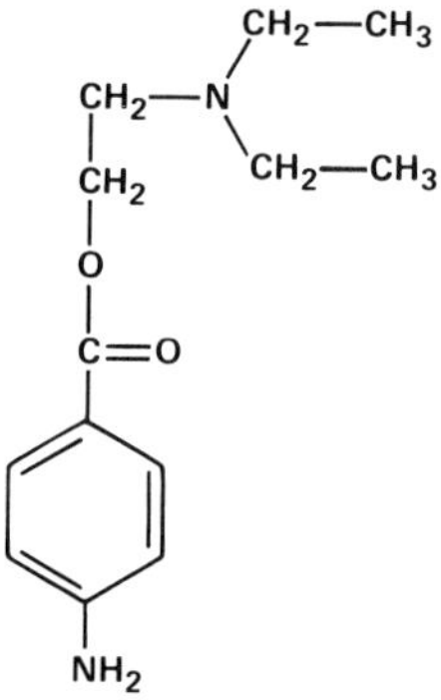

Tetracaine

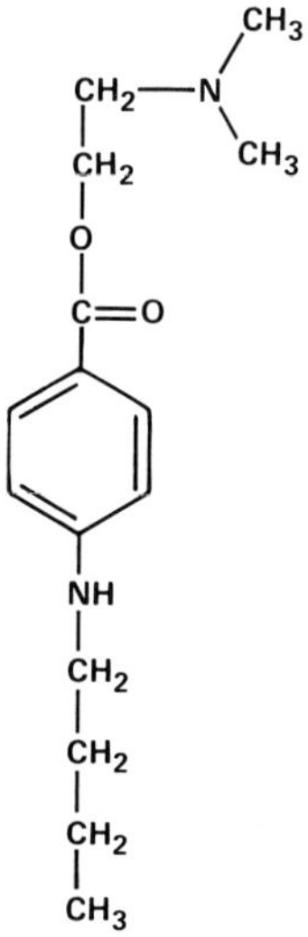

Dibucaine

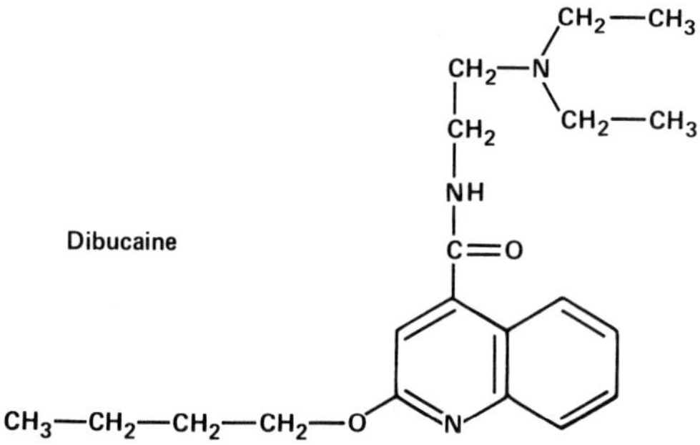

CH3—CH2—CH2—CH2—O

Figure 25.1   The structure of procaine and other local anesthetics.

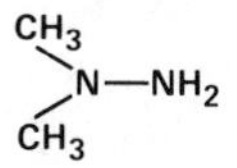

1,1-dimethylhydrazine

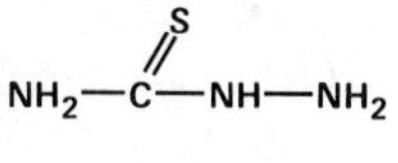

thiosemicarbazide

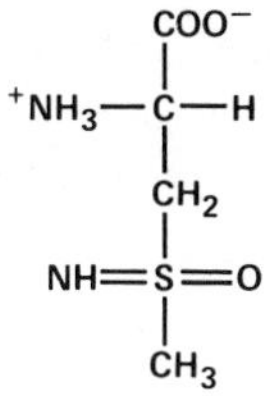

methionine sulfoximine

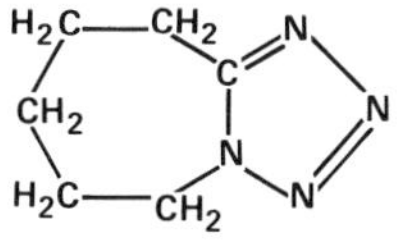

metrazole

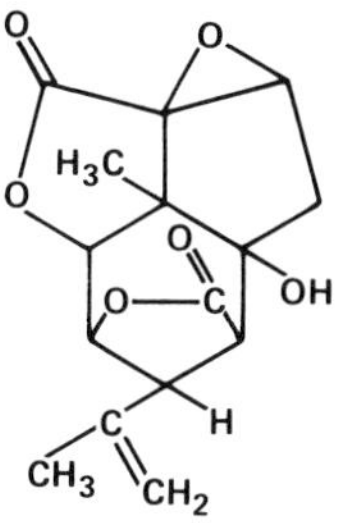

picrotoxin

Figure 25.2   The structures of various convulsant drugs.

acid, inorganic phosphate, and ADP increase markedly. Oxygen consumption by
the brain increases, and the acetylcholine content drops. As with the biochemical changes seen in anesthesia, the alterations seen during convulsions appear to
be consequences not causes.

   As mentioned in a previous section, the effects of some of the convulsants
have been linked to changes in the γ-aminobutyric acid content of the brain. The
carbonyl trapping agents (Fig. 25.2), such as the hydrazines and the semicarba-
zides, lower the γ-aminobutyric acid concentration by binding pyridoxal phosphate and thus inhibiting glutamate decarboxylase, the enzyme responsible for
the synthesis of γ-aminobutyric acid. No firm conclusion about the universality
of such a mechanism is possible, even in the case of the pyridoxal phosphate-
binding convulsants, since, under some conditions, the γ-aminobutyric acid levels

rise and convulsions still occur. Indeed, any number of chemical convulsants are
known which have no effect on γ-aminobutyric acid levels.

   Methionine sulfoximine (Fig. 25.2), which causes convulsions in all species
which have been tested, inhibits glutamine synthetase. It seems possible that the
convulsions are due to increased ammonia in the brain, which in turn is a consequence of the inability of the animal to remove ammonia through the formation
of glutamine. Even less is known about the mode of action of such convulsant
drugs as metrazole and picrotoxin (Fig. 25.2), except that they have relatively
minor effects on the respiration or the high-energy phosphate content of the
brain, and no apparent effect on γ-aminobutyric acid levels.

Anticonvulsants

The anticonvulsants are not general depressants. They do not depress overall
neuronal activity, but only the spikes or sudden bursts of high frequency and
high voltage associated with convulsions. Under certain conditions, hydroxylamine (Fig. 25.3) is an anticonvulsant. This action seems to be associated with
its ability to raise the cerebral γ-aminobutyric acid concentration by selectively
inhibiting γ-aminobutyric acid transaminase, which degrades γ-aminobutyric acid.
But, again, not all drugs which raise γ-aminobutyric acid levels are anticonvulsants, and most anticonvulsants have no effect on γ-aminobutyric acid levels.
Trimethadone, which is effective against petit mal, and diphenylhydantoin, which
counteracts grand mal, have no effect on γ-aminobutyric acid levels. Both drugs

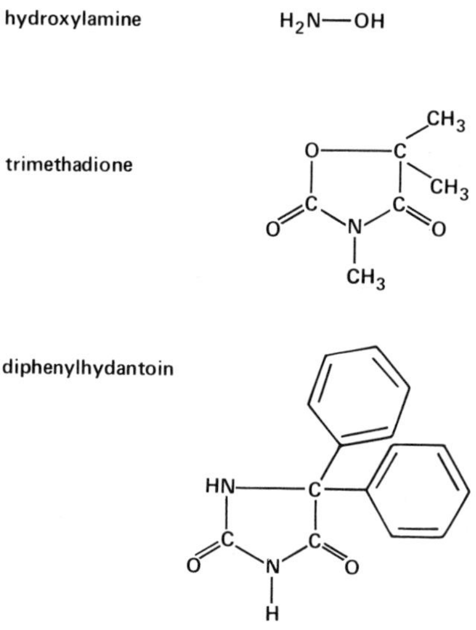

Figure 25.3 The structures of some anticonvulsant drugs.

inhibit the increased respiration caused by the electrical stimulation of brain
slices, but so do many drugs which are not anticonvulsants. Diphenylhydantoin
and other anticonvulsants alter the electrolyte balance of the brain; decreases of
25 to 35% in the sodium concentration of the brain of experimental animals have
been reported.

Major Tranquilizers

The tranquilizers or sedatives can be divided into two classes. The so-called major
tranquilizers or the neuroleptics include the phenothiazines, such as chlorpromazine (Fig. 25.4), and certain alkaloids, such as reserpine (Fig. 25.4). These produce sedation, without loss of consciousness, and decreased motor activity. Clinically, they have a calming effect during psychotic episodes. Subjectively, they
are said to produce indifference to the environment, drowsiness, and apathy [1,
8]. The biochemical actions of chlorpromazine are many and varied and it is not
clear which, if any, mediate the tranquilizing action. In high concentration chlorpromazine has profound effects on the energy-generating systems of the cell. It
inhibits respiration and glucose oxidation, and uncouples oxidative phosphorylation. Structural analogs which are inactive as tranquilizers also uncouple oxida-

chlorpromazine

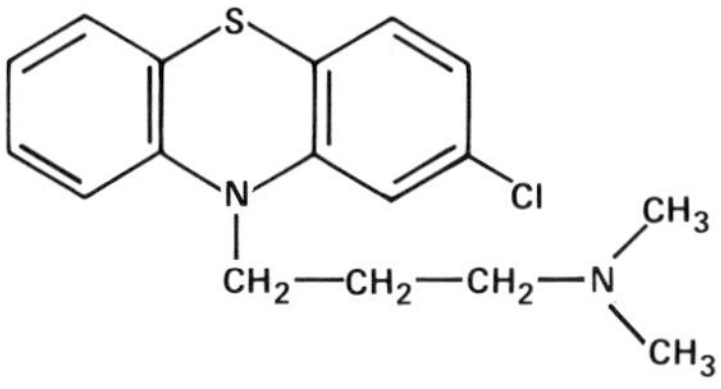

reserpine

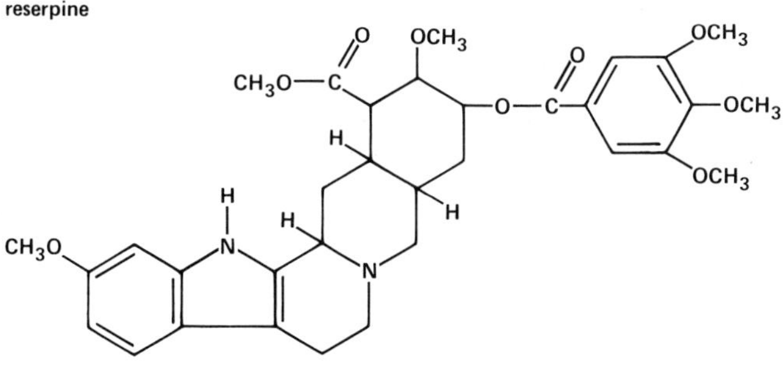

Figure 25.4   The structures of some of the major tranquilizers.

tive phosphorylation, so this is unlikely to be the mechanism by which chlorpromazine acts. Other reported effects include an inhibition of the incorporation of phosphate into certain phospholipids, and a competitive inhibition of the binding of the flavin cofactor to D-amino acid oxidase and other flavin enzymes. Some of the physiological responses to chlorpromazine are reflections of its hypothermic effect on intact animals. It is also well documented that chlorpromazine is a membrane stabilizer [9] and decreases membrane permeability. Although the basis of the pharmacological effect of chlorpromazine is not known, it is considered likely that its action is due to a membrane binding at a catecholamine site on the postsynaptic membrane, possibly to the dopamine receptor [1].

Reserpine, on the other hand, is better understood [10]. Its primary action is on the uptake and storage of monoamines. Reserpine is an alkaloid purified from the plant, *Rauwolfia*. Its therapeutic use as a tranquilizer is declining, although it is still used heavily as an antihypertensive. Reserpine inhibits the vesicular storage of monoamines [11] and, thus, releases amines from their storage sites in catecholaminergic neurons and in the adrenal medulla. It acts indiscriminantly since it also releases the indoleamine, serotonin, from its binding stores. But its action is probably due to release of the catecholamines since administration of dopa will counteract the sedative effect of reserpine, but 5-hydroxytryptophan is ineffective [1,12-14]. In any case, the mode of action of reserpine appears fairly clear; it prevents the binding of catecholamines to their vesicular sites, thus causing release of the amines and their subsequent destruction. The consequent inability of the neurons to store and release transmitter makes them unable to function normally and prevents their action.

Minor Tranquilizers

The minor tranquilizers include meprobamate and the diazepoxides (Fig. 25.5). These have actions similar to those of the major tranquilizers, but are less potent. They are not used as antipsychotics. They produce relaxation of both physical and mental faculties and may induce euphoria.

The barbiturates are considered minor tranquilizers although they will, in larger doses, produce anesthesia. The basis of their tranquilizing action is unknown, but they depress brain metabolism both in vivo and in vitro [1]. Certain barbiturates are also known to inhibit the electron transport chain [15], by preventing the reduction of ubiquinone and the cytochromes by reduced flavin. The barbiturates also induce one subset of the drug metabolizing enzymes found in the microsomes.

Alcohol

Alcohol is also a depressant. Its action seems closely related to the actions of the general anesthetics [16]. Moderate levels, on the order of 0.2% in the blood,

Meprobamate

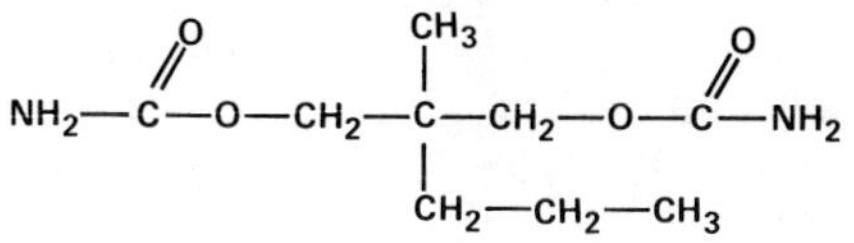

Chlordiazepoxide

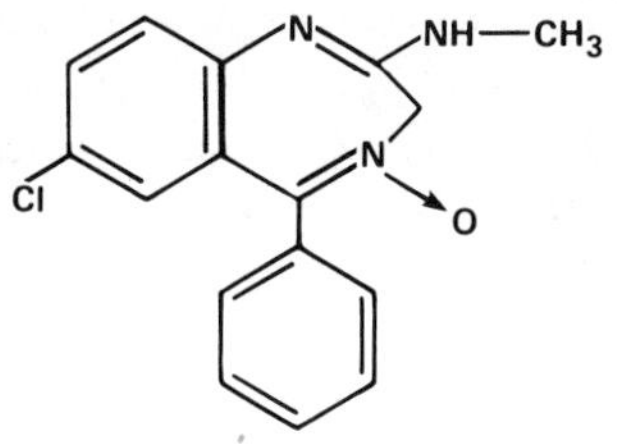

Phenobarbital

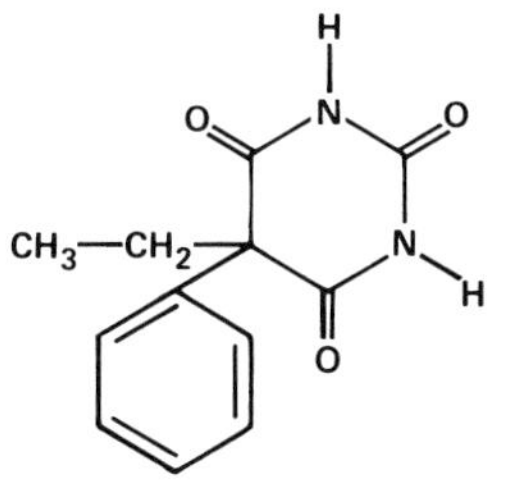

Figure 25.5   The structures of some of the minor tranquilizers.

have minimal metabolic effects. During heavy intoxication, when the blood contains 0.3% alcohol or more, the oxygen consumption of the brain decreases, as
does the glucose uptake. A compensating increase in cerebral blood flow is observed. More than 0.5% alcohol in the blood (0.11 M) is lethal. In vitro studies
are consistent with these effects; unstimulated brain slices show a depression of
respiration, but only when lethal concentrations of alcohol are applied. Slices
stimulated electrically or by high K^+ concentrations in the medium are more sensitive and respond to lower alcohol concentrations. Since comparable amounts
of alcohol have no effect on isolated mitochondria, it can be inferred that the
acute of ethanol is on the excitable membrane [17,18], as with the general
anesthetics, and that the decreased respiration and other metabolic effects are
due to a decrease in the activity of the neurons.

Although there are several reports which indicate that alcohol is not metabolized by the brain, others have found clear evidence for the presence of small amounts of alcohol dehydrogenase in neural tissue [19], and, indeed, for its inductive increase under certain conditions [20]. The acetaldehyde produced by the oxidation of alcohol can also be metabolized in brain.

There are a number of studies which show that γ-aminobutyric acid levels are altered by acute administration of alcohol [21-23]. Although there are some contradictory reports, most studies show that intoxicating concentrations of alcohol cause an increase in the amount of γ-aminobutyric acid in the brain. It is possible, then, that some, if not all, of the depressant effects of alcohol on the central nervous system are mediated by the inhibitory transmitter γ-aminobutyric acid. The mechanism by which alcohol increases the γ-aminobutyric acid content in the brain is unknown.

Antidepressants

As with some of the depressants such as reserpine, the pharmacological actions of many of the antidepressants appear to be mediated by their actions on the monoamine levels in the nervous system. The antidepressants, broadly, increase motor activity, and alertness, reduce appetite and sleeping time, and, in general, cause central sympathetic activation. Clinically, they may increase preexisting anxiety and similarly may exacerbate any psychotic tendencies already evident in the patient.

More specifically, amphetamine (Fig. 25.6), one of the most widely used of the stimulants, is a centrally acting excitant and a sympathomimetic agent [14]. It is a weak inhibitor of monoamine oxidase [1]. It also causes release of the monoamines [24] into the synaptic space of peripheral nerves and perhaps in the central nervous system as well, and inhibits the reuptake mechanism by which the monoamines are normally recaptured by the presynaptic neuron [25]. These various effects on the amine economy produce stimulation and euphoria. The ability of amphetamine to act on a given individual can be correlated with the amine levels in the brain of that individual at the time of administration. It is interesting, and puzzling, that amphetamine, although causing stimulation in normal individuals, frequently has a paradoxical calming effect on children classified as hyperkinetic.

A number of antidepressants are much stronger inhibitors of monoamine oxidase than is amphetamine and appear to act primarily through this inhibitor property. Iproniazide (Fig. 25.6) was the first of these to find wide use. Initially used as an antitubercular agent, it was observed to cause elation in tubercular patients. Since monoamine oxidase is a nonspecific enzyme and acts on all monoamines, the concentrations of all the monoamines increase when the enzyme is inhibited. Paradoxically, blood pressure in humans is decreased. The drug also potentiates the effects of alcohol. Its ability to induce the microsomal

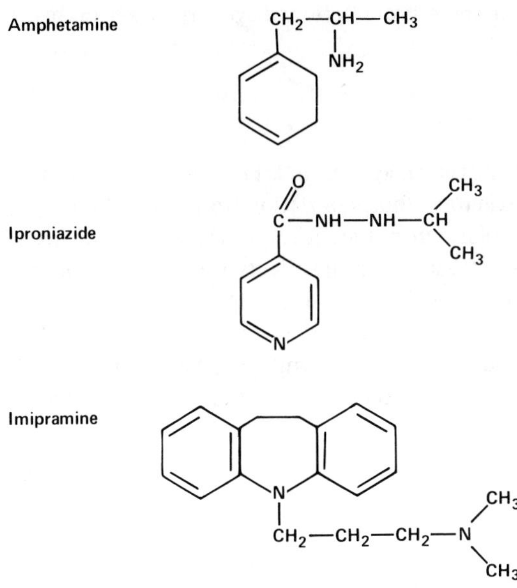

Figure 25.6   The structures of the common antidepressants.

drug-metabolizing enzymes can alter a patient's response to other drugs. It is not clear which of the monoamines is responsible for the alteration in mood produced by iproniazide.

The tricyclic antidepressants, of which imipramine is a representative (Fig. 25.6), have no effect on monoamine oxidase. Nevertheless, they appear to act through the mediation of the monoamines. Imipramine prevents the reuptake of the catecholamines and the indoleamines from the synaptic cleft and, hence, prevents inactivation of the amines. It is perhaps the most potent antidepressant available. This class of compounds is chemically related to the phenothiazines such as chlorpromazine, but the ring system contains an ethylene bridge rather than a sulfur and is known as dibenzazepine. The major site of action seems, then, on the reuptake mechanism at the neuronal membrane.

The action of lithium is described in Chap. 18. It is necessary only to recapitulate here that lithium is effective against mania and also used prophylactically against depression [26]. Lithium is thought to act by interfering with amine metabolism in the synaptic space. There is an alteration of the overall metabolism, perhaps in the rate of amine turnover [27-29]. Although lithium ion will substitute to a limited extent for the other monovalent cations, sodium and potassium, in various metabolic processes, and for sodium ion in nerve conduction, there is no evidence that these latter actions have any effect on behavior.

Analgesics

The analgesic action of the salicylates, such as aspirin, is not understood. It is known that they will uncouple oxidative phosphorylation from respiration. Their effects in brain slices are similar, but not identical, to those of low concentrations of dinitrophenol. Whether these metabolic effects are related to their analgesic properties is not known. More likely to be related is the recently observed interaction of the salicylates with prostaglandin synthetase. The initial observation that acetylsalicylic acid inhibits the enzyme [30] has been followed by more detailed studies [31,32]. These latter experiments show that the inhibition is caused by an acetylation of the enzyme, probably at the active site.

Morphine

The pharmacological action of the narcotic analgesics, such as morphine (Fig. 25.7), is well known [33]. When administered to humans they produce decreased respiration, decreased blood pressure, decreased gastrointestinal motility, and decreased awareness of pain. Accompanying these changes can be increased anxiety and euphoria, and sometimes convulsions. Repeated administration produces a tolerance in that there is a diminished pharmacological response to each succeeding dose. Repeated administration also produces a physical dependence manifest in the catastrophic effects of morphine withdrawal.

The wide variety of effects shown in various experimental systems has provided some interesting reading over the years, but no consensus as to the biochemical site of morphine action. Included among the many reports of effects are changes in K^+-stimulated oxygen uptake in brain slices [34], alterations in RNA synthesis in bacteria [35] and protein synthesis in brain [36,37], increases in radioactive phosphate incorporation into various phospholipids [38], release of catecholamines from adrenal [39], changes in serotonin turnover [40], competition for acetylcholine binding [41], release of corticosteroids [42], decrease of

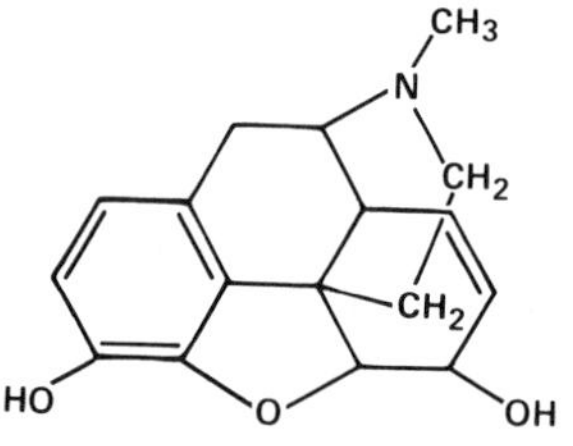

Figure 25.7 Morphine.

calcium levels in brain [43] , decrease in cerebral glycogen content [44] , and interference with dopamine-mediated transmission [45-47] . The fundamental problem is to decide which, if any, of these effects is the primary one, and, thus, the basis of the action(s) of the drug.

A remarkable amount of progress has been made in the last few years in studies on the morphine receptor [48] . Using radioactive ligands of high specific activity it has been possible to find and characterize the opiate binding sites in the brain [49-51] . The strategy has been to localize the receptor in membrane preparations using labeled agonists, and to evaluate the specificity of the binding by comparing the active and the inactive stereoisomers. Nonspecific binding is generally evaluated by measuring the binding of the labeled inactive stereoisomer or by adding large amounts of unlabeled carrier. Further proof of the significance of the data is obtained by comparing the affinities of a series of labeled opiates with the pharmacologic potency of the same series of opiates in the tissues being studied. If the strength of binding is in the same order as their potency, a convincing argument can be made for the physiological significance of the binding. Finally, the effect of known narcotic antagonists on the binding can be evaluated.

Through studies such as these the characteristics of the morphine receptor have been detailed [48] . It has been shown that the receptors have high affinity for both antagonists and agonists. Further, it has been shown that sodium ion has a marked influence on the binding of both antagonists and agonists. Low concentrations of sodium enhance the binding of antagonists and decrease the binding of agonists. While sodium alters the binding of agonists in vitro, the data imply that, in vivo, it is the agonists which interfere with sodium binding. That is, it seems likely that sodium is the ion whose conductance is altered by the opiates. Presumably, the opiates block ion channels in the postsynaptic cell, but there is evidence for the existence of presynaptic receptors as well.

The regional distribution of the opiate receptors in the central nervous system, as revealed by direct assay of various brain regions and by autoradiography, is consistent with several aspects of their action. Opiate receptors are found in the spinal cord, supporting the suggestion that a major part of their analgesic action is exerted in the cord. Also, receptors are found in the brainstem, highly localized in the nucleus controlling visceral sensory function; presumably this is the site at which the opiates induce nausea and vomiting. Also in the brainstem, the receptors are found in the accessory optic pathway, which has important connections with the hypothalamus and the pineal. Perhaps this is the site at which the opiates exert influence on endocrine functions. The localization of the receptors in the thalamus nuclei is consistent with the role of these nuclei in integrating the type of affective pain sensations which are most influenced by the opiates.

met-enkephalin

    Tyr-Gly-Gly-Phe-Met

leu-enkephalin

    Tyr-Gly-Gly-Phe-Leu   **Figure 25.8**   The sequences of the enkephalins.

Natural Analgesics

Since opiates do not occur naturally in the animal, the reason for the existence of an opiate receptor is not obvious. Put another way, the observation of the receptor led a search for the naturally occurring ligand. Such a ligand was reported in 1974 and 1975 [52-55] by several groups. The general term "endorphin" has been adopted to describe any endogenous opoid-like substance. The most widely studied of these two "enkephalins" derived from the brain. These are the pentapeptides, met-enkephalin and leu-enkephalin (Fig. 25.8). They are normally occurring morphine-like peptides which may be neurotransmitters or neuromodulators of pain. The distribution of the enkephalins in the brain parallels the distribution of the receptors. The enkephalins are enriched in the synaptic elements. They have great affinity for the morphine receptor, exhibiting binding constants of around 1×10^{-9} M. The synthetic enkephalins are weak exogenous analgesics; there is some evidence that met-enkephalin has addictive properties [56]. Also, there is evidence for the presence of a naturally occurring analgesic ligand in the blood, and this has been given the name "anodynin," but its structure is as yet not known.

One of the most interesting observations is that β-endorphin, the precursor peptide of met-enkephalin (Fig. 25.9) is itself part of the sequence of a larger molecule, β-lipotropin, a 91-amino acid peptide made in the pituitary. β-Lipotropin has fat-mobilizing activity. Also found in this molecule is β-melanotropin, which has a role in skin pigmentation. There are data to suggest that the β-endorphin has a greater affinity for the receptor, a greater analgesic action, and a longer duration of action than the enkephalins themselves.

A most provocative lead as to the mechanism of action of the opiates has come from the observation that morphine inhibits adenylate cyclase in homogenates of neuroblastoma-glioma hybrid cells [57,58]. Further, morphine inhibits cyclic AMP (cAMP) accumulation in intact cells [58,59] and in rat brain homogenates [60]. It has also been shown that upon prolonged contact between these cells in culture and morphine, the levels of adenylate cyclase increase by some compensatory mechanism and the levels of cAMP in the cells return to normal [61]. These data have been interpreted as a model for narcotic tolerance, since the levels of cAMP in the cells are now, in the presence of morphine, the same as they were in the absence of morphine. The system also lends itself to an understanding of the phenomenon of narcotic withdrawal, since upon removal

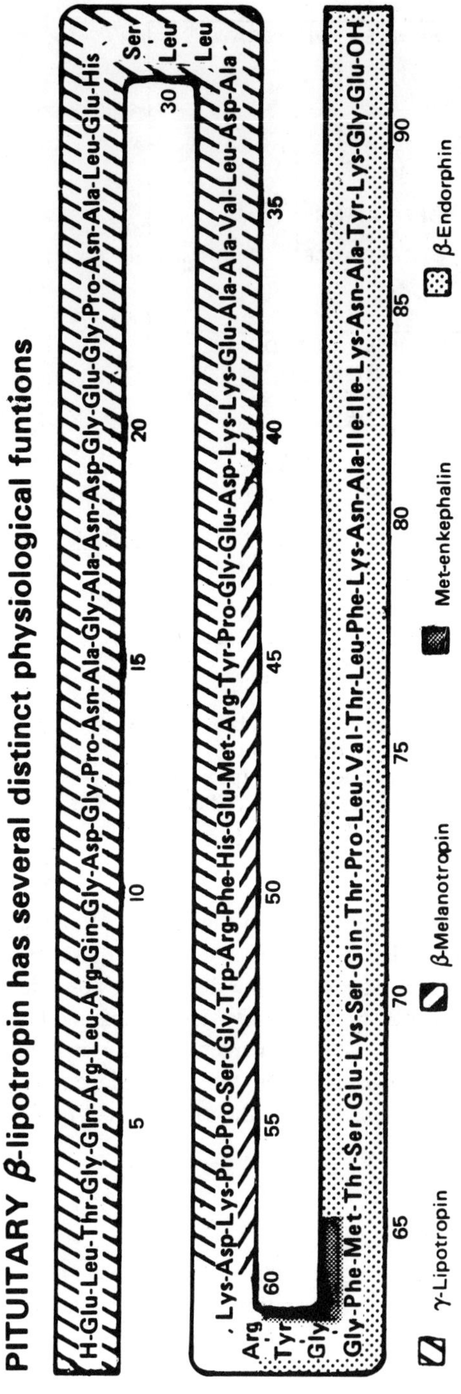

Figure 25.9   The structural relationships between β-lipotropin, β-endorphin, and met-enkephalin.

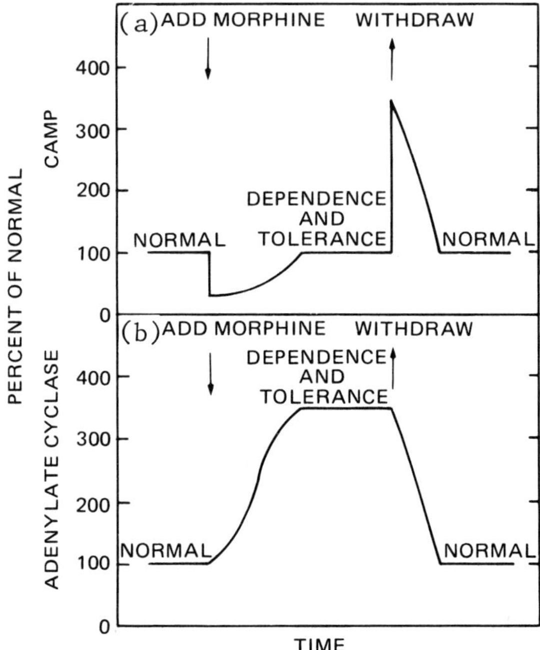

Figure 25.10   A model of the role of adenylate cyclase regulation in the development of morphine tolerance and dependence. (a) The effects of morphine upon cAMP levels; (b) the effects of the opiate upon adenylate cyclase activity as a function of time. [Figure taken from S. K. Sharma, W. S. Klee, and M. Nirenberg, *Proc. Natl. Acad. Sci., USA* 72:3092 (1975).]

of the morphine or its displacement from the receptor by an antagonist, the inhibition of adenylate cyclase is relieved and the synthesis of abnormally high levels of cAMP persists for some time due to the now higher levels of adenylate cyclase in the cell (Fig. 25.10). Whether this model has relevance to the true physiological effects of the narcotic analgesics is still somewhat open to question. It is known, however, that the endogenous opiates, such as met-enkephalin, will produce exactly the same effects in the hybrid cells [62,63].

Hallucinogens

Perhaps the most intriguing and puzzling of the neurochemical effectors are the hallucinogens, or the psychodysleptics [1,64]. As a group they cause in-

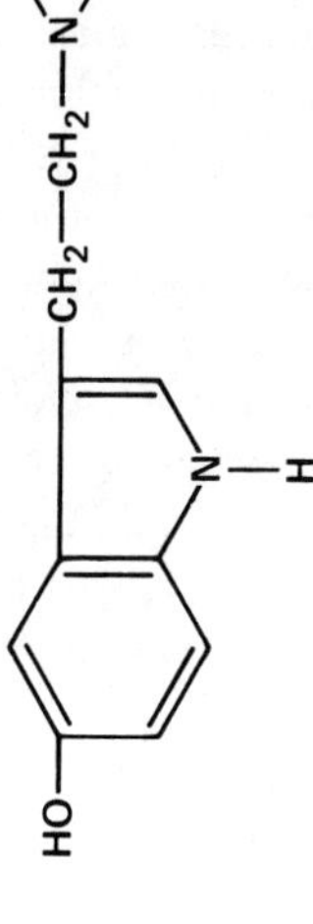

Figure 25.11   The structures of the major hallucinogens.

creased motor activity, some autonomic effects, and catatonia. Clinically, they produce marked changes in mood, perception, and thought processes; any pre-existing psychotic disturbance may be aggravated. Subjectively, they are said to cause distortion of sensory perception, color vision, and hallucinations.

There are at least three classes of hallucinogens about which information is available. The phenylalkylamines, such as mescaline (Fig. 25.11), are similar to the catecholamines in structure and are required in relatively large amounts to produce psychoactive effects [64]. Δ^9-Tetrahydrocannabinol, the major psychoactive principle in marijuana, is representative of the nonnitrogenous hallucinogens (Fig. 25.11). This compound is very lipid soluble and accumulates in brain, lung, and liver after administration to animals [65,66]. It is found in neurons in association with the lipoproteins. The pronounced lipid solubility has given rise to suggestions concerning its effects on the neuronal membrane and to investigations on the effects of Δ^9-tetrahydrocannabinol on lipid metabolism. So far, no consistent theory is available to explain its hallucinogenic action.

The most potent of the hallucinogens are included among the indole derivatives (Fig. 25.11). In this group are the dialkyltryptamines, psilocybin, bufotenine, and D-lysergic acid diethylamide (LSD), perhaps the most potent of all. These materials are effective in microgram quantities and produce profound psychological and perceptual changes. The dimethyltryptamines are relatively short acting and LSD relatively longer acting. Lysergic acid diethylamide produces excitment, hypermotility, and hyperthermia. The most often reported biochemical finding has been that LSD causes a decrease in serotonin turnover in the brain [1,64]. This finding is consistent with the structural relationship between the hallucinogen and serotonin, and also with studies which indicate that LSD competes with serotonin for certain binding sites in the central nervous system [67,68]. Along with the effects on serotonin turnover, and the frequently observed rise in serotonin concentration [69,70], some studies have also reported decreases in the catecholamines, norepinephrine and dopamine [71,72]. Also, there have been reports of actions of LSD on protein and RNA synthesis [73,74]. But these latter effects have not been uniformly observed, and it is considered likely at this time that the site of LSD action is at the locus of the serotoninergic neurons.

References

1. T. L. Sourkes, in *Basic Neurochemistry* (R. W. Albers, G. J. Siegel, R. Katzman, and B. W. Agranoff, eds.), Little, Brown, Boston, 1972.
2. L. J. Mullins, in *Handbook of Neurochemistry,* Vol. VI (A. Lajtha, ed.), Plenum, New York, 1971.
3. W. D. Dettbarn, in *Handbook of Neurochemistry,* Vol. VI (A. Lajtha, ed.), Plenum, New York, 1971.

4.   M. B. Feinstein, *J. Gen. Physiol. 48:*357 (1964).

5.   M. B. Feinstein and M. Paimre, *Biochim. Biophys. Acta 115:*33 (1965).

6.   M. P. Blaustein and D. Goldman, *Science 153:*429 (1966).

7.   M. P. Blaustein, *Biochim. Biophys. Acta 135:*653 (1967).

8.   S. Gabay, in *Handbook of Neurochemistry,* Vol. VI (A. Lajtha, ed.), Plenum, New York, 1971.

9.   P. S. Guth and M. A. Spirtes, *Int. Rev. Neurobiol. 7:*231 (1964).

10. P. A. Shore, in *Handbook of Neurochemistry,* Vol. VI (A. Lajtha, ed.), Plenum, New York, 1971.

11. C. O. Rutledge and N. Weiner, *J. Pharmacol. Exp. Ther. 157:*290 (1967).

12. R. Degkwitz, R. Frowein, C. Kulenkampff, and U. Mohs, *Klin. Woshenschr. 38:*120 (1960).

13. C. R. Creveling, J. Daly, T. Tokuyama, and B. Witkop, *Biochem. Pharmacol. 17:*65 (1968).

14. J. J. Schildkrat and E. S. Gershon, in *Handbook of Neurochemistry,* Vol. VI (A. Lajtha, ed.), Plenum, New York, 1971.

15. J. H. Quastel, in *Metabolic Inhibitors II* (R. M. Hochster and J. H. Quastel, eds.), Academic, New York, 1963.

16. H. Walgren, in *Handbook of Neurochemistry,* Vol. VI (A. Lajtha, ed.), Plenum, New York, 1971.

17. J. H. Mendelson, in *The Biology of Alcoholism* (B. Kissin and H. Beglerter, eds.), Plenum, New York, 1971.

18. H. Kalent, A. E. LeBlanc, and R. J. Gibbins, in *Biological Basis of Alcholism* (Y. Israel and S. Mardines, eds.), Wiley-Interscience, New York, 1971.

19. N. H. Raskin and L. Sokoloff, *J. Neurochem. 17:*1677 (1970).

20. N. H. Raskin and L. Sokoloff, *J. Neurochem. 22:*427 (1974).

21. H. M. Hakkinen and E. Kulonen, *Biochem. J. 78:*588 (1961).

22. H. M. Hakkinen and E. Kulonen, *Biochem. J. 105:*261 (1967).

23. I. A. Sytinsky, B. M. Guzikov, M. V. Gomanko, V. P. Eremin, and N. N. Konovalova, *J. Neurochem. 25:*43 (1975).

24. R. J. Ziance and C. O. Rutledge, *J. Pharmacol. Exp. Ther. 180:*118 (1972).

25. J. T. Coyle and S. H. Snyder, *J. Pharmacol. Exp. Ther. 170:*221 (1969).

26. M. Schou, in *Handbook of Neurochemistry,* Vol. VI (A. Lajtha, ed.), Plenum, New York, 1971.

27. H. Corrodi, K. Fuxe, T. Hokfelt, and M. Schou, *Psychopharmacologia 11:*345 (1967).

28. D. N. Stern, R. R. Fieve, N. H. Neff, and E. Costa, *Psychopharmacologia 14:*315 (1969).

29. J. J. Schildkrat, M. A. Logue, and G. A. Dodge, *Psychopharmacologia 14:*135 (1969).

30. W. L. Smith and W. E. M. Lands, *J. Biol. Chem. 246:*6700 (1971).

31. G. J. Roth, N. Stanford, and P. W. Majerus, *Proc. Natl. Acad. Sci. USA 72:*3073 (1975).

32. G. J. Roth, N. Stanford, J. W. Jacobs, and P. W. Majerus, *Biochemistry 16:*4244 (1977).

33. D. H. Clouet, in *Handbook of Neurochemistry,* Vol. VI (A. Lajtha, ed.), Plenum, New York, 1971.

34. A. E. Takemori, *Science 133:*1018 (1961).

35. E. J. Simon and D. van Pragg, *Proc. Natl. Acad. Sci. USA 51:*877 (1964).

36. D. H. Clouet and M. Ratner, *Brain Res. 4:*33 (1967).

37. D. H. Clouet and M. Ratner, *J. Neurochem. 15:*17 (1969).

38. M. Brossard and J. H. Quastel, *Biochem. Pharmacol. 12:*766 (1963).

39. A. K. Ray, M. Mukherj, and J. S. Ghosh, *J. Neurochem. 15:*875 (1968).

40. E. L. Way, H. H. Loh, and F. H. Shen, *Science 162:*1290 (1968).

41. J. Schuberth and A. Sundwall, *J. Neurochem. 14:*807 (1967).

42. R. K. MacDonald, F. T. Evans, V. K. Weise, and R. W. Patnck, *J. Pharmacol. Exp. Ther. 125:*241 (1959).

43. H. L. Cardenas and D. H. Ross, *J. Neurochem. 24:*487 (1975).

44. C. J. Estler and H. P. T. Ammon, *J. Neurochem. 11:*511 (1964).

45. E. Costa, A. Carenzi, A. Guidotti, and A. Revuelta, in *Frontiers in Catecholamine Research* (E. Usdin and S. Snyder, eds.), Pergamon, New York, 1973.

46. L. Kischinsky and O. Hornykeiwicz, *Eur. J. Pharmacol. 19:*119 (1972).

47. S. Puri, C. R. Reddy, and H. Lal, *Res. Commun. Chem. Pathol. Pharmacol. 5:*389 (1973).

48. S. H. Snyder and R. Simantov, *J. Neurochem. 28:*13 (1977).

49. C. B. Pert and S. H. Snyder, *Proc. Natl. Acad. Sci. USA 70:*2243 (1973).

50. C. B. Pert, G. W. Pasternak, and S. H. Snyder, *Science 182:*1359 (1973).

51. C. B. Pert and S. H. Snyder, *Mol. Pharmacol. 10:*868 (1974).

52. L. Terenius and A. Wahlstrom, *Acta Pharmacol. Toxicol.,* Suppl. 1, *35:*55 (1974).

53. J. Hughes, *Brain Res. 88:*1 (1975).

54. J. Hughes, T. Smith, H. W. Kosterlitz, L. A. Fothergill, B. Morgan, and H. R. Morris, *Nature (London) 258:*577 (1975).

55. G. W. Pasternak, R. Goodman, and S. H. Snyder, *Life Sci. 16:*1765 (1975).

56. E. Wei and H. Loh, *Science 193:*1262 (1976).

57. W. A. Klee and M. Nirenberg, *Proc. Natl. Acad. Sci. USA 71:*3474 (1974).

58. S. K. Sharma, M. Nirenberg, and W. A. Klee, *Proc. Natl. Acad. Sci. USA 72:*590 (1975).

59. J. Traber, K. Fischer, S. Larzin, and B. Hamprecht, *Nature (London) 253:*120 (1975).

60. H. O. J. Collier and A. C. Roy, *Nature (London) 248:*24 (1974).

61. S. K. Sharma, W. A. Klee, and M. Nirenberg, *Proc. Natl. Acad. Sci. USA 72:*3092 (1975).

62. W. A. Klee and M. Nirenberg, *Nature (London) 263:*609 (1976).

63. A. Lampert, M. Nirenberg, and W. A. Klee, *Proc. Natl. Acad. Sci. USA 73:*3165 (1976).

64. S. I. Szara, in *Handbook of Neurochemistry,* Vol. VI (A. Lajtha, ed.), Plenum, New York, 1971.

65. M. Wahlquist, I. M. Nilsson, F. Y. Sandberg, S. Agurell, and B. Granstrand, *Biochem. Pharmacol. 19:*2579 (1970).

66. W. L. Dewey, D. E. McMillan, L. S. Harris, and R. F. Turk, *Biochem. Pharmacol. 22:*399 (1973).

67.   J. T. Farrow and H. van Vunakis, *Biochem. Pharmacol. 22:*1103 (1973).

68.   J. L. Bennett and G. K. Aghajanian, *Life Sci. 15:*1935 (1974).

69.   W. C. Stern, P. J. Morgane, and J. D. Brinzino, *Brain Res. 41:*199 (1972).

70.   M. G. Zeigler, R. A. Lovell, and D. X. Freedman, *Biochem. Pharm. 22:* 2183 (1973).

71.   D. X. Freedman, *Am. J. Psychiatr. 119:*843 (1963).

72.   R. C. Smith, W. O. Boggan, and D. X. Freedman, *Psychopharmacologia 42:*271 (1975).

73.   I. R. Brown, *Proc. Natl. Acad. Sci. USA 72:*837 (1975).

74.   L. Holbrook and I. R. Brown, *J. Neurochem. 27:*77 (1976).

26
THE CHEMISTRY OF MEMORY

The business of the brain is cognition and retention, thought and memory. Of these, memory seems the most amenable to definition and experimentation. Essentially nothing is known for sure about the chemistry of either process. Indeed, the concept that thought and memory can even be described biochemically is only one of a number of alternative possibilities. Some investigators suggest that memory could be encoded in the patterns of connectivity between neurons and reinforced by repeated use; others see memory in electronic rather than biochemical terms. Be that as it may, the recent unraveling of the way in which hereditary information is stored has led to a heightened inquiry into the way in which experience is encoded.

Experimental Approaches

Basically, there have been four different experimental approaches to this study. All are dependent on the supposition that thought and experience are stored and processed as chemical entities, or, at the very least, changes occur in the brain during the process of thought which can be described in chemical terms. A test of this basic assumption underlies all efforts in this field. There are both conceptual and experimental difficulties associated with each of these several approaches.

Indeed, there are some interesting conceptual problems associated with the whole study of memory and thought in an experimental system. The first of these is alluded to above, i.e., it is necessary to accept, first and foremost, that a biochemical approach can yield significant information in the study of cognition. Once it is accepted that cognition can be described in biochemical terms, it is necessary to accept that it can be studied in any system other than the human, i.e., that lower animals have the same types of learning and memory of which humans are capable. Finally, these two propositions accepted, comes the question of how to measure learning. Learning cannot be measured directly, but only as a change in performance. Performance, in turn, can be influenced by fright, stimulation, and so on, i.e., many things which are not learning. Can changes in behavior imposed upon lower animals by conditioning be defined as learning and, if so, what are to be used as controls? Thus, the experimental study of learning is immediately beset by arguable concerns, independent of the exact approach taken. As will be pointed out, the various experimental strategies, while giving interesting and useful information, are themselves each laced with debatable concerns.

The four different approaches can be described as follows. First, there has been a search for a permanent chemical change in the brains of animals which have been trained to perform a specific task. Second, there have been attempts to intervene in the learning or memorization of a specific task using pharmacological or biochemical agents with known mechanisms of action. Third, there are many experiments in which attempts have been made to transfer specific pieces of learned information by injecting brain preparations made from trained rats into the bodies or the brains of untrained animals. Finally, there have been some attempts to duplicate in experimental animals certain diseases which afflict humans and which interfere with normal learning processes. Each of these approaches will now be considered in detail.

Chemical Studies

The search for a permanent chemical change in the chemistry of the brain in response to the learning of specific tasks has occupied a number of laboratories for a number of years [1]. As prelude to the discussion of these studies, it should be stated that there is no question that both the rates of synthesis and the amounts of various macromolecules in the brain are altered during a learning experience. The problem is that they are also altered during any number of other physiological stimulations of a nonlearning nature. It has been shown, for example, that there are changes in the content of RNA in certain neurons following prolonged electrical stimulation, sustained motor exercise, visual stimulation, acoustic stimulation, or rotary manipulation [1]. It is, of course, difficult to design learning tests which do not include physical components such as exercise or stress. Accordingly, it is difficult to divorce the changes caused by such nonlearning portions of the experiment from those associated with the learning. Thus, the choice of appropriate control tasks is a major part of the technology of such studies, and a major source of the criticism of the conclusions drawn.

Nevertheless, there are almost certainly changes in rates of synthesis of various macromolecules in the brain due to specific learning experiences. Mice which were trained to jump to a platform to avoid shock incorporated substantially more precursor into brain RNA than so-called yoked controls, i.e., animals which were given injections, shocks, lights, buzzers, and handling, but were not trained [2,3]. The increased incorporation appeared to be into a heterogenous

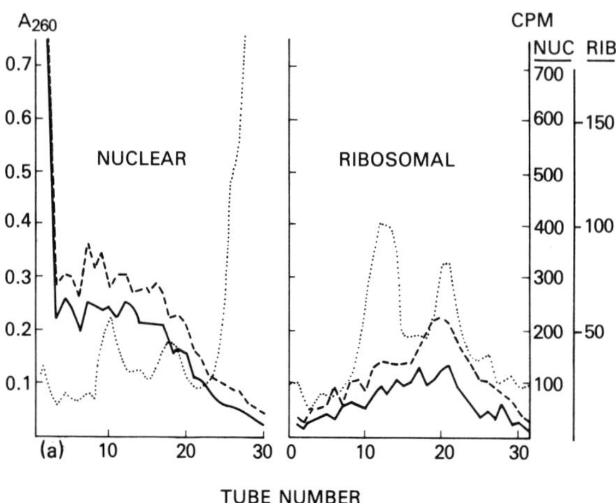

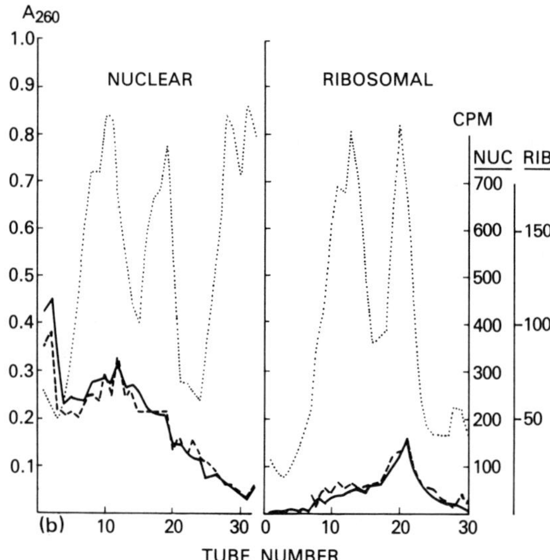

Figure 26.1 (a) Increased incorporation of precursors into brain RNA following a learning experiment. Sucrose gradient centrifugation of RNA from brains of trained and yoke control mice sacrificed immediately after training. One mouse was injected intracranially and intraperitoneally with [^{3}H] uridine, and another mouse similarly with [^{14}C] uridine. After 30 min, the former mouse was trained while the latter served as the yoke control. The training experience lasted 15 min, after which the brains were quickly removed and homogenized together. Ribosomes and nuclei were isolated, and RNA was extracted from each fraction. This RNA was centrifuged in a 5 to 15% sucrose gradient in 30 ml in the Spinco

population of RNAs (Fig. 26.1), and no new species of RNA were detected.
There was also no apparent difference in the base ratios of the RNA made [4].
Although there are other explanations possible, such as changes in the pool
sizes of the RNA precursors, it does appear that increased synthesis of RNA in
the brain accompanies the training experience.

Stimulation in the synthesis or the content of RNA should be reflected in
an eventual increase in protein synthesis and content. Indeed, there are many reports that learning experiences are accompanied by increases in incorporation of
amino acids into protein [5-8]. In fact, experiments in which goldfish must
learn to swim upright again after a float has been attached to their underside
have produced not only changes in overall protein synthesis, but increases in the
incorporation of amino acids into a few specific proteins in the brain [9]. Again,
it must be noted that many nonlearning situations produce changes in protein
synthesis in the brain [10,11].

The experiments of Hydén and his group were designed to uncover more
than simply a general change in the synthetic rates of macromolecules during
learning. After training an animal to perform a specific task, certain neurons in
the brain, thought to be concerned with the area of learning involved in the task,
were dissected by micromanipulation. The amount of RNA in these cells was
determined, and the base ratio of the RNA was examined.

Two different specific tasks were investigated. In the first, rats were
trained to climb a wire to obtain a food reward [12]. Following the training
the rats were killed and the RNA of certain neurons in the vestibular pathway
were removed and analyzed. Control animals were of two kinds: Some were untreated and others were rotated passively, but presumably had no learning experience. These latter animals were used to simulate the activity of the trained
animals. Changes in the content and composition of RNA in the chosen neurons
were noted. The amount of RNA in the neurons of the Deiter's nucleus in untreated controls was 683 pg per cell. In passively rotated controls the RNA content of comparable cells had risen to 722 pg. In the trained animals the RNA
content was 751 pg per cell. The base ratios of the total cellular RNA in the

---

Figure 26.1 (continued)
SW 25 rotor for 18 hr at 25,000 rpm, and 1-ml fractions were collected from the
bottom of the tube. Absorbance at 260 μm, ^{14}C, and ^{3}H were determined for
each fraction. The results were corrected for the ratio of ^{3}H:^{14}C in the UMP
pool. Solid line = ^{14}C, dashed line = ^{3}H, dotted line = absorbance at 260 μm.
(b) Sucrose gradient centrifugation of RNA from the livers of the trained and
yoke control mice sacrificed immediately after training. [Figure taken from J.
W. Zemp, J. E. Wilson, K. Schlesinger, W. O. Boggan, and E. Glassman, *Proc.
Natl. Acad. Sci. USA* 55:1423 (1966).]

three groups was similar; in the nuclear RNA, however, a difference in the base
ratios was reported. The untreated controls had adenine, guanine, cytosine, and
uracil contents of 21, 26, 32, and 21%, respectively. The rotated controls had
values of 21, 26, 31, and 22% for those bases, showing no difference from the
controls. The trained animals had values of 24, 27, 31, and 18%, which was
found to be significantly different than the controls.

In a second series of studies, animals were trained to reverse their handedness, i.e., to obtain food with the opposite paw to their normal preference [13,
14]. In this case the training was quite permanent; the shift in handedness took
4 days to impose and persisted for about 9 months following the training. Certain cells in the cortex said to be concerned with handedness were isolated and
analyzed. Comparable cells from the opposite side of the cortex were used for
controls. Cells from the "active" side had an average RNA content of 270 pg
per cell, while cells from the inactive or control side had 200 pg. In this case
also, an alteration, albeit small, was reported in the base composition of the RNA.

The general conclusion from this body of work is that specific types of
training lead to changes in the RNA content and base composition in specific
neurons in the brain. Later studies from this group show the expected changes
in amino acid incorporation into specific proteins in these neurons [15]. There
is no question that this work is visionary in concept, and ingenious in both design and execution. On the other hand, there is some doubt as to the accuracy
of the interpretation. The work has been highly visible and thoroughly discussed
[1]. Among other things which have been pointed out about the work, there
have been doubts expressed as to whether the results are due to the learning aspects of the tasks or to other facets of the training procedures. It is not clear
that the rotated controls duplicate the sensory input, the motivational factors,
or the motor activity of the trained animals. There are reservations about the
controls in the handedness experiments also, in that the two sides of the cortex
may not be equivalent. In none of these experiments have tissues other than
brain been looked at, either before or after training. The chemical results also
have been criticized because only very small numbers of animals were involved.
Perhaps the most telling of the reservations is that, because of the elaborate and
costly apparatus used, and the difficulties in the microanalyses, no other groups
have reported confirmation of the Hydén results. Aside from the problems of
interpretation, a judgment on the experiments themselves must await the repli-
cation of these studies in other laboratories.

Some of the vagaries in the search for permanent change following the
learning of a single task have been circumvented by a strategy in which experimental animals are raised in two totally different environments [16]. One,
termed the enriched condition (EC), involves the housing of 10 to 12 rats together in a large cage with a number of frequently changed stimulus objects
such as wheels, ladders, bottles, and so forth. The other, the so-called impover-

Table 26.1 Effects of Differential Experience on Occipital Cortex of S_1 Rats Kept in Enriched or Impoverished Conditions from 25 to 105 Days of Age

Measure	Percentage difference (EC > IC)	p	Number (EC > IC)
Weight	6.4	< 0.001	133/175
Total protein	7.8[a]	< 0.001	25/32
Thickness	6.3	< 0.001	45/52
Total acetylcholinesterase	2.2	< 0.01	102/171
Total cholinesterase	10.2	< 0.001	118/132
Total hexokinase	6.9[b]	< 0.01	17/21
DNA/mg	−6.1	< 0.001	4/23
RNA/mg	−0.7	−[c]	10/23
RNA/DNA	5.9	< 0.01	19/23
Number neurons	−3.1	−[c]	7/17
Number glia	14.0	< 0.01	12/17
Perikaryon cross-section	13.4	< 0.001	11.5/13

[a]Weight difference, 7.0% in these experiments.
[b]Weight difference, 5.5% in these experiments.
[c]Not significant.
Data taken from E. L. Bennett and M. R. Rosenzweig, in *Handbook of Neurochemistry,* Vol. VI (A. Lajtha, ed.), Plenum, New York, 1971.

ished condition (IC) has the animals housed individually without such stimulus objects. When the brains from these animals are compared after 80 days of such diverse environments there are a number of substantial differences (Table 26.1). The brains of the EC rats are bigger, their cortices thicker, there are more glia, and the levels of certain enzymes, such as cholinesterase, are higher. Recently, it has also been determined that the diversity of RNAs in the brains of the EC animals is greater [17,18]. It is felt that the increased stimulation experienced by the animals in the enriched condition leads to these permanent alterations in brain morphology and biochemistry. But in these experiments also, it has been difficult to accept the conclusion that the changes are due specifically to what would be defined in a human being as learning.

The general conclusion available from this first approach, then, is that during a specific training experience, or a large number of learning experiences, there are alterations, and perhaps permanent ones, in the chemistry of the brain, specifically in its macromolecules. These changes can be in the rates of synthesis, in the content, and even in the composition. What remains in question is the exact role of these changes in the training, and whether they are due to learning per se or to the other inseparable aspects of the training experience.

Inhibitor Studies

The second approach, and perhaps the one which has received the widest acceptance, is the use of drugs with known effect in other systems on the ability of
animals to remember a given task. For purposes of discussion it is necessary to
note that memory is generally divided into two phases, the short term and the
long term. The short-term memory is that which occurs shortly after the experience and lasts, depending on the species, anywhere from a few minutes to several
hours. No agents have been found which interfere with this kind of memory.
Neither actinomycin D, puromycin, nor cytosine arabinoside have any effect [1].
The process by which short-term memory is turned into long-term memory is
known as "consolidation," and there is some information as to the biochemical
basis of this process.

Studies in mice by Flexner et al. [19] and by others have shown that inhibitors of protein synthesis, such as puromycin [20], inhibit the ability of mice
to remember a training experience. The test involved training the animals to run
a Y maze and then injecting puromycin into several sites in the brain. If the antibiotic is administered within 1 to 3 days after the training, the animals, upon
later retest, will remember nothing of their experience. If the drug is given more
than 3 days later it has no effect on the retention. Agranoff has shown that the
effects of puromycin are also seen in goldfish trained to a specific discrimination
test [21,22]. Here the intervals involved are much shorter, but puromycin given
within 20 min after the training effectively blocks consolidation, whereas after
an hour or more comparable injections have no effect.

Acetoxycycloheximide, another inhibitor of protein synthesis, has been
studied by several groups. Flexner finds that even using acetoxycycloheximide
in concentrations sufficient to block more than 90% of the protein synthesis of
the brain, more inhibition than is caused by memory-blocking doses of puromycin, acetoxycycloheximide has no influence on memory. Indeed, it has been
shown that under certain conditions acetoxycycloheximide prevents the action
of puromycin on memory formation [23]. This surprising observation has been
explained as follows. The two inhibitors work in quite different fashion. Puromycin terminates the growth of the peptide chain, and puromycin peptides accumulate. Acetoxycycloheximide prevents protein synthesis at an earlier stage and
no puromycin peptides accumulate. Thus, perhaps the puromycin peptides themselves are the inhibitors. In support of this concept it has been observed that
puromycin peptides do indeed have remarkable persistence in the brain, remaining for several months, and, further, that injections of small amounts of saline
into the sites at which puromycin had been injected restores memory [24]. In
contrast, however, other workers have subsequently found that acetoxycycloheximide, injected into the brain of trained mice or of goldfish, does inhibit the
recall of a conditioned avoidance response quite effectively [25,26]. Thus, the
final word on inhibition by acetoxycycloheximide remains to be said.

More recently, another inhibitor of protein synthesis, anisomycin, has been studied [27,28]. This antibiotic has a relatively low toxicity compared to the others. It is also effective in inhibiting the formation of long-term memory.

Thus, inhibitors of protein synthesis inhibit the consolidation of experience into long-term memory. Inhibitors of RNA synthesis have the same effect. Actinomycin D cannot be used in mice because concentrations large enough to inhibit RNA synthesis more than 70% in the brain cause the animals to become quite ill. In goldfish, where the drug is less toxic, it appears to inhibit the formation of long-term memory [22]. Other inhibitors of RNA synthesis has been studied less thoroughly, but camptothecin and α-amanitin also prevent the formation of long-term memory.

In systems which are responsive to the memory-blocking effects of inhibitors of protein synthesis and RNA synthesis, inhibitors of DNA synthesis are without effect. Most investigators have employed cytosine arabinoside as a blocker of DNA synthesis, and in every case it has been found to be without effect on the consolidation process.

It seems clear, then, that the synthesis of protein and of RNA plays a role in the consolidation of long-term memory. The major question involves the specificity of such effects. It seems likely that the role of macromolecules is quite indirect. The function of the brain, after all, is to process and to store information. Since the synthesis of protein and of RNA is vital to the function of every cell, such processes may only permit the brain to serve its own unique function. It is possible, on the other hand, that the individual training tasks are coded and remembered in the form of proteins, but this seems unlikely. It is considered more plausible at the moment that consolidation is not a task-specific process. Perhaps it is simply a general response in the sense of remembering whatever just happened. The consolidation process would then contain no information; in the same sense, the developing fluid for a photograph has no relevance to the material appearing on the picture. But protein synthesis and RNA synthesis may be necessary for such a consolidation, and inhibitors of these processes might block it in the same way that electroconvulsive shock is known to block or disrupt consolidation. It is interesting that certain central nervous system stimulants facilitate retention. Strychnine, picrotoxin, and caffeine are in the category. Indeed, certain of these stimulants, including the amphetamines, are known to counteract the action of puromycin [29-31]. The mechanism of the stimulant action is not known.

Transfer of Learning

The experiments described above provide the most consistent and widely accepted line of evidence on the biochemical processes underlying memory. The third approach, the so-called "transfer of learning" experiments are, on the other hand, the most controversial and widely disputed studies in this area. It was reported by McConnell et al. in 1959 [32] that *Planaria,* a species of flatworm, could be trained, and that the memory of the training was preserved in both the head and the tail, because if the trained worm was cut in two and each portion allowed to regenerate a new animal, each new animal remembered the training. Experiments in other laboratories indicated, in addition, that regeneration of the tail portion in the presence of ribonuclease prevented the animal from remembering the training in that portion of the worm [33]. It was then reported that the information or training could also be conveyed if the trained animals were ground up and fed to other, untrained worms [34,35]. Although there has been substantial hesitation about this work, and even some reports that the basic premise, that *Planaria* can be trained, may be suspect [36], the experiments were widely discussed and immediately extrapolated into other experimental situations.

A large number of experiments were then done by various groups in which laboratory animals, primarily rats, were trained to a specific task. The animals were then killed and their brains homogenized; the brains were either used directly or subjected to some preliminary fractionation. Finally, the homogenates or the fractions thereof were administered to naive animals, and the ability of those animals to perform the same task was measured. A remarkable number of successes have been reported. A partial list of those behaviors which are reportedly transferred by such techniques include maze performance [37-41], Skinner box bar pressing [42,43], active [44,45], and passive [46,47] avoidance learning, conditioned reflex [48,49], and habituation to sound [50]. Unfortunately, in spite of all the successes, there are an equal number of reported failures [51-55]. Indeed, the experiments are so significant, if true, that other investigators have taken great pains to replicate meticulously the conditions of each study and repudiate it. Initially, many of the successful reports with fractionated preparations from brain led to the conclusion that the "transfer factors" were RNA. This conclusion was weakened by demonstrations that RNA injected intraperitoneally did not enter the brain intact [56,57], even though intraperitoneal injections of brain preparations were reported to be effective in transfer of learning studies. More recently, as discussed below, the weight of evidence has been that if the transfer phenomenon is real, the materials carrying the information are peptides.

When it was felt that RNA had some effect on memory, yeast RNA was fed to elderly patients in an attempt to improve their memory [58-61]. It was reported that the ingestion of gram quantities of yeast RNA was indeed effective. This work, however, has been challenged on the basis that the effects of the ingestion could be nonspecific nutritional or pharmacological influences, rather than the result of a direct effect on mental functioning itself.

The most elaborate and extensive transfer of learning experiments, and the most controversial in a controversial field, have been done by Ungar [62-64]. He and his collaborators reported some years ago that tolerance to morphine could be transferred by the injection of brain extracts from morphine-tolerant animals into naive animals [65]. From those studies the work progressed to a more formal learning task which involved the training of rats to avoid the dark, a response contrary to their normal preference. Material from the brains of the trained rats was then injected into untrained rats, and they also exhibited a fear of the dark [66], while extracts from naive animals had no effect. The principle from the brains of the trained animals was isolated and purified and is reported to be a pentadecapeptide (Fig. 26.2) [67]. This peptide, named "scotophobin" from the Greek words meaning fear and darkness, has been produced synthetically [68], and is reported to have comparable effects in other species [69,70]. The work on scotophobin has been vigorously debated, and criticized on the basis of both the behavioral experiments [71] and the chemical characterization [72]. Be that as it may, there is now a large and increasing body of work on these task-specific peptides, albeit largely from a single research group, that of Ungar. Ameletin, from the Greek word for indifferent, is a hexapeptide (Fig. 26.3) isolated from the brains of rats habituated to a loud and sudden sound, and is further stated to convey such habituation to naive animals [50,73]. More recently, the brains of goldfish taught to discriminate between the colors green and blue have been extracted. A peptide was extracted from the brains of the fish trained to avoid blue and prefer green, the so-called BA (blue-avoiding) peptide; from the other fish came the GA peptide. These peptides both contain 13 amino acids, including a common internal sequence of eight of the amino acids (Fig. 26.4) [74]. In spite of the widespread scepticism surrounding this

Ser-Asp-Asn-Asn-Gln-Gln-Gly-Lys-Ser-Ala-Gln-Gln-Gly-Gly-Tyr-NH$_2$

Figure 26.2   The structure of scotophobin.

pyro-Glu-Ala-Gly-Tyr-Ser-Lys

Figure 26.3   The structure of ameletin.

BA   p-Glu-Ile-Gly-Ala-Val-Phe-Pro-Leu-Lys-Try-Gly-Ser-Lys

GA   Nac Lys-Gly-Gln-Ile-Ala-Val-Phe-Pro-Leu-Lys-Tyr-Gly-Ser

or

Nac Lys-Gly-Ala-Val-Gln-Ile-Phe-Pro-Leu-Lys-Tyr-Gly-Ser

Figure 26.4   The structures of the BA and GA peptides.

research there are a substantial number of corroborating studies. It is stated in
a recent paper by Tate and coworkers [74] that the "reports of successful bioassays, suggesting the formation of learning-induced substances in the brain, have
been published to date from at least 42 laboratories in over 180 publications."

There is, indeed, increasing evidence that peptides may play an important,
if less specific role in learning and retention. It appeared later in the transfer of
learning studies that the materials being transferred were not RNAs but probably
peptides [39,40,42,43]. Clearly there are peptides which have general activating
or depressant action on the brain. deWied has shown, more specifically, that
adrenocorticotropic hormone (ACTH) retards the loss of the memory (extinction)
of a conditioned avoidance [75-78]. Further, the peptide enhances learning in
hypophysectomized animals. The N-terminal pentapeptide of ACTH is active in
this regard, as are other peptides. It seems that the influence of peptides on specific processes in the brain, from the opiate peptides, through the learning acti-
vators, and perhaps even the task-specific peptides, is an area just coming into
focus. Probably the transfer of specific pieces of information does not occur,
but behavioral modification does seem to take place after the injection of brain
extracts into untrained animals. The distinct possibility exists, however, that
this modification is due to nutritional or pharmacological effects rather than to
information transfer.

Animal Models of Mental Retardation

The fourth approach is an attempt to produce permanent alterations in learning
ability in experimental animals, by reproducing in them known human genetic
abnormalities. If it is possible to obtain such a model, studies on it might lead
to an understanding of the lesion in biochemical terms and thus, perhaps, to an
understanding of the normal functioning of the brain through an understanding
of its malfunction. The search for such a model has centered largely around
phenylketonuria. The reasons for this are that the disease is well characterized
biochemically and clinically; the enzyme lesion is known, as are the metabolic
sequelae. What is not understood is how a loss of this liver enzyme leads to the
catastrophic loss of mental capability and the most severe retardation.

Attempts to create an animal model of phenylketonuria began a number
of years ago in the group directed by Waisman. These experiments involved the
feeding of high levels of phenylalanine to mature rats, and the measurement,
during the treatment, of certain behavioral parameters [79]. It is now clear,
from the further work done by this group and by others, that such treatment
was chemically and developmentally inaccurate. The treatment produced in-
appropriate amino acid alterations in the blood and intoxication of a temporary
nature, and whatever behavioral alterations were observed proved to be quite
reversible. Since that time further attempts at producing such a model have been
sporadic and noncomprehensive.

The rediscovery a few years ago of the amino acid analog, p-chlorophenyl-alanine [80], and its identification as an inhibitor of tryptophan and phenyl-alanine hydroxylases, provided a tool by which it seemed possible to produce a more satisfactory model [81]. This analog, which is a weak competitive inhibitor of phenylalanine hydroxylase in the test tube, is a potent and irreversible inhibitor in vivo [82]. The mechanism of its action is as yet unknown, but it in some way inhibits the enzyme irreversibly, and the return of enzyme activity over the next several days can be slowed by the administration of inhibitors of protein synthesis [82]. These observations indicate that the return of activity is due to the synthesis of new enzyme protein. The mechanism of action of p-chlorophenylalanine on tryptophan hydroxylase appears to be similar.

An attempt has been made over the last several years, to produce an acceptable model of phenylketonuria using p-chlorophenylalanine [83,84]. At the beginning, it seemed appropriate to establish some criteria by which the adequacy of a model could be judged, as follows:

1. The model animals should have the appropriate biochemistry. That is, they should have low levels of phenylalanine hydroxylase activity in the liver, a high phenylalanine level in the blood, a high phenyl-alanine/tyrosine ratio in the blood, and the appropriate phenylketones in the urine.
2. The model animals should exhibit global changes in behavior, and those changes should overlay those seen in the children. That is, in addition to difficulties in problem solving, the animals should be hyperactive, irritable, perhaps fearful, and should have seizures.
3. The animals should show a consistent pathology, and this should include microcephaly, gliosis, and a delayed myelination.
4. Finally, the lesion should be permanent and remain even long after the termination of the chemical insult.

A simple protocol was adopted in which the animals were given p-chloro-phenylalanine to inhibit the enzyme, and phenylalanine to create the appropriate amino acid imbalance in the blood. This regimen was begun on the day of birth and continued until day 21.

The first experiments were designed to determine if the animals, during the treatment period, had the appropriate characteristics. It was shown that, indeed, the enzyme was severely inhibited. The blood amino acid picture also seemed to be correct. The phenylalanine levels were raised, not as much as in the children, but quite substantially. More important, the phenylalanine/tyrosine ratio was elevated, perhaps 20-fold, since the blood tyrosine level did not rise. Brain serotonin levels were decreased. The animals did not experience spontaneous convulsions, but their seizure thresholds, as measured by their susceptibility to chemical convulsants, were lowered by the treatment [85]. This suggested that they had a higher cerebral excitability. Furthermore, the increased

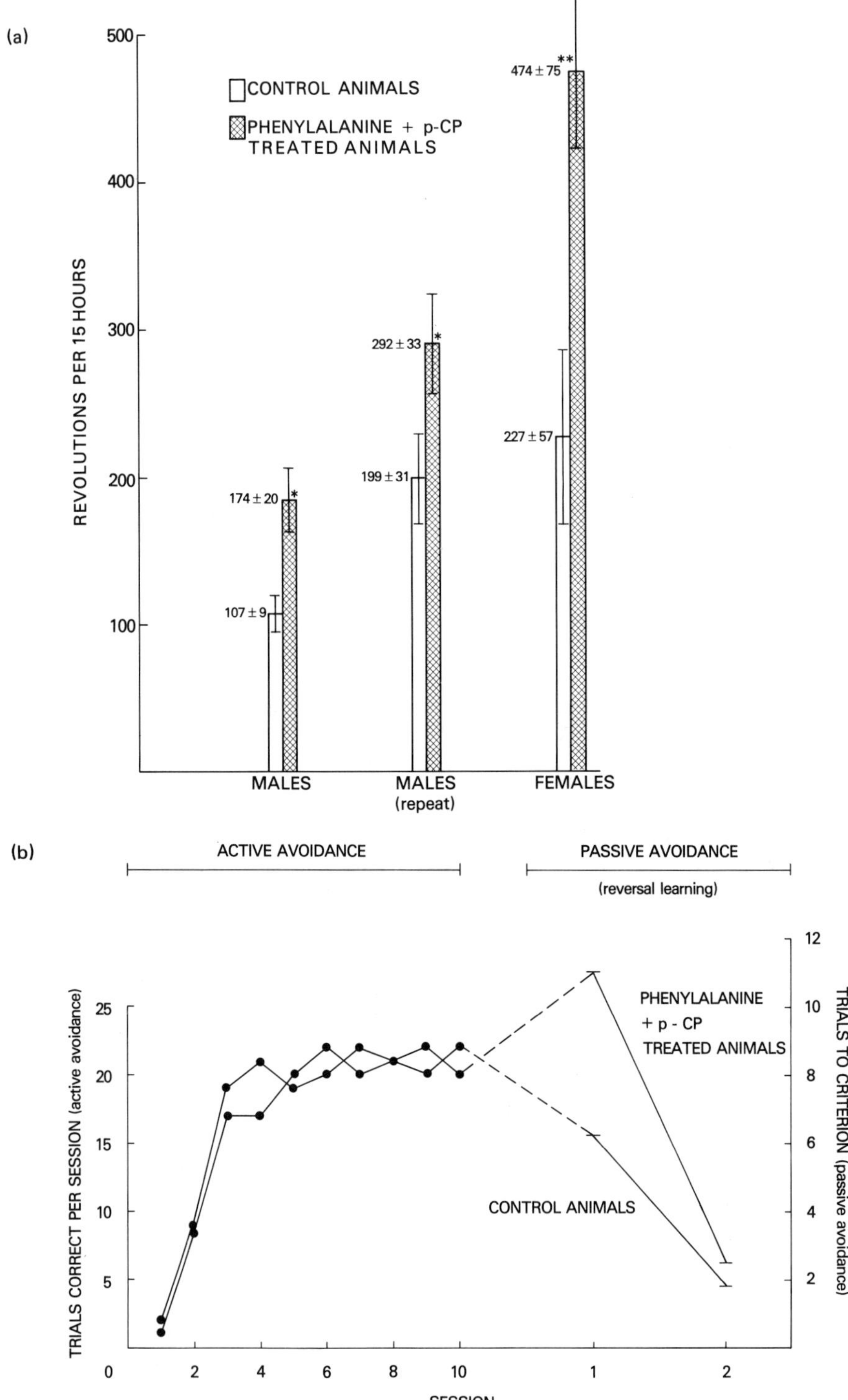

(a)
500
400
300
200
100
REVOLUTIONS PER 15 HOURS
CONTROL ANIMALS
PHENYLALANINE + p-CP TREATED ANIMALS
474 ± 75 **
292 ± 33 *
199 ± 31
227 ± 57
174 ± 20 *
107 ± 9
MALES
MALES (repeat)
FEMALES
(b)
ACTIVE AVOIDANCE
PASSIVE AVOIDANCE
(reversal learning)
25
20
15
10
5
TRIALS CORRECT PER SESSION (active avoidance)
12
10
8
6
4
2
TRIALS TO CRITERION (passive avoidance)
PHENYLALANINE + p - CP TREATED ANIMALS
CONTROL ANIMALS
0
2
4
6
8
10
1
2
SESSION

excitability was age-dependent, as appears to be the case in the phenylketonuric children. The movement of the treated animals also seemed somewhat different, being more sudden and jerky than those of comparable controls. Finally, their developmental milestones were delayed.

Thus, the treatment of the young animals seemed to produce a phenylketonuria-like condition. In order to determine if this early treatment was reflected in the later behavior of the animals they were kept for 6 to 12 months for recovery. Testing of these adults showed that they were hyperactive by at least two different measures and did exhibit some, albeit subtle, learning problems (Fig. 26.5) [83,84]. They seemed, on the one hand, more aggressive both by casual observation and by the test of mouse killing and, on the other, more fearful, in that they could not be motivated to perform in a reward-conditioning setting, but remained crouched in a corner when presented with the need to make right-left choices.

It seemed then that the behavioral changes were permanent. Equally permanent were certain changes in brain structure. First, even though the body weights of these adult animals were indistinguishable from those of the controls, they remained somewhat microcephalic. They had thinner cerebral cortices and fewer cells. Detailed study of the optic nerves of treated animals at various stages of development [86] revealed a myelin deficit and an associated gliosis which was not only permanent, but appeared to be increasing in severity as the animal matured, even months after the end of the treatment.

There are, then, areas of the biochemistry, the neuropathology, and the behavior of these animals which parallel the observations on phenylketonuric chil-

---

Figure 26.5   The activity (a) and the learning ability (b) of p-chlorophenylalanine and phenylalanine treated rats. (a) Overnight activity wheels. The symbol I represents SEM. Asterisks (*) refer to statistically significant differences between experimental and control groups: one asterisk (*), $p < 0.025$; two asterisks (**), $p < 0.01$. (b) Active avoidance training, followed by passive avoidance of the active response. Left: the scores of male subjects over 10 sessions of active avoidance in a two-way shuttlebox. Each session consisted of 30 trials. There is no significant difference between groups on active avoidance testing. Right: the response of these same subjects (excluding phenylalanine) to passive avoidance of the learned active avoidance response. The saline-treated and p-chlorophenylalanine + phenylalanine treated animals now split into two distinct groups; the statistically significant difference between groups is maintained when the subjects are retested 1 week later. The results of the first passive avoidance sessions are 6.6 ± 1.1 trials for controls and 11.2 ± 1.8 trials for p-chlorophenylalanine + phenylalanine treated rats (standard error given). The results of the repeat passive avoidance session are 1.7 ± 0.2 trials for controls and 2.6 ± 0.3 trials for treated. [Data taken from A. Andersen and G. Guroff, *Proc. Natl. Acad. Sci. USA 69:*863 (1972); A. Andersen, V. Rowe, and G. Guroff, *Proc. Natl. Acad. Sci. USA 71:*21 (1974).]

dren. There are, however, a number of things which are seen in the animals which do not parallel the course of the disease. First, there is a marked mortality among the treated animals, not seen in the children. Then, there are cataracts in the treated animals [87] which do not seem to be characteristic of phenylketonuria. Finally, the changes observed in these animals are quite subtle, whereas the changes in the children are catastrophic.

More important than the things which are seen and should not be there, are the things which are not seen which should be there. Upon careful study of the animals during treatment, phenylketones were not found in the urine [88]. Phenylketones, reminiscent of these excreted by phenylketonuric children, can be produced, but the animals must be treated with much more phenylalanine than the standard regimen includes. The most obvious interpretation of these data is that the behavioral changes in the animals are not caused by the accumulation of the most characteristic products of the disease.

Fundamentally, then, this model is not yet completely satisfactory [89]. The cataracts, the lack of phenylketones, and the subtlety of the changes, among other things, indicate that, so far, no reliable information about the biochemical cause of phenylketonuria can be obtained here. Whether this model can be made more accurate in the future is yet another question. All that can be said at this point is that it has been possible, using a specific and discrete chemical treatment, to produce a defined, reproducible, and permanent behavioral alteration in rats which has some resemblance to phenylketonuria in children.

It seems possible that no satisfactory model of human mental retardation will ever be produced in an experimental animal, because the animal does not have the behavioral or cognitive repetoire, by definition, to express such a lesion. It remains, however, a reasonable approach both to the problems of aberrant neural function in disease, and to the study of normal cognitive processes in the healthy brain.

Thus, in spite of substantial effort, little is known about the biochemistry of cognition and retention. The experiments are complicated by the vagaries of trying to measure learning in experimental animals. At present, the best evidence is that the synthesis of macromolecules is involved in the functioning of the cells which participate in learning and remembering. But the exact substrate for the storage and retrieval of specific pieces of information is still unknown. Whether the memory program is inherent in any biochemical substance or in structure at some higher level of organization and integration is still an open question.

References

1.   E. Glassman, *Ann. Rev. Biochem. 38:*605 (1969).
2.   J. W. Zemp, J. E. Wilson, K. Schlesinger, W. O. Boggan, and E. Glassman, *Proc. Natl. Acad. Sci. USA 55:*1423 (1966).

3.   L. Adair, J. E. Wilson, J. Semp, and E. Glassman, *Proc. Natl. Acad. Sci. USA* *61:*606 (1968).

4.   J. W. Zemp, J. E. Wilson, and E. Glassman, *Proc. Natl. Acad. Sci. USA 58:* 1120 (1967).

5.   H. Hydén and P. W. Lange, *Proc. Natl. Acad. Sci. USA 65:*898 (1970).

6.   I. B. Levitan, G. Ramirez, and W. E. Mushynski, *Brain Res. 47:*147 (1972).

7.   P. P. G. Bateson, G. Horn, and S. P. R. Rose, *Brain Res. 84:*207 (1975).

8.   M. Hershkowitz, J. E. Wilson, and E. Glassman, *J. Neurochem. 25:*687 (1975).

9.   V. E. Shashoua, *Science 193:*1264 (1976).

10.   H. Waelsch and A. Lajtha, *Physiol. Rev. 41:*709 (1961).

11.   K. Richardson and S. P. R. Rose, *J. Neurochem. 21:*521 (1973).

12.   H. Hydén and E. Egyhazi, *Proc. Natl. Acad. Sci. USA 48:*1366 (1962).

13.   H. Hydén and E. Egyhazi, *Proc. Natl. Acad. Sci. USA 52:*1030 (1964).

14.   H. Hydén and P. W. Lange, *Proc. Natl. Acad. Sci. USA 53:*946 (1965).

15.   H. Hydén and P. W. Lange, in *Handbook of Neurochemistry,* Vol. VI, (A. Lajtha, ed.), Plenum, New York, 1971.

16.   E. L. Bennett and M. R. Rosenzweig, in *Handbook of Neurochemistry,* Vol. VI (A. Lajtha, ed.), Plenum, New York, 1971.

17.   L. Uphouse and J. Bonner, *Dev. Psychobiol. 8:*171 (1974).

18.   L. Grouse, B. K. Schrier, E. Bennett, M. Rosenzweig, and P. G. Nelson, *J. Neurochem. 30:*191 (1978).

19.   L. B. Flexner, J. B. Flexner, and R. B. Roberts, *Science 155:*1377 (1967).

20.   J. B. Flexner, L. B. Flexner, and E. Stiller, *Science 141:*57 (1963).

21.   B. W. Agranoff, in *Basic Neurochemistry,* 2nd ed. (R. W. Albers, G. J. Siegel, R. Katzman, and B. W. Agranoff, eds.), Little, Brown, Boston, 1976.

22.   B. W. Agranoff, in *The Nervous System,* Vol. 1 (D. B. Tower, ed.), Raven, New York, 1975.

23.   L. B. Flexner and J. B. Flexner, *Proc. Natl. Acad. Sci. USA 35:*369 (1966).

24.   L. B. Flexner and J. B. Flexner, *Proc. Natl. Acad. Sci. USA 60:*923 (1968).

25.   S. H. Barondes and H. D. Cohen, *Proc. Natl. Acad. Sci. USA 58:*157 (1967).

26.   B. W. Agranoff, *Science 158:*1600 (1968).

27.   J. F. Flood, M. R. Rosenzweig, E. L. Bennett, and A. E. Orme, *Physiol. Behav. 10:*555 (1973).

28.   L. R. Squire and S. H. Barondes, *Brain Res. 66:*301 (1974).

29.   S. H. Barondes and H. D. Cohen, *Proc. Natl. Acad. Sci. USA 61:*923 (1968).

30.   R. G. Serota, R. B. Roberts, and L. B. Flexner, *Proc. Natl. Acad. Sci. USA 69:*340 (1972).

31.   M. E. Hall, K. Schlesinger, and E. Stamm, *J. Neurochem. 4:*353 (1976).

32.   J. V. McConnell, A. L. Jacobson, and D. P. Kimble, *J. Cell. Comp. Physiol. 52:*1 (1959).

33.   W. C. Corning and E. R. John, *Science 134:*1363 (1961).

34.   J. V. McConnell, *J. Neuropsychiatr.,* Suppl. 1, *3:*42 (1962).

35.   R. A. Westermann, *Science 140:*676 (1963).

36. E. L. Bennett and M. Calvin, *Neurosci. Res. Prog. Bull. 2:*3 (1964).
37. E. J. Fjeringstad, Th. Nissen, and H. H. Roigaard-Peterson, *Scand. J. Psychol. 6:*1 (1965).
38. T. Nissen, H. H. Roigaard-Peterson, and E. J. Fjeringstad, *Scand. J. Psychol. 6:*265 (1965).
39. F. Rosenblatt and R. G. Miller, *Proc. Natl. Acad. Sci. USA 56:*1423 (1966).
40. F. Rosenblatt and R. G. Miller, *Proc. Natl. Acad. Sci. USA 56:*1683 (1966).
41. W. B. Essman and G. M. Lehrer, *Fed. Proc. 26:*263 (1967).
42. F. Rosenblatt, J. T. Farrow, and S. Rhine, *Proc. Natl. Acad. Sci. USA 55:* 548 (1966).
43. F. Rosenblatt, J. T. Farrow, and S. Rhine, *Proc. Natl. Acad. Sci. USA 55:* 787 (1966).
44. D. J. Albert, *Neuropsychologia 4:*79 (1966).
45. D. H. Malin, A. M. Bolub, and J. V. McConnell, *Nature (London) 233:*211 (1971).
46. R. Gay and A. Raphelson, *Psychon. Sci. 8:*369 (1967).
47. G. Ungar, L. Galvan, and R. H. Clark, *Nature (London) 217:*1259 (1968).
48. S. Reinis, *Achvitas Nervosa Superior 7:*167 (1965).
49. F. Rosenblatt, J. T. Farrow, and W. F. Herblin, *Nature (London) 209:*46 (1966).
50. G. Ungar and C. Oceguera-Navarro, *Nature (London) 207:*301 (1965).
51. C. G. Gross and F. M. Carey, *Science 150:*1749 (1966).
52. M. W. Gordon, G. G. Deanin, H. L. Leonhardt, and R. H. Gwynn, *Am. J. Psychiatr. 122:*1174 (1966).
53. L. D. Hutt and L. Elliott, *Psychon. Sci. 18:*57 (1970).
54. A. Goldstein, P. Sheehan, and J. Goldstein, *Nature (London) 233:*126 (1971).
55. W. L. Byrne, D. Samuel, E. L. Bennett, M. R. Rosenzweig, E. Wasserman, A. R. Wagner, F. Gardner, R. Galembos, B. D. Berger, D. L. Margules, R. L. Fenichel, S. Stein, J. A. Corson, H. E. Eneses, S. E. Choroxer, C. F. Holt, III, P. H. Schiller, L. Chiepettg, M. E. Jarvik, R. C. Leaf, J. D. Dutcher, Z. P. Horovitz, and P. L. Carlson, *Science 153:*658 (1966).
56. S. Sved, *Can. J. Biochem. 43:*949 (1965).
57. M. Luttges, T. Johnson, C. Brick, J. Holland, and J. McGaugh, *Science 151:* 834 (1966).
58. D. E. Cameron, *Am. J. Psychiatr. 114:*943 (1958).
59. D. E. Cameron, L. Solyom, and L. Beach, *Neuro-Psychopharmacol. 2:*351 (1961).
60. D. E. Cameron and L. Solyom, *Geriatrics 16:*74 (1961).
61. D. E. Cameron, V. A. Kral, L. Solyom, S. Sved, B. Waenrib, C. Beaulieu, and H. Enesco, in *Macromolecules and Behavior,* (J. Gaito, ed.), York University, Toronto, ACC Div. Meredith Publishing, Toronto, 1966.
62. G. Ungar, *Life Sci. 14:*595 (1974).
63. G. Ungar, *Biochem. Pharmacol. 23:*1553 (1974).
64. G. Ungar, *Int. Rev. Neurobiol. 17:*37 (1975).

65. G. Ungar and M. Cohen, *Int. J. Neuropharmacol. 5:*183 (1966).

66. G. Ungar, L. Galvan, and R. H. Clark, *Nature (London) 217:*1259 (1968).

67. G. Ungar, D. Desidero, and W. Parr, *Nature (London) 238:*198 (1972).

68. W. Parr and G. Holzer, *Hoppe-Seylers Z. Physiol. Chem. 352:*1043 (1971).

69. R. C. Bryant, N. H. Santos, and W. L. Bryne, *Science 177:*635 (1972).

70. D. H. Malin and H. N. Guttman, *Science 178:*1219 (1972).

71. A. Goldstein, *Nature (London) 242:*60 (1973).

72. W. W. Stewart, *Nature (London) 238:*202 (1972).

73. S. R. Burzynski, *Anal. Biochem. 70:*359 (1976).

74. D. F. Tate, L. Galvan, and G. Ungar, *Pharmacol. Biochem. Behav. 5:*441 (1976).

75. B. Bohus and D. deWied, *Science 153:*318 (1966).

76. D. deWied, *Am. J. Physiol. 207:*255 (1964).

77. D. deWied, in *The Neurosciences,* 3rd Study Program (F. O. Schmitt and F. G. Worden, eds.), MIT Press, Cambridge, 1974.

78. D. deWied, B. Bohus, I. Urban, T. van Wimersma Greidanus, and W. Gispen, in *Peptides: Chemistry, Structure, Biology* (R. Walters and J. Meienhofer, eds.), Science Publishing, Ann Arbor, 1975.

79. V. J. Polidora, R. F. Cunningham, and H. A. Waisman, *J. Comp. Physiol. Psychol. 61:*436 (1966).

80. B. K. Koe and A. Weissman, *J. Pharm. Exp. Ther. 154:*499 (1966).

81. M. A. Lipton, R. Gordon, G. Guroff, and S. Udenfriend, *Science 156:*248 (1967).

82. G. Guroff, *Arch. Biochem. Biophys. 134:*610 (1969).

83. A. Andersen and G. Guroff, *Proc. Natl. Acad. Sci. USA 69:*863 (1972).

84. A. Andersen, V. Rowe, and G. Guroff, *Proc. Natl. Acad. Sci. USA 71:*21 (1974).

85. J. W. Prichard and G. Guroff, *J. Neurochem. 18:*153 (1971).

86. L. Avins, G. Guroff, and T. Kuwabara, *J. Neuropathol. Exp. Neurol. 34:*178 (1975).

87. V. D. Rowe, S. Zigler, A. E. Andersen, J. B. Sidbury, and G. Guroff, *Exp. Eye Res. 17:*245 (1973).

88. V. D. Rowe, H. M. Fales, J. J. Pisano, A. E. Andersen, and G. Guroff, *Biochem. Med. 12:*123 (1975).

89. G. Guroff, in *Clinical and Biochemical Medicine* (S. Kumar, ed.), Pergamon, Oxford, 1978.

RECENT DEVELOPMENTS

The inevitable lag in the publishing process is not compatible with the rapid progress in neurobiology. Within the 18-month period required to put this book together, not to mention the time between the writing of the various sections and their submission in final form, the accumulation of new information and the opening of new fields of investigation has proceeded at a normal, rapid pace. This chapter attempts to deal with the problem.

Included here are short sections on subjects which, from an admittedly subjective view, appear to have progressed to the point that omitting them from such a book would be ridiculous. The chapter is not intended to be exhaustive, nor to deal with information in all the areas discussed in the book. Notwithstanding the title "Recent Developments," there are some findings described here that are not so recent but which seem to have come to fruition, perhaps, within the last couple of years.*

Axonal Flow

The basic concept of this process, the nourishment of the nerve endings by an orderly progression of materials down the axon, has not been challenged. Nevertheless there is a new appreciation that the opposite process, retrograde axonal flow, is also of crucial importance for the appropriate development of the neuron. Especially for the signaling of growth processes by trophic factors such as nerve growth factor, the retrograde progress of information from target organ to neuronal cell body has assumed a more prominent place in the conceptual framework of studies in this area.

Two other important concepts about this process have crystallized in the past few years. First, the dissection of the various rates of axonal flow into at

*The authors referenced in this chapter do not appear in the Author Index.

least five different classes and the identification of the component proteins have
suggested that there is a functional meaning to the various classes. That is, proteins transported together may subserve a single specific function in the cell.
Second, the proteins transported may be altered during the transport. That is,
the processing known to occur in the cell, changing the initial gene product to a
finished, functional protein, may happen during the axonal transport of the
protein.

The dissection of the various transport rates has been possible through the
work of several groups. Foremost among these, perhaps, have been the investigations of Willard and of Lasek and their colleagues. Using retinal ganglion cells
from the rabbit, Willard [1] observed more than 40 peptides being transported.
He was able to divide these into five groups based on their rates of flow. The
rates of flow and the proteins or organelles thought to participate in them are
given in the table below.

Group	Flow rate (mm/day)	Protein or organelle
I	>240	Possibly membrane-associated proteins
II	34–68	Mitochondria
III	4–8	Myosin
IV	2–4	Tubulin, myosin, and actin
V	0.7–1.1	Proteins of the neurofilaments

Lasek and his colleagues, concentrating on slow axoplasmic flow in retinal
ganglion cells from the guinea pig, have identified two components, SC_a at 0.2–
0.5 mm/day and SC_b at 2 mm/day. In the slowest moving component, SC_a,
there are five polypeptides [2], a triplet comprised of neurofilament proteins,
and two subunits of tubulin. In SC_b there are 20 polypeptides and all are different from the five seen in the slowest component. Among the 20 is actin [3],
whose identification has been accomplished partially through its unusual affinity
for DNase I. Also in this component, probably, is tropomyosin. Actin does not
occur in SC_a, nor is it seen among the proteins of the fast components. Myosin,
on the other hand, has been found in groups III and IV [1].

Thus, in spite of some differences in the rates measured in different systems, and the consequent differences in the assignment of specific proteins to
the various groups, it seems clear that several discrete components exist and that
proteins transported at one rate do not overlap into other groups. Further, it
may be that the proteins in the various groups are of related function.

Clearly, in certain cases, the protein that leaves the cell body is not the
same when it reaches the terminal. It has been suspected for some time that
the peptides oxytocin and vasopressin and their respective carrier proteins, the

neurophysins, were produced in the cell from a single, large protein precursor. Evidence substantiating this suspicion has been presented by Gainer, Sarne, and Brownstein [4]. By injecting [^{35}S] cysteine adjacent to the supraoptic nucleus they obtained the labeling of two proteins with molecular weights on the order of 20,000. As these proteins moved down the axon they were processed by the removal of more basic regions of the structure and gave rise to the neurophysins. Since the oxytocin and vasopressin portions would be more basic than the overall protein, the authors concluded that the two original proteins represented high molecular weight precursors to neurophysin and oxytocin, and neurophysin and vasopressin, respectively. Supporting evidence has come from studies with Brattleboro rats [5]. These rats, genetically unable to produce vasopressin, produce just one of the 20,000 molecular weight species found in controls.

Myelin Proteins

Information has accumulated on the presence of glycoproteins in myelin. Although the glycoprotein content of myelin is small, there is some suggestion that it may have an important role in the initial glial-axonal interaction in the formation of the myelin sheath. The continuing work of Quarles, Brady, and their colleagues has led to the identification of a glycoprotein of 100,000-110,000 molecular weight in highly purified myelin [6-8]. This myelin-associated glycoprotein (MAG) has found originally only in the central nervous system, although other, lower molecular weight glycoproteins are present in myelin from peripheral nerves. The glycoprotein can be labeled with various carbohydrate precursors, such as fucose, glucosamine, and N-acetylmannosamine, as well as with sulfate. Antibody against this protein has been prepared [9], and this antibody reacts with components in the cytoplasm of oligodendroglia even before myelination begins. These studies have also indicated that the myelin-associated glycoprotein occurs in peripheral myelin, notwithstanding the earlier inability to detect it through more direct observation. The function of the myelin-associated glycoprotein is unknown, as are the exact functions of the other, more abundant proteins of myelin. It is of interest, however, that the glycoprotein is apparently altered in some way in the early stages of multiple sclerosis.

Peptides

There has been a virtual explosion in research on neuroactive peptides [10,11]. Such peptides, now numbering 20 or more, are generally molecules which occur in the body at sites other than those associated with endocrine or neuroendocrine function. Many are found in the intestine, which may indicate that the intestine and the central nervous system have some common precursor in the embryo. In the brain they are found in the neurons and specifically in the nerve terminals. Biosynthetically they are usually produced from some larger precursor through proteolytic processing. Among them are small peptides and large peptides. Some are thought to be neurotransmitters impinging directly on target neurons. Others are apparently neuromodulators acting more diffusely to moderate the response of the neuron to other transmitters. Some are excitatory, others inhibitory. Although none participate in the synaptic events at a large portion of central nervous system synapses, probably less than 1% in each case, they already outnumber all previously identified nonpeptide central nervous system transmitters. The enormous outpouring of research on the subject suggests there will be peptides in profusion discovered in the immediate future.

A simple listing of the peptides now found in the brain and thought to have transmitter-like functions would take a good portion of a page. The first, historically, was substance P. This molecule, described in earlier sections of this book, was characterized as a putative neurotransmitter several years ago. It has been found in neurons by immunohistochemistry, in the synaptosomal portion by inspection of the relevant synaptosomes, and in specific portions of the brain using radioimmunoassay techniques in punch samples of the central nervous system. Substance P is found in the substantia nigra, in the interpeduncular nucleus, and in several hypothalamic nuclei. It is also found in spinal sensory ganglion cells with substance P terminals in the spinal cord, and these terminals have been observed to release substance P. It is of interest that the neurons containing substance P also seem to contain, in some cases, serotonin [12], providing some further evidence against the now shaky Dale hypothesis of one neuron-one transmitter.

The enkephalins appear to be neurotransmitters or neuromodulators. Neurotensin, a peptide causing vasodilation, hypotension, and many other physiological effects, has also a potent analgesic effect and is, indeed, 1000 times more potent than the enkephalins. It is found in the hypothalamus and the basal ganglia and is localized in the synaptosomal portion of neurons in these areas. The vasoactive intestinal peptide (VIP), initially found in peripheral nerves of the wall of the intestine, has now been identified in cortical neurons. Given peripherally, it causes vasodilation and stimulates glycogenolysis and lipolysis. Centrally it is found in the synaptosomal portion of the relevant neurons and, as are most transmitters, is released by a Ca^{2+}-dependent mechanism. Cholecystokinin, an intestinal hormone causing gallbladder contraction and pancreatic secretion, has now been found in cortical neurons and in white matter. Its function here is unclear, but it is localized in neurons, and only in neurons in the cortical areas. Bradykinin causes slow contraction of the ileum and elicits cardiovascular responses. It has now been found in brain extracts, and bradykinin neurons have been identified in the hypothalamus. Angiotensin II is a vasoconstrictor and stimulates the release of aldosterone. Angiotensin II neurons have been identified in the central nervous system, although their exact anatomical distribution is still not known. Thyrotropin releasing hormone (TRH) is found, of course, in the hypothalamus, and its action there is pretty well understood. But 80% of the total TRH is present outside the hypothalamus, primarily in the thalamus, the cerebrum, and the brainstem. Its function in these other sites is unknown, but it is present in the synaptosomal portions of the cell, and it is released in a Ca^{2+}-dependent manner. Somatostatin, found in the stomach, intestine, and pancreas, has been identified as the hypothalamic principal inhibiting the release of growth hormone. Only 25% of the somatostatin present in the central nervous system is in the hypothalamus, major portions occurring also in the spinal cord. Luteinizing hormone releasing factor has potent depolarizing action on sympathetic ganglia. Indeed, this list of actions by peptide neurotransmitters could continue for pages. Suffice it to say that, in addition to the peptides mentioned above, there is reason to consider the following also as putative or at least candidate neurotransmitters: ACTH, bombesin, carnosine, β-endorphin, gastrin, oxytocin, prolactin, and vasopressin.

Carnosine

Among the many peptides listed above, perhaps the best characterized is the smallest, carnosine. Through the work of Margolis and his colleagues it has become clear that this peptide is most likely the transmitter of the olfactory chemoreceptor neurons [13]. The original studies, in which denervation was found to have no effect on the content of the well-described transmitters, e.g., norepinephrine, acetylcholine, γ-aminobutyric acid, suggested that the transmitter in the olfactory system was none of these. Inspecting the other constituents of the tissue showed that following denervation a dansyl chloride-reactive spot disappeared. Identification of this spot as carnosine (β-alanyl-L-histidine) suggested that this peptide might be the transmitter. This suggestion was supported by the finding that the carnosine content of the tissue was high (2 nmol/mg) and that the enzyme responsible for its synthesis, carnosine synthetase, was higher in the olfactory bulb than in any other part of the brain.

Evidence accumulated over the last few years indicates that, indeed, carnosine fulfills most, if not all, of the criteria generally accepted as proof of transmitter function. Thus, it is present presynaptically, and the enzyme involved in its synthesis is also present. Further, the enzyme involved in the degradation of carnosine, carnosinase, is also present, but since its concentration is not altered by either deafferentation or by denervation, it is thought to be extraneuronal. There are high-affinity uptake systems for the accumulation of the precursors, β-alanine and histidine, and there are high-affinity receptors present for the binding of carnosine. The relevant electrophysiology does not appear to have been done, nor is there any evidence to suggest that physiological changes in olfactory function are reflected in alterations of transmitter (carnosine) synthesis or release. Nevertheless, the data now available strongly suggest that carnosine is the transmitter in this system.

The most recent studies [14,15] have focused on the characteristics of the carnosine binding site, the presumed receptor. Using olfactory bulb membranes it has been shown that the binding of carnosine is sterically and structurally specific. It shows saturation, and the binding is fully reversible. The affinity constant for binding is less than 1 μM and the binding is reduced after deafferentiation. The binding of carnosine, then, is similar in nature to the binding of other transmitters to their receptors and provides more encouragement for the concept that carnosine is a transmitter. Overall, these studies describe what is probably the most discrete and convincing evidence that a peptide can serve as transmitter in any of the many peptidergic systems.

Benzodiazepine Receptor

The minor tranquilizers Librium and Valium have been the most widely used prescription drugs in the world for several years. It has been estimated that some 30% of the population of this country have taken these drugs at one time or another. Their mechanism of action is as yet unknown. Recent studies have concerned the nature of the receptor for these compounds which fall into the chemical class benzodiazepine.

The receptors have been studied by now-conventional techniques [16-20]. High specific activity radio-labeled benzodiazepine compounds have been prepared, and binding of these materials to brain particulates has been measured. The receptors have been found in the highest concentration in cortical areas, in intermediate levels in striatum and cerebellum, and in low amounts in spinal cord. They are neuronal in location, and the binding of the various benzodiazepines to these receptors correlates well with their pharmacological potency.

There is evidence that the action of the benzodiazepines is related in some way to the action of γ-aminobutyric acid (GABA) [21]. The pharmacological actions of these compounds are compatible with such a concept. That is, the benzodiazepines are anxiolytic, muscle relaxant, and anticonvulsant, all properties which might be due to a potentiation of GABA neurons. Recently [22], it has been found that GABA also potentiates benzodiazepine binding.

Following the reasoning which proved so profitable in the uncovering of the endogenous analgesics, i.e., the search for the endogenous ligand of the morphine receptor, several groups have searched for the endogenous ligand of the benzodiazepine receptor [23,24]. Isolation of materials from tissue preparations which competed with labeled benzodiazepines led to the isolation of hypoxanthine and inosine. These two purines, although they have low affinity, seem to be endogenous ligands for this receptor. They are specific for the receptor, although adenosine and guanosine and some of the methylated xanthines also bind weakly. There is some evidence [25] that they have the same physiological action as the benzodiazepines, but their action seems weaker. While the benzodiazepines totally prevent the seizures evoked by pentylenetetrazole, inosine merely increases seizure latency, i.e., partially antagonizes the seizures.

Thus, inosine and hypoxanthine appear to be endogenous ligands for the benzodiazepine receptor. Their rather low affinity, however, leaves some doubt as to whether they are the only ligand. So, the search continues in some laboratories for a peptide or protein ligand in tissues. Nevertheless, the finding that purines can act as ligands for tranquilizer receptors is surprising and provocative. Although the connection is not yet obvious, the possibility of purinergic neurons in the central nervous system, and the purine lesions in the Lesch-Nyhan syndrome, provide an intriguing framework for these recent findings.

Nutrition and Transmitter Synthesis

It has long seemed axiomatic that the brain is insulated by physiological and biochemical mechanisms from alterations in the environment. It has seemed even more certain that the ability of the brain to function depends upon its communication via neurotransmitters and that, of all the properties of the brain, this communication should be the most protected. It has been, therefore, surprising to find this concept challenged. The challenge has come through a series of studies by Wurtman and Fernstrom and their colleagues [26]. Basically, these studies suggest that, under a number of different conditions, the synthesis of various transmitters is regulated by the availability of precursors from the blood which, in turn, is regulated by the nutritional state of the animal at the time.

In this view, neurotransmitter synthesis reflects the availability of neurotransmitter precursor for the following reasons. First, the initial enzymes in the synthetic pathway to certain transmitters are rate limiting and not saturated with their substrates at normal cellular concentrations. Secondly, there is no storage of these precursor substrates in the brain. Finally, there is a known fluctuation of the circulating levels of these substrates.

The first studies on this subject concerned the synthesis of serotonin. Clearly, the enzyme tryptophan hydroxylase is not saturated at usual cellular levels of tryptophan. Indeed, the K_M for tryptophan is about 6×10^{-5} M, which is on the order of the blood concentration of tryptophan. In support of the calculations it is known that administration of tryptophan raises the levels, in turn, of blood tryptophan, brain tryptophan, and brain serotonin. Further, it is known that administration of tryptophan increases the excretion of serotonin metabolites. So it appears clear that increased availability of tryptophan in the blood increases the synthesis of the transmitter, serotonin. There is some evidence that increased serotonin synthesis increases the functional aspect of serotonin transmission, but such evidence is not easy to come by.

Incidental to these studies, perhaps, have been investigations on the factors which intervene between blood levels of an amino acid and brain levels of this same amino acid. In order to explain some rather anomolous nutritional effects on the levels of serotonin synthesis, a series of investigations was done on the relation between the uptake of tryptophan into the brain and the levels of other amino acids in the blood. The outcome of these studies has been the solidification of the concept that the uptake of an amino acid into the brain, and its metabolism as well, is regulated by the levels of other amino acids in the blood that may compete for transport. In the case of tryptophan it was clearly shown that the levels of tryptophan in the brain reflect not just the levels of tryptophan in the blood, but also the levels of the group of large neutral amino acids, tyrosine, phenylalanine, leucine, isoleucine, and valine, which compete with tryptophan for uptake. Thus, even though a given meal may raise the levels of tryptophan in the blood, if the levels of these competing amino acids also rise, the level of tryptophan in the brain may go down.

The synthesis of acetylcholine also appears to be under some nutritional control. The K_M for choline acetyltransferase is 400 μM, well above actual brain choline levels. Indeed, acetylcholine levels increase after the administration of choline [27]. There is also some evidence that the transsynaptic induction of tyrosine hydroxylase, thought to be due to the release of presynaptic acetylcholine on the postsynaptic catecholamine neurons in the central nervous system, can result from the ingestion of choline. This experiment would suggest that even the functional consequences of transmitter release can follow the administration or ingestion of the neurotransmitter precursor.

There are some conditions in humans thought to be due to insufficient release of acetylcholine. Prominent among these is tardive dyskinesia. The administration of choline, then, which may increase the synthesis of acetylcholine, has been used in this condition, with some reported utility. A more felicitous treatment seems to be the administration of lecithin as a choline source. This device for administration of choline circumvents the unpleasant side effects of choline ingestion and appears, in any case, more effective in raising blood choline levels [28]. An evaluation of the possible clinical effectiveness of this treatment is as yet premature, but there appear to be some encouraging preliminary studies.

The synthesis of catecholamines has also been studied. Here the case is not so persuasive. Certainly under some conditions, e.g., when the metabolism of the catecholamines is blocked pharmacologically, there seems to be some effect of precursor availability on catecholamine transmitter synthesis. But in the normal animal this does not seem to be the case. This is not too surprising, since the first and rate-limiting enzyme, tyrosine hydroxylase, appears to be saturated at normal cellular concentrations of tyrosine, so fluctuations in blood or brain concentrations would not seem to have any effect. So far, accordingly, there is not much evidence that nutritional state affects the synthesis of catecholamines.

That there is some effect of precursor availability on the synthesis of certain transmitters is undeniable. That this increased synthesis produces increased transmitter function seems possible or even likely. In spite of the widespread resistance to these ideas, the studies on this subject have opened a new avenue of thinking.

The Role of Cyclic AMP in Synaptic Transmission

The Greengard hypothesis [29], that cAMP and the attendant phosphorylation or membrane proteins mediate the slow hyperpolarizing response, the catecholamine-induced inhibitory postsynaptic potential (IPSP) in superior cervical ganglia, and the more general concept, that cyclic nucleotides function as "second messengers" in the neurotransmission of the nervous system, have come under heavy attack within the last several years. The evidence against the direct involvement of cAMP as an intermediate here takes many forms [30,31]. Basically, however, many groups have been unable to provide evidence for the two principal predictions of the hypothesis. These are, first, that application of the suspected second messenger, namely, cAMP, should mimic the effects of the transmitter it is presumed to mediate. Second, treatments that affect the synthesis or breakdown of the second messenger ought to appropriately alter the response being studied.

In the first case, a number of groups [32-35] have been unable to obtain any alteration in the membrane permeability or the membrane potential upon direct application or even by intracellular administration of cyclic AMP. Equally unsuccessful have been attempts in most laboratories to obtain the postulated slow depolarizing response (EPSP) upon the application of cyclic GMP, although there is one report of a depolarization caused by cGMP administration [36]. In the second case, the administration of blockers of adenylate cyclase such as prostaglandin, which should lower cAMP levels, or of phosphodiesterase inhibitors such as papaverine, which should raise cAMP levels, should have consistent effects on the synaptic potential. Prostaglandin had no such effects, and papaverine had an effect opposite to that predicted, although some phosphodiesterase inhibitors did potentiate the inhibitory potential. Thus, the outcome of these several experiments does not support the second messenger role for cyclic AMP.

On the other hand, whether or not cAMP is a mediator, the data relating cAMP concentrations to transmitter release are unquestioned and strongly suggest that cAMP plays some role in the events surrounding or following transmitter release. Accordingly, investigations on the actions of cyclic AMP in postsynaptic neurons have proceeded. It has been observed [37] that the action of cAMP in synaptosomes leads to the phosphorylation of two membrane-associated proteins by $[^{32}P]$ ATP. These two proteins are of molecular weights 86,000 and 80,000 and apparently have the same isoelectric point. They are both susceptible to degradation by collagen. One of the proteins in neuron specific and found only in nervous tissue, and is enriched in synaptic areas. Ca^{2+} causes the same two proteins to be phosphorylated, but it is said that the sites of phosphorylation are different from those increased by cAMP. This action of cAMP in stimulating phosphorylation is entirely consistent with the actions of cAMP on protein kinases in other systems. What relation these phosphorylations have to transmitter function remains to be seen. Right or wrong, we have learned a lot from the Greengard postulate.

2-Deoxyglucose Methodology

Perhaps the most bountiful development in the past several years has been the elucidation by Sokoloff and his colleagues [38-42] of the 2-deoxyglucose method for assessing energy metabolism of the brain and its coupled functional correlates. Basically, the technique assumes that functional activity in a portion of the brain will be coupled to an increase in energy metabolism, and further, that this energy metabolism will be reflected in increased uptake of glucose. Then, by using a labeled analog of glucose, 2-deoxyglucose, which can be phosphorylated by hexokinase but not metabolized further, and looking at its increased uptake via autoradiography, a functional map of the entire central nervous system can be obtained.

[^{14}C]-2-Deoxyglucose is injected intravenously. After 45 min the animal is decapitated and the brain is frozen. The plasma and tissue glucose levels are monitored and 20-μm sections of the brain are prepared and examined by autoradiography. By densitometric evaluation and the appropriate computer manipulations, the cerebral glucose utilization can be calculated.

Although some better resolution can be obtained using tritiated material, there is not sufficient resolution obtained to assess the uptake by specific cells or processes. This is due to the innate diffusability of 2-deoxyglucose. As yet no remedy for this limitation is in sight.

Nevertheless the applications of this method are remarkable in scope. Initially, simple experiments were reported comparing the patterns in conscious and anesthetized rats and monkeys. Now, however, any number of functional parameters have been explored. Various odors light up the olfactory bulb; obstruction of the ears diminishes the activity in the auditory pathways. In the visual cortex it has been possible to dissociate active from inactive ocular dominance columns by injecting 2-deoxyglucose into monkeys in which one eye has been removed or covered. Energy metabolism has been explored in epileptogenic foci, in exercise, in various pharmacological treatments, and after the administration of hormones. The most astounding demonstration of the sensitivity of the technique, perhaps, is the ability to discern a hot spot of 2-deoxyglucose uptake after manipulation of a single whisker on the injected rat.

In a modification of the methodology, it is possible to inspect energy parameters in human brains by computerized axial tomography. This application to the clinic requires a derivative that gives rise to X-rays via positron emissions. Such a derivative, 2-fluoro[^{18}F]-2-deoxyglucose, is available and has been applied to the study of energy metabolism in the living human brain.

By any criteria, the 2-deoxyglucose method is one of the most impressive of contemporary developments and one that promises large rewards in both the basic research on energy metabolism and the clinical research and diagnosis of brain malfunction.

References

1. M. Willard, *J. Cell Biol. 75:*1 (1977).
2. P. N. Hoffman and R. J. Lasek, *J. Cell Biol. 66:*351 (1975).
3. M. M. Black and R. J. Lasek, *Brain Res. 171:*401 (1979).
4. H. Gainer, Y. Sarne, and M. J. Brownstein, *Science 195:*1354 (1977).
5. M. J. Brownstein and H. Gainer, *Proc. Natl. Acad. Sci. USA 74:*4046 (1977).
6. R. H. Quarles, J. L. Everly, and R. O. Brady, *J. Neurochem. 21:*1177 (1973).
7. J. M. Matthieu, R. O. Brady, and R. H. Quarles, *Brain Res. 86:*55 (1975).
8. J. M. Matthieu, R. H. Quarles, J. F. Poduslo, and R. O. Brady, *Biochim. Biophys. Acta 392:*159 (1975).
9. N. Sternberger, R. H. Quarles, Y. Itoyama, and H. deF. Webster, *Proc. Natl. Acad. Sci. USA 76:*1510 (1979).
10. L. L. Iversen, *Trends in NeuroSciences* (Ref. Ed.), *1:*15 (1978).
11. S. H. Snyder and R. B. Innis, *Ann. Rev. Biochem. 48:*755 (1979).
12. V. Chan-Palay, G. Jonsson, and S. L. Palay, *Proc. Natl. Acad. Sci. USA 75:*1582 (1978).
13. F. L. Margolis, *Trends in NeuroSciences* (Ref. Ed.), *1:*42 (1978).
14. J. D. Hirsch, M. Grillo, and F. L. Margolis, *Brain Res. 158:*407 (1978).
15. J. D. Hirsch and F. L. Margolis, *Brain Res. 174:*81 (1979).
16. R. F. Squires and C. Braestrup, *Nature (London) 266:*732 (1977).
17. C. Braestrup and R. F. Squires, *Proc. Natl. Acad. Sci. USA 74:*3805 (1977).
18. H. Mohler and T. Okada, *Science 198:*849 (1977).
19. R. S. L. Chang and S. H. Snyder, *Eur. J. Pharmacol. 48:*213 (1978).
20. M. Williamson, S. M. Paul, and P. Skolnick, *Nature (London) 275:*551 (1978).
21. M. Karobath, *Trends in NeuroSciences 2:*116 (1979).
22. M. Karobath and G. Sperk, *Proc. Natl. Acad. Sci. USA 76:*1004 (1979).
23. P. Skolnick, P. J. Marangos, F. K. Goodwin, M. Edwards, and S. M. Paul, *Life Sci. 23:*1473 (1978).
24. T. Asano and S. Spector, *Proc. Natl. Acad. Sci. USA 76:*977 (1979).

25. P. Skolnick, P. J. Syapin, B. A. Paugh, V. Moncada, P. J. Marangos, and S. M. Paul, *Proc. Natl. Acad. Sci. USA 76:*1515 (1979).

26. R. J. Wurtman and J. D. Fernstrom, *Biochem. Pharm. 25:*1691 (1976).

27. E. L. Cohen and R. J. Wurtman, *Life Sci. 16:*1095 (1975).

28. M. J. Hirsch and R. J. Wurtman, *Science 202:*223 (1978).

29. P. Greengard, *Nature (London) 260:*260 (1976).

30. I. Kupfermann, *Ann. Rev. Neurosci. 2:*447 (1979).

31. F. F. Weight, J. A. Schulman, P. A. Smith, and N. A. Busis, *Fed. Proc. 38:* 2084 (1979).

32. J. P. Gallagher and P. Shinnick-Gallagher, *Science 198:*851 (1977).

33. N. J. Dun, K. Kaibara, and A. G. Karczmar, *Science 197:*778 (1977).

34. N. A. Busis, F. F. Weight, and P. A. Smith, *Science 200:*1079 (1978).

35. H. Kobayashi, T. Hashiguchi, and N. S. Ushiyama, *Nature (London) 271:* 268 (1978).

36. T. Hashiguchi, N. S. Ushiyama, H. Kobayashi, and B. Libet, *Nature (London) 271:*267 (1978).

37. W. Sieghart, J. Forn, and P. Greengard, *Proc. Natl. Acad. Sci. USA 76:* 2475 (1979).

38. L. Sokoloff, *Trends in NeuroSciences* (Ref. Ed.), *1:*75 (1978).

39. C. Kennedy, M. H. DesRosiers, J. W. Jehle, M. Reivich, F. Sharpe, and L. Sokoloff, *Science 187:*850 (1975).

40. L. Sokoloff, M. Reivich, C. Kennedy, M. H. DesRosiers, C. S. Patlak, K. C. Pettigrew, O. Sakurada, and M. Shinohara, *J. Neurochem. 28:*897 (1977).

41. R. C. Collins, C. Kennedy, L. Sokoloff, and F. Plum, *Arch. Neurol (Chic.) 33:*536 (1976).

42. M. Reivich, D. Kulil, A. Wolf, J. Greenberg, M. Phelps, T. Ido, N. Cosella, J. Fowler, E. Hoffman, A. Alavi, P. Som, and L. Sokoloff, *Circul. Res. 44:*127 (1979).

AUTHOR INDEX

Numbers in brackets are reference numbers and indicate that an author's work is referred to although the name is not cited in the text. Italic numbers give the page on which the complete reference is listed.

G

SUBJECT INDEX

Numbers in boldface indicate that a structure is given on that page. F and T indicate figure and table, respectively.